IMMUNOLOGY: Essential and Fundamental

THIRD EDITION

Sulabha Pathak, PhD
Homi Bhabha Fellow
Department of Biological Sciences
Tata Institute of Fundamental Research
Mumbai, India

Urmi Palan, MPhil
Faculty Member
Department of Microbiology
Ramnarain Ruia College
Mumbai, India

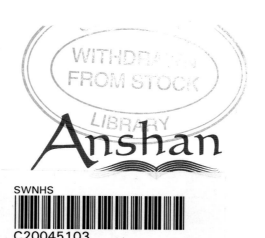

Anshan

IMMUNOLOGY
Essential and Fundamental 3E

Published in the UK by:

Anshan Ltd
11a Little Mount Sion
Tunbridge Wells
Kent. TN1 1YS

Tel: +44 (0) 1892 557767
Fax: +44 (0) 1892 530358

e-mail: info@anshan.co.uk
web site: www.anshan.co.uk

ISBN: 978 1 848290 33 4

British Library Cataloguing in Publication Data
A catalogue record for this book is available from the British Library.

Cover Design: Emma Randall
Cover Image: Science Photo Library
Copyediting and editorial inputs: Gauri Pathak
Illustrations: Zain Enterprises
Preface illustration: Scanning electron micrograph of a lymphocyte migrating into the high endothelial venule in a lymph node (1000×), a gift from Dr. W. van Ewijk, Leiden University, The Netherlands.

Icons

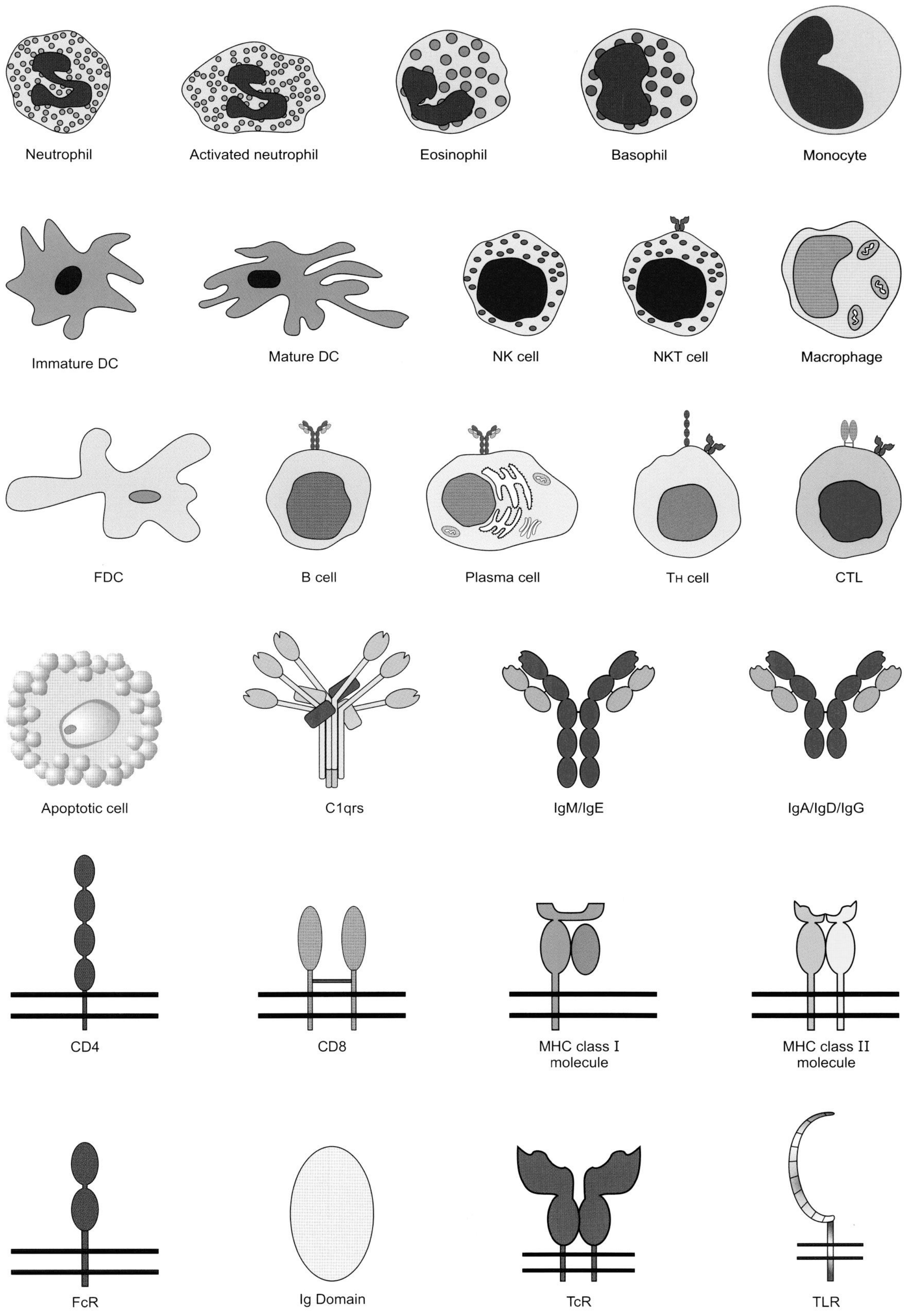
Neutrophil
Activated neutrophil
Eosinophil
Basophil
Monocyte
Immature DC
Mature DC
NK cell
NKT cell
Macrophage
FDC
B cell
Plasma cell
TH cell
CTL
Apoptotic cell
C1qrs
IgM/IgE
IgA/IgD/IgG
CD4
CD8
MHC class I molecule
MHC class II molecule
FcR
Ig Domain
TcR
TLR

Preface to the Third Edition

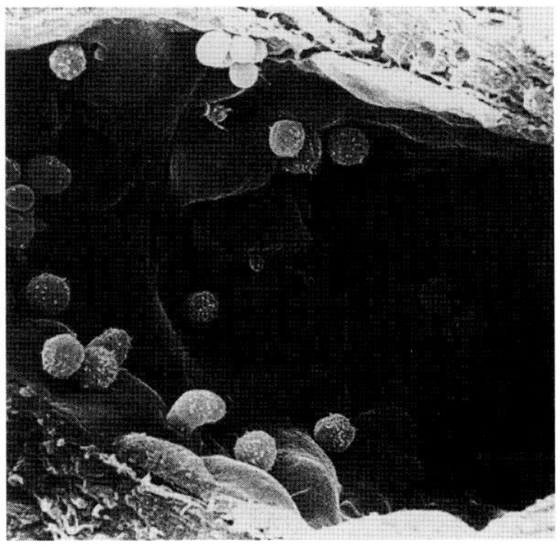

Immunology is a vast and rapidly growing field that touches almost all aspects of human health and wellbeing. In this third edition, Dr Pathak, with her international research exposure, and Ms Palan, with her extensive academic experience, collaborate yet again to bring students the latest in the field of immunology in a format that is non-intimidating and both comprehensive and comprehensible.

The authors have aimed at making this book about ideas and concepts rather than purely about facts and minutiae. *Immunology: Essential and Fundamental* views immunology in the context of everyday human life. Metaphors linking the content to popular culture and daily life help the student to think beyond the course material to view the subject holistically and in context. Throughout, this book endeavours to remind the reader that although immunology is an extremely specialised field, it must be viewed in the context of the human body, the individual, and sociocultural milieu. It takes us beyond immunology as a compartmentalised field to place it in context and enrich our understanding of the big picture.

Immunology: Essential and Fundamental also encourages students to question the obvious and to explore the implications of the material they learn. The 'sidetracks' present supplementary information, real-life implications, clinical uses, or research consequences for the interested student. They also engage students by providing tantalizing glimpses of possible future applications and suggesting avenues for further research, adding depth to their grasp of the subject.

The third edition, now in colour, expands and builds upon the popular features of the second edition, such as the summary boxes, sidetracks, and chapter acronym lists. Besides incorporating developments made in the field since the second edition, the authors have reorganised some of the earlier matter to foster understanding. The figures have been substantially enhanced to aid comprehension. An entirely new chapter focusses on infectious agents and the human immune system's responses to them. Moreover, mindmaps and flowcharts provide schematic organisation of data and visual overviews of processes. This edition of *Immunology: Essential and Fundamental* is an excellent tool for novices to the field who wish to explore the diverse aspects of immune function and for educators who wish to spark their students' interests and imaginations.

Acknowledgements: We are deeply grateful to all our colleagues and students for their invaluable feedback and insights. We must also thank Gauri Pathak for allowing us to unabashedly tap her as a resource and for all her efforts and creative inputs. A debt of gratitude is owed to Chetan Bandekar, Ali Fezi and his team (Murali and Suresh) for patiently and cheerfully entertaining our requests with regard to the illustrations. Finally, we also thank the Publishers for their efforts in making this four-colour book a reality.

S Pathak, PhD
U Palan, MPhil

November, 2010

Contents

1 Introduction

It's the eye of the tiger, it's the thrill of the fight
Risin' up to the challenge of our rival
And the last known survivor stalks his prey in the night
And he's watchin' us all in the eye of the tiger

Face to face, out in the heat
Hangin' tough, stayin' hungry
They stack the odds 'til we take to the street
For we kill with the skill to survive

Risin' up, straight to the top
Had the guts, got the glory
Went the distance, now I'm not gonna stop
Just a man and his will to survive

— Survivor, *Eye of the Tiger*

1.1 Introduction

Any organism, with its easy supply of nutrition and moisture, is an attractive target for parasites — a parasite is an organism that spends a significant portion of its life in or on a living host organism, harming the host. Parasitism is the most common lifestyle on earth; by some estimates, parasitic species may outnumber nonparasitic species by a ratio of four to one. Parasites attempt to gain a foothold in host organisms; hosts fight back. The host and the parasite are constantly engaged in a struggle for dominance, and any living system, whether unicellular or multicellular, has defensive mechanisms to curb the invader and eliminate it. The more complex the host, the more complicated the system of defence. The collection of mechanisms within an organism that protect it against potential threats is the *immune system*. It consists of organs, tissues, cells, and proteins that act in concert to confer protection. *Immunity* is the ability to defend against infectious agents; *immunology* is the study of all aspects of host defence (including adverse consequences). *Immune response* or *immune reaction* refers to the production of soluble factors and cells that defend the body against potentially dangerous biological and chemical agents. These agents, in turn, are referred to as *immunogens* or *antigens*.

Immunology, like most other fields, has its own lexicon (or jargon). The terminology aims to convey scientific principles and definitions in context. Although essential to immunological study, these terms may discourage a fresh learner. Trying to figure out where to start studying (and understanding) immunology is a bit like trying to figure out the chicken or egg dilemma. To understand immunology, you need to know the terminology, and to understand the terminology, you need a basic knowledge of immunology. In this introductory chapter, we have attempted to introduce some terms and immunological concepts for easy comprehension, necessitating oversimplification. The oversimplification will be rectified in later chapters. The aim of this book is to foster the understanding of the human immune system and immune responses. However, our knowledge of the human immune system often stems from experiments in mice. Hence, though the focus of this book is the human immune system, it includes substantial information about the murine system as well.

An immune response is the result of a perceived threat to an animal's well-being[1]. The immune system has a two-pronged strategy to defend the human body — a general mechanism to keep intruders out (the innate system) and a specific mechanism to deal with those that get through (the adaptive system; Table 1.1). This is in some sense similar to the multitiered defence of medieval castles. A peripheral area stocked with predators discouraged interlopers from entering castles. Armed soldiers — often issued with sketches of known criminals or enemy agents — were also deployed around the grounds to watch out for unlawful entrants (Figure 1.1). The key word is multitiered — the innate and adaptive immune systems do not operate in isolation but are well-integrated parts of a single system of host defence. They participate in the host defence in tandem. They can, and do, communicate with and regulate each other. The two systems are treated as separate entities for easier understanding (and because scientists like to separate and organise everything into tiny little niches or boxes).

[1] This is a classic example of circular logic. An *immunogen* is that which elicits an immune response in a healthy individual; an *immune response* is the result of introduction of an immunogen. The immune system responds to immunogenic challenge by producing soluble proteins and cells capable of specifically recognising and reacting with the immunogen.

1.2 The First Line of Defence

The innate immune system is the body's watchdog. All living systems, including unicellular organisms, have some kind of defensive mechanism. Even the 'lowly'

Table 1.1 Features of Innate and Adaptive Immune Systems

Features	Innate immune system	Adaptive immune system
Found in	All metazoans, plants, and even unicellular organisms	Jawed vertebrates
Major effectors		
Soluble	Antimicrobial peptides and proteins (including complement)	Antibodies
Cellular	Epidermal layers, phagocytes, natural killer cells, dendritic cells, mast cells, and granulocytes	$CD8^+$ T cells
Recognition system	Limited number of nonclonal receptors present on a variety of cell types	Highly diverse, clonally distributed antigen receptors found on B and T lymphocytes
	Receptors specifically recognise molecules or molecular patterns associated with pathogens	Highly specific receptors that can discriminate between even minor differences in molecular structure; receptors recognise a wide variety of substances that may or may not be associated with pathogens
Response time upon first encounter	Seconds to hours	Days
Memory responses	Absent	Present; second or subsequent encounters result in rapid and more intense responses

bacteria have a mechanism for defending the integrity of their DNA. Unmethylated foreign DNA (such as phage DNA) is destroyed by bacterial restriction enzymes; the host DNA remains protected from digestion by these enzymes because it is methylated at specific sequences. Similarly, bacteria (and mammals) have specific systems that destroy double-stranded RNA (a signature molecule of some viruses). These systems did not primarily evolve for host defence. Rather, the existing systems were put to defensive use during evolution. The innate immune system in humans is, needless to say, much more complicated.

We discuss the innate immune system in detail in chapters 2 and 3. Chapter 4 outlines what the immune system considers potential threats and how it recognises these threats (also see the sidetrack *Us and Them*). The innate immune system is the first one to encounter a threat and respond to it. It consists of multiple layers of security measures. First are the physical and chemical barriers designed to keep intruders out. The system is also armed with antimicrobial substances (present in the blood and body secretions) to eliminate pathogens that may breach these barriers before they gain a foothold in the body. Chief of these substances is a family of proteins called complement, which we describe in chapter 3. Complement gets activated by (amongst other things) the intruder's outer surface. Its activation can result in lysis of the intruding cell. Even as the pathogen tries to escape the chemical attack, other innate immune mechanisms swing into action. Chemical signals released as a result of the pathogen's metabolic activities and/or tissue

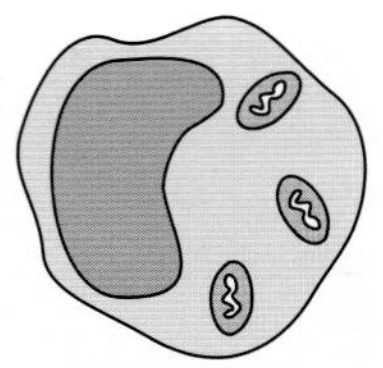

Macrophage

trauma cause an influx of fluids and leukocytes to the site of intrusion. Some of these leukocytes are scavenger cells that normally phagocytose (internalise and degrade) dead and dying cells. When phagocytic cells sense intruders, they get activated and phagocytose the aggressor. Activated phagocytes also liberate a large number of enzymes and chemicals capable of killing the pathogen. Phagocytic cells have another important role — that of alerting the adaptive immune system of the intrusion and of triggering the adaptive immune response. The phagocytes are an excellent example of how existing systems were put to defensive use during evolution. Phagocytosis is a means of obtaining nutrition in unicellular organisms. However, in multicellular organisms, it has been transformed into a mechanism of protection, and phagocytic cells have come to have a key role in both innate and adaptive immune responses.

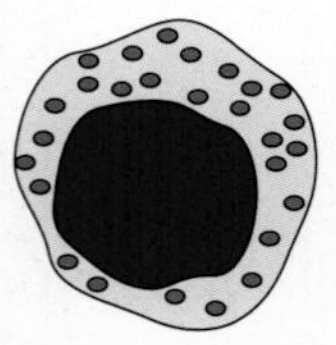

Natural killer cell

Natural killer cells are a type of lymphocytes belonging to the innate immune system. A distinctive feature of these cells is the presence of poison-filled cytotoxic granules. These cells are activated by the chemical signals released by phagocytes. Natural killer cells recognise infected body cells. These infected cells are like spies working for the enemy and they must be eliminated. Activated natural killer cells release the cytotoxins contained in their granules directly at infected cells, killing them and the pathogens inside. In this manner, innate defence mechanisms collectively try to overwhelm and kill the pathogen. Cells of the innate immune system also release chemical signals that activate the cells of the adaptive immune system.

1.3 Specific Defence

In case initial measures fail and the pathogen establishes a focus of infection, cells of the innate immune system recruit the adaptive immune system into the battle. This adaptive arm of immunity is also called *specific immunity*, as it seems to track specific threats and responds to repeat assaults with increasing ferocity (i.e., it appears to have a memory of past encounters). Chapters 5 through 12 describe aspects of the adaptive immune system. As mentioned earlier, all living systems have an innate system of host defence. The adaptive system seems to have developed suddenly in jawed vertebrates. This does not mean that lampreys and hagfish (jawless vertebrates) and other 'lower' animals or plants do not have an immune system that can track specific threats. On the contrary, some insects, sea urchins, lampreys, and even some plants have molecules that are associated with specific antipathogen defence. However, the diversity of such molecules is limited when compared to the mammalian system, and the structure of these molecules is also different.

1.3.1 Organisation of the Adaptive Immune System

The tissues and cells of the adaptive immune system are called the lymphoid system. This system consists of organs, such as the thymus, bone marrow, spleen, and lymph nodes, as well as loose patches of tissues distributed throughout the body. We describe the lymphoid system in chapter 5. The adaptive immune system, like any defensive organisation, has a hierarchical structure. At the helm are the thymus and bone marrow. They are involved in producing lymphocytes — the immune system's armed guards and the effectors and executors of adaptive immunity. Local branches of the adaptive immune system — the lymph nodes — are distributed throughout the body. The spleen may be regarded as regional military headquarters, where all blood-borne threats are dealt with. The thymus and bone marrow are lymphocyte training camps. Those trained in the bone marrow are called B lymphocytes; those trained in the thymus are T lymphocytes (described in chapters 8 and 9, respectively). During the training process, lymphocytes are equipped with detectors to help them recognise pathogens. The lymphocytes become licensed to kill, maim, or otherwise damage anything that dares enter the body without permission. The lymphocytes are thus equipped to respond to immunogenic challenges.

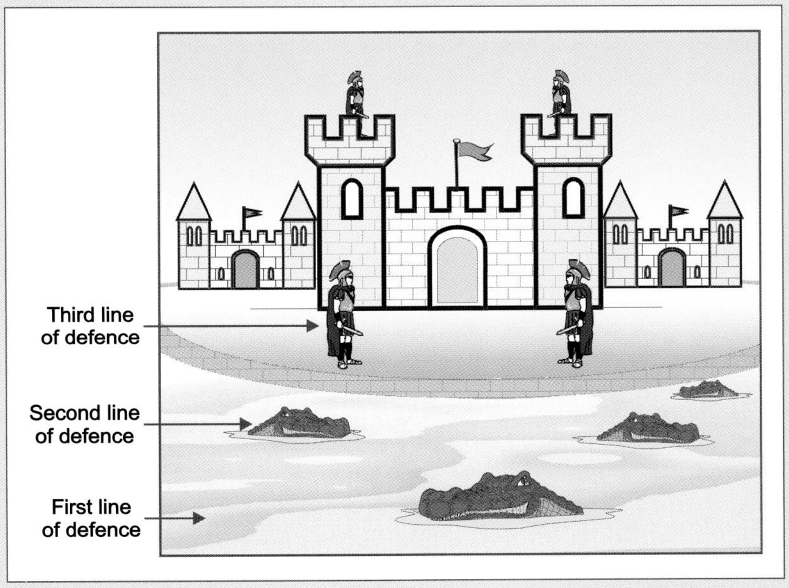

Figure 1.1 The body is a well-defended fortress with different components having designated defensive functions. Physical and chemical barriers, consisting of intact skin, moist mucous membranes, complement, and antimicrobial substances in the blood and body secretions act as the first line of defence. They form a bulwark against the entry and establishment of pathogens. If this bulwark is breached, phagocytic cells and cytotoxic natural killer cells form the second line of defence. Phagocytes rapidly engulf and ingest any pathogens that enter the body; natural killer cells release cytotoxins to kill intruders. Lymphocytes of the adaptive immunity are the third line of defence. They specifically recognise and destroy intruders. This specific defence is a two-pronged strategy consisting of free-roaming antibodies and of cells that defeat the microbe in one-on-one combat.

1.3.2 The Recognition Molecules

Cells of the adaptive immune system are equipped with special recognition molecules. A recognition system implies two parts — an identity tag and a detector for that tag. All cells of the body sport identity tags. Lymphocytes have detectors which allow them to recognise cells without appropriate tags. These detectors (called antigen receptors) are described in chapter 6. The genetic mechanisms involved in their development are also explained in that chapter. We like to think of the lymphocyte, with its antigen receptor, as Cinderella's prince with her glass shoe. The lymphocyte tries to find cells with the correct fit. In this search, lymphocyte receptors briefly engage with prospects. If there is a fit, cells enter a prolonged relationship. Chemical messages are exchanged, resulting in lymphocyte activation. If the fit is not good enough, the lymphocyte moves on.

The two classes of lymphocytes differ in their detectors. B lymphocytes have receptors that directly recognise and bind antigens, that is, intruders or their metabolic products. B cells' recognition of antigens is similar to a soldier's recognising of invaders by their external appearance, for example, by their uniforms. Antigen detection by T cells is more complicated. T cell detectors cannot recognise antigens by themselves; antigens have to be loaded on special molecules, called MHC molecules. These molecules are described in chapter 7. MHC molecules can be thought of as flagpoles; antigens function as flags. Depending upon the fabric available in the cell, different flags are mounted on these MHC molecules. Self-cells load fragments derived from self-proteins, declaring to T cells that they remain faithful and thus escape immune attack. Conversely, infected cells load fragments of proteins derived from the infecting pathogen, and recognition of such MHC:peptide complexes by T cell detectors triggers an immune response. Thus, unlike B cells, T cells recognise invaders based on the flags these infected cells carry. Both the flag (antigens) and the flagpole (MHC molecules) are important in T cell recognition. The process of identification requires physical interaction between the detector (antigen receptor), the flag (fragments derived from the intruder), and the flagpole (MHC molecules). Cells that carry a different flagpole (i.e., express a nonself MHC molecule), such as transplanted cells, are immediately attacked by T cells[2].

MHC molecules come in two varieties. MHC class I molecules are present on all nucleated cells of the body and announce these cells as belonging to the body, that is, as self-cells. MHC class II molecules have a more restricted distribution — they are usually expressed by cells of the immune system. Other cells start expressing them only when they sense danger. Two types of T cells are recognised on the basis of the MHC molecules they interact with. Those that recognise antigen loaded on MHC class I molecules express CD8 molecules and are called $CD8^+$ T lymphocytes. These are killer T cells (called cytotoxic T lymphocytes) that kill target cells by releasing toxins. T cells that recognise antigen loaded on MHC class II molecules express the CD4 molecule; they are $CD4^+$ T cells. These cells help B cells, phagocytic cells, and $CD8^+$ T cells mount immune responses. Such $CD4^+$ T cells are usually referred to as *T helper cells*. $CD4^+$ T cells are also involved in regulating immune responses and ensuring that the responses themselves are not deleterious. $CD4^+$ T lymphocytes are thus multifunctional cells involved in assisting B cells, regulating immune responses, and killing infected self-cells. CD1 molecules are structurally related to MHC class I molecules except that they can be loaded with lipid, not peptide, antigens. T cells recognising antigen loaded on CD1 molecules may be $CD4^+CD8^+$ or $CD4^-CD8^-$. The lipid-specific T cells are similar to peptide-specific T cells in that they are also involved in protecting from infections and in regulating the immune response.

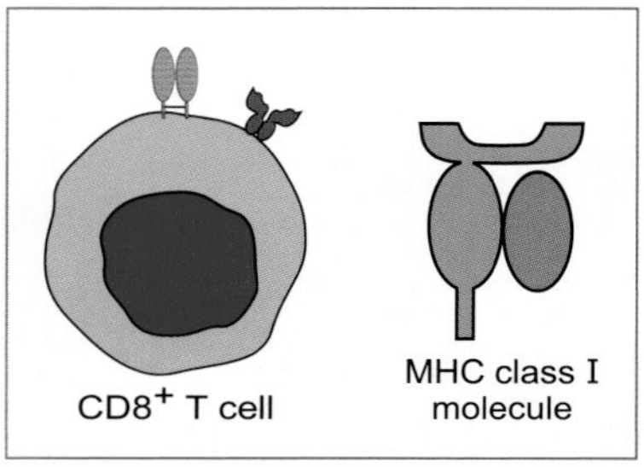

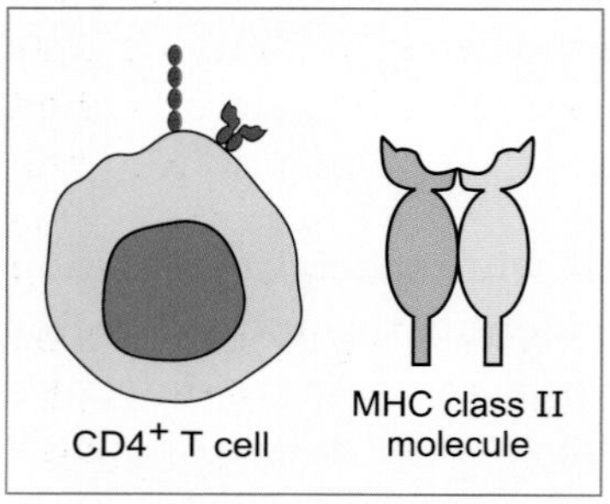

[2] Interestingly, cells that stop expressing MHC class I molecules — e.g., because of a viral infection — are also attacked. However, this attack comes from natural killer cells, not T cells.

1.3.3 The Weaponry

The adaptive immune system's destructive weapons are described in chapters 10 and 11. With these weapons, an adaptive immune response can damage or kill trespassers. The first weapons in the adaptive immune system's arsenal are antibodies (or immunoglobulins), which specifically bind and neutralise the pathogen (e.g., virus or bacteria) or its products. Immunoglobulins are comparable to free-roaming pre-programmed guided missiles that can combine with and destroy specific targets. Five classes of immunoglobulins are recognised in humans, based on their structures and functions. The binding of some classes of antibodies to a pathogen can activate complement and result in lysis of the target cell. Alternatively, antibodies can prevent binding of the pathogen to self-cells. Given that viruses have to bind to host cells before infection can occur, coating viral particles with antibodies can prevent further infection. Immunoglobulins are also important in neutralising toxins and enzymes of the pathogen; they combine with the active site of the toxin or enzyme, interfering with its activity. Antibody-coating also hastens removal and destruction of target cells, as coated cells are also more easily taken up by phagocytic cells.

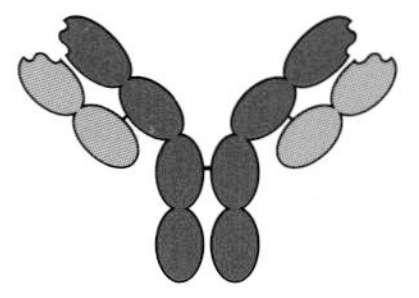

Immunoglobulin

Nevertheless, immunoglobulins cannot reach pathogens inside infected cells. In these cases, the best way to eliminate the pathogen is to kill the infected cell. Hence, activation of the immune system also results in the formation of $CD8^+$ cytotoxic T lymphocytes that specifically recognise and kill infected cells. These cells have toxin-filled granules that cause lysis of target cells. Toxins liberated by cytotoxic T lymphocytes result in the formation of pores in the target cell's plasma membrane, resulting in its death. This is not the only way that cytotoxic T cells kill target cells; they can also induce target cell death by inducing it to commit suicide.

1.4 The Immune Response

To best understand the principles and players of the immune system, let us follow an attempted intruder in its journey. When a pathogen falls on intact skin, it cannot cross that barrier and is eventually shed with the outer layer of dead skin cells. A pathogen trying to gain entry by way of the respiratory system will be likely to suffer the same fate — reflexes, such as coughing or sneezing, will ensure that the pathogen is thrown out. If, however, these barriers have already been breached (e.g., because of a wound), the pathogen may enter. Once inside, the pathogen will try its best to settle in its niche and proliferate (establish a focus of infection). The pathogen will encounter a plethora of antimicrobial substances in the blood and tissue fluids that will try to thwart its efforts. Phagocytic cells will get activated by the chemical signals released by the pathogen as well as those transmitted by damaged tissue cells. Activated phagocytic cells are potent killing machines. They also activate natural killer cells. These cells will try to kill the pathogen and infected tissue cells in an attempt to prevent spread of the infection/pathogen. Meanwhile, the immune system cannot afford to idly wait for the conclusion of the pathogen–innate immune system tussle. Some cells of the innate immune system become messengers for the adaptive immune system. These innate immune cells that recruit the adaptive immune system are called antigen-presenting cells. Prominent among these are the dendritic cells that are distributed throughout the body.

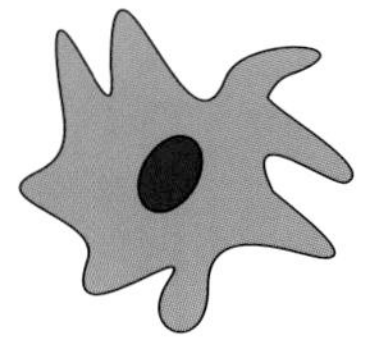

Immature dendritic cell

Dendritic cells and phagocytic cells will capture the pathogen or its products and carry them to the nearest branch of the adaptive immune system, the lymph node. *En route*, they will digest the pathogen and load peptide fragments derived from the pathogen onto MHC molecules and display them on their cell surfaces. The recognition of the MHC:peptide complex by T cells will occur in lymph nodes. T cell recognition of intruding pathogen-derived protein fragments loaded on MHC class I and class II molecules is the first in a chain of events that result in targeted action against the pathogen.

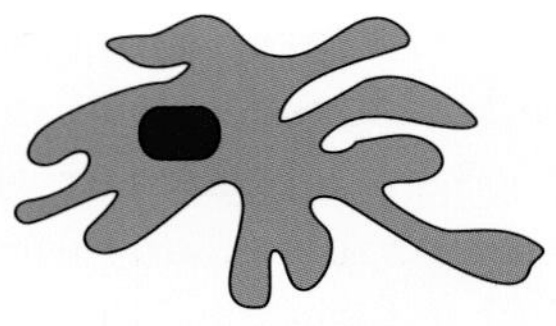

Mature dendritic cell

As mentioned, the adaptive immune system uses a two-pronged approach to neutralise invaders — a humoral response that consists of specific proteinic missiles that bind target cells and a cytotoxic cell-mediated response that kills infected cells.

- **Humoral immune response**: Lymphocytes that have never seen their antigens are called naïve or virgin cells. The recognition of MHC class II:peptide complexes by naïve $CD4^+$ T cells activates them. The activated T cells immediately start proliferating. They slowly metamorphose into T cells that are capable of helping other lymphocytes in fighting against the intruder (i.e., they become T helper cells). B cells also join this developing response. Unlike T cells, B cells need not wait for news from messenger cells. Their receptors directly recognise strange molecules (or antigens). In this way, B cells recognise not only the pathogen but also enzymes, toxins, and other metabolites produced by the pathogen. B cells get stimulated upon encountering the antigen, but the mere recognition of the pathogen (or its product) is often insufficient to activate them. They need signals from activated T helper cells and proliferate only after receiving appropriate T cell help. The proliferating B cells eventually differentiate and give rise to two types of cells — plasma cells and memory cells.

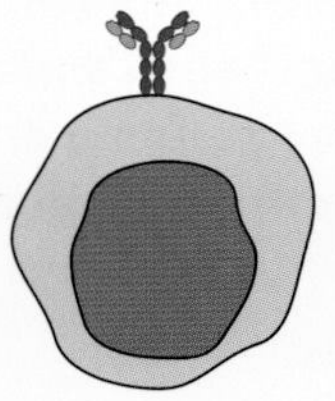

B cell

  - **Plasma cells** are virtual weapons factories. They synthesise and secrete several thousand immunoglobulin molecules per second. These antibodies find their way into tissue fluids and blood. As these immunity-conferring molecules are found in the blood, tissue fluids, and secretions, this form of immunity is called humoral immunity.

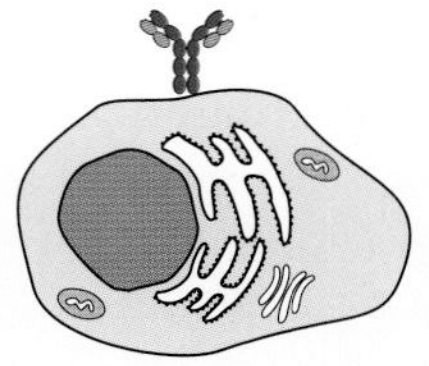

Plasma cell

  - **Memory cells**[3] are long-lived progeny of activated B cells that are essentially similar to their mother cells, that is, they are capable of proliferating and differentiating in response to the particular antigen that activated the mother cell. Memory cells have a greater sensitivity to the specific antigen they recognise.

- **Cell-mediated adaptive immune response**: Cytotoxic $CD8^+$ T cell receptors can recognise pathogen-derived peptides loaded on the MHC class I molecules that are expressed on messenger cell surfaces. The first step in $CD8^+$ T lymphocyte activation is the recognition of class I MHC:peptide complexes on the antigen-presenting cell. Just as in B cells, recognition alone is not enough to activate $CD8^+$ T cells. They too require T helper cell signals for activation and memory cell generation[4]. Once activated, cytotoxic T lymphocytes proliferate

[3] During the immune response, activated T helper cells, B cells, and cytotoxic T lymphocytes give rise to descendants that survive. Because the progeny have already encountered their antigen, they are no longer *naïve* but are termed *memory cells.*

[4] This two-signal system is similar to the dual-key system of some bank vaults. Two distinct keys, kept with separate people, are simultaneously required to unlock such vaults. The two-signal system is a common feature of immunology. It is believed to have evolved to avoid accidental activation of the immune system.

and differentiate. They produce two types of daughter cells — effector cells capable of killing their targets and memory cells that can differentiate to effector cells upon antigen encounter. The effector cells have cytotoxin-filled granules. They recognise infected cells because of their expression of MHC class I molecules loaded with pathogen-derived peptides. Once their receptor is engaged by the appropriate MHC class I:peptide complex, cytotoxic T cells direct their cytotoxic cargo at the infected cell, causing its lysis.

Immunoglobulin molecules and cytotoxic T lymphocytes (ably assisted by T helper cells) kill the pathogen. Phagocytic cells mop up the tissue debris resulting from all this activity, and the body is then free to heal and to repair the damage caused by the infection and the ensuing immune response. As the lymphocytes have also proliferated during the immune response, the numbers of memory cells are higher. In the course of antigen-induced proliferation, their signalling pathways, genetic make-up, and antigen receptors (in the case of B cells) change subtly. When the memory cells encounter a similar pathogen again, their response, called a *memory response*, is much faster than that of the naïve cell.

1.5 Checks, Balances, and Imbalances

Given that our environment is full of potentially harmful chemical and biological agents, the immune system must get activated a staggering number of times. It is obvious that the immune system, with its potent killing instruments, must be well-regulated. The immune system has evolved mechanisms that allow it to discriminate between intruders and the body's own cells (self-cells). These mechanisms allow the immune system to concentrate its powers on the former while being tolerant of the latter. The mechanisms of immune tolerance are discussed in chapter 12. Chapters 13 and 14 discuss specific examples of how the immune system concertedly acts for the defence of the body. Chapter 13 outlines how the immune system deals with the various infectious agents it encounters. It also lays out examples of a few infections to illustrate the strategies used by the immune system to defeat the pathogen and those used by the pathogen to subvert the immune system. Chapter 14 is devoted to cancer and the immune response to transformed self-cells.

When the immune system starts attacking self-cells, the result is a spectrum of disorders called *autoimmune diseases*. The mechanisms and possible causes of autoimmune disorders are discussed in chapter 15. The immune system may occasionally become hypersensitive and react to innocuous environmental substances. Chapter 16 describes such hypersensitive disorders and their effects. Finally, the application of the principles of self and nonself recognition and tolerance induction in transplantation of tissues and organs are discussed in chapter 17.

This book attempts to introduce students to fundamental principles and aspects of immunology. We have included some emerging concepts and ideas essential to understanding the subject. A careful reading of this book will not only introduce students to the world of immunology, but it will also equip them with the terms and theories crucial to understanding research in the field.

Us and Them

The immune system has a difficult mandate. It must destroy pathogens and infected cells while sparing healthy self-cells. *Per definition*, the immune system must have discriminatory powers to distinguish the self from the nonself. A model was needed to explain how the immune system decides when, how, and by what means to respond. Until recently, an explanatory theory based on Burnet's *Clonal Deletion Hypothesis* was generally accepted. This *Self–Nonself Theory* of immune recognition postulates that the immune system defines *self* as that which is present early in life; anything that comes later is deemed *nonself*. The immune system is trained to tolerate the self while identifying and attacking the nonself. The theory proposes that lymphocytes capable of responding to self-antigens are deleted in the primary lymphoid organs (i.e., the thymus and the bone marrow) during the process of lymphocyte maturation. There is ample experimental evidence to show that such deletional mechanisms are indeed active in both the thymus and bone marrow. Since B cells require T cell help to respond to most antigens, the thymus is thought to be more important in the deletional process. According to the Self-Nonself Theory, the body is populated with mature lymphocytes that only react to foreign antigens. Self-cells are spared by the immune system, because surviving lymphocytes are incapable of reacting to antigens expressed on self-cells. Although a straightforward and elegant model, this theory takes a very simplistic view of immunology. The basic assumption of the Self–Nonself Theory is that the thymus has a sampling of all the antigens expressed in the body. Yet, it is well established that many new proteins are expressed during puberty and maturation. The theory fails to explain how self-reactive cells to these proteins can be eliminated during early development. It also fails to explain the observation that autoreactive antibodies and cells are present in healthy adult mice and in humans and yet fail to cause autoimmune diseases. The fact that the foetus is not being rejected by its mother or that her body does not mount an immune response to her lactating breasts or her milk can also not be explained by this theory.

The last decade has seen the emergence of two theories that help explain what the immune system considers dangerous and why. The first is the *Infectious Nonself Hypothesis* or *Stranger Theory* of the late Charles Janeway; the second is the *Danger Hypothesis* of Paula Matzinger. Both rely on the foundation of the Self–Nonself Theory — they accept that deletional mechanisms eliminate most self-reactive cells. Both these theories differ from the Self–Nonself Theory in that they shift the onus of self–nonself discrimination away from the adaptive immune system. The Infectious Nonself Theory lays greater emphasis on the innate immune system's role in initiating an immune response. It proposes that the innate immune system of vertebrate animals recognises pathogens (the infectious nonself) by virtue of the host's ability to recognise the conserved products of microbial metabolism that are essential but unique to the microbe (and not expressed by host cells). The recognition strategy is based on invariant structures produced by and unique to all microorganisms. It suggests that recognition of these pathogen-associated molecular patterns allows the innate immune system to discriminate between the infectious nonself and the non-infectious self. In the absence of pathogens, antigen-presenting cells remain quiescent (not activated) and fail to activate T cells, even if the MHC:peptide complexes they display are recognised by T cells. The recognition of pathogen-associated molecular patterns activates antigen-presenting cells. Such activated cells can efficiently stimulate T cells and trigger an immune response. The discovery of a number of receptors that recognise pathogen-associated molecular patterns (see chapter 2) has helped strengthen the Infectious Nonself Hypothesis.

The Danger Hypothesis suggests that the immune system is more concerned with damage than with foreignness and is called into action by alarm signals from injured tissues rather than by recognition of nonself molecules. It postulates that antigen-presenting cells are activated by danger/alarm signals from injured cells exposed to pathogens, toxins, mechanical injury, etc. Antigen presentation by such activated cells results in an immune response, whereas antigen presentation in the absence of the danger signals results in tolerance. The discovery of endogenous nonforeign alarm signals, such as mammalian double stranded DNA, heat shock proteins, and interferon-α, has lent support to the Danger Hypothesis. The model takes the burden of triggering the immune response away from the immune system (whether adaptive or innate) altogether. Instead, it believes in tissue power. It assumes that when tissues are healthy, they induce tolerance; when they are distressed, they trigger an immune response. Debates about whether the immune response is triggered by the innate immune system or by a cross-talk between tissues and cells of the immune system are still raging and hopefully the picture will become clearer in the coming years.

2 Innate Immunity

I am alone,
Gazing from my window to the street below
On a freshly fallen silent shroud of snow.
I am a rock,
I am an island.

I've built walls,
A fortress steep and mighty,
That none may penetrate.

— Simon and Garfunkel, *I Am a Rock*

2.1 Introduction

Abbreviations

ADCC:	Antibody-Dependent Cell-mediated Cytotoxicity
ANCA:	Anti-Neutrophil Cytoplasmic Antibodies
APC:	Antigen-Presenting Cell

Immunity is a state of having sufficient biological defences to avoid infection, disease, or other unwanted biological and chemical invasions. It entails the recognition and elimination of infectious organisms. In mammals, two integrated systems achieve this goal — the innate and the adaptive systems. The key word is integrated. The two systems influence, communicate with, and regulate each other. Innate immunity, also referred to as nonspecific[1] or nonadaptive immunity, is developmentally the older of the two. It is the first and immediate response to an infectious challenge. An animal's first exposure to an invader results in a complex and interrelated series of events, which only after 4-7 days culminates in an effective specific response. During this time, protection is primarily limited to the largely nonadaptive component of immunity. In the past, the innate immune system was naïvely assumed to be simple, like a coat of armour that helped hold pathogens at bay. We should have known better. Evolution has ensured that *nothing* in human biology is simple! Innate immunity is now known to be involved in the triggering and regulation of adaptive immunity, besides being the first and second line of defence. It is thus critical to an individual's well-being. Physical and emotional stress, chronic diseases, old age, and malnutrition can affect the innate immune system, predisposing a person to ill health (see the sidetrack *Staying Fighting Fit*). To summarise:

- Innate immunity is the most primitive form of defence, present in all metazoans and some plants.
- It consists of molecules and cells that recognise specific, conserved constituents of microorganisms, that is, it acts against any potential threat to the body.
- The receptors involved in threat recognition are germ-line encoded, meaning that their specificity is genetically determined. Unlike the antigen receptors of adaptive immunity, these receptors are not clonally distributed; identical receptors are present on all clones of a given cell type.
- The innate immune system can be mobilised within hours of contact with a potential pathogen.
- It does not bestow lasting protection, and the degree of resistance conferred remains unchanged even after repeated challenges by the same agent.

Innate immunity is the combined effect of several factors. These factors are broadly divided into two groups — the physiological and chemical barriers that prevent entry and establishment of pathogens and the cells that are actively involved in their elimination. The two are, of course, intimately linked.

2.2 Physiological and Chemical Barriers

These are relatively passive players of innate immunity.

- **Epithelial surfaces**[2]: The body surface, which is in constant contact with the outside environment, is covered by the epithelium. Intact healthy skin, consisting of an outer layer of keratinised dead cells covering the epidermis, is virtually impenetrable to most organisms. The repeated shedding of skin's cells further helps dislodge organisms from its surface. Mucous membranes present in areas not covered by skin have a damp, wet surface that acts as an efficient trapping agent. Ciliated cells present at some sites (for example, in the respiratory tract) help sweep away any foreign particles that may gain entry. Besides acting

[1] Given the fact that the recognition systems involved are actually very specific for certain signature molecules of microorganisms, it is time to reconsider the 'nonspecific' designation!

[2] The epithelium is a diverse group of tissues that covers or lines nearly all body surfaces, cavities, and tubes. It functions as an interface between different biological compartments. The epithelium is separated from underlying supporting tissue by a basement membrane consisting of myriad glycoproteins. Epithelial layers provide physical protection and containment. They are also involved in organ-specific transport. Often, one surface of an epithelial cell (called the *apical surface*) comes in contact with a non-sterile environment. Hence, these cells have special junctions called *tight junctions* that join them and do not allow movement of fluids across cells. Various epithelia are classified according to the arrangement and shapes of the cells present. The endothelium, by contrast, is the inner layer of capillaries, blood vessels, lymphatics, etc., and is an essentially sterile environment.

as physical barriers, epithelial cells lining the skin, gastrointestinal tract, and bronchi express antimicrobial peptides. These peptides prevent pathogens from adhering to epithelial surfaces. Incidentally, production of such antimicrobial peptides is found in almost all living systems — from amoebae to frogs to mammals — and the peptides have a similar structural design and mechanism of action.

Staying Fighting Fit: Nutrition, Exercise, Stress, and Immunity

It makes intuitive sense that lifestyle choices directly influence immunity — both innate and adaptive. Recent studies provide scientific backing for this intuitive wisdom.

Diet and immunity have long been considered to be intertwined by ancient medical traditions and popular wisdom. Systematic studies show that nutrient deficiencies impair immune responses and can lead to frequent and severe infections, especially in children. Protein–energy undernutrition results in a reduction in numbers and functions of T cells and phagocytic cells. Secretory IgA (Immunoglobulin A) antibody response is also impaired. In addition, levels of many complement components are reduced. Other nutritional factors — especially vitamins A, B_6, C, E, and trace elements, such as Cu, Fe, Se, and Zn — have been found to be important to optimal immune function. Their mode of action is, unfortunately, not well understood.

- **Vitamin A** deficiency impairs innate immunity by impeding the normal regeneration of mucosal barriers damaged by infection and by diminishing the function of neutrophils, macrophages, and NK (Natural Killer) cells. Vitamin A is required for adaptive immunity and plays a role in the development of both **T H**elper (T_H) cells and B cells. In particular, vitamin A deficiency diminishes T_H2 cell-mediated antibody responses.
- **Vitamin D** is important in suppressing autoimmune disorders. It is thought to suppress inflammatory T cell activity by stimulating TGF-β and IL-4 production.
- **Vitamin E** is a potent antioxidant with the ability to modulate immune functions. Deficiency of this vitamin leads to a downward trend in most immune parameters. Vitamin E plays an important role in the differentiation of immature T cells in the thymus. Its deficiency leads to decreased differentiation of immature T cells, and it results in an early decrease of cellular immunity with spontaneous aging in hypertensive rats. In animals, vitamin E supplementation induces the early reversal of thymic atrophy following X-ray irradiation. Vitamin E is important for effective maintenance of the immune system, especially in sick and elderly people. Decreased cellular immunity caused by aging or the development of HIV/AIDS is markedly improved by a diet high in vitamin E.
- **Zn** deficiency is associated with profound impairment of cell-mediated immunity. It results in reduced $CD4^+$ and $CD8^+$ T cell numbers and decreased chemotaxis of phagocytes. In addition, thymic function is also affected, as levels of Zn-dependent thymic hormones markedly decrease with Zn deficiency.
- **Cu**, like Zn, is critical to the proper functioning of the innate and adaptive immune systems. Research has shown that even marginal Cu deficiency is

CNS:	Central Nervous System
CR:	Complement Receptor
DC:	Dendritic Cell
ECF-A:	Eosinophil Chemotactic Factor-A
ECP:	Eosinophil Cationic Protein
EDN:	Eosinophil-Derived Neurotoxin
EPO:	Eosinophil Peroxidase
FcR:	Fc Receptor
G-CSF:	Granulocyte-Colony Stimulating Factor
GM: -CSF	Granulocyte Macrophage-Colony Stimulating Factor
HPA: axis	Hypothalamus-Pituitary-Adrenal axis
ICAM:	Intercellular Adhesion Molecule
IFN:	Interferon
Ig:	Immuno-globulin
ITAM:	Immuno-receptor Tyrosine-based Activating Motif
ITIM:	Immuno-receptor Tyrosine-based Inhibitory Motif
KIR:	Killer cell Ig-like Receptor
LAK: cell	Lymphokine-Activated Killer cell
LBP:	LPS-Binding Protein
LFA:	Leukocyte Function associated Antigen
LPS:	Lipopolysaccha-ride
MBL:	Mannose-Binding Lectin
MBP:	Major Basic Protein
MCP:	Macrophage Chemotactic Protein
M-CSF:	Macrophage-Colony Stimulating Factor
MIP:	Macrophage Inhibitory Protein

linked to IL-2 deficiency and is likely to reduce T cell proliferation. In case of severe Cu deficiency, the number of neutrophils in human peripheral blood is reduced. The neutrophils' ability to generate the superoxide anion and kill ingested microorganisms is decreased in both overt and marginal Cu deficiency.

- **Arachidonic acid-derived eicosanoids**[3] modulate the production of pro-inflammatory and immunoregulatory cytokines. Overproduction of these cytokines is associated with septic shock and chronic inflammatory diseases, such as arthritis and multiple sclerosis. The n-3 **P**oly**u**nsaturated **F**atty **A**cids (PUFAs), eicosapentaenoic acid, and docosahexaenoic acid, all found in fish oils, suppress production of arachidonic acid-derived eicosanoids. Thus, dietary fats rich in n-3 PUFAs have the potential to alter cytokine production. Several human studies show that supplementing the diet of healthy volunteers with n-3 PUFAs results in the reduced *ex vivo* production of IL-1, IL-6, TNF-α, and IL-2 by peripheral blood mononuclear cells. Animal studies indicate that dietary fish oil reduces responses to endotoxins and pro-inflammatory cytokines, resulting in increased survival; such diets have been beneficial in some models of bacterial challenge, chronic inflammation, and auto-immunity.

Exercise is believed to increase immunity. Moderate exercise seems to increase resistance to upper respiratory tract infections, irrespective of age. Repeated strenuous exercise, on the other hand, suppresses immune function. After intense long-term exercise, the immune system is characterised by concomitant impairment of the cellular immune system and increased inflammation. Strenuous exercise seems to decrease the number of circulating lymphocytes and their proliferation, and it suppresses innate immunity. The levels of secretory IgA antibodies in saliva are lowered simultaneously with high levels of circulating pro-inflammatory and anti-inflammatory cytokines. Conversely, exercise has been found to attenuate changes in the immune system that are related to aging, provided the exercise is long-term and of a sufficient volume to induce changes in body weight and fitness. Only such exercise can improve immunity, especially in the old.

Stress is known to affect immunity in different ways. There is continuous direct and indirect chemical communication between the neuroendocrine and immune systems through neurotransmitters and cytokines. Although growth hormones are involved in the activation of phagocytic cells, glucocorticoids severely impair the phagocytic and cytotoxic activities of neutrophils and macrophages. Their capacity to produce ROI (**R**eactive **O**xygen **I**ntermediates), induce NOS (Nitric Oxide Synthase), and secrete lysosomal enzymes in response to activation is substantially reduced. The oxidative burst of phagocytes is also inhibited by epinephrine and β-endorphins. Stressed individuals are therefore likely to have impaired innate immune responses and be more susceptible to infections.

Aging is associated with a generalised reduction in immunity and increased inflammatory activity, reflected by increased circulating levels of TNF-α, IL-6, cytokine antagonists, and acute-phase proteins *in vivo*. Epidemiological studies suggest that chronic low-grade inflammation in aging promotes an atherogenic profile and is related to age-associated disorders (e.g., Alzheimer's disease, atherosclerosis, and type 2 diabetes) and enhanced mortality risk. Therefore, the aging process is suggested to be linked to dysregulated production of inflammatory cytokines. *In vivo*, infectious models show delayed termination of inflammatory activity and a prolonged fever response in elderly people, suggesting that the acute-phase response is altered by aging.

NK cell:	Natural Killer cell
NO:	Nitric Oxide
NOS:	NO Synthase
PAF:	Platelet Activating Factor
PAMP:	Pathogen-Associated Molecular Pattern
PDGF:	Platelet-Derived Growth Factor
PECAM-1	Platelet Endothelial Cell Adhesion Molecule-1
PMN:	Polymorpho-nuclear granulocyte
PRR:	Pattern Recognition Receptor
PUFA:	Polyunsaturated Fatty Acids
RNI:	Reactive Nitrogen Intermediates
ROI:	Reactive Oxygen Intermediates
TH cell:	T helper cell
TLR:	Toll-Like Receptor
TNF:	Tumour Necrosis Factor

[3] Eicosanoids derive their name from Greek *eicosa*, meaning 20. They are a class of lipid mediators that have 20 carbon fatty acid derivatives with a wide variety of biological activities. Four main classes of eicosanoids are recognised — prostaglandins, prostacyclins, thromboxanes, and leukotrienes.

They are amphipathic (i.e., have clusters of hydrophobic and cationic amino acids) in nature and target the negatively charged molecules on the microbial membranes outer surface. Peptides with a significant role in innate defence include:

- **Defensins and cathelicidins**, two major families of antimicrobial peptides, are found in a variety of tissues. They may be expressed either constitutively or induced by the action of pro-inflammatory cytokines[4]. In addition to their antimicrobial action, both cathelicidins and defensins possess chemotactic properties for various phagocytic leukocytes, immature **D**endritic **C**ells (DCs), and lymphocytes. Thus, they seem to have a role in alerting, mobilising, and amplifying innate and adaptive antimicrobial immunity.
- **Dermicidin** is a protein specifically and constitutively expressed in sweat glands. It is secreted into sweat and transported to the epidermal surface. Here it undergoes proteolysis to generate a broad-spectrum antimicrobial peptide that remains active over a wide pH range and in high salt concentrations.
- **Melanin** is the pigment responsible for skin colour. It protects against damage caused by exposure to UV rays, also found in sunlight. It is therefore important in the prevention of UV-induced skin cancer; having less melanin in your skin provides less protection from the damaging effects of sunlight. Recent research shows that melanin is also a potent antibacterial and antifungal agent[5].

- **Adverse conditions in the digestive tract**: The digestive tract is constantly exposed to a wide variety of microorganisms; therefore, a number of protective factors operate there. The pH of the stomach (pH 1-2) helps eliminate a large fraction of organisms entering the digestive tract. A reversal of pH (acid to alkaline), caused by the presence of bile in the gut, helps eliminate many acidophilic organisms that survive the stomach. The surfactant property of bile also aids in microbial destruction.
- **Normal flora**: Microbes colonise all exposed surfaces of the body and live in semi-peaceful co-existence with the body. To protect its ecological niche, normal flora produces many antimicrobial factors, such as colicin (gastrointestinal tract), long chain fatty acids (skin), and acidity (vagina). Moreover, the sheer unavailability of space prevents potential pathogens from colonising areas occupied by normal flora, whom they compete with for nutrients. Normal flora, thus, unintentionally helps defend the body.
- **Microbicidal factors in tissues and blood**:
  - **Lysozyme**, present in tears and saliva and other secretions, such as those of the nose and tissues, is effective against Gram-negative organisms.
  - **Lactoferrin**, a ubiquitous and abundant constituent of human external secretions, chelates iron and makes it unavailable to bacteria.
  - **Complement** consists of a complex group of heat-labile proteins (C1-C9) found in blood. It is important in inflammation, opsonisation[6], activation of macrophages, etc. Complement components aid phagocytosis by immobilising the organism on the macrophage surface and then activating the process of ingestion and degradation of the internalised organism. Complement activation also leads to the lysis of target cells (see chapter 3). The C3a and C5a released by this activation are chemotactic and help recruit phagocytes to the site of invasion. Complement deficiency can lead to increased susceptibility to infections by Gram-negative organisms, as well as to autoimmune diseases[7].
  - **Properdin**, a γ globulin that can activate complement, is found to exert haemolytic, bactericidal, and virucidal actions (see section 3.4).

[4] Cytokines are soluble glycoproteins produced by a variety of cells. They are multipotent (i.e., have multiple effects) and play a major role in the initiation, regulation, and course of the immune response. Cytokines also affect a variety of other processes. Those that promote inflammation include **T**umour **N**ecrosis **F**actor-α (TNF-α), interleukins such as IL-1, IL-6, and **Inter**fero**n**-γ (IFN-γ). See appendix II.

[5] Beware, users of antiperspirants and 'fairness creams'!

[6] Certain proteins such as acute-phase proteins, antibodies, and complement components can interact with target cells and coat them. These proteins are called *opsonins*. Specific receptors on the cell surfaces of phagocytes recognise opsonins. The opsonin-coated particle therefore gets bound to the phagocyte's cell surface. Opsonins enhance engulfment and digestion of target particles by phagocytes; the process is called *opsonisation*.

[7] Autoimmune diseases are caused when the immune system attacks self-tissues and organs.

Iron Politics: Fe and Innate Immunity

During the febrile response, most bacteria have a decreased ability to synthesise their own Fe chelators (siderophores), and they therefore need an iron source. As iron is needed for the bacterial electron transport chain, its unavailability inhibits the growth of most bacteria. During infection, the body makes considerable metabolic adjustments to deny microorganisms access to iron. Much of this is due to the production of a defence chemical called leukocyte-endogenous mediator, which lowers blood iron content and reduces iron absorption from the gut. The amount of Fe in plasma declines with a concomitant increase in Fe storage in the form of ferritin. Synthesis of Fe chelators, such as lactoferrin and transferrin, which trap Fe for use by human cells while making it unavailable to most microbes, is increased. Available Fe is further decreased by transporting lactoferrin to common sites of microbial invasion, such as the mucous membranes. Recent evidence indicates that the chelating activity of lactoferrin causes a characteristic twitching in underlying cells, dislodging bacteria and preventing them from forming biofilms at the environmental interface. By contrast, transferrin gains entry into tissues during inflammation, causing a local decline in iron concentrations. However, some bacteria (such as *Neisseria gonorrhoeae, N. meningitidis*, and *Haemophilus influenzae*) have receptors for human lactoferrin and transferrin, and they can utilise iron that is bound to these compounds.

- **Interferons**[8] (IFN-α, IFN-β, and IFN-γ) are potent immunomodulatory cytokines that also link adaptive and innate immunity. They are especially important in host defence against viral infections.
  - IFN-α and IFN-β, also called type I IFNs, are produced by various cell types upon exposure to viruses. A type of DC, called plasmocytoid DC, is the major source of these IFNs. IFN-α does not denote a single cytokine, but is a family of closely related proteins, whereas IFN-β is a single protein. The binding of these IFNs to receptors on infected cells induces the formation of molecules that interfere with viral replication. Type I IFNs also activate **N**atural **K**iller (NK) cells. In certain viral infections, they potentiate the production of IFN-γ, by itself a potent immune stimulator. In addition, they induce MHC[9] class I expression on uninfected cells, making them resistant to the action of NK cells (section 2.3.2).
  - IFN-γ (type II IFN) is produced predominantly by T lymphocytes and NK cells. It enhances macrophage activity and triggers DC maturation. DCs are crucial in the triggering of an adaptive response. IFN-γ also increases expression of MHC molecules by all types of cells and thus augments the activation of the adaptive immune response. Apart from its role in inflammation and the potentiation of NK cell activity, IFN-γ also plays a crucial role in the development and functioning of a subset of helper T cells known as TH1 cells (see section 9.5.1).

- **Other physiological nonspecific defence mechanisms**: Although often diagnosed and treated as deleterious reactions, diarrhoea, vomiting, and sneezing are all designed to dislodge microorganisms and therefore constitute a part of the innate defence mechanism. If uncontrolled, these physiological processes may prove harmful. The IL-1 (**I**nter**l**eukin-1), IL-6, **T**umour **N**ecrosis **F**actor-α (TNF-α), and lymphotoxin produced by damaged tissue cells, activated macrophages, lymphocytes, and brain cells affect the hypothalamus — the centre of temperature control — resulting in pyrexia (fever). Increase in body

[8] These cytokines are called *interferons* because they interfere with the replication of viral RNA or DNA.

[9] MHC stands for **M**ajor **H**istocompatibility **C**omplex. These proteins are crucial in triggering the adaptive immune response and are discussed in chapter 7.

temperature is also a defence mechanism — many pathogens are sensitive to or grow poorly at higher temperatures[10] while phagocytosis is enhanced at elevated temperatures. Thus, fever can be a beneficial response, provided it does not rise too high (above 104°F, 40°C) and is of short duration.

- **Pattern Recognition Receptors or molecules (PRRs)**: Molecules on the cell envelope of microorganisms form unique geometric patterns and are often referred to as PAMPs (**P**athogen-**A**ssociated **M**olecular **P**atterns). Such PAMPs become excellent targets for the host system because they are indispensable for microbial survival, are unique to those microbes, and are found on a large group of microbes. The LPS (**Lipo**poly**s**accharide) of Gram-negative bacteria is one of the best examples of PAMPs. Host molecules capable of recognising and reacting with these molecular patterns are termed **P**attern **R**ecognition **R**eceptors. Thus, the innate immune system seems to be designed to focus on a few highly conserved structures in a large group of potential pathogens rather than on recognising individual organisms. In contrast to the antigen receptors of adaptive immunity (discussed in chapter 6), PRRs are expressed on effector and noneffector cells of innate immunity. Also, their expression, unlike immunoglobulins or Igs, is not clonal — each type of PRR expressed by a particular cell type has identical specificity. PRR engagement by a molecular pattern on the pathogen also triggers the effector cell to perform its function immediately, rather than after proliferation. The concentration of some of these receptors increases dramatically in the acute-phase reaction (section 2.4.2). Three broad functional classes of PRRs are recognised: secreted, endocytic, and signalling PRRs.
  - **Secreted PRRs** function as opsonins; the binding of these molecules to microbial cell walls tags them for phagocytosis and results in their lysis by the complement cascade.
    - **C-reactive protein**, so named because it was found to bind the C-protein (a phosphorylcholine moiety) of pneumococci, can recognise and bind to molecular groups found on the cell walls of various bacteria and fungi (e.g., phosphocholine residues, lipoproteins, monophosphate esters, and histones). It can also activate the complement cascade. It is thought to be a prototype pattern recognition molecule. The presence of this protein in serum is a recognised indicator of pathological activity.
    - **LBP** (**L**PS-**B**inding **P**rotein) binds LPS, the major outer surface membrane component of Gram-negative bacteria. It is an extremely strong stimulator of innate immunity in species ranging from insects to humans. LBP is an acute-phase protein with the ability to bind and transfer bacterial LPS. LBP binds LPS and then transfers the complex to cell-surface CD14[11], a receptor on macrophages and B cells. CD14 engagement triggers the inflammatory response through the engagement of **T**oll-**L**ike **R**eceptors (TLRs), described later in this section. LBP is also thought to be involved in alcohol-induced liver injury.
    - **Collectins** are a group of PRRs that are structurally similar to the complement component C1q, itself a PRR. Their name derives from their collagen domain, which is connected to a lectin[12]-binding domain. Collectins recognise distinct but overlapping carbohydrate ligands, such as mannose, N-acetylglucosamine, and L-fucose, found on microbial cell walls (Table 2.1). They do not bind galactose and sialic acid, two carbohydrates commonly found on mammalian cell walls. The binding

[10] In 1927, Wagner-Jauregg was awarded the Nobel Prize for having discovered malariotherapy for curing neurosyphilis. As part of this therapy, neurosyphilis patients were inoculated with the malarial parasite *Plasmodium vivax*. The resultant high temperatures caused the death of the spirochete that caused syphilis. The patients were eventually treated with quinine to get rid of the malaria.

[11] The term CD (**C**luster of **D**ifferentiation) is a historical term used to define cell-surface molecules recognised by a given set of antibodies of a single specificity. As new molecules were discovered and defined by scientists, they were given numbers reflecting the order of discovery. Thus, the name CD14 indicates that it was the 14th molecule so defined and has no bearing on the molecules' function. Determining the function was an altogether different task from discovering the molecule. Some functions are still to be determined. Appendix I details the CD molecules mentioned in this book.

[12] Lectins are a group of proteins that specifically bind or crosslink carbohydrates; C-type lectins or Ca-dependent lectins require Ca^{2+} for binding.

specificity of these proteins is due to the shape of their respective binding pockets, which help distinguish subtle alterations in topology. Collectins include C1q, ficolins, conglutinin, and **M**annose-**B**inding **L**ectin or **P**rotein (MBL/MBP[13]). Both C1q and MBL opsonise bacteria, enhance bacterial clearing by phagocytes, and activate complement (see chapter 3). MBL is a serum protein that selectively recognises and binds to mannose or N-acetylglucosamine on microbial surfaces.

- **Endocytic PRRs** are present on a phagocyte's surface. These receptors mediate the uptake and delivery of pathogens into lysosomes.
  - **The mannose receptor** is a member of the Ca-dependent lectin family. It specifically recognises carbohydrates with a large number of mannoses or fucoses. Besides initiating phagocytosis, engagement of this receptor leads to the secretion of cytokines like TNF-α, IL-1, IL-6, and IL-12.
  - **Scavenger receptors** are a broad group of receptors expressed by myeloid cells (macrophages and DCs) and certain endothelial cells. They bind and internalise microorganisms and their products, including lipoteichoic acids of Gram-positive bacteria, LPS of Gram-negative bacteria, intracellular bacteria, and CpG DNA[14]. They are involved in lipid metabolism and bind modified low-density lipoproteins.
- **Signalling receptors** recognise PAMPs and activate signal transduction pathways (i.e., they transmit the signal generated by the binding of the ligand to the receptor across the cell membrane and thus activate a series of reactions; see chapter 6 for a better understanding of signal transduction).
  - **CD14**: It is expressed on phagocytic cells, although a soluble form is also found in serum. It recognises LPS and other cell wall components of Gram-negative and Gram-positive bacteria[15]. CD14 does not have a cytoplasmic domain. Hence, it requires TLRs (see below) for signal transduction. The binding of LPS or other bacterial constituents to CD14 results in the activation of a number of genes involved in innate and adaptive immune responses.
  - **Toll receptors**: The first Toll receptor was identified in *Drosophila* and was shown to be involved in embryonic development. It was later found to be involved in the immune responses of adult flies as well. Homologues of Toll were also identified in humans, and these homologues are referred to as **T**oll-**L**ike **R**eceptors (TLRs). To date, 11 members of the TLR family have been reported. They play a key role in the induction of immune and inflammatory responses to pathogens. TLRs recognise conserved products of microbial metabolism, such as LPS, peptidoglycan, and lipoteichoic acids (Figure 2.1). They are expressed by a variety of cells, including macrophages, immature DCs, vascular endothelial cells, and intestinal epithelial cells. Activation of TLRs induces the expression of a variety of cytokines and costimulatory molecules[16] that are also crucial to initiation of adaptive responses.
  - **Integrin family members**: Although known to be adhesion molecules that bind cells, some members of the integrin family also function as PRRs. Heterodimers of CD11b/CD18 (Mac-1 or Complement Receptor 3 — CR3) and CD11c/CD18, present on macrophages and neutrophils, can recognise microbial products, such as LPS, lipophosphoglycans, and β-glucans, and activate the phagocytes.

[13] MBL is the preferred name, as it helps avoid confusion between the mannose-binding protein and the major basic protein of eosinophils.

[14] Unmethylated CpG dinucleotides are very common in bacteria but highly under-represented in vertebrates. Not surprisingly, PRRs that can recognise these bacterial oligonucleotides have evolved in vertebrates. CpG DNA is thus a very potent immune stimulator.

[15] LBP is required for the binding and transfer of LPS to CD14.

[16] Costimulatory molecules are expressed on cells of the immune system. Engagement of these molecules is a prerequisite for the initiation of adaptive responses. See chapters 6, 8, and 9 for more information.

Figure 2.1 TLRs play a key role in the induction of immune and inflammatory responses to pathogens. These leucine-rich receptors recognise a variety of products of microbial and host origins. Some, such as TLR2, associate with other TLRs (e.g., TLR1 or TLR6; represented as TLRx) and are expressed on cell surfaces. TLR2 heterodimers can specifically bind products of microbial cell walls (LPS, peptidoglycan, yeast cell walls). CD14, also expressed on cell surface, recognises and binds LPS, but requires TLR4 for signal transduction. TLR5 is expressed on the basolateral surfaces of intestinal epithelial cells that are not exposed to the outside environment. These receptors are engaged only when flagellated bacteria penetrate epithelial surfaces. Engagement of TLR5 activates pro-inflammatory gene expression. TLR9, essential for responses to bacterial or viral DNA and CpG DNA, is not expressed on the surface but is found in the lysosomes of host cells. Microbial DNA has to be internalised to trigger immune activation through TLR9.

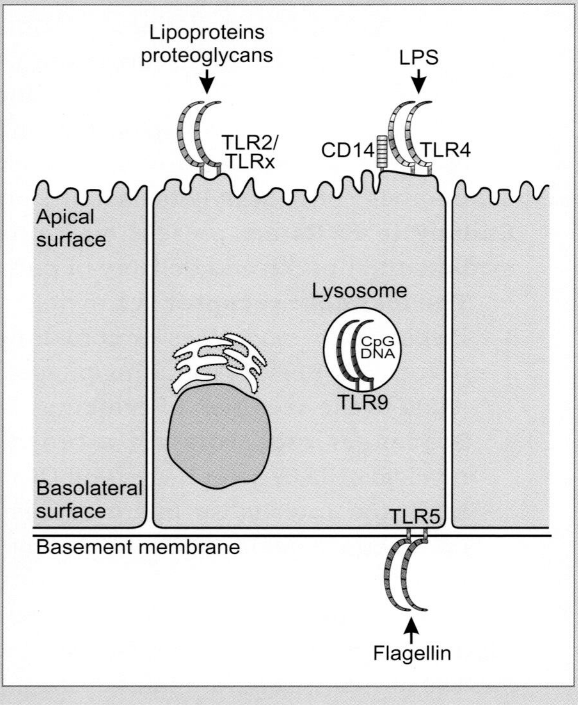

2.3 The Cellular Players

Several cells are actively involved in the elimination of pathogens. These include phagocytic cells, which internalise and kill microbes, and cytolytic cells, which bring about their lysis. Also included are some lymphocytic cells. Although originally thought to be involved solely in adaptive immunity, three subsets of lymphocytes are now known to participate in innate immune responses. DCs, which are a large and diverse family of leukocytes, are also considered part of the innate immune system. Their sole function is in triggering the adaptive immune response.

2.3.1 Phagocytic Cells

Phagocytosis is a means of obtaining nutrition in unicellular organisms. However, it was adapted and transformed into a protective mechanism in multicellular organisms. Thus, although multiple cell types (such as epithelial cells and endothelial cells) have limited phagocytic capacity, they are inefficient in phagocytosis — that is not their principal function. True phagocytic cells are extremely efficient at internalisation and degradation; scavenging is one of their main functions, and it is these cells that we consider here. Phagocytes are found in almost all body tissues and are strategically located to maximise encountering and trapping of foreign particles. Originally thought to merely be scavengers that mop up after lymphocytes, phagocytic cells have now been established to have a key role in host defence and maintenance of normal tissue structure and function. Their major functions are:

❒ To **engulf**, **internalise**, **and destroy foreign particles** (especially large particles >0.5 μm)

Besides clearing infectious agents, phagocytes clear senescent (old) cells and

Table 2.1 Common Human PRRs

PRR	Type of PRR	PAMPs recognised	Function
C-reactive protein	Secreted	Molecular groups found on cell walls of a variety of bacteria and fungi, including phosphocholine moieties, lipoproteins, monophosphate esters, and histones	Promotes opsonisation and activates the complement cascade
LBP	Secreted	LPS	Binds LPS and transfers the complex to CD14
Collectins	Secreted	Bind glucose, L-fucose, mannose, N-acetylmannosamine, and N-acetyl-glucosamine residues on microbial cell walls C1q binds IgG or IgM antibodies MBL recognises mannose or N-acetyl-glucosamine residues Ficolins recognise N-acetylglucosamine, N-acetylgalactose, and L-fucose	C1q, MBL, and ficolins opsonise bacteria, enhance their clearance, and activate complement
Mannose receptor	Endocytic	Carbohydrates with many mannoses or fucoses	Initiates phagocytosis; promotes secretion of pro-inflammatory cytokines by macrophages
Scavenger receptors	Endocytic	Bind microorganisms and their products, including lipoteichoic acids, LPS, and CpG DNA, and low-density lipoproteins	Initiate internalisation and may alter cell morphology
CD14	Signalling receptor	Binds LPS and other cell wall constituents of Gram-positive and Gram-negative bacteria	With TLRs, activate several genes involved in innate and adaptive immune responses
TLRs	Signalling receptors	Recognise conserved products of microbial metabolism, such as LPS, peptidoglycan, and lipoteichoic acids	Activation induces the expression of a variety of cytokines and costimulatory molecules crucial to the initiation of adaptive immune responses

cellular debris from tissues. Activated phagocytes have potent microbicidal activity and are therefore important in cell-mediated immunity (see chapter 11).

- To **secrete biologically active compounds**
  Phagocytes secrete numerous hydrolytic enzymes, complement components of the classical and alternative pathway, plasma proteins, coagulation factors, oxygen metabolites, metabolites of the arachidonic pathway, etc. Thus, they are crucial not only in host defence but also in tissue regeneration (Table 2.2). Macrophages secrete TNF-α, IL-1, and IL-6 — cytokines involved in the inflammatory response. These cytokines are termed *endogenous pyrogens*, as they are responsible for elevation of body temperature. These cytokines also induce the synthesis of acute-phase proteins by liver hepatocytes (section 2.4.2).
- To **present antigen** (i.e., act as **A**ntigen-**P**resenting **C**ells; APCs)
  Macrophages internalise pathogens or other antigens, breakdown their proteins/lipids to small fragments, and then load and display the fragments on MHC/CD1 molecules (chapter 7). Recognition of the peptide:MHC (or lipid:CD1)

Table 2.2 Macrophage Products and Their Functions

Function	Product
Microbicidal	
*ROI**	Superoxide anion, H_2O_2, hydroxyl radicals, singlet oxygen, chloramines
*RNI**	NO*, nitrates, nitrites
Bioactive peptides	Defensins, cathelicidins, glutathione
Enzymes	Neutral proteases, acid hydrolases, lysozyme
Tissue acting	
Tissue damaging	ROI, RNI, TNF-α*, neutral proteases
Tissue remodelling	Elastase, collagenase, hyaluronidase, growth factors (fibroblast growth factor, angiogenesis factor, TGF-α*, TGF-β)
Pyrogenic	IL-1, TNF-α, IL-6
Inflammation regulators	
Cytokines	IL-8, IFN-γ, erythropoietin, MIP-1*, MIP-2, MIP-3, G-CSF*, GM-CSF*, M-CSF*, and PDGF*
Metabolites of the arachidonic pathway	**P**rosta**g**landins (PGE2, PGE2α), prostacyclin, thromboxane A2, **L**euko**t**rienes (LTB4, LTC4, LTD4, LTE4)
Coagulation factors	Thromboplastin; factors V, VII, IX; prothrombin; plasminogen activator and its inhibitor
Complement components	C1–C5; properdin; Factors B, D, I, P, H
Proteinases	Plasminogen activator, neutral proteases (collagenase, elastase, angiotensin convertase, stromelysin), and acid proteases (cathepsin D, L)
Acid hydrolases	Lysozyme, lipases, glycosidases, ribonucleases, phosphatases
Protease inhibitors	α2-macroglobulin, α-antitrypsin, plasmin, collagenase inhibitor, plasminogen activator inhibitor
Stress proteins	Heat shock proteins
Bioactive peptides	Glutathione

*G-CSF — **G**ranulocyte-**C**olony **S**timulating **F**actor, GM-CSF — **G**ranulocyte **M**acrophage-**C**olony **S**timulating **F**actor, M-CSF — **M**acrophage-**C**olony **S**timulating **F**actor, MIP — **M**acrophage **I**nhibitory **P**rotein, NO – **N**itric **O**xide, ROI — **R**eactive **O**xygen **I**ntermediates, RNI — **R**eactive **N**itrogen **I**ntermediates, PDGF — **P**latelet-**D**erived **G**rowth **F**actor, TGF — **T**ransforming **G**rowth **F**actor

complex by T lymphocytes results in these cells' activation and culminates in an adaptive immune response.

- To **retain** the **antigen** for a prolonged period of time

  This function helps avoid overwhelming the adaptive immune system. The phagocytes thus have a major role in augmenting and controlling the adaptive immune response.

i. Macrophages and related cells: Macrophages are actively phagocytic cells that arise from bone marrow stem cells or circulating monocytes. These cells display considerable heterogeneity and are relatively long-lived (average life: 6-16 days). They are found in connective tissues, at other sites of possible pathogenic challenge (such as the gastrointestinal tract and lungs) and in organs (such as the liver and spleen) that take care of old and dying cells (Table 2.3). Mature macrophages can localise to different tissues and become fixed macrophages (called resident

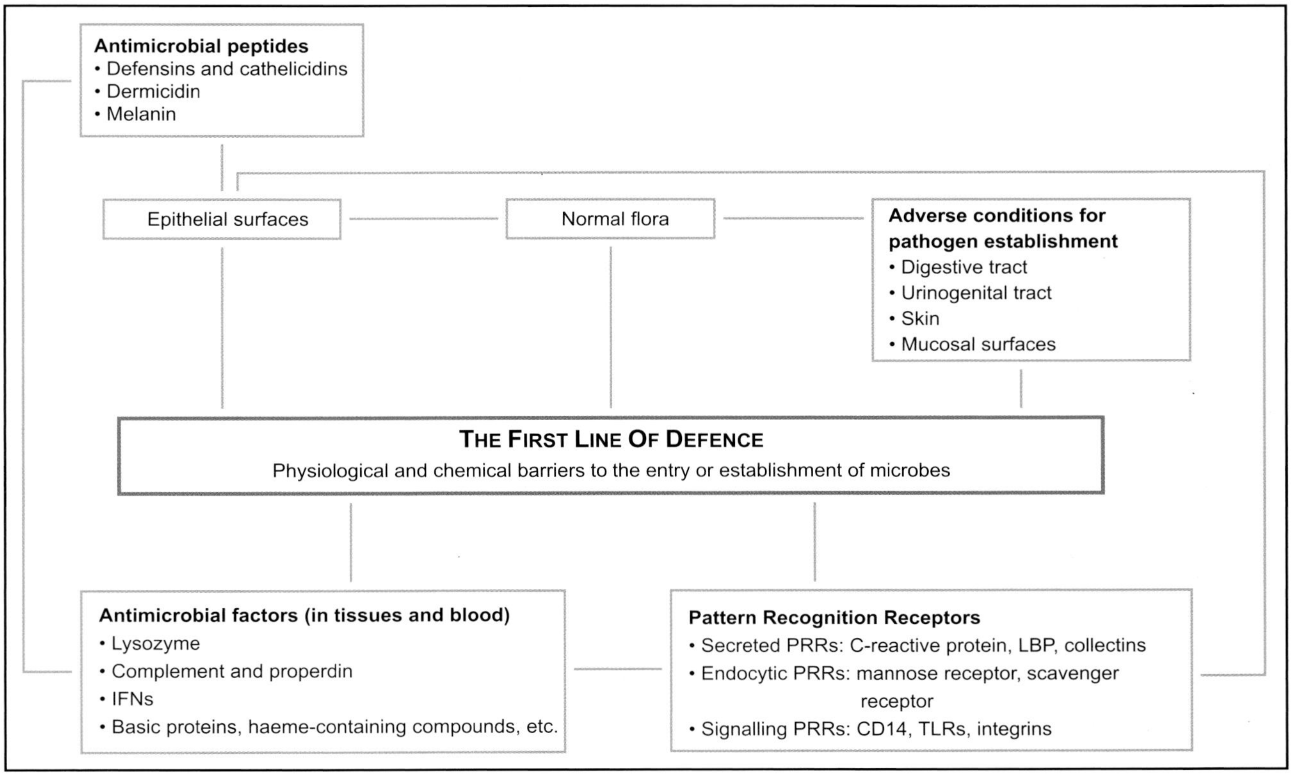

Mind Map 2.1 The first line of defence

macrophages) or remain motile and become free wandering macrophages. Tissue macrophages proliferate locally, but their numbers can also be augmented by the influx of blood monocytes. Normal tissue macrophages are immunologically quiescent. They have low O_2 consumption and do not secrete cytokines. Cytokines, such as IFN-γ or TNF-α, produced in response to invasion, activate them. **Activated macrophages express a variety of cell receptors and secrete an array of antimicrobial agents and cytokines, turning them into potent mediators of innate immunity**.

Macrophages and monocytes express a number of specialised receptors and proteins on their cell surfaces. The number and type of receptors expressed at any point of time depends upon the state of activation (quiescent *versus* activated), location, life cycle, and local milieu. These receptors include PRRs, FcRs[17], and receptors for complement components, cytokines, chemokines[18], and growth factors. Like all nucleated cells of the body, they express MHC class I molecules. They also express MHC class II molecules that are involved in antigen recognition by $CD4^+$ T cells. The expression of these molecules is upregulated by activation or exposure to cytokines (such as TNF-α and IFN-γ) or microbial products (such as LPS). Important types of macrophages include:

- **Monocytes**: Promonocytes in the bone marrow give rise to blood monocytes. These form a circulating pool, from where they migrate to various organs and tissue systems to give rise to macrophages. Monocytes can also differentiate into immature DCs. Monocytes are thought to regularly migrate from the blood into tissues, where they differentiate into immature DCs or macrophages, depending upon the cytokine environment (Figure 2.2). Monocytes that enter a

[17] Antibodies (immunoglobulins) are Y-shaped structures that bind specifically to antigens by the two arms of the Y. The stem of the Y is called the Fc region. Many cell types express receptors for this region. These receptors are collectively called *FcRs* (chapter 10). Both, the type of cell and FcR, dictate the outcome of binding of the antigen–antibody complexes to the FcR.

[18] Chemokines are cytokines that attract phagocytic cells to the site of inflammation. Besides directing the migration of leukocytes along a concentration gradient, chemokines are important in their activation (see the sidetrack *Instant Chemistry*, chapter 9).

Table 2.3 Types of Macrophages/Monocytic Cells

Site	Type
Blood	Monocytes, neutrophils
Bone marrow	Promonocytes and monocytes
Brain	Microglial cells
Connective tissue	Histiocytes
Kidney	Mesangial macrophages
Liver	Kupffer cells
Lungs	Alveolar macrophages, interstitial macrophages
Lymph nodes	Resident and recirculating macrophages
Serous fluids	Pleural and peritoneal macrophages
Spleen	Splenic macrophages
Synovial cavity	Synovial cells

site of inflammation have variable fates. They may become resident macrophages, transform into epithelioid cells, fuse with other macrophages to become multinucleated giant cells, or die.

- **Kupffer cells**[19]: Macrophages are most densely concentrated in the lungs and liver. Both organs contain a heterogeneous population of macrophages. Kupffer cells are the liver's resident macrophages, situated at locations that expose them to foreign material entering the liver *via* the blood stream. They are among the first phagocytic cells to encounter the antigen after it is absorbed from the intestinal lumen. Kupffer cells not only digest foreign matter but they also remove a large number of dead self-cells and can therefore trigger an autoimmune response. Fortunately, Kupffer cells express low levels of the MHC class II molecules which are needed to activate $CD4^+$ T lymphocytes, and they are situated in such a way that they are not easily accessible to naïve T cells (i.e., T cells that have never encountered antigen). Kupffer cells are activated by various bacterial stimuli, including LPS. Activated Kupffer cells express MHC class II molecules and can present antigen; they also secrete cytokines that can activate liver NK cells. The liver leukocytes and the adaptive immune responses that they induce play a crucial role in defence against bacterial infections and haematogenous tumour metastases.
- **Alveolar macrophages**: The lungs have the body's largest epithelial surface area in contact with the external environment. As a consequence, they need an active and elaborate defence system. Alveolar macrophages, with interstitial macrophages, are part of this defence. Alveolar macrophages line the alveoli of the lungs and rapidly clear inhaled particulates, such as pollutant particles, allergens, and microorganisms. Apart from other receptors, these macrophages also express receptors for IgA, an antibody involved in mucosal immunity.

[19] First identified by German scientist von Kupffer — thankfully, the system of naming cells or molecules after scientists has been abandoned!

ii. Polymorphonuclear granulocytes (PMNs): PMNs are produced in the bone marrow and are short-lived cells that do not differentiate further. These granulocytes

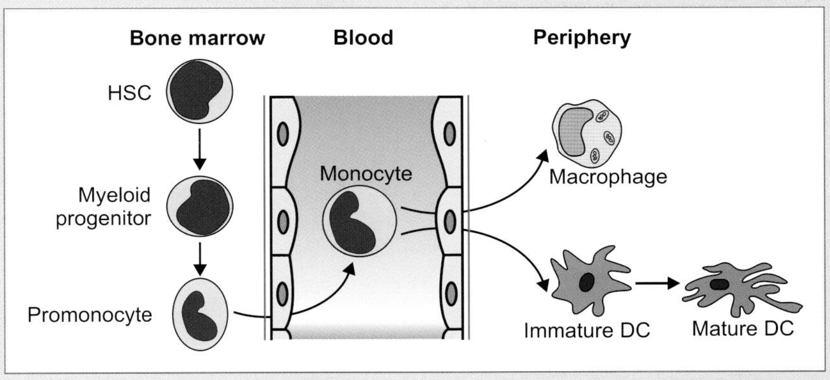

Figure 2.2 Monocytes arise from haematopoietic stem cells and can differentiate to macrophages or DCs. Haematopoietic stem cells (HSC) in the bone marrow give rise to CD34$^+$ common myeloid progenitor cells that further give rise to promonocytes. Monocytes arise from the promonocytes and circulate in blood. Monocytes can further differentiate to macrophages or DCs, depending upon the cytokine milieu. Cytokines, such as GM-CSF, promote differentiation to DC lineage, whereas exposure to M-CSF promotes differentiation to macrophages.

comprise upto 70% of the leukocytes in normal blood. The mature granulocyte contains a multilobed nucleus and many cytoplasmic granules. PMNs have an important role in acute inflammation. They accumulate rapidly at the site of inflammation. Acting in concert with complement and antibodies, PMNs are the major defence against microorganisms. Based on their morphology and staining reactions, PMNs are divided into neutrophils, eosinophils, and basophils.

- **Neutrophils**: Human neutrophils have a multilobed nucleus and a highly granular neutrophilic cytoplasm. They are derived from the bone marrow and represent about 90% of circulating granulocytes and 60 to 65% of blood leukocytes[20]. When released into circulation, these cells are in a quiescent state and have a half-life of only 4-10 hours before they marginate and enter tissue pools. They survive there for 1-2 days. Cells of the circulating and marginated pools interchange with each other. Senescent neutrophils undergo apoptosis, or programmed cell death (Figure 2.3), prior to their removal by macrophages. Being a highly mobile cell type responding to various stimuli, neutrophils have a major role in inflammation and are among the first cells recruited at the site of tissue damage. **Their primary function is phagocytosis**.

  Neutrophil granules are storage depots for mediators of innate host defence. These granules are generated during cell differentiation, and their contents are available on demand as soon as a pathogen is encountered. On the basis of function and enzyme content, human neutrophil granules are divided into three types — azurophil granules, specific granules, and small storage granules. Individual granule populations can also be characterised biochemically and morphologically (e.g., azurophil granules are larger and contain more electron-dense material than specific granules). Their function is not just to provide enzymes for hydrolytic substrate degradation but also to kill ingested bacteria and to secrete contents extracellularly to regulate various physiological and pathological processes, including inflammation.

  Neutrophil granules contain antimicrobial or cytotoxic substances, neutral and acid proteases, acid hydrolases, and a pool of membrane receptors (Table 2.4). Azurophil granules function predominantly in the intracellular milieu (in the phagolysosomal vacuole), where they are involved in the killing and degradation of microorganisms[21]. Myeloperoxidase, present in azurophil granules, is a critical enzyme in the conversion of hydrogen peroxide (H_2O_2) to hypochlorous acid. Along with H_2O_2 and a halide cofactor, it forms

[20] The bone marrow of a normal adult produces 10^{11} neutrophils per day; this number can increase to 10^{12} during infection.

[21] The latest research indicates that human neutrophil elastase may be a key host defence protein that targets bacterial virulence proteins. It may be especially important in immunity to enteric pathogens like *Shigella* spp.

Figure 2.3 Cell death can be of the necrotic or apoptotic type. Necrosis occurs when cells die from severe injuries (physical/chemical trauma). The cell membrane is damaged, and there are early changes in mitochondrial shape and function. The cell swells and ruptures, spilling its contents into surrounding tissue spaces and provoking an inflammatory response. Apoptosis is a more subtle form of cell death, often seen when cell death is physiologically determined or acceptable, such as during the morphogenetic death of cells during embryonic development, elimination of self-reactive clones, or induction of death in infected self-cells. In apoptosis, the cell shrinks and the cell membrane becomes ruffled and blebbed (resembling a bubbled surface). The nucleus shrinks and becomes dense, and its DNA undergoes fragmentation. Intracellular contents are not released; inflammation is not provoked.

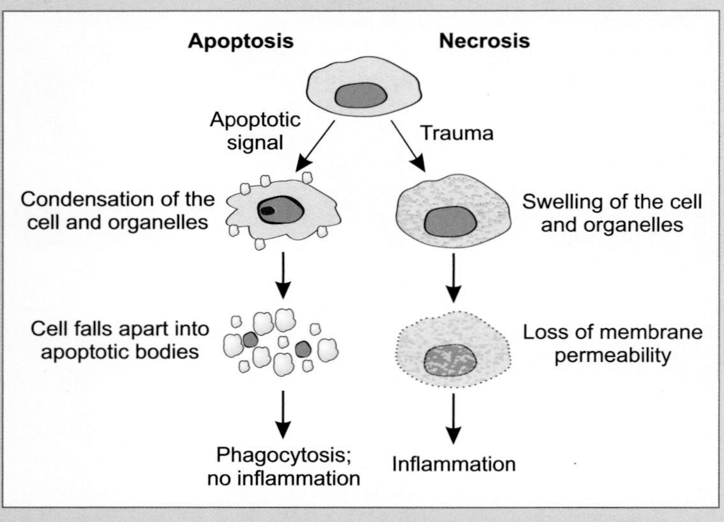

the most effective microbicidal and cytotoxic mechanism of leukocytes — the myeloperoxidase system[22]. Conversely, neutrophil specific granules are particularly susceptible to releasing their contents extracellularly, and they appear to have an important role in initiating inflammation. Specific granules represent an intracellular reservoir of various plasma membrane components, including cytochrome *b558*[23], receptors for complement fragment iC3b (CR3, CR4), laminin, and formylmethionyl peptides. Granule constituents may also participate in adaptive immune responses. For example, defensins and cathelicidins can stimulate IL-8 production and mobilise various types of phagocytic leukocytes, immature DCs, and lymphocytes. These peptides can also cause mast cell degranulation, helping in the release of more bioactive mediators (see below). Thus, neutrophils are critical in alerting, mobilising, and amplifying the innate and adaptive antimicrobial immunity of the host.

- **Eosinophils**: These PMNs take up the acidic component (eosine) of compound stains used in blood staining. They account for 2-5% of leukocytes in healthy, nonallergic individuals. Only a small fraction of the total eosinophil population is in circulation. Most eosinophils are in connective tissues immediately underneath the respiratory, gut, and urinogenital epithelia. Their numbers may shoot up dramatically in allergic states or worm infestations. Human eosinophils usually have a bilobed nucleus and many cytoplasmic vesicles. Apart from the usual receptors (for complement components, adhesion molecules, cytokines, chemokines, etc.), eosinophils also express FcεRI (the high-affinity receptor for the Fc region of IgE antibodies; section 10.7) and a low-affinity receptor for IgG antibodies. Eosinophils have also been shown to express MHC class II proteins under the influence of certain cytokines, and they can act as APCs under those conditions.

The large granules (vesicles) in a mature eosinophil are membrane-bound organelles with a core that differs in electron density from the surrounding matrix. They contain four cationic proteins that exert a range of biological effects on host cells and microbial targets. The **M**ajor **B**asic **P**rotein (MBP; not to be confused with MBL) is thought responsible for the eosinophils' cytotoxicity to parasites. It is demonstrable in the blood and sputum of patients with

[22] Myeloperoxidase is responsible for the characteristic green colour of pus.

[23] Cytochrome *b558* is a component of NADPH oxidase, the enzyme responsible for producing superoxide.

Table 2.4 Neutrophil Granules

Attributes	Azurophil granules	Specific granules	Small storage granules
Antimicrobial	Myeloperoxidase Lysozyme Bacterial permeability-increasing protein* Defensins Cathelicidins Azurocidin	Lysozyme Lactoferrin	—
Proteolysis:			
Neutral proteases	Elastase Cathepsin G Proteinase 3	Collagenase Complement activator	Gelatinase Plasminogen activator
Acid proteases	Cathepsin B Cathepsin D	—	Cathepsin B Cathepsin D
Hydrolysis	β-d-glucuronidase α-mannosidase Phospholipase-α2	Phospholipase-α2	β-d-glucuronidase α-mannosidase
Membrane receptors	—	Complement receptors 3 & 4 Laminin receptor N-form-met-peptide receptor$	—
Miscellaneous	Chondroitin sulphate	Cytochrome *b558* Monocyte chemotactic factor Histaminase Vitamin B_{12} binding protein	Cytochrome *b558*

*Bacterial permeability-increasing protein is a member of the perforin family and is highly toxic to Gram-negative bacteria.

$N-formylmethionine is an amino acid exclusive to bacterial protein synthesis; the receptor is a detector of bacterial protein synthesis.

hypereosinophilia. Other cationic proteins in the large granules are **E**osinophil **C**ationic **P**rotein (ECP), **E**osinophil-**D**erived **N**eurotoxin (EDN), and **E**osinophil **P**er**o**xidase (EPO). These cationic proteins are important not only in host defence against helminthic parasites but also in tissue dysfunction and damage in eosinophil-related inflammatory and allergic diseases. The large granules also contain histaminase, arylsulphatase, and large amounts of peroxides. Like neutrophils, eosinophils possess small granules that house enzymes, such as arylsulphatase, acid phosphatase, and gelatinase. These cells are also capable of synthesising mediators upon degranulation. Hence mediators contained in the granules are called *preformed*, whereas those synthesised just before release are called *newly synthesised*. Newly synthesised mediators include prostaglandins, leukotrienes, and cytokines, all of which augment the inflammatory response, activate epithelial cells, and help recruit more phagocytes and leukocytes. The cytokines secreted by eosinophils include IL-3, IL-5, IL-6, IL-8, GM-CSF, TNF-α, and both TGF-α and TGF-β. Eosinophils also synthesise IL-4 when stimulated with chemokines.

The immune system uses two strategies to destroy potential threats: those threats that can be engulfed are internalised, digested, and finally destroyed; those that are too large to be engulfed are subjected to an array of cytotoxins

To Eat or Not to Eat, *That* is the Question...

Recognition and ingestion of self-cells by phagocytes is crucial in protecting tissues from the toxic contents of dying cells and avoiding an inflammatory response. Just how phagocytes distinguish between viable and dying cells is slowly becoming clear. Because apoptosis involves radical changes in the cell, intracellular molecules normally not expressed on the cell surface gain access to the surface, enabling selective recognition of apoptotic cells by phagocytes. Apoptotic cells are known to flip their cell membrane, exposing phosphatidylserine residues that are normally present on the inner leaflet of the cell membrane, on the cell surface. The dead or dying cells thus deliver positive signals to the phagocyte, that is, they express signals absent in healthy, viable cells. There has recently been evidence for negative signalling as well, such as the active repulsion of phagocytes by viable cells. The molecule CD31 (or PECAM-1) is implicated in this process. CD31 is involved in a number of cellular interactions and lymphocyte migration. Engagement of CD31 expressed on viable cells by phagocytes was found to lead to phagocyte detachment and thus cell survival. Such detachment is disabled in dying cells. In apoptotic cells, CD31 engagement promotes tight binding and ultimate ingestion by the phagocyte instead.

and eventually walled off and contained. Neutrophils and phagocytic cells are the major players in the former strategy; eosinophils, basophils, and mast cells are the primary players in the latter. Thus, eosinophils are capable of phagocytosing, but only to a limited degree, as that is not their primary function. The neutrophil lysosomal enzymes act primarily on material engulfed in phagolysosomes; the eosinophil granule acts mainly on extracellular targets, such as parasites. **Eosinophils degranulate when triggered by an appropriate stimulus**. Perhaps because of their cytotoxic potential, eosinophils are under tight regulatory controls. Few eosinophils are normally present in circulation, and these do not express FcεRI. Only activated eosinophils express FcεRI and become primed for their protective function. Crosslinking of the FcεRI receptors by the IgE–antigen complexes leads to activation of the cell membrane and finally degranulation of the cell (Figure 2.4). During degranulation, the vesicles (the granules) fuse with the cell membrane and their contents are released outside the cell. Through this simple strategy, eosinophils use antimicrobial substances stored in granules to incapacitate large targets that cannot be phagocytosed. **This degranulation mechanism may be important in defence against helminth infestations**; infected tissues show degranulated eosinophils adhering to helminths. Because of their role in allergic reactions, doubts regarding eosinophils' exact roles in such protection still persist.

Eosinophils also participate in hypersensitivity reactions through the two lipid inflammatory mediators, LTC4 and **P**latelet **A**ctivating **F**actor (PAF). Both mediators contract airway smooth muscles, promote mucous secretion, alter vascular permeability, and elicit eosinophil and neutrophil infiltration. MBP, on the other hand, can stimulate histamine release from basophils and mast cells. It also stimulates EPO release from mast cells. In this way, once stimulated, eosinophils can serve as a local source of specific lipid mediators and induce mediator release from mast cells and basophils. The cationic proteins of eosinophils, by themselves, can also contribute to acute allergies and inflammation, especially in the late-phase of an allergic reaction (see section 16.2). Paradoxically, like many cells of the immune system, eosinophils can

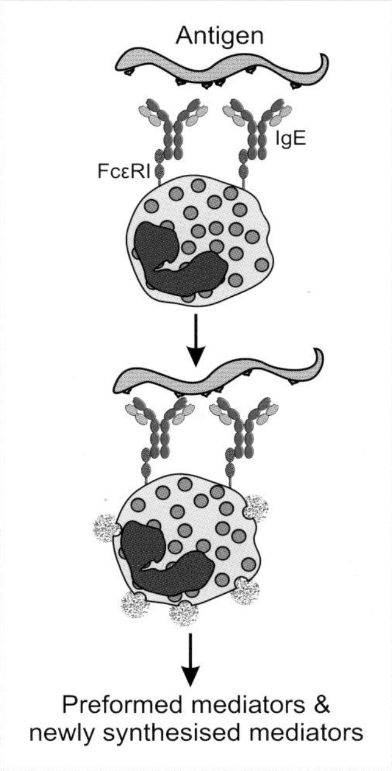

Figure 2.4 Crosslinking of FcεRI-bound IgE antibodies by multivalent antigen results in eosinophil degranulation. Eosinophils are considered important in defence against helminth infestations. Crosslinking of IgE antibodies bound to the high-affinity IgE receptors (FcεRI) expressed on eosinophils by epitopes on parasites results in the fusion of granule membranes with the cell membrane. Eosinophil granules are a rich depository of toxic metabolites, such as MBP, ECP, EDN, and EPO, and these granular contents are rapidly released into the external milieu. Degranulation also results in the synthesis and release of pro-inflammatory mediators, such as prostaglandins, PAF, and leukotrienes. Together, these mediators result in contraction of smooth muscles and mucous secretion — physiological responses important in ridding the body of the parasite. However, in hypersensitive persons, these mediators are responsible for the unpleasant manifestations of allergies.

also negatively regulate their effector function, and some of their granular contents can actually inactivate mediators of anaphylaxis. Thus, for example, arylsulphatase B can inactivate the mixture of LTC4, LTD4, and LTE4 that is responsible for some of the manifestations of allergies, phospholipase D destroys the platelet lytic factor, and histaminase degrades the histamine released by mast cell degranulation.

- **Basophils**: Basophils account for 0.5-1% of blood leukocytes. They contain granules that stain prominently with basic dyes because of the presence of large amounts of acidic proteoglycans. These membrane-bound granules are storehouses for preformed mediators, such as histamine, heparin, **E**osinophil **C**hemotactic **F**actor-**A** (ECF-A), chondroitin sulphate, and neutral proteases. However, basophils contain only about one-fourth as much MBP as eosinophils and merely detectable amounts of EDN, ECP, and EPO. Basophils express FcεRI and receptors for complement components (C3a and C5a). **Crosslinking of FcεRI by antibody–antigen complexes leads to basophil degranulation**[24]. Basophils can also degranulate directly upon contact with certain antigens, independent of IgE antibodies. Membranes of the granules fuse with the cell membrane and release preformed mediators to the external milieu. Antibody-mediated degranulation also results in the release of newly synthesised mediators, such as LTC4, PGD2, cytokines (TNF-α, TGF-β, IL-1, IL-4, IL-5, IL-6, and IL-13), and chemokines. Thus, IgE-mediated degranulation leads to the release of a host of pharmacologically active mediators. These mediators cause the adverse symptoms of allergy (inflammation, vasodilation, increased vascular permeability, smooth muscle contraction, etc.). On the other hand, cytokines and chemokines play an important role in inflammation, adaptive immune responses, and tissue remodelling. Like eosinophils, basophils are thought to be important in antihelminth defence. Their capacity to rapidly release large quantities of IL-4 and IL-13 suggests a role in initiation of TH2 responses.

[24] Although FcεRI crosslinking is the most studied mode of inducing degranulation, it is by no means the only one. Factors that induce degranulation include tissue injury caused by high temperature, irradiation, or trauma; toxins and venoms; and proteases or cationic mediators released by eosinophils or neutrophils.

The Neutrophil Paradox

Although neutrophils are essential to host defence, they have also been implicated in the pathology of many chronic inflammatory conditions. A large influx of neutrophils in the blood of people with acute bacterial infections or in the lungs of patients with adult respiratory distress syndrome causes widespread tissue damage because of the oxidants and hydrolytic enzymes released from activated neutrophils; it may eventually lead to death. The activation of neutrophils by immune complexes in synovial fluid contributes to the pathology of rheumatoid arthritis. Chronic activation of neutrophils may also initiate tumour development; ROI generated by neutrophils damage the DNA, and the released proteases can promote tumour cell migration. Oxidants of neutrophil origin have also been shown to oxidise low-density lipoproteins which are then more effectively bound to the cell membrane of macrophages through specific scavenger receptors. Uptake of these oxidised low-density lipoproteins by macrophages is thought to initiate atherosclerosis. Cytoplasmic constituents of neutrophils are known to give rise to specific **A**nti-**N**eutrophil **C**ytoplasmic **A**ntibodies (ANCA) in susceptible individuals. ANCA are antibodies against enzymes that are found mainly within the azurophil granules of neutrophils. These antibodies are closely related to the development of systemic vasculitis and glomerulonephritis. Neutrophils also release a large number of hydrolases besides ROI. Under normal conditions, hydrolytic damage to host tissue, and therefore, chronic inflammatory conditions, are kept at bay by the presence of antioxidants and protease inhibitors in the blood. If the antioxidants and antiproteases are overwhelmed or excluded from tissues, there can be severe damage. Oxidative stress may initiate tissue damage by reducing the concentration of extracellular protease inhibitors until it is below the level required to inhibit released proteases. High levels of activated neutrophils have been found in people with hypertension, Hodgkin's disease, inflammatory bowel disease, psoriasis, sarcoidosis, and septicaemia.

Mast cells are sessile cells similar to basophils, and they are widely distributed in the body's connective tissues. They have granules containing preformed mediators and express receptors for IgE antibodies. Although they have a role in innate immune responses, especially against parasites and viruses, their role has been more intensively investigated in hypersensitivity reactions. They have therefore been discussed in chapter 16.

iii. Movement of phagocytic cells: Cells of the immune system circulate as nonadherent cells in the blood and lymph, and when necessary, they migrate as adherent cells in tissues. Rapid transition between adherent and nonadherent states is crucial to immune surveillance and responsiveness. This capability to switch allows them to efficiently protect the body from infectious agents. Although we describe only the migration of phagocytic cells in and out of blood vessels below, other cells of the immune system (such as DCs and lymphocytes) also move in an essentially similar fashion through blood vessels or tissues.

Phagocytes are capable of chemotactic movement and move towards the source of chemotaxin by cytoplasmic streaming. Common chemoattractants include LTB4, LPS, histamine, PAF, peptides having N-formylmethionine, chemokines (e.g., IL-8), complement component C5a, and, to a much lesser degree, C3a. In order to enter various tissues, circulating monocytes and neutrophils must first adhere to and then cross the endothelial lining of blood vessels. Migration in host tissues consists of three steps — rolling, adhesion, and extravasation.

- **Rolling**: Normally, leukocytes are transported by the blood stream and weakly collide with and probe the surfaces of endothelial cells lining blood vessels.

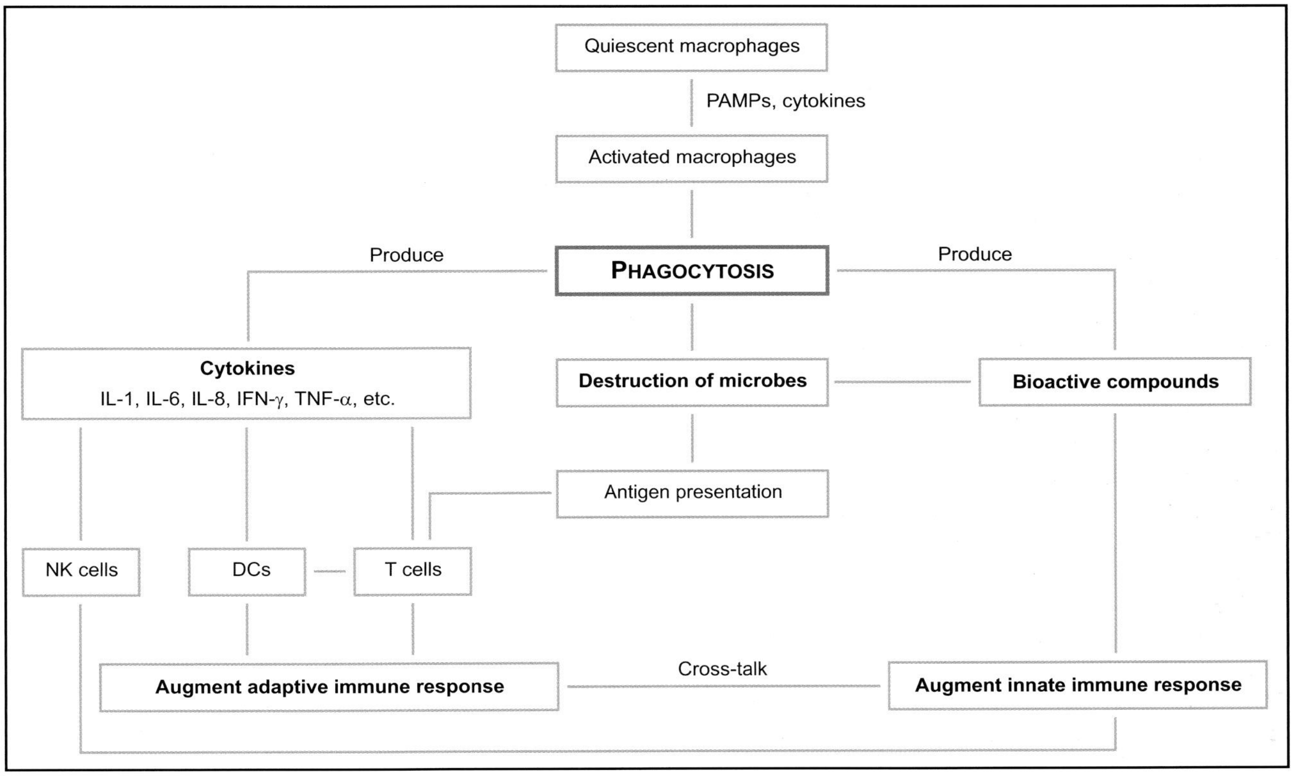

Mind Map 2.2 Phagocytosis

Firmer interaction is required for emigration. Blood flow rate is slowest in post-capillary venules, and it is here that this interaction takes place. At sites of inflammation, dilation of vessels reduces blood flow rate, increasing chances of collision between the leukocyte and endothelium. A set of adhesion molecules known as selectins initiate the first transient contact of the leukocyte with the endothelium. Selectins bind to transmembrane glycoproteins on the surface of the interacting cells and come in three types — L-selectin, P-selectin, and E-selectin. L-selectin is expressed by leukocytes and transiently interacts with carbohydrates on the endothelial cells in the initial interaction. Normally, endothelial cells do not express selectins but store them in granules. Exposure to LTB4, C5a, or histamine released by local trauma leads to the release of P-selectin to the cell surface. The next to appear is E-selectin. Together, these molecules allow a loose attachment between the endothelium and the phagocyte's cell-surface glycoproteins. With the force of the bloodstream, this leads to rolling (also called *margination*) of leukocytes along the endothelium.

- **Adhesion**: Under the influence of chemotaxins released by platelets (such as IL-8 and PAF) phagocytes upregulate the expression of another set of adhesion molecules — integrins — that orchestrate leukocyte trafficking. This family of adhesion molecules includes LFA-1 (**L**eukocyte **F**unction associated **A**ntigen-**1**; a heterodimer of CD11a/CD18), Mac-1 (CD11b/CD18), ICAM-1 (**I**nter**c**ellular **A**dhesion **M**olecule-**1**) and ICAM-2. The resting endothelium does not express ICAM-1, and it expresses only low levels of ICAM-2. Exposure to TNF-α induces the expression of ICAM-1 and upregulates the expression of ICAM-2. LFA-1 and Mac-1 on leukocytes bind the ICAM-1 and ICAM-2 expressed on endothelial

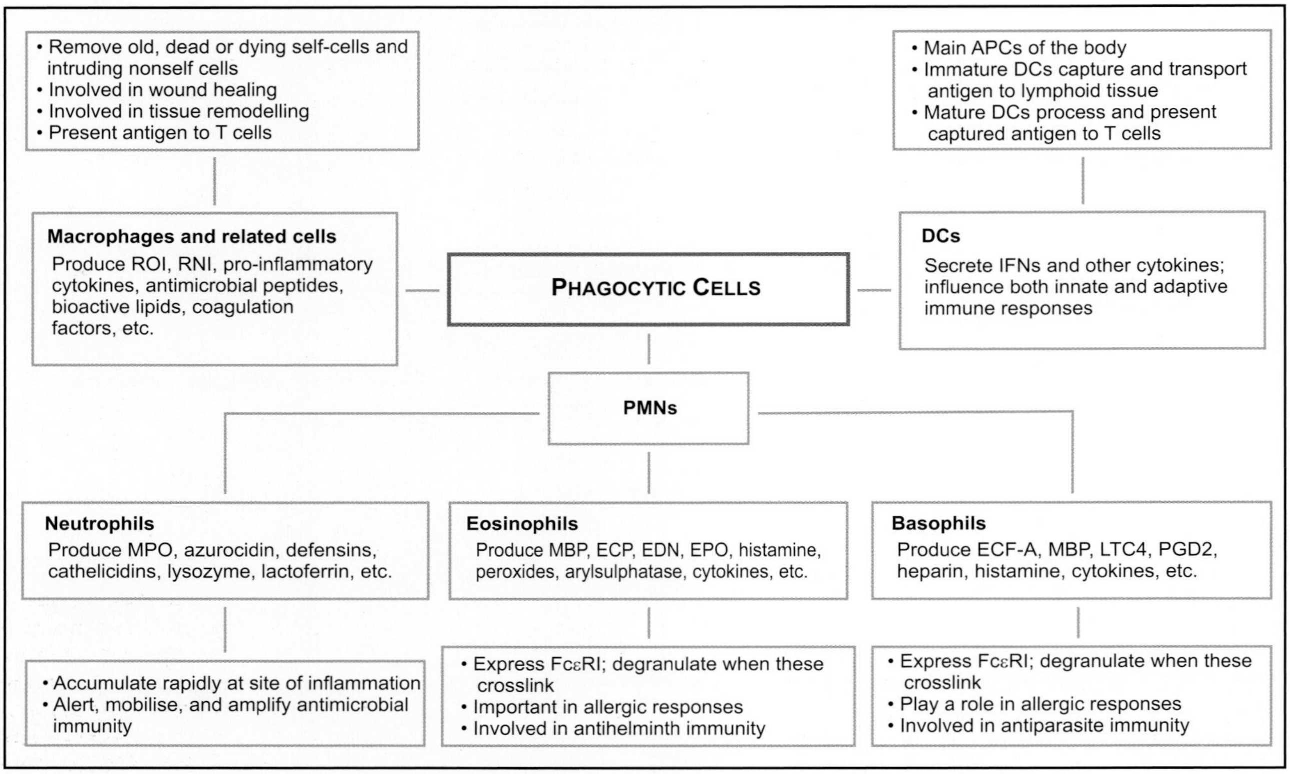

Mind Map 2.3 Phagocytic cells

cells. These adhesion molecules arrest rolling leukocytes and allow them to stick to the endothelium. New adhesion contacts are formed in the direction of movement, and adhesion is reduced at the trailing end.

❒ **Extravasation**: Phagocytes move towards endothelial cell–cell junctions because of integrin-mediated adhesive interactions. PECAM-1[25] — expressed both by leukocytes and intracellular junctions of endothelial cells — helps the phagocyte squeeze through the junction and cross the endothelial wall (crossing of the endothelial wall is termed *extravasation* and movement through the wall is called *diapedesis*). The phagocytes initially extend a pseudopod between the endothelial cells while maintaining nuclei and granules on the lumenal side. Eventually, the whole cell migrates across the endothelial boundary. Transmigration across the endothelium to the basal endothelial membrane is rapid (within minutes of attachment) and one way. Phagocytes accumulate briefly between the basement membrane and endothelial cells before entering connective tissue. Proteases secreted by phagocytes help in digestion of proteins of the basement membrane (Figure 2.5).

iv. Phagocytosis: Discovered by Metchnikoff in 1884, *phagocytosis* is a process by which a cell ingests and destroys particulate substances. Phagocytosis by macrophages is crucial for uptake and degradation of not only pathogens but also senescent, apoptotic, and damaged tissue cells. Phagocytosis is an integral part of development, tissue remodelling and wound healing, innate and adaptive immune responses, and inflammatory processes. It occurs in three phases:

❒ **Attachment**: Attachment of the microorganism to the phagocyte is the first step in phagocytosis. Thanks to the variety of receptors they express, phagocytes

[25] PECAM-1 (**P**latelet **E**ndothelial **C**ell **A**dhesion **M**olecule-1; CD31) is homophilic, that is, its ligand is also CD31.

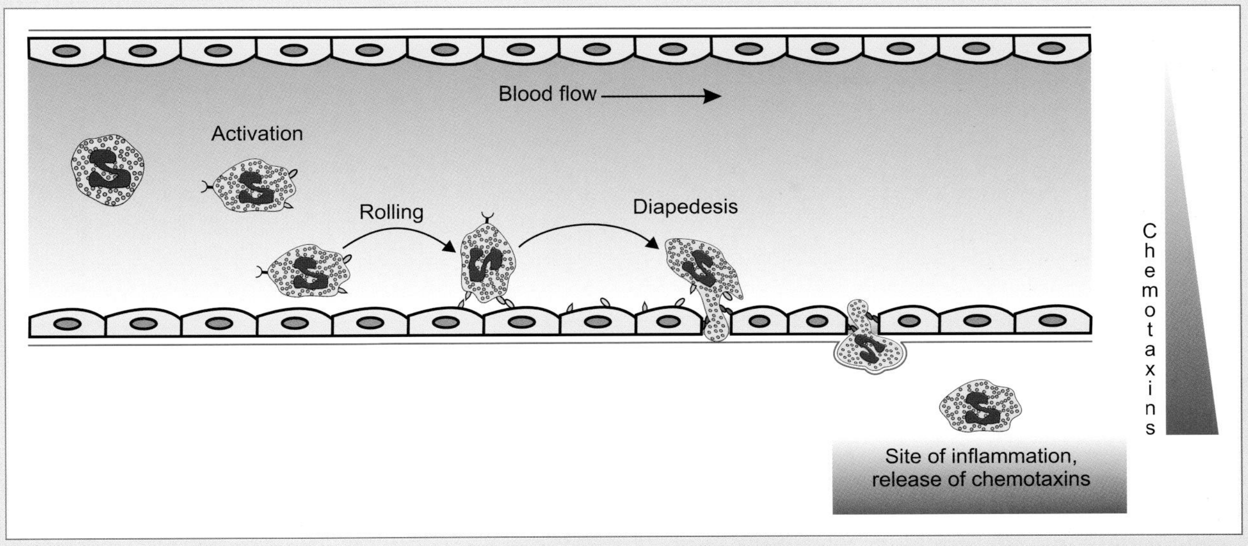

Figure 2.5 Phagocytes cross the endothelium and enter the site of infection by a multistep process initiated by the adhesion molecules that are expressed by activated endothelial cells. The normal endothelium is nonadhesive and interacts fleetingly with leukocytes as they are transported across the endothelium by blood flow. Tissue injury induces the expression of selectins on endothelial cells, allowing firmer interaction with glycoproteins on the phagocytic cell surface. This interaction is not firm enough to anchor the phagocyte to the endothelium but slows it down and causes it to roll across the endothelial surface. New adhesion contacts are formed in the direction of movement while adhesion is reduced at the trailing end. Expression of ICAM-1 on the endothelium and its ligands (e.g., LFA-1) on the phagocytes allows stronger interaction. The phagocyte adheres to the endothelium and move towards intracellular junctions. CD31, expressed by endothelial cells and phagocytes, helps the cells squeeze through the junction towards the basement membrane, and it is called *diapedesis*. Matrix metalloproteases, expressed by the phagocyte, aid in the digestion of proteins of the basement membrane. Once outside the blood vessel, the leukocyte migrates as a result of the chemokine concentration. These chemokines are released by the interaction of tissue cells or macrophages upon initial encounter with the pathogen, and they bind to proteoglycans in the tissue's extracellular matrix. They form a matrix-associated concentration gradient, along which leukocytes migrate to the focus of infection.

can attach themselves to many organisms. Besides PRRs, the FcRs and CRs expressed by phagocytes are important to attachment. Opsonisation of an organism by complement components and antibodies or expression of adhesion molecules on the phagocytic cell membrane greatly enhances attachment.

- **Pseudopod formation**: The phagocytic membrane is activated by the attachment and leads to pseudopod formation and eventual engulfment of the organism. The formation and fusion of pseudopods involves the same proteins that are involved in muscle movement — actin and myosin. When the pseudopods fuse, the microorganism is internalised into a phagosome. It then matures through a series of fusion and fission events. Finally, lysosomes fuse with the mature phagosome to form a phagolysosome[26].
- **Digestion**: Within the confines of the phagolysosome, the microorganism is subjected to a battery of bactericidal factors and is killed by two distinct mechanisms.
  - **The oxygen-dependent mode** of intracellular killing is dependent on cellular glycolysis and is a by-product of a marked increase in metabolic activity, termed the metabolic or respiratory burst, that accompanies phagocytosis (Figure 2.6). In the metabolic burst, cells consume a large

[26] Several intracellular pathogens, such as *Mycobacterium tuberculosis* or *Chlamydia* spp., subvert the system and survive by interfering with phagolysosome formation (see chapter 13).

Inner Lining: Endothelium

The cells lining body surfaces and blood vessels are not passive observers in protection and immunity. Endothelial cells, by their capacity to express adhesion molecules and cytokines, are intricately involved in inflammatory processes. During immune injury, endothelial cell activation by inflammatory cytokines stimulates leukocyte adhesion to the endothelium. The endothelium is transformed from an anticoagulant surface to a pro-coagulant one, and this results in the release of vasoactive mediators, cytokines, and growth factors. Endothelial cells have been shown to express cytokines, such as IL-1, IL-5, IL-6, and IL-8; several colony-stimulating factors (G-CSF, M-CSF, and GM-CSF); and the chemokines **M**acrophage **C**hemotactic **P**rotein-**1** (MCP-1) and RANTES. Processes as diverse as hypoxia or bacterial infection can induce the expression of these cytokines. IL-1 and TNF-α, produced by infiltrating inflammatory cells can also induce endothelial cells to express several of these cytokines and adhesion molecules. There seems to be cross-talk between inflammatory cells and the endothelium, and this may be critical to the development of chronic inflammatory states.

The escalating inflammatory processes set off by pathogen entry must also be diffused. Otherwise, the pro-coagulant endothelium will allow indiscriminate cell adhesion, blocking blood vessels. Endothelial cells have a role in this process as well. Cytokine activation of endothelial cells in the later stages of the immune response results in increased TGF-β synthesis. This cytokine has immunosuppressive functions and is found to inhibit E-selectin expression, hindering leukocyte adhesion and transmigration. TGF-β also influences the mechanisms of vascular remodelling and tissue healing. Thus, endothelial-derived cytokines may be involved in a variety of processes, such as haematopoiesis, cellular chemotaxis and recruitment, bone resorption, coagulation, acute-phase protein synthesis, and inflammatory response regulation.

amount of oxygen, resulting in increased H_2O_2 production. Other bactericidal oxidising agents, such as the superoxide anion (molecular oxygen having an extra electron; O_2^-), singlet oxygen ($^1O_2^*$ molecules with high-energy electron), and hydroxyl radicals (OH^* together called **R**eactive **O**xygen **I**ntermediates or ROI), are also produced. The enzyme myeloperoxidase, also present in neutrophils, increases damage by catalysing the peroxidation of numerous surface molecules on microorganisms in the presence of the toxic oxygen metabolites.

- **The oxygen-independent mechanism** of killing includes:
  - **A progressive decrease in pH and the action of hydrolytic enzymes**, such as cathepsins, phosphatases, phospholipases, glycosidase, lysozyme, and arylsulphatase, that bring about the digestion of cell walls or outer coverings of pathogens. Ribonucleases, lipases, and proteases further digest intracellular contents.
  - **Nitric Oxide** (NO), one of the most important weapons in the macrophage arsenal, is effective in the immediate vicinity of phagocytes, as well as at a considerable distance from them, as it diffuses easily across cellular barriers. The antimicrobial activity of NO is thought to be due to its mutagenic activity. Moreover, the RNI (**R**eactive **N**itrogen **I**ntermediates) formed during NO synthesis are also bactericidal.
  - **Peptides**, such as defensins and cathelicidins, that are microbicidal.
  - **Lactoferrin**, which binds Fe^{2+}, making it unavailable to microbes.

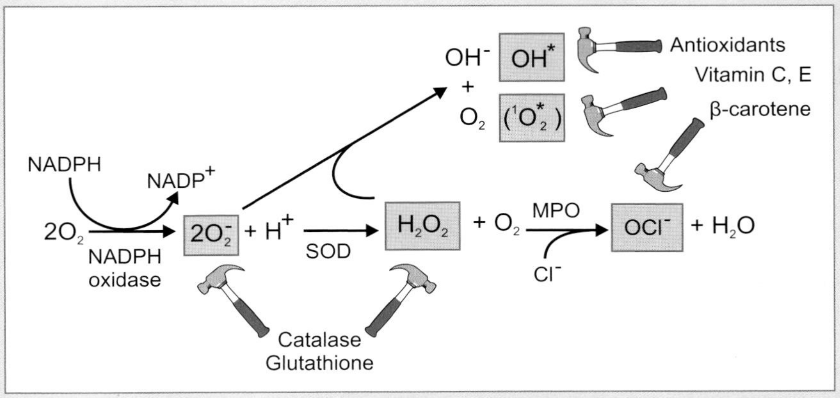

Figure 2.6 Phagocytosis causes a respiratory burst that is accompanied by a transient increase in oxygen consumption and which results in the production of superoxide anions (O_2^-), hydrogen peroxide, singlet oxygen ($^1O_2^*$), hydroxyl radical (OH^*), and hypohalite (OCl^-) by macrophages and neutrophils. The respiratory burst is a part of the oxygen-dependent mechanism of killing that allows the formation of highly toxic but short-lived metabolites and free radicals. They are generated by NADPH oxidases and other enzymes present in the lysosomes of the phagocytes. The critical intermediate is the highly reactive superoxide which can be converted to the equally damaging H_2O_2 by the action of superoxide dismutase (SOD). In the presence of chlorides, H_2O_2 is converted to hypochlorite by the enzyme MPO found in neutrophils. The toxic metabolites can also cause extensive host tissue damage and need to be rapidly neutralised. Antioxidants, such as β-carotene and vitamins C and E, help in scavenging the free radicals, whereas catalase and glutathione help convert H_2O_2 and superoxide anions to water.

Eventually, the microorganism is digested (Figure 2.7). Portions of the microbial cell are displayed on the surface of the phagocyte in the context of MHC molecules, setting the stage for T cell stimulation. If the ingested particle is too large, the phagocyte cannot digest it. The whole particle-containing lysosome is then thrown out of the cell by exocytosis. The corrosive lysosomal brew is damaging to bystander tissue cells. Such exocytosis is linked to some of the harmful effects of hypersensitivity. When antigens or microorganisms effectively resist the microbicidal activity of macrophages, the resulting chronic inflammation leads to the formation of a characteristic nodular mass called *granuloma* (section 2.4.1).

2.3.2 Lymphocytic Cells

Lymphocytes have a large nucleus and a coarse, dense cytoplasm. Most immunological research originally focused on antibodies produced by B cells and the effector functions of T cells, resulting in the perception that lymphocytes were the instruments of adaptive immune responses only. In recent years, some lymphocytes have been shown to be involved in innate responses.

i. NK cells have an instrumental role in innate immune responses against bacterial, viral, and parasitic pathogens. They play a crucial role in suppressing tumour metastasis and outgrowth and are called *natural killer cells* because they demonstrate their cytolytic activity even in normal animals not exposed to an infectious agent. These large granular lymphocytes constitute about 15% of peripheral blood lymphocytes and are a heterogeneous group of cells arising from the bone marrow that are characterised by their large granular morphology and ability to lyse target cells. They do not express Igs and T cell receptors and hence are distinct from both B and T lymphocytes. The NKT cells (a subset of NK cells),

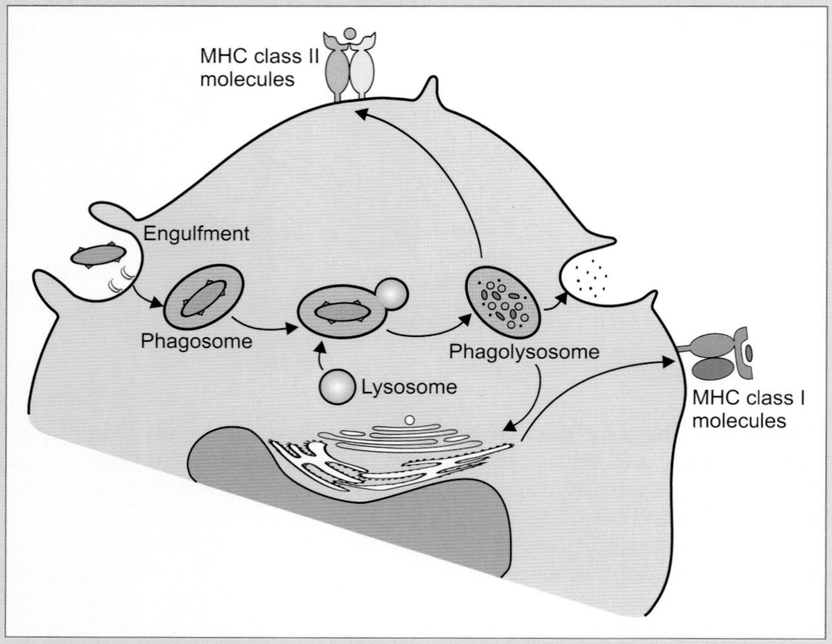

Figure 2.7 Engagement of receptors on the phagocyte surface by ligands on the microbe results in phagocytosis and the eventual destruction of the pathogen; it also triggers the adaptive immune response. The activated phagocytic cell membrane rapidly forms pseudopodia and engulfs the pathogen. The pathogen is internalised in a membrane-bound vesicle termed the phagosome. The phagosome fuses with one or more highly acidic internal vesicles called lysosomes that are rich in enzymes and toxic metabolites required for destruction of the microbe. The pathogen is eventually digested in the phagolysosome, and indigestible material is eliminated by exocytosis. Fragments derived from the pathogen are loaded onto MHC molecules and displayed on the cell surface. They trigger the adaptive immune response.

express T cell receptors and are discussed in chapter 9. Unlike T cells, NK cells can mature in the absence of the thymus and can be found even in animals lacking this organ. Many NK cells originate from precursors in the foetal liver (6–8 weeks), although they may also develop from immature thymocytes.

The grouping and nomenclature of NK cells can be confusing, and for simplicity, only human NK cells will be considered here. Human NK cells are divided into two subsets on the basis of CD56 expression — CD56^{bright} and CD56^{dim}. The functional significance of CD56 expression is not understood, and a murine homologue of CD56 is yet to be discovered. Generally speaking, CD56^{dim} cells are more cytotoxic than CD56^{bright} cells and are consequently more granular. NK cells express a number of other receptors and markers[27] on their cell surfaces, although all the markers need not be present simultaneously on the same cell. Thus, different clones of NK cells express different subsets of markers. These include CD16 (a kind of FcR; FcγRIII), a variety of chemokine receptors, and cytokine receptors (e.g., for IL-1, IL-2, IL-10, IL-12, and IFN-γ). NK cells expressing CD16 can target antibody-coated pathogens and participate in ADCC (**A**ntibody-**D**ependent **C**ell-mediated **C**ytotoxicity). These NK cells, called K cells, will be dealt with later, in the chapter on cell-mediated immunity.

NK cells express both activatory and inhibitory receptors (Figure 2.8) and may be visualised as constantly being in a tug of war regulated by these opposing signals. Inhibitory receptors monitor the target cells for normal expression of host proteins. Their cytoplasmic domains have **I**mmunoreceptor **T**yrosine-based ***I**nhibitory* **M**otifs (ITIMs[28]). The best characterised inhibitory receptors bind classical or nonclassical MHC class I molecules that are expressed by almost all nucleated cells of the body. Signals delivered by the inhibitory receptors are often dominant over signals emanating from activating receptors with ITAMs (**I**mmunoreceptor **T**yrosine-based ***A**ctivating* **M**otifs) and prevent NK cell effector functions. **A balance of signals generated by these inhibitory and activatory receptors**, along with engagement of various adhesion (ICAM-1, LFA-3, etc.) and costimulatory molecules,

[27] The cells of a specific lineage express a number of molecules on their cell surfaces. The presence of such a molecule (or set of molecules) identifies the cell type. Such molecules that are present (or absent) from a particular cellular type are termed *markers*. Markers aid the characterisation of, understanding of the function and biology of, and eventually isolation of various cell types. Although stem cells do not express any of the markers distinguishing the different lineages they give rise to, they express distinct markers.

[28] ITIMs and ITAMs are universal motifs found in the cytoplasmic domains of molecules involved in signal transduction. We discuss them in chapter 6.

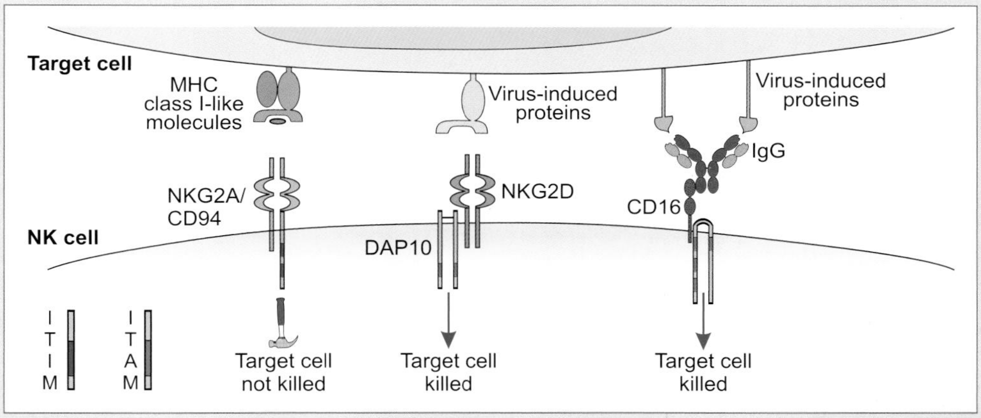

Figure 2.8 A balance of signals generated by inhibitory and activatory receptors governs NK cell cytotoxicity. NK cells express a variety of receptors for ligands on target cells. Activatory receptors have ITAMs in their cytoplasmic tails and result in NK cell activation. Inhibitory receptors have ITIMs in their cytoplasmic tails, and signals generated by their recognition are dominant over activatory signals. The recognition of MHC class I or MHC class I-like molecules expressed on self-cells by inhibitory receptors, such as heterodimers of NKG2A/CD94, therefore allows the target cell to escape. Activating receptors, such as NKG2D, recognise ligands induced by viral infection. Viral infections often result in downregulation of MHC class I molecule expression. This results in the lack of an inhibitory signal and allows NK cell-mediated lysis of the virally infected cell. Some activating receptors do not have a cytoplasmic tail and transmit the activating signal through bridging molecules (called *adaptor molecules*), such as DAP10 or DAP12. Many NK cells also express CD16, a low-affinity IgG antibody receptor. Crosslinking of CD16-bound IgG by an antigen transmits an activating signal and may result in the lysis of the target cell.

governs the activation of NK cell cytotoxicity. Thus, the fate of the target cell is determined by this mixture of signals. If the activatory signals are stronger than the signals generated by inhibitory receptors, the target cell is killed; otherwise, it will be spared.

Inhibitory receptors are critical in ensuring that self-cells are protected from the destructive potential of NK cells. Downregulation of either MHC class I synthesis or the molecules' export to the surface is a mechanism used by many viruses to escape recognition by cytotoxic T cells (e.g., cytomegalovirus). In this respect, the nonclassical MHC molecule HLA-E is of special importance. It binds members of the CD94/NKG2 family of NK receptors. It presents peptides from the signal sequences of other HLA-molecules in the cell and is an indicator of cellular MHC class I synthesis. Incorrect processing of MHC class I molecules within a cell suppresses the display of HLA-E:peptide complexes at the cell surface, reporting to NK cells the impairment of intracellular processing that is essential for antigen presentation and making the cell susceptible to NK cell-mediated lysis. Activating receptors, on the other hand, recognise the carbohydrates, sulphated proteoglycans, or glycoproteins expressed on cell surfaces. Viral infection often results in an alteration in the glycosylation pattern of cellular proteins, and these receptors therefore allow NK cells to detect such changes. Activating signals, in the absence of inhibitory receptor engagement, allow NK cell activity to be targeted at infected and/or transformed cells that may otherwise escape the specific killing mechanisms of the adaptive immune system.

A large number of inhibitory and activatory receptors have now been identified. However, their nomenclature and description can be very confusing. The best studied of these include the following:

NO Laughing Matter: NO and RNI

NO and RNI are produced by immune system cells (DCs, NK cells, mast cells, and phagocytic cells), as well as other cells involved in the immune reaction (e.g., endothelial cells, epithelial cells, vascular smooth muscle cells, fibroblasts, keratinocytes, and hepatocytes). NO is derived from the amino acid L-arginine by the enzymatic activity of NOS. It has multiple roles. Although some of its effects (such as its antimicrobial activity) are well documented, others (such as its immunosuppressive effects) are very poorly understood. NO and RNI are potent molecules that have local effects at the site of synthesis as well as effects at sites further away because of rapid and unhampered diffusion. The most obvious function of NO and RNI is antimicrobial action; they can kill or reduce the replication of pathogens, such as bacteria, fungi, and viruses. NO or RNI combine with DNA, causing mutations. They also cause alterations in proteins because of S-nitrosylation or tyrosine nitration, and disrupt enzymes by disrupting S–Fe clusters or haeme groups or through peroxidation of membrane lipids. NO may combine with O_2^- to form peroxynitrite ($ONOO^-$), a potent antibacterial molecule. Local arginine depletion caused by NO synthesis in host tissues can also inhibit growth or induce parasite death. NO inhibits tumour cell growth and may also induce their death; however, many tumours seem to express NOS, and NO seems to help their survival. NO can influence leukocyte chemotaxis by modulating chemokine production and/or functioning as an intracellular messenger in chemokine signalling pathways. NO downregulates endothelial expression of adhesion molecules and hence can significantly affect the rolling and transmigration of leukocytes. Thus, NO seems to govern a broad spectrum of processes. These include the differentiation, proliferation, and apoptosis of immune cells; the production of cytokines and adhesion molecules; and the synthesis and deposition of extracellular matrix components.

- **K**iller cell **I**g-like **R**eceptors (KIR): Some of these receptors recognise MHC class I molecules (HLA-A and -B) and are inhibitory receptors, whereas others recognise nonself proteins and are activatory.
- C-type lectin superfamily receptors that are heterodimers of NKG2 and CD94 in humans; Ly49 is their murine counterpart. This family of receptors also consists of activatory and inhibitory receptors.

Engagement of activatory receptors, combined with exposure to cytokines, such as IFN-α, IFN-β, and IL-12, activates NK cells. IFNs not only activate cells but also enhance the activated cells' cytolytic capacity. **Activated NK cells bring about the lysis of target cells by cell-to-cell interaction** through the release of granzymes and perforin contained in the lysosomes or by inducing the cell to apoptose. Perforin inserts itself in the target cell membrane and polymerises to form transmembrane channels. These channels lead to the leakage of low-MW cellular contents while allowing an influx of water. Granzymes are a family of enzymes that promote DNA fragmentation and cell death (see chapter 11). Apart from their direct cytolytic activity, NK cells secrete IFN-γ and other cytokines in significant amounts upon stimulation. IFN-γ production is an important function of NK cells. It induces NOS (**NO S**ynthase) and promotes production of ROI and RNI. It is also known to activate macrophages and cause DC maturation, triggering the adaptive immune response. NK cells can also secrete TNF-α, which has antiviral and immunoregulatory properties. The cytolytic functions of NK cells have been suggested to be uniquely used in particular compartments. In the case of cytomegalovirus infection, for instance, perforin-dependent NK cell effects may be more important in the spleen, whereas IFN-γ production may be more important for liver defence.

Up in Smoke and Down the Drain: Cigarette Smoke and Innate Immunity

Tobacco contains more than 4,500 compounds in particulate and vapour phases. These compounds contain at least five known human carcinogens and many toxic agents, including CO, NH_3, acrolein, acetone, benzopyrenes, nicotine, hydroquinone, and NO. The carcinogenic action of many of these compounds has been linked to DNA damage and/or increased cellular proliferation. Many of these compounds are at much higher concentrations in sidestream smoke (released into the air during burning) than in mainstream smoke (inhaled by the smoker).

The lungs are equipped with nonspecific and specific defence mechanisms to protect against the environmental pathogens they are exposed to. Alveolar macrophages and other monocytes are the lungs' most important innate immune mechanisms. Smoking increases the number of alveolar macrophages by several folds. These cells express increased levels of lysosomal enzymes and elastase, both of which are thought to damage connective tissue and parenchymal cells of the lung, greatly increasing risks of bronchitis and emphysema. Alveolar macrophages from smokers are functionally impaired and show a decreased ability to phagocytose and kill pathogens. They also secrete considerably lower levels of pro-inflammatory cytokines. These cytokines are crucial for early responses to immunological challenges and upregulation of local host defences. The NK cells of smokers also show reduced activity against tumour cells. Chronic exposure to cigarette smoke increases the frequency of spontaneous tumours in laboratory animals, which could partially be attributed to impaired NK cell function. To summarise, smoking produces various morphological, physiological, biochemical, and enzymatic changes in macrophages and NK cells, which might impair antibacterial defences, regulatory activity, and inflammatory responses in the lungs, eventually leading to lung pathogenesis.

The **L**ymphokine-**A**ctivated **K**iller (LAK) cells, one more class of lymphocytic cells derived from NK cells, specialise in tumour cell lysis. These cells have the capacity to induce apoptosis, and their cytotoxic potency is enhanced by cytokines, such as IL-1, IL-2, IL-6, IFNs, and TNF-α. Although LAK cells represent a very minor subset, they are important in tumour immunity because they can stimulate T cell and eosinophils. LAK cell therapy has been shown to induce partial and complete remissions in up to 20% of patients with malignant melanoma and renal cell carcinoma.

ii. $\gamma\delta$ T lymphocytes is a special subset of T cells that differs from the conventional T cells of the adaptive immune response in two respects — these lymphocytes do not express CD8 or CD4 surface molecules, and they do not recognise peptide:MHC complexes. They express antigen receptors of a very limited diversity that recognise target antigens directly. Most of the well-defined ligands of $\gamma\delta$ T cells fall into a group of self-molecules potentially capable of indicating cellular stress or molecules that are expressed by tissue cells only in response to infection, for example, heat shock proteins, MIC-A, and MIC-B (see chapter 7). $\gamma\delta$ T cells are therefore thought to be unique in that they are activated by self-molecules expressed as a consequence of infection rather than by pathogen-associated molecules alone. Recent reports show that these cells can also act as APCs and present antigens to conventional T cells. They are **considered important in conferring immunity against infections in the intraepithelial sites**, **where they are found in substantial numbers**. Although many studies now suggest that $\gamma\delta$ T cells have a prominent role in the innate response against pathogens and tumours, information about underlying mechanisms remains scarce. Mice deficient in $\gamma\delta$ T cells have exaggerated responses

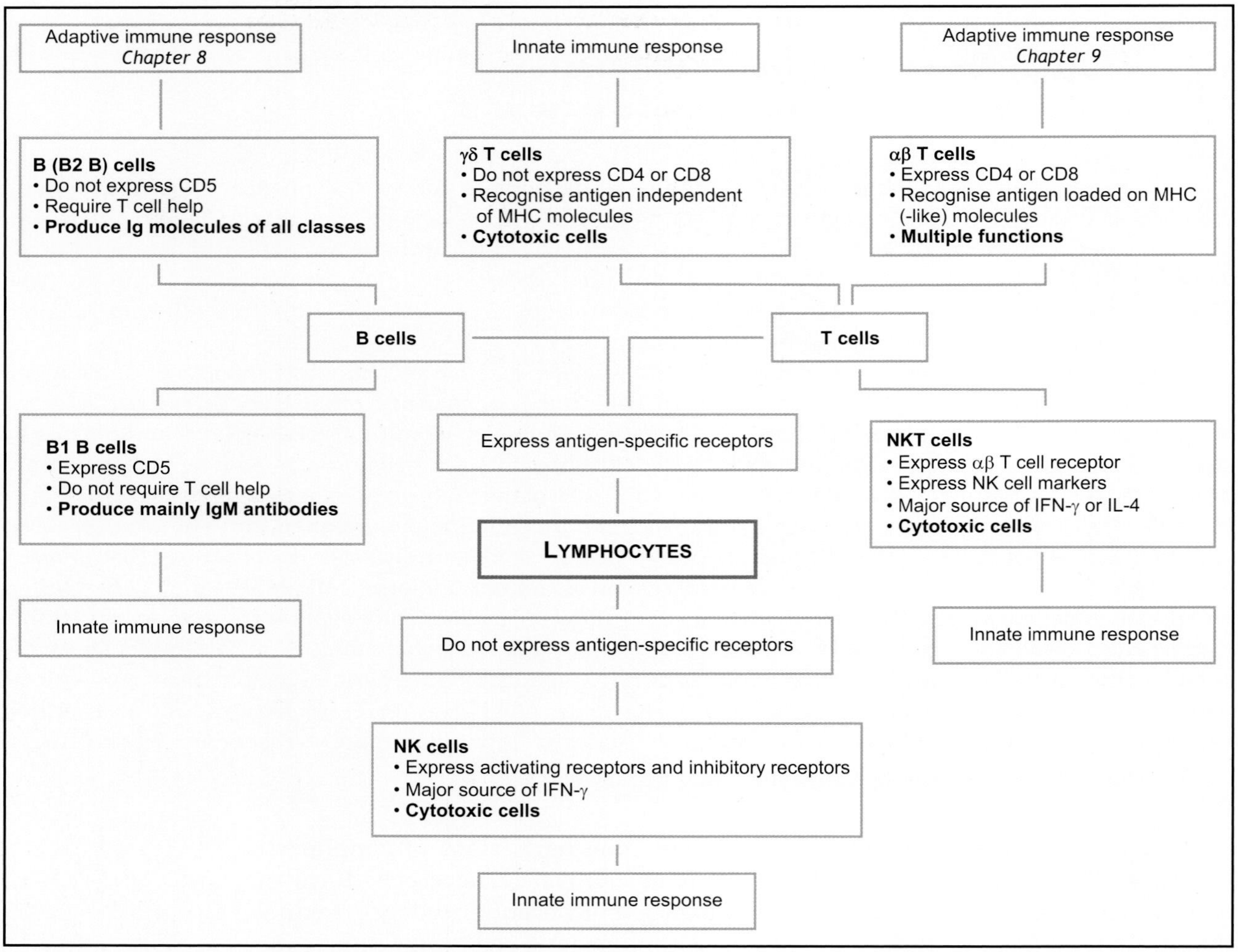

Mind Map 2.4 Lymphocytes

to pathogens and self-tissues, suggesting that these T cells may play a role in modulating immune responses. γδ T cells are cytotoxic; they kill target cells by means of perforin and granzymes. They also secrete cytokines upon activation and may modulate immune responses through these molecules.

iii. B1 cells are a set of B lymphocytes (<5%) that **develop in the foetal/neonatal animal and express the CD5 molecule**. They are termed B1 B cells since they are the earliest B cell type to develop. B lymphocytes that develop in adult bone marrow are, according to this system of nomenclature, of the B2 type. B1 B cells are thought to contribute to the production of serum Igs and *natural antibodies*, which are antibodies produced in the absence of specific immunogenic challenge and are found even in germ-free animals fed on an antigen-free diet. Typically, B1 cells do not need T lymphocyte help to mount an immune response. In adult animals, B1 cells proliferate in the peritoneal and pleural cavities and seem to interact with self-antigens and bacterial antigens found in the gut flora. They differentiate to antibody-producing plasma cells rapidly in the first stages of immune response. By contrast, it is almost a week before conventional B cells differentiate and start producing antibodies. The antibodies produced by these cells are predominantly of the IgM type and play a role in defence against bacteria and viruses. They are thought to be especially important in mucosal immunity and are often regarded to be part of a primitive mechanism bridging innate and adaptive immunity.

2.3.3 DCs

DCs is a term used to describe a heterogeneous group of cells residing in most areas of the body. DCs derive their name from their peculiar morphology — immature and mature DCs show the presence of dendrites or veils (i.e., long processes). Immature DCs may be formed directly from haematopoietic stem cells of the bone marrow or from other cell types, such as monocytes. Such cells that can differentiate to DCs are often called pre-DCs. DCs (whether pre- or immature) are efficient sentinels of the immune system. Immature DCs are highly mobile cells and colonise lymphoid and nonlymphoid tissues extensively (see chapter 7). They have a role in both innate and adaptive immune responses, besides being a major source of type I IFNs. They are therefore important in antiviral defence and NK cell activation. The TNF-α and NO produced by DCs are thought to be important in antimicrobial defence, especially in the early hours after infection. Immature DCs also double up

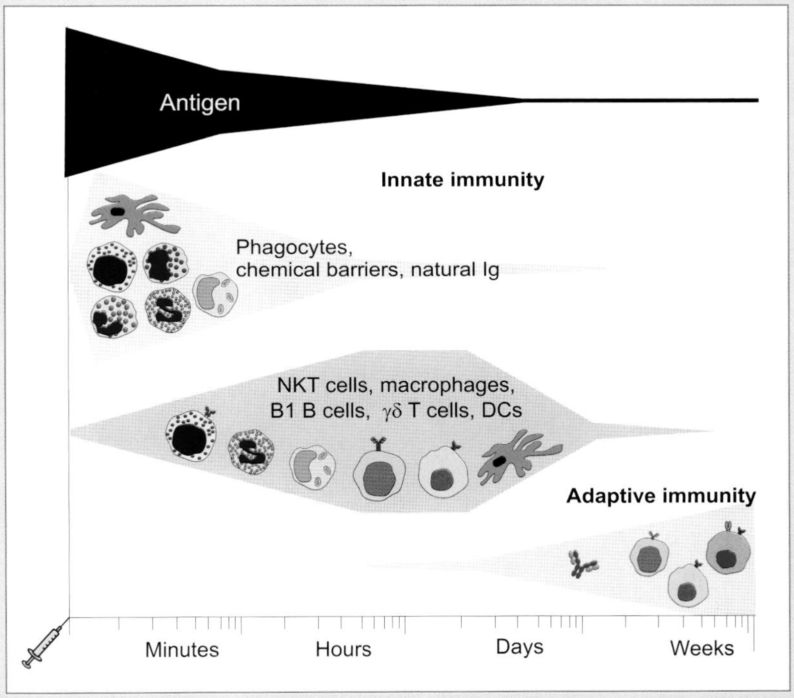

Figure 2.9 Components of the innate immune system act at different stages of the innate immune response to effectively meet the antigenic challenge and remove most of the antigen before the adaptive immune system swings into action. Complement, natural antibodies, and antimicrobial factors of the innate immune system act within minutes of microbial invasion to restrict the invader. The cellular component of the innate system is the next to join the battle. Neutrophils are one of the first cells to enter the site of infection. With mast cells and eosinophils, they rapidly recruit macrophages and DCs into the developing response. Actively phagocytic cells try to limit spread of the infection while killing infectious agents by releasing antimicrobial substances. Meanwhile, lymphocytic components of innate immunity (B1 B cells, γδ T cells, and NKT cells) also join the developing response. Days, and even weeks, elapse before the development of the adaptive response; it is triggered only after the capture and transport of the antigen to the lymphoid organs. The lymphocytic compartment of innate immunity bridges this gap, ensuring optimal transition while aiding antigen removal. (adapted from Martin & Kearney, *Current Opinion in Immunology*, 2001, 13:195).

as 'lookouts' of the immune system that continuously scout the body for signals of potential danger and then trigger adaptive immune responses. Consequently, **immature DCs are highly endocytic cells that capture and internalise intruders that have breached the innate barriers**. PRRs expressed by immature DCs allow them to recognise and capture such intruders. They efficiently trap and transport the captured antigen to lymphoid tissue. Contact with microbes or their products triggers DC maturation. **Mature DCs interact with T lymphocytes in the lymphoid tissue and activate them**, functioning as links between innate and adaptive immune responses. DCs and their role in triggering adaptive immune responses are discussed in chapter 7.

2.4 The Innate Immune Response

So far, we have discussed the cells and cellular processes of the innate immune response. This section describes the chain of events following an encounter with a pathogen to give a better perspective on the entire process (see Flow chart 2.1). Epithelial barriers act as a first line of defence to keep the microbe out. If the pathogen succeeds in breaching these barriers, it is immediately recognised by phagocytes in subepithelial connective tissues. These cells bind and phagocytose the pathogen. The pathogen is then subjected to an arsenal of antimicrobial substances in the phagolysosome, resulting in its destruction. If only a few microbes have crossed the epithelial barrier, this is the end of the encounter. If the innate immune system fails in overcoming the microbial challenge, a focus of infection is established, and both innate and adaptive immune responses are brought into action. As a result of interaction with the pathogen, the recruited phagocytes secrete a number of cytokines. The combined effect of the release of antimicrobial molecules and cytokines starts a series of events, culminating in an inflammatory response. This response occurs within hours of microbial intrusion. Some phagocytic cells (DCs, macrophages, eosinophils) load fragments of ingested pathogens onto MHC class II molecules and display them on their cell surfaces. The recognition of these peptide:MHC complexes by T lymphocytes triggers the adaptive immune response. The primary adaptive immune response takes between 4-7 days for generation, but later encounters result in faster responses (Figure 2.9).

2.4.1 Inflammation

Inflammation can be defined as a localised, protective event, elicited by injury, that serves to destroy, dilute, or wall off injurious agents and injured tissue. The inflammatory response is triggered by chemical signals released upon tissue injury. Appropriate cells are immediately recruited to the site to kill any microbes that have gained entry and infected host cells. Inflammatory mediators stimulate endothelial cells to express proteins that trigger blood clotting in local small vessels. This checks pathogen spread. Tissue surrounding the injury/intrusion is liquefied to halt spread of the microbe. In the final phase of the response, the tissues damaged by the microbe or the host response are repaired. The inflammatory response is analogous to a quick central government response which rushes paramilitary and military forces to the site of disturbance to control unrest and begin damage control.

The inflammatory response is a well-orchestrated event caused by the interaction of several inflammatory mediators. Inflammation is initiated in response to a traumatic, infectious, post-ischemic, toxic, or autoimmune injury[29] (Figure 2.10).

[29] To witness an inflammatory response, use your fingers to slap your inner forearm. The traumatised area immediately responds by turning red. It will also feel warmer and painful to touch — the pain ought to dissuade repetitions.

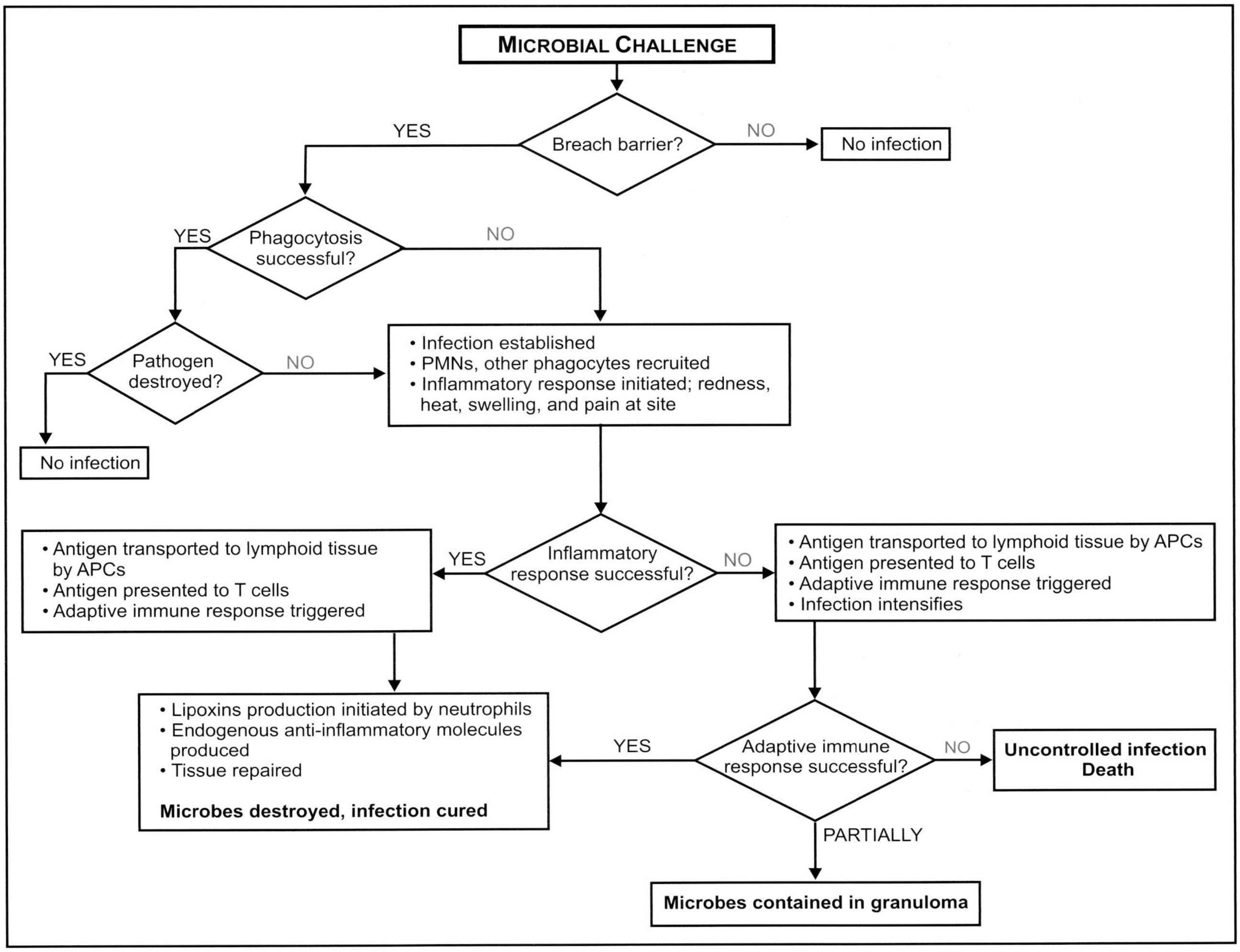

Flow Chart 2.1 Events following pathogen encounter

The bioactive peptides released by neurons and heat shock proteins or mitochondrial proteins released by dying cells trigger cytokine production by tissue cells. The pharmacological mediators released by mast cells and chemokines and the cytokines released by macrophages cause the endothelium to express adhesion molecules, resulting in an influx of PMNs. The endothelium also becomes leaky, leading to the influx of fluids to the site of inflammation. The released cytokines and pro-inflammatory mediators activate macrophages and neutrophils as well, turning them into virtual killing machines. Thus, there are three components of an acute inflammatory response — vascular changes resulting in increased flow and adhesion; increased vascular permeability; and increased leukocyte margination, migration, and activation. Inflammatory mediators include:

- **Chemokines**: These are a superfamily of polypeptides that cause chemotaxis, increased cellular adhesion, and leukocyte activation. Chemokines thus control the movement of phagocytic cells. As described before, multiple cell types involved in the innate response secrete chemokines.
- **Cytokines**: Cells of the innate immune system (especially phagocytes) secrete several cytokines in response to infection. These cytokines promote inflammation (i.e., are pro-inflammatory; examples include IFN-γ, TNF-α, IL-1, and IL-6).

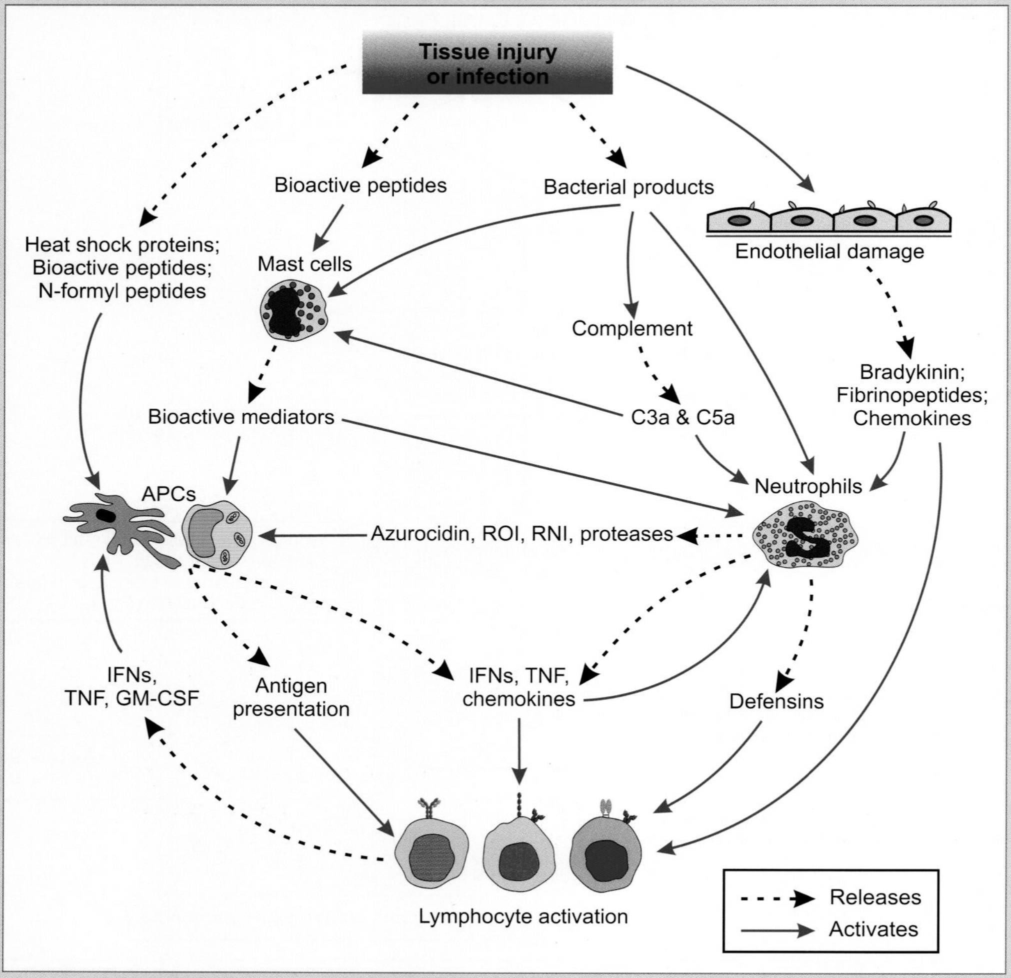

Figure 2.10 A complex interplay of factors is responsible for initiation of the inflammatory response. The inflammatory response is a well-orchestrated, protective event involving multiple cell types. Each cell type is activated by chemical signals released by other cell types, and in turn, releases factors that activate or recruit still other cell types. Chemical signals released by tissues injured either by physical trauma or bacterial invasion trigger the inflammatory response. Infection results in activation of the complement cascade and release of chemotactic C3a and C5a. The release of bradykinins and fibrinopeptides by damaged endothelial cells and the release of chemokines at the site of injury result in neutrophil recruitment. Neutrophils are the first cells to arrive at the site of injury. They express PAMPs and are activated by bacterial infection. Bioactive peptides released because of tissue injury also activate mast cells in the vicinity, causing their degranulation. Pharmacological mediators released by mast cells further recruit neutrophils, immature DCs, and macrophages. Heat shock proteins, neuropeptides, and intracellular contents, released by dead or dying tissue cells or the N-formyl peptides that are signature molecules of bacterial metabolism, activate immature DCs and phagocytic cells through engagement of PRRs. ROI, RNI, and azurocidin (amongst others) produced by neutrophils enhance activities of macrophages and immature DCs. Antigen capture, transport, and presentation by DCs activate naïve T cells. Defensins liberated by neutrophils recruit memory T lymphocytes to the developing response. The phagocytic cells and lymphocytes cross-regulate each other's activation through cytokines. Thus, through a well-coordinated response, the body rids itself of the challenging agent and simultaneously arms the adaptive immune system for future challenges (adapted from Nathan, *Nature,* 2002, 420:346)

They have local effects (chemotaxis, activation of the endothelium, increased vascular permeability, recruitment and activation of cells to the site of infection, etc.) and systemic effects (pyrexia, induction of the acute-phase response, increased metabolism). One of their systemic effects is an increase in the circulating leukocytes (the extra leukocytes are called in from the bone marrow) and the recruitment of immature DCs to the site of infection. DCs are potent APCs that help prepare for the next phase of defence — the adaptive immune response.

- **Components of the complement cascade**: C3a and C5a, released as a result of complement activation, are potent mediators of inflammation. The binding of these molecules to receptors on mast cells results in their degranulation, releasing histamine and other pharmacologically active mediators. These mediators are partly responsible for the observed increase in capillary diameter (vasodilation), increased phagocyte motility, and an increased influx of fluids. This influx allows antibodies, enzymes, and phagocytes to enter the site of inflammation.
- **Plasma mediators**: Endothelial damage starts a cascade in the kinin enzyme system that leads to the formation of bradykinin. This is a potent vasoactive peptide that increases vascular permeability, causes smooth muscle contraction, and induces pain. Damage to a blood vessel starts the fibrinogen cascade to help form a clot, minimise blood loss, and isolate the damaged area. An offshoot of this cascade is the formation of fibrinopeptides, which act as chemoattractants and further increase vascular permeability.
- **Lipid inflammatory mediators**: Phagocytes release a variety of molecules in response to infectious challenge, including metabolites of the arachidonic pathway, such as leukotrienes, prostaglandins, PAF, and thromboxane. Leukotrienes cause smooth muscle contraction and are potent chemoattractants; prostaglandins induce chemotaxis and vasodilation. Thromboxane causes platelet aggregation and blood vessel constriction.

Thus, a **characteristic inflammatory response** results in

- **Increased blood supply** to, and resultant reddening of, the affected area (this reddening is also referred to as *erythema*).
- **Increased capillary permeability** that causes swelling, or *oedema*, which is partially a result of the retraction of endothelial cells that normally line blood vessels. The increased permeability allows larger molecules, normally incapable of penetrating the endothelium, to reach the site of infection/ injury. This enables various soluble mediators of immunity to reach the affected site.
- **Migration of leukocytes** to the affected area, which is facilitated by newly expressed adhesion molecules on the endothelial cells that line the blood vessels.
- **Local experience of pain and an increase in temperature** that is partly due to increased blood supply and partly due to the effect of the release of different chemicals (cytokines, chemokines, etc.) in the area.
- **Possible loss of function** of the affected area, caused by the inflammatory response.

This achieves a two-fold objective.

- It limits the exposure of tissue to the external environment (limiting the entry of potentially harmful agents).
- It walls off the damaged area so that any harmful agent that may have entered cannot easily spread to the rest of the body.

Acute inflammation (inflammation that lasts from a few hours to a few days) is generally associated with a systemic response known as the acute-phase reaction, and it is normally self-limiting. Chronic inflammation (inflammation that persists beyond 10-14 days) develops because of pathogen persistence or chronic activation of the immune system, as happens with *Mycobacterium tuberculosis* infection, autoimmune disorders, or some cancers. Chronic inflammation leads to the accumulation and activation of macrophages. This chronic macrophage activation also activates fibroblasts. The activated fibroblasts produce collagen, leading to fibrosis at the inflamed site. Formation of scar tissue (e.g., in the joints) can interfere with normal function and precipitate diseases. Chronic inflammation can also lead to granuloma formation. Granuloma is a characteristic nodular mass consisting of a central area of activated macrophages surrounded by activated lymphocytes (chapters 13 and 16). In a way, this is a protective response because the damaging pathogen is effectively isolated from the rest of the tissue.

Prolonged inflammation becomes deleterious. The dysregulated production of pro-inflammatory cytokines, such as TNF-α and IL-1, can lead to acute systemic shock, resulting in capillary leakage, lowering of blood pressure, and organ failure. It is often fatal. Counter-regulatory molecules produced in the later stages of the inflammatory response help restore immunological equilibrium. **The resolution of inflammation is a highly controlled and coordinated process that suppresses pro-inflammatory gene expression, leukocyte activation, and migration.** It is followed by the enhanced clearance of cell debris and apoptotic cells by phagocytes. Molecules and signals involved in this process are still being elucidated. The process of resolution begins with the death and removal of microbes and their products by macrophages. Neutrophils initiate the process by triggering production of anti-inflammatory lipoxins by tissue cells.

- **Lipoxins** are extremely potent endogenous lipid anti-inflammatory mediators. They seem to act early in the process of resolution.
  - They inhibit neutrophil and eosinophil activation and migration.
  - Lipoxins suppress the release of superoxide anions by neutrophils.
  - They dampen genes involved in adhesion molecule expression and activation.
  - They inhibit the secretion of chemotactic IL-8.
- Macrophages secrete a **serine protease inhibitor** that is expressed late after exposure to microbial products or cytokines. It suppresses ROI production and elastase secretion by neutrophils. It also promotes tissue healing.
- Systemic production of **endogenous anti-inflammatory molecules**, such as adrenaline, noradrenaline, and 5-hydroxytryptamine, reduce vascular leakage. This reduces inflammation, as reduced vascular leakage stops fresh neutrophils from entering the site.
- **cAMP**, which is induced by several hormones, inflammatory mediators, and cytokines, has a central role in dampening the inflammatory response. It has multiple functions.
  - It suppresses the release of histamine and leukotrienes from mast cells, monocytes, and neutrophils.
  - It inhibits the secretion of cytokines and NO from macrophages.
  - It inhibits the release of lysosomal enzymes and ROI from neutrophils.
- **Glucocorticoids**, produced by the adrenals, further help by inducing the expression of anti-inflammatory proteins, such as annexin-1.
- **Cyclopentenone prostaglandins** are anti-inflammatory prostaglandins with multiple functions. They are found in the resolution phase of inflammation.

INFLAMMATION

- It is a localised, protective event elicited by injury.
- Inflammation serves to destroy or wall off injurious agents and injured tissue.
- The inflammatory process includes
  - Tissue-based response triggered by recognition of microbial penetration
  - Recruitment of macrophages, PMNs, and DCs, and the unleashing of their molecular arsenal
  - Dispatching of APCs to the draining lymph node
  - Liquefaction of surrounding tissues to prevent microbial spread
  - Healing of tissue damaged by the infectious agent or the host response
- Inflammation is characterised by
  - *Rubor* (redness) due to increased blood supply
  - *Calor* (heat) caused partly by the influx of blood and partly by the release of cytokines and chemokines
  - *Tumour* (swelling) due to increased capillary permeability fluid influx
  - *Dolor* (pain) due to the release of bioactive peptides by neuronal cells
  - Loss of function of the affected organ or part
- The inflammatory response is induced by pro-inflammatory mediators, such as
  - Chemokines (IL-8, MCP, RANTES)
  - Cytokines (TNF-α, IL-1, IL-6, IFN-γ, TGF-β)
  - Components of the complement cascade (C3a, C5a)
  - Plasma mediators (bradykinin, fibrinopeptides)
  - Lipid mediators (metabolites of the arachidonic pathway)
- The resolution of inflammation is promoted by
  - Serine protease inhibitor, released by macrophages
  - Endogenous anti-inflammatory mediators, such as adrenaline and noradrenaline
  - cAMP
  - Annexin-1
  - Lipoxins and cyclopentenones
  - TGF-β

  - Cyclopentenone prostaglandins decrease the expression of adhesion molecules by endothelial cells.
  - They inhibit macrophage/monocyte activation and migration.
  - They decrease the expression of NOS by macrophages.
  - Along with lipoxins, cyclopentenone prostaglandins promote leukocyte apoptosis and their uptake and clearance by phagocytic cells.
- The uptake of apoptotic cells by phagocytes induces the phagocytic cells to release the immunosuppressive cytokine TGF-β, promoting tissue repair.

2.4.2 Acute-Phase Reaction

This is the set of immediate inflammatory responses initiated by PRR engagement that localises spread of infection and enhance systemic resistance to infection. The acute-phase response is a systemic **response initiated by a sudden rise in circulating cytokines, such as IL-1**, **IL-6**, **and TNF-α**. These cytokines act on the brain, the neuroendocrine system, and other tissues and organs, leading to fever and profound hormonal and metabolic changes. The **H**ypothalamus–**P**ituitary–**A**drenal (HPA) axis is then activated and serves as the primary regulator of immune and inflammatory reactions. Insulin, glucagon, and catecholamine levels rise. HPA activation also leads to the release of glucocorticoids by adrenal glands. Bone marrow activity and leukocyte function are high, so the number of circulatory leukocytes increases dramatically. The liver initiates rapid production of acute-phase proteins, with a parallel decrease in the production of other proteins, such as albumin and transerythrin[30]. Acute-phase proteins are a group of ~20 proteins which include α1 acid-glycoprotein, serum amyloid A and P, α2 macroglobulin, PRRs (C-reactive protein, LBP, MBL), fibrinogen, some complement components, enzyme inhibitors, and anti-inflammatory proteins whose serum concentration increases several hundred- to thousand-fold within 24-48 hours. Many acute-phase proteins equip the body for antimicrobial defence because they bind a broad range of pathogens. They also cause increased synthesis of pro-inflammatory cytokines (IL-1, IL-6, TNF-α) in the brain, resulting in pyrexia. Some acute-phase proteins are serine/ cysteine protease inhibitors responsible for inactivating the proteolytic enzymes secreted during the inflammatory response. Along with anti-inflammatory proteins, they help in limiting inflammation.

The thymus is dramatically affected by the acute-phase response, which is accompanied by profound neuroendocrine and metabolic changes. The most striking effect of glucocorticoids on the immune system is the induction of apoptosis in the thymus. In concert with glucocorticoids, elevated catecholamine levels selectively suppress immune responses. This temporary suppression of specific immunity might serve to protect the body from the possible generation of adverse immune reactions. The acute-phase reaction may be considered an emergency response; it represents a switch in host defence, from the adaptive immune response, which is slow to develop and is commanded by the thymus and T lymphocytes, to a less specific but more rapid and intense reaction. Acute-phase proteins therefore provide enhanced protection against microorganisms and modify inflammatory responses by affecting cell trafficking and mediator release.

[30] The acute-phase response is an expensive, energy-consuming event; this explains why fever is often accompanied by exhaustion.

Mind Your Immunity

The two major 'adaptive systems' of the body — the brain and the immune system — are involved in bi-directional communication (or cross-talk) that allows homeostasis (*status quo* maintenance). Neuroendocrine regulation of immune function is also essential for survival during stress or infection and in modulating immune responses in inflammatory diseases. The brain, or more specifically the **C**entral **N**ervous **S**ystem (CNS), regulates the immune system through two key pathways — the HPA axis and the sympathetic nervous system. The CNS can also regulate the immune system *via* peripheral nerves, through the release of neuropeptides and locally produced corticotrophin-releasing hormone.

The main components of the HPA axis are the paraventricular nucleus in the hypothalamus, the anterior pituitary gland, and the adrenal glands. Corticotrophin-releasing hormone, secreted by the paraventricular nucleus, stimulates the release of adrenocorticotropin hormone by the anterior pituitary gland. This induces the adrenal glands to produce and secrete glucocorticoids in plasma. Cells of the immune system express receptors for glucocorticoids. The binding of glucocorticoid to its receptor modulates multiple gene expressions. Glucocorticoids have an anti-inflammatory effect on the immune system, and they

- suppress expression of pro-inflammatory cytokines, such as IL-1, IL-2, IL-6, IL-8, IL-12, TNF-α, and IFN-γ;
- upregulate expression of anti-inflammatory cytokines, such as IL-10 and IL-4;

- decrease neutrophil and macrophage migration by repressing the expression of adhesion molecules, such as ICAM-1 and E-selectin on endothelial cells, and by decreasing expression of chemokines like IL-8, RANTES, and MCPs;
- suppress synthesis of pro-inflammatory mediators, such as prostaglandins and NO, at the site of inflammation; and
- promote the rapid transport of the anti-inflammatory molecule annexin-1 from the cytoplasm to the cell surface and upregulate its synthesis.

The catecholamines norepinephrine (also called noradrenaline) and epinephrine are the principal end-products of the sympathetic nervous system. The sympathetic nervous system modulates local immune responses *via* these neurotransmitters. Noradrenergic sympathetic nerve fibres run from the CNS to primary and secondary lymphoid organs. These nerve terminals make synapse-like connections with neighbouring immune cells by releasing norepinephrine. B cells, T cells, NK cells, neutrophils, and macrophages express adrenoreceptors that bind norepinephrine. Locally released norepinephrine or circulating epinephrine released by the sympathetic nervous system can affect lymphocyte trafficking, circulation, and proliferation. They also modulate cytokine production and the functional activity of different lymphoid cells. B and T cell differentiation and proliferation is influenced by norepinephrine. Microbes or their products can influence the turnover of norepinephrine in lymphoid organs. Both epinephrine and norepinephrine were originally thought to have immunosuppressive effects. Recent data suggests that they may be immunostimulatory or immunosuppressive, depending upon the location and stage of development of target cells.

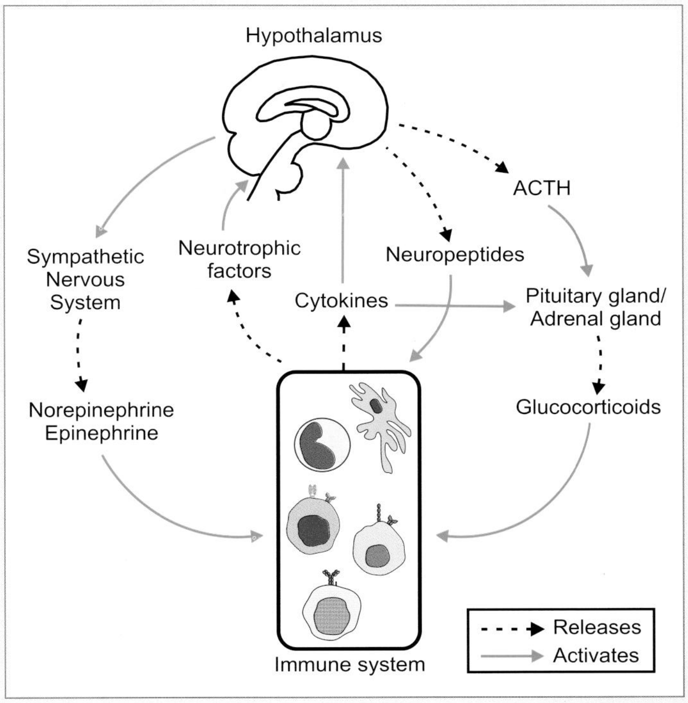

Figure 2.S1 The CNS — through the HPA axis and sympathetic nervous system — and the immune system communicate with and influence each other. The neuroendocrine system regulates functioning of the immune system through the release of ACTH (adrenocorticotropin hormone), glucocorticoids, epinephrine, and norepinephrine. Similarly, cytokines and neurotrophic factors secreted by the immune system, influence neuroendocrine functioning.

The immune system signals the CNS through cytokines. Cytokine receptors are expressed on cells of the CNS, peripheral nerves, and ganglia, and cytokines can influence neuronal survival, growth, and differentiation. Through their receptors, cytokines can also influence peripheral nerve activity and neurotransmitter release. Thus, neuronal cells respond to IL-2 treatment by enhanced neurite growth *in vitro*. IL-1, IL-2, IL-6, and TNF-α have all been shown to influence norepinephrine release. They signal the brain to trigger activation of the HPA axis and sympathetic nervous system (Figure 2.S1).

3 Complement

Domino motion jump starts when we touch
The blackout approaching
Here it comes now, wish me luck
It's all over, it's all over, it's all over in a flash
I can't remember,
What have I done now?

Go, go, faster wider
More more get it down ya

— Imogen Heap, *Glittering clouds*

3.1 Introduction

The term *complement* is used to describe a complex group of about thirty heat-labile, sequentially interacting proteins and glycoproteins found in the blood, plasma, and cell surfaces of all vertebrates. Complement was originally thought to be a single component that *complemented* an antibody's antibacterial activity. Immunology has come a long way since then; it is now well established that complement components and their receptors are important in a variety of interrelated physiological activities, including the augmentation of innate and adaptive immune responses and apoptotic cell clearance (section 3.7).

Complement activation is a cascading[1] and sequential process. Each protein in the cascade activates the next one. Many components exist as proenzymes, that is, as inactive or nearly inactive precursors of enzymes. They require conversion to the active form by proteolytic cleavage. When activated, several become serine proteases[2] that cleave and activate the next component in the cascade. The **key event in complement activation is the formation of the enzyme C3-convertase**, which cleaves C3. This activation can occur through three different pathways. The downstream events following C3 cleavage — termed *the terminal pathway* — can result in the lysis of the target cell[3]. The three pathways that activate C3 are

- the classical pathway,
- the MBL (**M**annose-**B**inding **L**ectin) pathway, and
- the alternative pathway.

The nomenclature of complement proteins is far from logical. The letter C, followed by a number, is used to designate all classical pathway proteins. Unfortunately, numerical insanity reigns supreme; the numbering is not sequential but in the order of discovery. Native molecules are labelled C1-C9, and the products formed are named by adding lower case letters (C4a, C3b, etc.). The smaller fraction is generally labelled *a*, while the larger is designated *b* — C2 is an exception. C2a is the larger fragment; C2b, the smaller. Proteins of the alternative pathway are named *Factors*, followed by a letter. The proteins of the MBL pathway are MASPs (for **M**annose-binding lectin-**A**ssociated **S**erine **P**roteases), followed by a number.

Complement components other than C1 are primarily synthesised by liver cells, although other cells, such as tissue macrophages, monocytes, fibroblasts, epithelial and endothelial cells, adipocytes, and astrocytes, also synthesise them. Even neurons are involved in their synthesis. C1 is primarily synthesised by the epithelium of the gastrointestinal and urinogenital tracts. Inflammation increases the synthesis of complement components, probably as a result of the IL-1 and IFN-γ produced during inflammation. Complement proteins appear in the blood during foetal development before circulating antibodies. This probably reflects complement's evolutionary history — phagocytic cells and complement were major defence mechanisms for vertebrates before they developed antibodies.

3.2 The Classical Pathway

The origins of the complement system have now been traced near to the beginnings of multicellularity. Although the classical pathway has a prominent role in mammalian immune system, the earliest complement-activation pathway was probably similar to the MBL pathway, and C1q, the first component of the

Abbreviations

APC:	Antigen-Presenting Cell
C1EI:	C1 Esterase Inhibitor
C1INH:	C1 Inhibitor
C4bp:	C4-Binding Protein
CR1:	Complement Receptor 1
CR2:	Complement Receptor 2
DAF:	Decay-Accelerating Factor
FDC:	Follicular Dendritic Cell
FHL-1:	Factor H-Like protein-1
HRF:	Homologous Restriction Factor
MAC:	Membrane Attack Complex
MA-p19:	MBL-Associated protein of 19 KD
MASP:	Mannose-Binding Lectin-Associated Serine Protease
MBL:	Mannose-Binding Lectin
MCP:	Membrane Cofactor Protein
MCP-1:	Macrophage Chemoattractant Protein-1

[1] It is like a waterfall — increasing in strength as it proceeds. This is achieved by built-in activation where each activated component has the potential to activate many molecules of the next component.

[2] The complement system consists of many proteases. Ten of these — C1s; C1r; the closely related MASP-1, -2, and -3; C3/C5-convertases; and Factors B, D, and I — are serine proteases, that is, have serine in their active sites.

[3] Erythrocyte lysis was one of the first observed effects of complement activation and is still used in some diagnostic tests (see appendix IV).

RCA: Regulators of Complement Activation
SCID: Severe Combined Immunodeficiency
SLE: Systemic Lupus Erythematosus
sMAP: Small MBL-Associated Protein

classical pathway, predates immunoglobulins. In species like lampreys that lack immunoglobulins, C1q activates the complement pathway by binding to MASPs (serine proteases that trigger the MBL pathway).

The first step in the classical pathway is the binding and activation of C1. The prototypical activator of C1 is an antibody bound to its homologous antigen. However, not all Ig isotypes are capable of binding C1. In humans, IgG1, IgG2, IgG3, and IgM antibodies can activate the complement cascade.

Not Just the Sum of Its Parts: Multiple Functions of Complement Components

Complement components were believed to be limited to a role in inflammation and immunity. However, studies investigating the components' distinct expression profiles in various tissues and at varying developmental stages found evidence for other functions.

- Complement components have a role in bone and cartilage development. They promote the differentiation of mononuclear progenitors to osteoclasts. The distribution patterns of C3, Factor B, Factor H, C5, C9, and properdin suggest a potential role for these components in cartilage-bone transformation, matrix degradation and bone remodelling, and vascularisation.
- The presence of almost all complement components and membrane regulators has been documented in the epithelial and vascular tissues lining the entire female reproductive tract. This prominent expression of DAF (**D**ecay-**A**ccelerating **F**actor), MCP (**M**embrane **C**ofactor **P**rotein), CR1 (**C**omplement **R**eceptor **1**), and CD59 in reproductive epithelia and sperm surfaces was thought to protect these tissues from autologous complement activation. These regulators are now proposed to have a role in maintaining foetomaternal tolerance during early pregnancy. There is also increasing evidence that the biosynthesis of several complement components and receptors in the reproductive tract is subject to fine hormonal regulation. This synthesis follows stage-specific expression patterns during the menstrual cycle, suggesting a role in normal reproductive processes. Some tantalising evidence points to a role of complement components in fertilisation, but further research is needed to clarify this.
- Complement components have also been implicated in organ regeneration. C3 has been known to be involved in limb regeneration in urodeles (amphibians such as axolots and newts). In mammals, recent findings note that, in addition to TNF-α and IL-6, C5a and its receptor C5aR are also important to regeneration of the liver (one of the few mammalian organs capable of regeneration). Research suggests that the complement system has an important immunoregulatory role in hepatic growth and homeostasis.
- C1q and MBL modulate cytokine production in human peripheral blood monocytes, leading to the suppression of the pro-inflammatory IL-1.
- C1q and its phagocytic receptor, C1qRp, have been implicated in early haematopoietic development. Recent experiments suggest that the coengagement of CD46 (MCP) with T cell antigen receptor causes greater and more sustained proliferation of T cells than is achieved by stimulation through CD28/T cell antigen receptor (see chapter 6). This suggests a role for complement regulators in T cell biology as well.
- C1q is expressed at synapses throughout the brain only during the period when selective synapse pruning is shaping brain structure. Studies show that C1q helps tag weak or inappropriate synapses for elimination during normal development. Recent evidence also points to a role for C1q in neurodegenerative diseases. The improper activation of C1q and the resulting cascade is thought to be a common step in various neurodegenerative diseases.

3.2.1 Binding and Activation of C1

C1 consists of three subunits — **C1q**, **C1r**, **and C1s** — that are held together by Ca^{2+} ions. C1q is a 400 KD protein made of six subunits, with each subunit consisting of three polypeptide chains. When assembled, the polyprotein resembles a bunch of six tulips or pods (Figure 3.1). C1r and C1s are smaller (83 KD) proteins that have a catalytic domain and an interacting domain. They can act as serine proteases because of the catalytic domain. The interacting domain facilitates their interaction with each other or with C1q. They exist in serum as an S-shaped $C1r_2s_2$ tetramer. Binding to C1q changes the S-configuration to that of the figure 8 and yields the $C1qr_2s_2$ complex (Figure 3.1).

- **C1q binds to the CH2 domain[4] of IgG or the CH4 domain of the IgM** molecule *via* the tulip heads. Native Ig molecules have a relaxed conformation, and hence, C1q cannot access the binding site. Antigen binding causes a conformational change in the Ig molecule that exposes the binding site. C1q recognition and binding is stabilised mainly by electrostatic charges and hydrophobic interactions. As binding of each individual pod is dependent on intrinsically weak interactions, multiple pods must be engaged for activation. A doublet of antigen-bound IgG molecules (two molecules lying side by side) is therefore necessary for C1q activation. A single IgM molecule can activate C1q provided more than one subunit of the pentamer binds the antigen. C1q activation is not exclusively dependent upon antigen–antibody interaction. **A variety of non-Ig molecules activate C1**. These include acute-phase proteins (e.g., C-reactive protein), certain bacterial glycolipids, monosodium urate crystals, LPS, nucleic acids, and chromatin.

[4] Ig molecules are made of two chains — light and heavy. Each of these chains has domains with differing functions, as explained in section 10.4.

Wasted by Waste: C1q and Autoimmunity

C1q can bind to monocytes, PMNs, B lymphocytes, platelets, endothelial cells, and fibroblasts, suggesting the existence of protein receptors. Although identifying these receptors (C1qRs) was difficult because of the sticky nature of C1q, a number of receptors have been identified.

- Two intracellular proteins that may be found on the surfaces of damaged or apoptotic cells have been shown to be C1qRs. One is calreticulin, a chaperone protein, and the other, a mitochondrial matrix protein. They bind both C1q and MBL. After binding to C1q, and the protein-C1q complex associates with CD91, a signalling molecule that drives apoptotic cell phagocytosis.
- Another C1q binding molecule has recently been found to be expressed on phagocytic cells and is termed C1qRp (C1qR of phagocytosis).
- CR1 (CD35) has also been shown to bind C1q.

C1q helps tag damaged or apoptotic cells and aids in their removal by phagocytes. These damaged tissue cells are thought to be major autoantigen sources. Deficiency in complement components is a predisposing factor for autoimmune diseases; C1q deficiency carries the greatest risk. C1q seems to be especially important in the development of SLE (**S**ystemic **L**upus **E**rythematosus), an autoimmune disease with severe and multiple manifestations. There is an almost 100% correlation between C1q deficiency and SLE development, although a lack of C4 can also cause the disease. This has led to the hypothesis that SLE is the result of an antibody response to self-antigens driven by the nonclearance of necrotic and apoptotic cells.

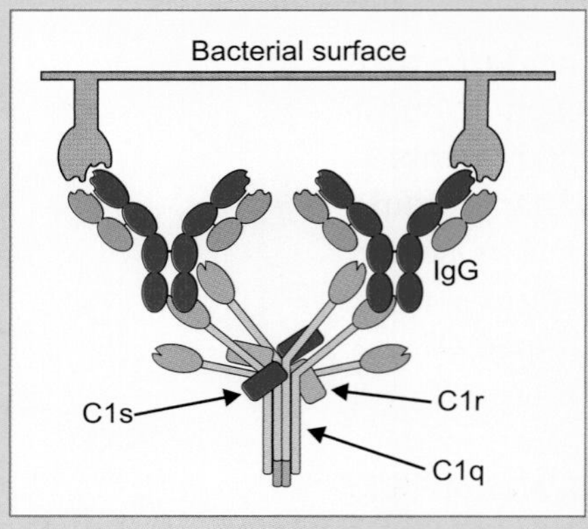

Figure 3.1 C1, the first component of the classical pathway, consists of three subunits — C1q, C1r, and C1s — held together by Ca^{2+} ions. C1q consists of six subunits and resembles a bunch of tulips. One unit of C1q associates with two units each of C1r and C1s. At least two subunits of C1q must be engaged for C1q activation. Activated C1q activates C1s and later C1r and converts them both to serine proteases. C1s cleaves C2 to trigger the classical complement pathway.

- Conformational change caused by **the binding of C1q** to a receptive surface **leads to the autoactivation of C1r** and its conversion to a serine protease. **Activated C1r cleaves C1s** and converts it to a serine esterase.

3.2.2 Formation of Classical Pathway C3-convertase

Both C4 and C2 are substrates of activated C1s. They can also be cleaved by the MBL pathway. Thus, the classical and the MBL pathways converge at this point.

- **Activated C1s cleaves C4** to form a larger C4b and smaller C4a[5] fragment; C4a is released to the fluid, while C4b binds to a site in the molecule's vicinity.
  - Several C4 molecules can be cleaved by one molecule of activated C1s — this is the first amplification step in the cascade.
  - C4 cleavage exposes a highly reactive thioester bond on the C4b molecule. This bond allows covalent binding of C4b to a site in the immediate vicinity of the molecule, such as to a cell membrane near the site of activation or to the C1qrs complex itself. If C4b is not rapidly covalently bound, the thioester bond is hydrolysed irreversibly, inactivating the molecule. This mechanism ensures that complement action remains localised to the surfaces of target cells.
- **C2 binds to C4b** in the presence of Mg^{2+}; **C4b cleaves the bound C2**[6] to yield a small C2b fragment and a larger C2a one.
  - In contrast to C4b bound C2, free C2 is weakly susceptible to the proteolytic action of C1s. Free C2a formed by this cleavage cannot bind C4b and has no role in the complement cascade.
  - The released C2b is a prokinin that becomes biologically active upon enzymatic alteration by plasmin.
  - The activation of C2b leads to the accumulation of fluids and oedema at the site of complement activation.
- **C2a remains bound to C4b to form the classical pathway C3-convertase** — C4b2a (Figure 3.2).
  - The C4b2a complex is unstable and dissociates with a half-life of less than five minutes.
  - Two plasma proteins, Factor H and **D**ecay-**A**ccelerating **F**actor (DAF), interact with this complex to accelerate dissociation. The dissociated C4b is then cleaved to a biologically inactive form by Factor I (section 3.8).

[5] Although initially thought to be bioactive, no known function has been observed for human C4a. Research has also failed to identify its receptor.

[6] The naming of C2 fragments is a classical case of confusion. The large active fragment of C2 was originally designated C2a. It was later proposed that the nomenclature be reversed and brought in line with other components (i.e., rename the old C2a as C2b). However, the proposal has now been laid to rest, and it has been (finally) decided to continue with older nomenclature.

3.2.3 C3 Cleavage

C3 is a key molecule of the complement cascade. Evolutionarily, it arose about 700 million years ago, long before the appearance of Igs. Cleaving of C3 represents a major amplification step in the complement cascade. Probably the most versatile protein of this system, it is central to all three pathways of complement activation. It is the most abundant of all complement components (1.2 gms/litre in blood) and is pivotal to complement-mediated immune functions.

- **The classical pathway C3-convertase**, **C4b2a**, **cleaves C3** into a small C3a and a larger C3b fragment (Figure 3.3). Over 200 C3 molecules are split by a single molecule of C3-convertase, representing a major amplification loop.
- The smaller C3a, a multipotent molecule, is released in the fluid phase. It is called an *anaphylatoxin* as it mimics some effects of anaphylactic shock (see chapter 16).
- **The C3b fragment closest to the C4b2a complex binds it** to form the C5-convertase of the classical cascade — C4b2a3b. Not all C3b molecules produced by C3 cleavage bind C4b2a, and they end up with different fates.
  - C3, like C4, has a reactive thioester bond that is exposed on cleavage and allows C3b to be covalently bound to appropriate surfaces; the cleavage of covalently bound C3b by serum enzymes results in the formation of several products (iC3b, C3c, C3d(g), etc.) that are bioactive. These bioactive products exert their effect through receptors found on a variety of cells.

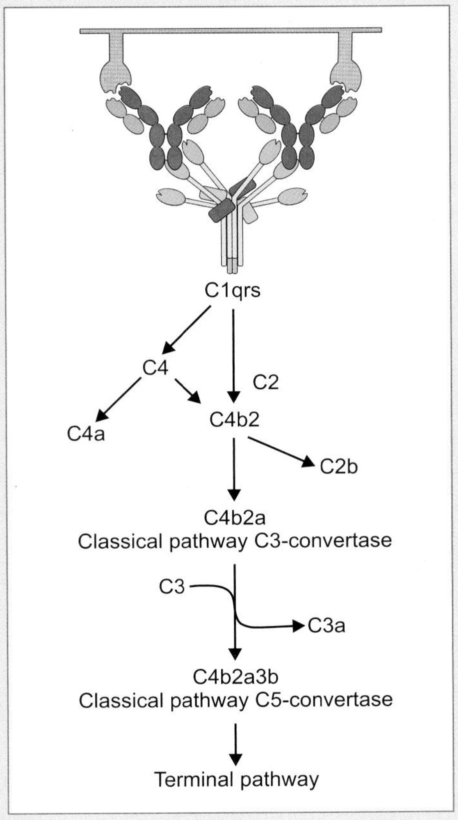

Figure 3.2 The classical pathway of complement activation is triggered when two subunits of C1q bind adjacent antigen-bound Ig molecules. C1q activation sequentially activates C1r and C1s. Activated C1s cleaves C4; C4a is released in the fluid phase, and C4b binds C2. C1s also cleaves C4b-bound C2. C2b, a bioactive molecule, is released to the fluid phase, while C2a remains bound to C4 to yield C4b2a — the classical pathway C3-convertase. The enzyme can rapidly cleave many molecules of C3. The bioactive C3a is released to the fluid phase. C3b binds the C4b2a complex to yield C4b2a3b — the classical pathway C5-convertase. With the formation of this proteolytic complex, the cascade enters the terminal pathway.

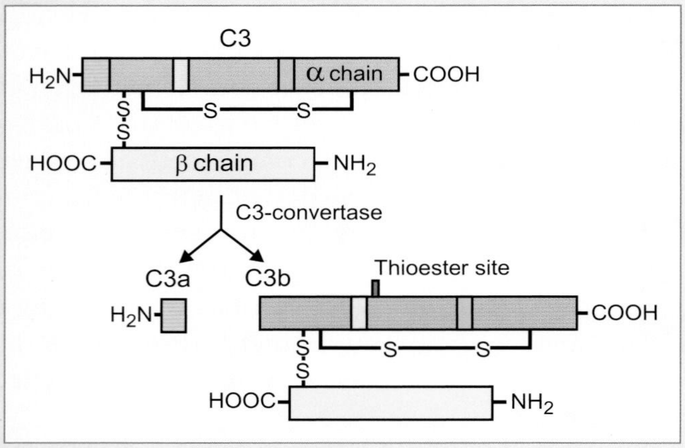

Figure 3.3 Structural features of C3 allow it to become the most versatile protein of the complement cascade. Human C3 is a 185 KD glycoprotein consisting of α and β chains held together by a single disulphide bond and noncovalent forces. C3-convertase cleaves C3 to C3a and C3b. The larger C3b consists of a fragment of the α chain and the whole β chain. Native C3 has an internal thioester bond that is hidden inside a hydrophobic pocket and exposed only in the C3b fragment. This thioester bond has a half-life of 100 μs and can transiently participate in transacylation reactions with hydroxyl groups in its neighbourhood. Unlike the native C3, C3b expresses multiple binding sites for various complement components including C5; properdin; factors B, H, and I; and MCP (CD46). Binding of these proteins leads to either amplification of the cascade or inactivation of C3b.

- Some C3b molecules may be deposited on membrane surfaces in the vicinity of the reaction and may be inactivated or further cleaved. The carbohydrate environment and the nature of the surface to which C3b is attached largely dictate whether inactivation or further amplification of the reaction occurs.
- Some C3b molecules are released in the fluid phase and bind to and opsonise microbes in their vicinity.
- Some C3b molecules may bind Factor B and activate the alternative pathway. By contrast, C3b molecules that bind Factor H become susceptible to Factor I, which catabolises them to inactive products.
- Unbound C3b molecules are rapidly inactivated by hydrolysis.

❐ **With the formation of C5-convertase, the cascade enters the terminal pathway.** C4b2a3b complex binds C5 and makes it susceptible to the activity of C2a. This C5-convertase of the classical pathway can hydrolyse both C3 and C5. Binding of additional C3b molecules increases its affinity for C5.

3.3 The Lectin Pathway of Complement Activation[7]

Mounting evidence supports the importance of this relatively newly discovered pathway in innate immunity — in some individuals, MBL deficiency is associated with repeated and severe infections. MBL pathway proteins are homologues of classical pathway components. The pathway is initiated when MBL binds to carbohydrates.

❐ **MBL**: A C-type lectin, MBL requires Ca^{2+} for binding to carbohydrates. It is a member of the collectin family of proteins, which includes C1q and MBL, amongst others. Members of the collectin family have a collagen-like region and a carbohydrate recognition domain. **MBL can bind to an array of carbohydrate structures on the surfaces of microbes** (yeasts, bacteria, viruses, and parasitic protozoa). It recognises molecular patterns formed by repetitive mannose or N-acetylglucosamine residues that are present on microbial surfaces. MBL therefore preferentially recognises glucans, lipophosphoglycans, or glycoinositol phospholipids with mannose, glucose, fucose, or N-acetylglucosamine as terminal hexoses. Such patterns are either absent or present in limited amounts on mammalian cell walls. Instead, mammalian cell walls have structures that terminate with sialic acid and are therefore protected from MBL damage.

[7] Consensus has yet to be reached on the nomenclature of this pathway. Although it is referred to as the lectin-binding pathway, this is perhaps not the best name, as 'lectin-binding' can be presumed to imply that any lectin can activate complement. It would be more appropriate to call it the MBL pathway.

CLASSICAL COMPLEMENT PATHWAY

C1	Consists of C1q, C1r, and C1s; binding of two C1 molecules triggers the classical pathway
C4	Activated C1s cleaves C4
C2	C2 binds to C4b; C1s cleaves C2 to form the classical pathway C3-convertase — C4b2a
C3	C3-convertase cleaves C3; C3b binds to C4b2a to yield the classical pathway C5-convertase — C4b2a3b
C5	C5 is cleaved by C5-convertase
C6 & C7	C6 and C7 sequentially attach to C5b; C5b67 gets inserted into lipid bilayer
C8	β subunit of C8 binds to C5b67; both C8α and C8β get inserted into the lipid bilayer
C9	C5b678 causes polymerisation of C9 and formation of MAC; MAC lesions allow the outward passage of small molecules and inward passage of water, leading to cell lysis

- **MASPs**: Three serine proteases are found to be associated with MBL, and they are called *MASPs* (**M**BL-**A**ssociated **S**erine **P**roteases-1, -2, and -3). The association between MBL and MASPs is Ca^{2+} dependent. MASP-1 and -2 are homologous to C1r and C1s. These serine proteases normally exist as zymogens and have to be converted to their active forms. MASP-3 is an alternatively spliced version of MASP-1.
- **MA-p19** (**M**BL-**A**ssociated **p**rotein of **19** KD) or **sMAP** (**s**mall **M**BL-**A**ssociated **P**rotein): This nonprotease associated with the MBL pathway is a truncated form of MASP-2. Its exact role in the MBL pathway is unclear.
- **Ficolins**: These lectins contain a collagen-like structure and recognise carbohydrates (N-acetylglucosamine) *via* a fibrinogen-like structure. Complexes of ficolins and MASPs also activate the MBL pathway.

As MBL and MASPs are homologues of C1q, C1r, and C1s, the MBL and classical pathways are very similar. The **MBL pathway generates C3-convertase of the classical pathway** (Figure 3.4).

- **The binding of MBL or ficolins** to microbial surfaces **activates the serine protease MASP-2**; it cleaves C4 to C4a and C4b.
- **C2** binds to C4b and **is cleaved by MASP-1 or MASP-2**, forming the classical C3-convertase C4b2a.
- Further cleavage of C3 and the formation of C5-convertase occur by the classical pathway, and it sets the terminal pathway in motion.
- MASP-1 by itself has been shown to cleave C3. The low levels of cleaved C3 required for the alternative pathway may be formed through the binding of MBL to receptive surfaces.

3.4 The Alternative Pathway of Complement Activation

Components of the alternative pathway bypass the initial sequences of the classical pathway and directly cleave C3. The alternative pathway[8] can function in the absence of IgM or IgG antibodies and is spontaneous. It provides a means of nonspecific resistance to infection without the participation of antibodies. Like the MBL pathway, the alternative pathway provides a true first line of defence against several infectious agents. The alternative pathway is thought to be a prototype of the classical pathway and it may have developed later in evolution through gene duplication after or along with the development of antibodies. **LPS, teichoic acids of Gram-positive bacteria, zymosan of yeast cell walls**[9], **and the surface**

[8] In a way, *alternative pathway* is a misnomer, a hangover from the era when only two pathways were known to activate complement — the classical and the nonclassical (i.e., alternative).

[9] *Zymosan* is an insoluble preparation from yeast cell walls. It is a mixture of glucans, mannans, proteins, chitins, and lipids. Several key functional effects related to zymosan can be ascribed to its major component β-glucan. A *zymogen*, on the other hand, is a pro-enzyme.

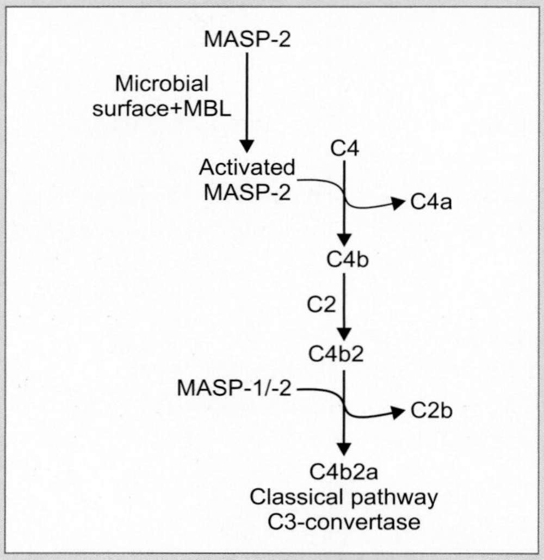

Figure 3.4 The MBL pathway of complement activation is homologous to the classical pathway. The MBL pathway is activated by the binding of MBL to microbial surfaces. This binding activates MASP-2. MASP-2 cleaves C4. C4a is released in the fluid phase; C4b binds C2. MASP-1 and MASP-2 cleave the bound C2 to form the classical pathway C3-convertase C4b2a.

components of some animal parasites are capable of activating complement by the alternative pathway. Igs that cannot fix complement by the classical pathway (IgE, IgA) can also activate the alternative pathway.

Serum proteins important in the initiation and progress of the alternative pathway are:

- **Factor B** (C3 proactivator) is a zymogen serine protease very similar to C2. It is even produced by a gene closely linked to the C2 gene (Figure 7.7). Factor B is cleaved into Ba & Bb. Ba has no known biological activity, but Bb combines with and cleaves C3.
- **Factor D**, found in serum in trace amounts, is a serine protease resembling activated C1s. An active enzyme, it can only cleave Factor B complexed with C3b. It is ineffective on free Factor B.
- **Properdin** (from Latin *perdere* — to destroy) is the protein that led to the discovery of the alternative pathway. It stabilises C3-convertase of the alternative pathway.
- **Factor H** is a regulator of the alternative pathway.

The **formation of a few C3b molecules is the first step in the alternative pathway**. The thioester bond in native C3 tends to hydrolyse spontaneously at a very slow rate. Trace amounts of C3b can thus be found in normal serum and can activate the alternative pathway under appropriate conditions.

- The cell-bound or fluid-phase **hydrolysed C3** ($C3(H_2O)$[10]) **binds Factor B** in the presence of Mg^{2+} ions; binding of B to C3b/$C3(H_2O)$ exposes a site on Factor B that is recognised by Factor D.
  - If C3b or $C3(H_2O)$ attach to the surfaces of host cells, the molecules are quickly inactivated by sialic acids present on most mammalian cell surfaces.
  - Microbial cell walls generally lack sialic acids, and deposition of C3b or $C3(H_2O)$ can result in the switching on of the alternative pathway.
  - C3b formed by the classical or MBL pathway may act as the focus for the formation of alternative pathway convertase.
- **Factor B is cleaved by Factor D** to Ba and Bb. Ba is released, leaving behind the C3bBb/C3b (H_2O)Bb complex — the C3-convertase of the alternative pathway.

[10] $C3(H_2O)$ is also referred to as C3i.

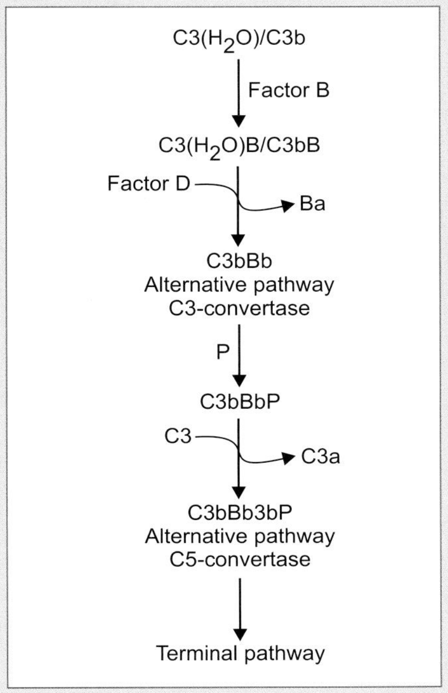

Figure 3.5 Multiple proteins are involved in the initiation and progress of the alternative pathway of complement activation. C3b or C3(H_2O), formed as a result of hydrolysis of C3, binds Factor B and makes it susceptible to the action of Factor D. Factor D cleaves Factor B to release Ba and yield C3bBb, the alternative pathway C3-convertase. The binding of properdin stabilises the enzyme and allows the formation of further C3b molecules. The association of an additional molecule of C3b with this complex yields the alternative pathway C5-convertase. The classical and alternative pathways merge beyond this point and share the terminal pathway.

- The complex is inherently unstable and is stabilised by properdin.
- C3bBbP is an efficient C3-convertase that cleaves many molecules of C3 (Figure 3.5). This step represents an amplification loop.

❒ An **additional molecule of C3b associates with C3bBbP to form the C5-convertase** of the alternative pathway (C3bBb3bP). Once the C5-convertase is formed, the two pathways converge into the terminal pathway. C5 cleavage and the membrane attack stage are set into motion, ultimately resulting in cell lysis.

3.5 Terminal Pathway

Whatever the mode of formation of C5-convertase, its formation sets the terminal pathway in motion. Complement components C5-C9 involved in the terminal pathway are called MAC (**M**embrane **A**ttack **C**omplex) components. Except for C5, which is a structural homologue of C3 and C4, the remaining components of the terminal pathway are nonenzymatic, hydrophilic proteins that display hydrophobic conformation upon binding. The proteins C6, C7, the α and β subunits of C8, and C9 are all structurally and genetically related.

❒ **C5-convertase** (whether C4b2a3b or C3bBb3bP) **binds C5 and cleaves it** to smaller C5a and larger C5b fragments; **C5b binds C6**.
- C5a, an extremely potent anaphylatoxin (see section 3.6 and Mind map 3.2), is released in the fluid phase.
- C5b is highly unstable and is inactivated within two minutes unless bound by C6.

❒ **C5b6 binds one molecule of C7**, causing a conformational change and transiently exposing a hydrophobic lipid-binding site; the C5b67 complex gets inserted in a membrane.

- Up to this point, the reaction occurs in the fluid phase because of its hydrophilic nature; with the binding of C7, the complex becomes hydrophobic.
- In the absence of a membrane to insert into (as happens on surfaces of immune complexes), the C5b67 complex dissociates from the surface.
- Under rare circumstances, the dislodged complex attaches to neighbouring cells, causing lysis of these bystander cells.

❒ The next step in MAC formation is the binding of C8 to the complex. **C8 binds to C5b67 *via* its β subunit; C8α and C8β are inserted in the lipid bilayer** (Figure 3.6).

- C8 is an oligomeric protein consisting of three subunits — α, β, and γ. C8α and γ form a disulphide-linked dimer, and C8β is noncovalently associated with that dimer.
- The α subunit of C8 has multiple binding sites. It has sites for binding C8β and γ. A third site binds C9 and directs the incorporation of this component in MAC. It also has a site for the regulatory protein CD59 which inhibits the formation of MAC. A lipid-binding site is exposed once C8α binds to the complex.
- C8β has a binding site for C5b67 as well as one or more sites that can bind to the target cell membrane.
- The role of C8γ in MAC formation remains unclear.
- The complex C5b678 can cause moderate membrane damage, especially in non-nucleated cells, such as RBCs. High-density lipoproteins in serum, as well as polyanionic agents, such as heparin, block the interaction of the complex with the cell membrane. Histones and protamines enhance its lytic activity.

❒ **C5b678 facilitates the binding and polymerisation of multiple molecules of C9 and the formation of MAC** — an annular ring structure of 12-18 molecules of C9.

- Binding of one molecule of C9 initiates the process of C9 oligomerisation.
- After at least 12 molecules are incorporated in the complex, a discrete channel is formed. The size of the channel varies according to the number of C9 molecules incorporated into the structure.
- During polymerisation, C9 undergoes a hydrophilic-amphophilic transition; the polymerised C9 is hydrophobic and can insert itself into the lipid bilayer.
- The completed MAC has a hydrophobic external surface and a hydrophilic internal channel of about 7-10 nm diameter.

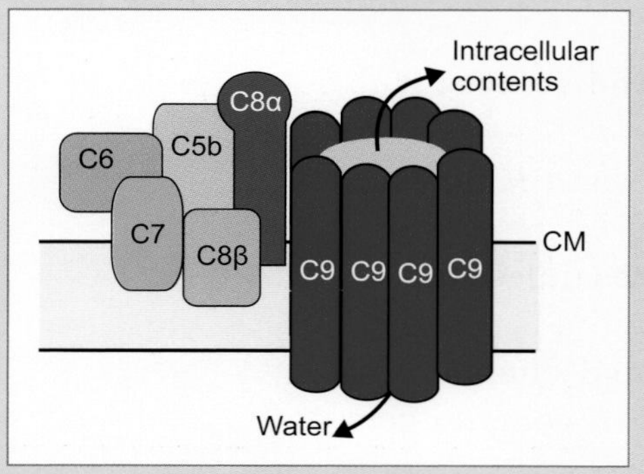

Figure 3.6 MAC is formed by C9 polymerisation, and C5b678 facilitates this formation. C5b, formed by the action of C5-convertase, binds C6 and C7. Binding of the C5b6 complex to C7 causes a conformational change in the molecule and exposes a hydrophobic site, allowing it to be inserted in the cell membrane. C8 binds to C5b67. The α and β subunits of C8 also get inserted in the lipid bilayer. The C5b678 complex promotes the polymerisation of C9 and results in the formation of an annular structure. MAC allows free exchange of electrolytes and water through its channel, destroying the osmotic stability of the cell and causing its lysis.

- MAC allows the free exchange of electrolytes and water through its channel, destroying the osmotic stability of the cell and causing its lysis.
- MAC damage is restricted to cells on which initial events of the cascade have occurred; bystander cells escape because of the instability of MAC (it quickly loses its cytotoxicity).
- Nucleated cells try to limit complement damage by exocytosing or endocytosing the area of the membrane that has MAC complexes.

❒ Suboptimal concentration of MAC on the cell membrane of nucleated cells is not lytic. Instead, it induces proto-oncogenes, activates the cell cycle, and enhances cell survival. It also inhibits apoptosis.

3.6 The Anaphylatoxins

C3a and C5a formed by the cleavage of C3 and C5 respectively, mimic some of the symptoms of anaphlactic shock (smooth muscle contraction, vasodilation, mast cell degranulation, etc.). They are therefore termed *anaphylatoxins*. Of the two, C5a is more potent; it is effective at concentrations of 10^{-12} M. Receptors for C3a and C5a (C3aR and C5aR) are expressed on mast cells, neutrophils, eosinophils, basophils, and monocytes. In macrophages, eosinophils, and neutrophils, anaphylatoxins can trigger an oxidative burst. Both these molecules can also profoundly affect eosinophils. They regulate the synthesis of eosinophil cationic protein, increase eosinophil adhesion to endothelial cells, and influence eosinophil chemotactic migration. Both C3a and C5a can cause basophil and mast cell degranulation and histamine release by these cells. C3a can stimulate serotonin release from platelets in guinea pigs and modulate the synthesis of IL-6 and TNF-α by B lymphocytes and monocytes. C3a can also induce mucous secretions by

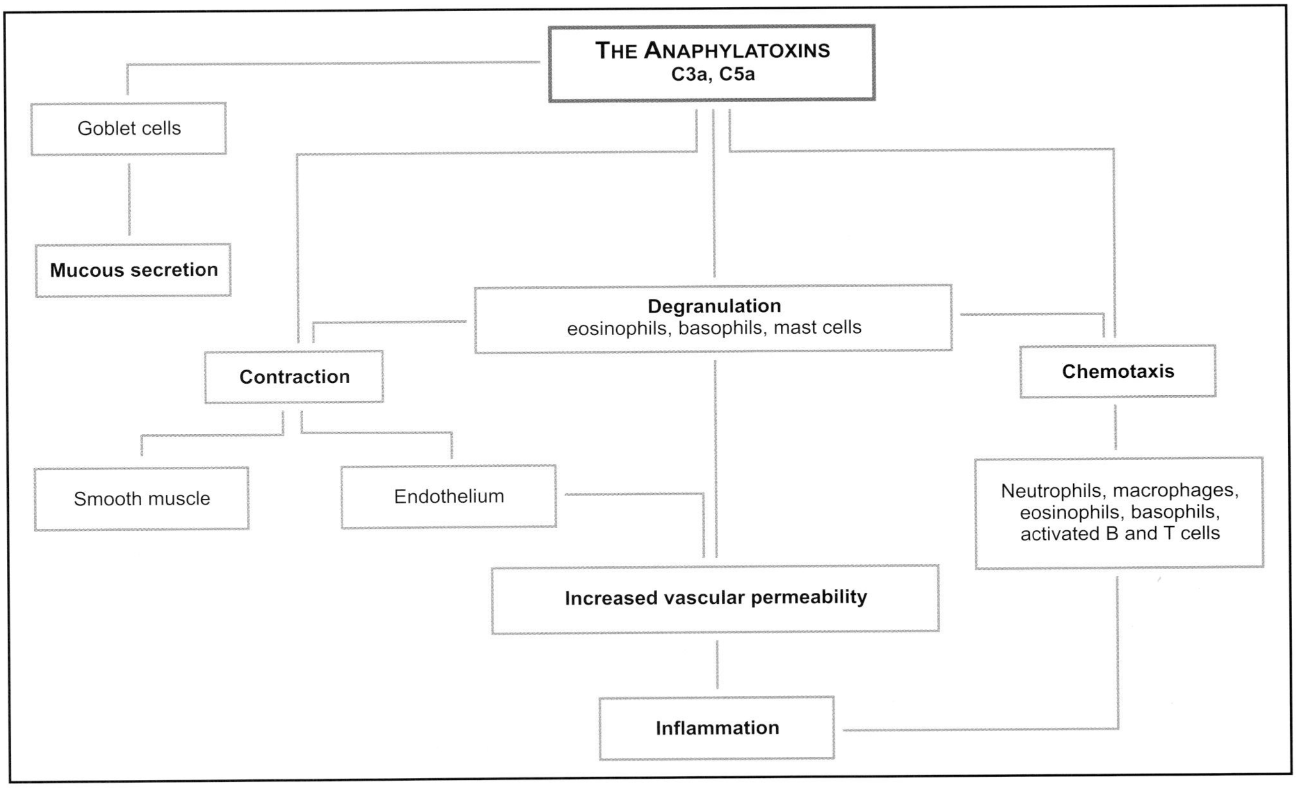

Mind Map 3.1 The anaphylatoxins

tracheal goblet cells[11] *in vitro.* C5a is a powerful chemoattractant for macrophages, neutrophils, basophils, mast cells, and activated B and T lymphocytes. It induces the production and secretion of leukotrienes[12] and IL-1 by macrophages.

C3 and C5 cleavage results in a terminal arginine at the carboxy-end. The properties of both C3a and C5a are critically dependent on this terminal arginine. Carboxypeptidase N, present in serum, can remove the terminal arginine from these molecules and seems to be especially important in inactivating C3a. Recent evidence shows that another enzyme, carboxypeptidase R (also called carboxypeptidase U), is the key enzyme in the inactivation of C5a. C3adesarg, formed by removal of the terminal arginine, cannot bind to C3aR and is thus devoid of biological activity. C5adesarg can, however, bind to C5aR; hence, it is capable of exerting considerable functional activity. Loss of arginine leads to a loss of C5a spasmogenic activity, even though chemotactic and other neutrophil-stimulating effects are retained, albeit at a reduced level. Being such potent molecules, anaphylatoxins are implicated in a number of diseases, including septic shock, asthma, immune complex disease, and delayed-type hypersensitivity.

3.7 Complement in Health and Disease

Complement plays a major role in innate and adaptive immune responses. The role of complement is not limited to immune responses (see the sidetrack *Not Just the Sum of its Parts*). Not surprisingly, deficiencies in any of the components or regulators of complement can lead to a number of diseases (Table 3.1).

❒ **Complement components are integral to innate immune defence.** Complement promotes

➢ **Opsonisation**

The binding of complement-coated target cells to CRs expressed on phagocytic cells promotes their phagocytosis and destruction. C3b is especially important in this process, although C4b also has a role. Interestingly, complement component C5a dramatically increases the number of such receptors on phagocytic cells.

➢ **Macrophage activation and chemotaxis**

C3a and C5a, released as a result of complement activation, can enhance macrophage activity.

➢ **Lysis of the target cell**

Gram-negative bacteria and many enveloped viruses are susceptible to complement-mediated lysis. Gram-positive bacteria and nucleated cells, however, are resistant to complement-mediated damage.

➢ **Virus neutralisation**

Complement aids in the process of virus neutralisation in multiple ways that are independent of the neutralising effect of antiviral antibodies.

- It causes the lysis of most enveloped viruses.
- Complement components coat virus particles and physically interfere with their attachment to host cells.
- Component C3b facilitates the aggregation of viruses and their eventual phagocytosis.

➢ **Augmentation of inflammatory responses**

Complement has a pro-inflammatory effect.

➢ C3a and C5a are vasodilators and cause an influx of fluids to the site of complement activation; C2b increases fluid accumulation.

[11] Tracheal goblet cells are cells lining the trachea. They are responsible for mucous secretions — when you have a bad cold, these cells are working overtime. ☺

[12] Leukotrienes are by-products of the arachidonate pathways. Leukotrienes induced by complement components include prostaglandin E_2, leukotriene B4, and thromboxane. They have a range of effects which are similar to those of histamine.

- C3a, C5a, and the complex C5b67 increase adhesion of monocytes and neutrophils to vascular endothelial cells and their extravasation and migration towards the site of complement activation.

❒ **Complement augments adaptive immune responses**.

- **Complement aids in antigen-processing and antigen presentation**. The coating of the antigen by complement allows it to be more efficiently internalised by APCs (**A**ntigen-**P**resenting **C**ells).

The Heart of the Matter: Complement and Heart Disease

Complement is implicated in both atherosclerosis and ischemic heart disease. *Atherosclerosis* is derived from the Greek *athero,* meaning gruel or paste, and *sclerosis,* meaning hardness. Narrowing and hardening of the coronary arteries results in restricted blood flow to the heart muscle and inadequate oxygen supply to the heart. Ischemia is defined as an insufficient supply of blood to an organ, usually caused by a blocked artery, and it can lead to cardiac arrest. Atherosclerosis seems to be a consequence of a chronic inflammatory process induced by the activation of macrophages, complement, and T lymphocytes. Research indicates that complement may have an important role in both the initiation and progression of atherosclerosis.

❒ Continuous activation of complement seems to be an active part of the atherosclerotic process.
❒ There is evidence that autoantibodies against lipoproteins are deposited in the arterial walls and may lead to complement activation.
❒ Cholesterol, especially enzymatically modified LDL cholesterol, can activate complement by the alternative pathway.
❒ Cellular debris and subcellular particles of the arterial walls also activate complement.
❒ Chronic complement activation has multiple effects on arterial walls.
❒ It acts as a pro-inflammatory stimulus.
❒ Its activation releases chemoattractants, such as C5a and MCP-1 (**M**acrophage **C**hemoattractant **P**rotein-**1**), which recruit monocytes to the activation site.
❒ It induces cell injury and lysis, which further potentiate the inflammatory response.
❒ Sublytic assembly of MAC (**M**embrane **A**ttack **C**omplex) on smooth muscle cells and endothelial cells can induce the activation and proliferation of these cells, contributing to fibrosis of the arterial wall.

As explained, atherosclerotic narrowing of arteries can lead to ischemia. Tissues and organs deprived of oxygen are often severely damaged when revascularised (i.e., when blood circulation is surgically re-established). This phenomenon is known as ischemia/reperfusion injury. Part of the damage is due to exposure of the hypoxic (ischemic) tissue to oxygen when reperfused — the sudden increase in oxygen results in the formation of reactive oxygen radicals. A considerable body of evidence suggests that complement plays a key role in furthering damage during reperfusion and therefore contributes to the pathophysiology of ischemic heart disease. Experimental models of acute myocardial infarction and autopsy specimens taken from acute myocardial infarction patients demonstrate that complement is selectively deposited in infarction areas. Furthermore, the inhibition of complement activation or depletion of complement components prior to myocardial reperfusion has been shown to reduce complement-mediated tissue injury in numerous animal models. The exact pathway of complement activation in ischemia is, however, unclear. There is evidence suggesting that naturally occurring IgM antibodies may have a role in this damage. Therapeutic approaches to prevent ischemic injury include —

❒ administration of C1INH (**C1 inh**ibitor; a naturally occurring inhibitor of the classical and MBL pathways of complement activation),
❒ use of anti-MBL antibodies to control the lectin pathway, and
❒ administration of recombinant soluble CR1, or peptides derived from it, to mop up the C3b, C4b, and C1q released during ischemic injury.

Table 3.1 Diseases Associated with Complement Deficiencies

Deficiency	Associated disease
C1	Hypogammaglobulinaemia Severe combined immunodeficiency (SCID) Systemic lupus erythematosus (SLE) Glomerulonephritis
C2 and C4	Recurrent bacterial infections Organ-nonspecific autoimmune diseases
C3	Severe infections; susceptibility to pyogenic infections Decreased antibody repertoire Immune complex disease (glomerulonephritis, vasculitis, etc.)
C5	Increased susceptibility to infection Liver fibrosis in mice
C6, C7 and C8	Increased susceptibility to *Neisseria* infections
Properdin	Meningococcal infections
Factor D	Respiratory infections
C1INH	Hereditary angioedema Autoimmune diseases
DAF and CD59	Haemolysis and thrombosis, which result in paroxysmal nocturnal haemoglobinuria
Factor I	Recurrent bacterial infections
CR1	Immune complex disease
CR3	Recurrent infections
MBL	Recurrent pyogenic infections and a failure to thrive in young children

- **It enhances B cell activation**. Coligation of the CR2 (CD21) and B cell antigen receptor by complement coated antigen–antibody complexes reduces, 100- to 1000-fold, the amount of antigen required for B cell activation.
- Complement components are **necessary for the localisation of antigen–antibody complexes** on **F**ollicular **D**endritic **C**ells (FDCs) in germinal centres[13]. Such localisation is essential for memory B cell development.

❒ **Complement aids waste disposal**. Complement opsonises apoptotic cells and immune complexes, hastening their clearance by phagocytes.

- C1q and MBL bind apoptotic cells and aid in their clearance.
- Erythrocytes bind opsonised cells/pathogens/immune complexes *via* CR1 and target them to the spleen and liver for disposal.

[13] Germinal centres are structures found in secondary lymphoid organs and are described in chapters 5 and 8. FDCs are specialised cells present in these structures. They express receptors for complement components and Ig. They also trap antigen–antibody complexes on their cell surfaces for prolonged periods.

3.8 Regulation of the Complement Cascade

Complement activation has tremendous capacity for self-amplification. For example, C3-convertase can deposit approximately 1200 C3b molecules on the

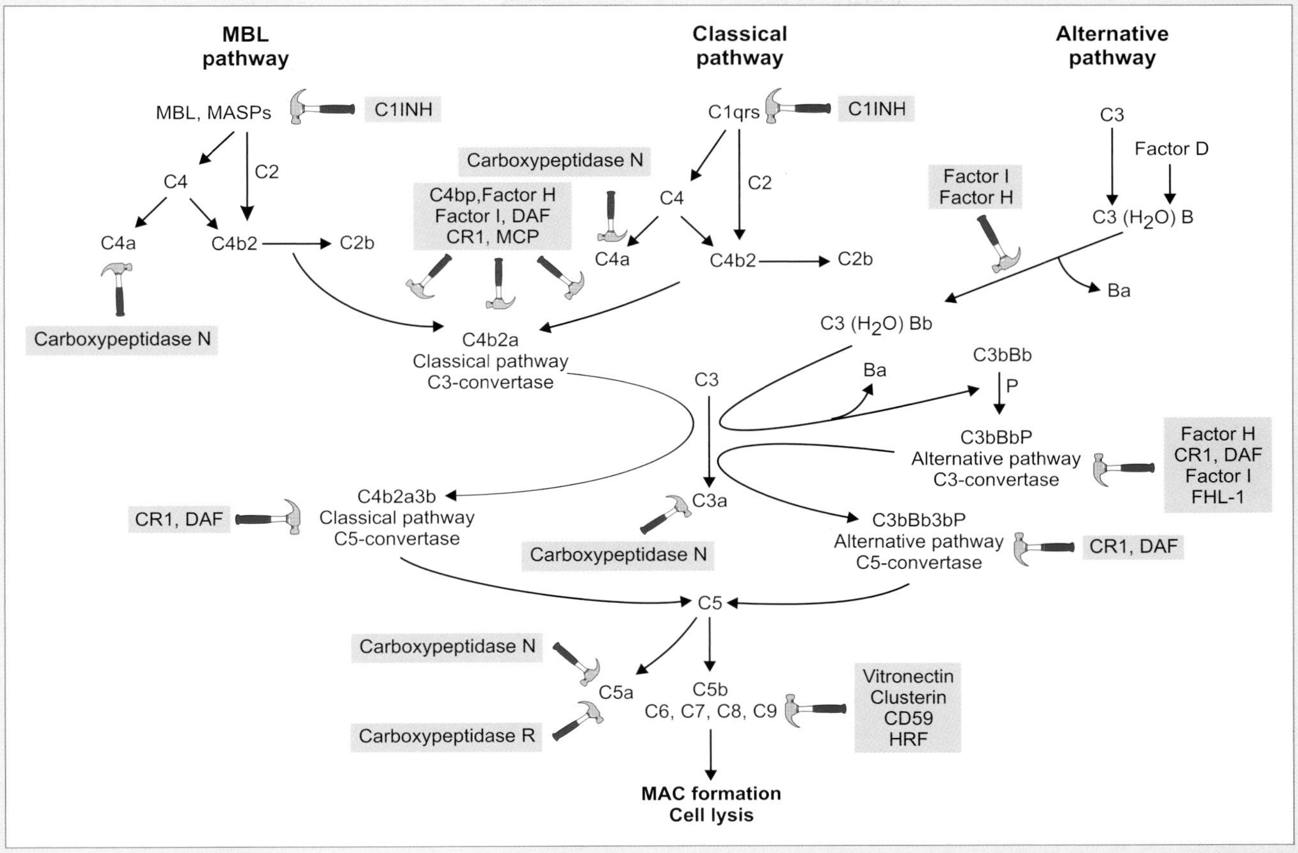

Figure 3.7 The complement cascade is regulated at almost every step by multiple regulatory proteins. The fluid-phase C1INH regulates activation of the MBL and classical pathways by regulating MASPs and C1. It thereby controls the triggering of both these pathways. Two homologous proteins, C4bp and Factor H, inactivate the C3-convertases of the classical pathway and alternative pathways, respectively. Factor I, CR1, DAF, and MCP aid this inactivation. In addition, FHL-1 inactivates the alternative pathway C3-convertase. Apart from regulating the C3-convertases, CR1 and DAF can also accelerate the inactivation of both classical and alternative pathway C5-convertases. The progress of the terminal pathway is inhibited by vitronectin and clusterin, which bind the C5b67 complex. C9 polymerisation is inhibited by CD59 and HRF. The complement cascade also liberates the multipotent molecules C3a and C5a. Carboxypeptidases N and R inactivate them and help prevent a runaway response. Carboxypeptidase N also inactivates C4a. Together, these members of the RCA family prevent damage of excess complement activation.

cell membrane near a single IgM molecule. C5-convertase is equally efficient in cleaving C5. Once switched on, the cascade has remarkable destructive potential. This is extremely dangerous as the system is nonspecific and does not discriminate between host cells and microbes. It is essential that activation of complement is focussed on the surface of the invading pathogen and that its deposition on normal cells or tissues is limited. This, and the fact that complement has multiple functions, necessitates its strict regulation. If this regulation goes awry, the complement system can be injurious to health. The complement pathway is therefore subject to dual regulation at both the intrinsic and extrinsic levels. Intrinsic regulation ensures that each activated component has a short half-life; it gets inactivated if it fails to attach to a cell membrane or to activate the next component. Extrinsic regulation provides for inhibitory molecules that can inactivate or degrade activated

Table 3.2 Complement-evasion by Microorganisms

Protein	Microorganism	Ligand
C1q	*E. coli*	C1q-binding protein
C3-convertases	*Staphylococcus spp.*	Staphylococcal complement inhibitor
Terminal pathway	*S. aureus* Group A streptococci	*SSL7,* binds C5 Streptococcal inhibitor of complement; binds the soluble C5b67 complex
Properdin	Herpes simplex virus	gC-1
CR2 (CD21)	Epstein-Barr virus	gp350/220
CR3	*M. tuberculosis* West Nile virus	C3 fragments deposited on cell surfaces
MCP (CD46)	*N. gonorrhoeae, N. meningitides* *Helicobacter pylori* Measles virus Human herpes virus 6 *Streptococcus pyogenes*	Pili Pili BabA protein Haemagglutinin Glycoproteins H, L, and Q M protein
DAF (CD55)	*E. coli* *Trypanosoma cruzi* Echovirus 7, coxsackie virus A21 HIV-1	Dr-like antigen, X adhesin gp 160, Trypomastigote-DAF Capsid Acquires host DAF
C4bp	*S. pyogenes* *N. gonorrhoeae* Vaccinia virus	Some M proteins Porin VCP
Factor H	*S. pyogenes* *N. gonorrhoeae* HIV-1	Some M proteins Lipo-oligosaccharides, Porin gp41, gp120
Factor H like-protein-1	*S. pyogenes*	Some M proteins
CD59	HIV-1 *Borrelia burgdorferi*	Acquires host CD59 80 KD surface protein

components. Thus, a number of proteins and regulatory mechanisms are actively involved in controlling excess complement activation[14]. Many pathogens subvert the defensive action of complement by expressing ligands that bind the regulatory components/receptors and use them to gain entry into cells (Table 3.2).

- **Inactivation of C1 and MASPs**: **C1 E**sterase **I**nhibitor (C1EI, also called C1INH[15]) inhibits the initiation of the classical pathway by forming a complex with C1r and C1s, causing them to dissociate from C1q. It also controls MASPs.
- **Inactivation of C3-convertases and C5-convertases**: C3- or C5-convertases of the human complement system are controlled by fluid-phase proteins and membrane proteins belonging to the RCA (**R**egulators of **C**omplement **A**ctivation) family. They inhibit convertases of the classical and the alternative pathways by causing their decay and/or by acting as cofactors for their degradation by Factor I — a serine protease that can cleave C4b and C3b.

[14] The exception is properdin. It is the only regulatory protein that actually stabilises alternative pathway C3-convertase.

[15] C1INH is a *serpin* — serine protease inhibitor. Serpins bind to and are cleaved by serine proteases. The cleaved serpin undergoes a massive allosteric change in its tertiary structure, acts as a chaperone, and targets the serpin:protease complex for degradation.

Table 3.3 Complement Receptors

Receptor (*Ligand*)	**Characteristics**
CR1 or CD35 *(C3b, iC3b, C4b)*	Found on erythrocytes, neutrophils, eosinophils, mononuclear phagocytes, mast cells, FDCs, B cells, and some T cells Promotes opsonisation, stimulates phagocytosis, helps in the clearance of apoptotic cells and immune complexes, and promotes immune adherence Blocks/promotes cleavage of C3/C5-convertase by Factor I
CR2 or CD21 *(major ligand: C3b; can bind iC3b, C3c, C3dg)*	Expressed by B cells, thymocytes, and FDCs Part of B cell antigen coreceptor complex; engaging this receptor makes the B cell 100 times more sensitive to the antigen Receptor for the Epstein-Barr virus
CR3 or CD11b/CD18 *(iC3b)*	Found on monocytes, macrophages, neutrophils, eosinophils, FDCs, NK cells, and K cells Facilitates extravasation of neutrophils Stimulates phagocytosis
CR4 or CD11c/CD18 *(iC3b)*	Closely related to CR3; expressed by monocytes and macrophages — especially tissue macrophages, neutrophils, NK cells, and K cells
CR5 *(C3d)*	Expressed by neutrophils and platelets
C1qR *(C1q)*	Expressed on B cells, macrophages, monocytes, platelets, and endothelial cells Helps bind immune complexes to phagocytes Aids apoptotic cell clearance
C3aR and C5aR *(C3a and C5a respectively)*	Expressed on mast cells, monocytes, macrophages, neutrophils, basophils, and T cells Small G protein-coupled receptor Responsible for biological manifestations of anaphylatoxins

- **Classical C3-convertase decay**: Three proteins, C4bp (**C4**-**b**inding **p**rotein), DAF (CD55), and CR1 (CD35) cause the decay of C3-convertase of the classical pathway by binding and displacing C2a from the C4b2a complex. C4bp is a soluble protein; DAF and CR1 are membrane bound. Apart from disrupting the already formed C3-convertase, these proteins can also block its formation by binding C4b and allowing its cleavage by Factor I. Factor I cleaves C4b into two fragments — the larger C4c is released, and the smaller C4d remains attached to the activated surface. CR1 and C4bp act as cofactors in this cleavage. Another member of the RCA family, **M**embrane **C**ofactor **P**rotein (MCP; CD46 — no relation of Macrophage Chemoattractant Protein), also acts as a cofactor for the cleavage of C4b by Factor I, but it cannot cause the decay of C3-convertase (Figure 3.7). Both DAF and MCP are expressed on a wide variety of cells and are thought to protect these cells from complement attack.
- **Alternative C3-convertase decay**: CR1, DAF, and a homologue of C4bp (called Factor H) cause the decay and degradation of alternative pathway

C3-convertase (C3bBb). Factor H competes with Bb for the binding site on C3b. The surface to which C3b is attached determines which of these two ultimately combines with C3b. Certain surfaces like those containing polysaccharides are called *activator surfaces* because they favour the binding of Factor B to C3b. Others, such as heparin, favour the binding of Factor H, even when the C3b is bound to an activator surface. The inactivation of C3b by Factor I also requires Factor H, CR1, or MCP. An alternatively spliced variant of Factor H, FHL-1 (**F**actor **H**-**L**ike protein-**1**), can also cause the decay of alternative C3-convertase besides acting as a cofactor in its degradation.

- **C5-convertase disruption**: C5-convertase of the classical and alternative pathway can similarly be disrupted by CR1 or DAF. These regulators bind C3b in the enzyme complexes. The bound C3b is then susceptible to cleavage by Factor I. Another trypsin-like enzyme in the serum can cleave C3b (called C3bi) to form a larger C3c, which is released, and a smaller C3d, which remains attached to the cell surface.

❒ **Inactivation of the terminal pathway**:

- S protein (vitronectin) and clusterin (also named apolipoprotein J), both found in serum, are fluid-phase inhibitors of the terminal pathway. They compete with membrane lipids for metastable binding sites on C5b67. They bind the complex in the fluid phase and prevent it from binding to cell membranes. S protein allows the binding of C8 and C9 to Cb567 but prevents C9 polymerisation. The bound complex retains its hydrophilic character and is therefore unable to insert itself into membranes.

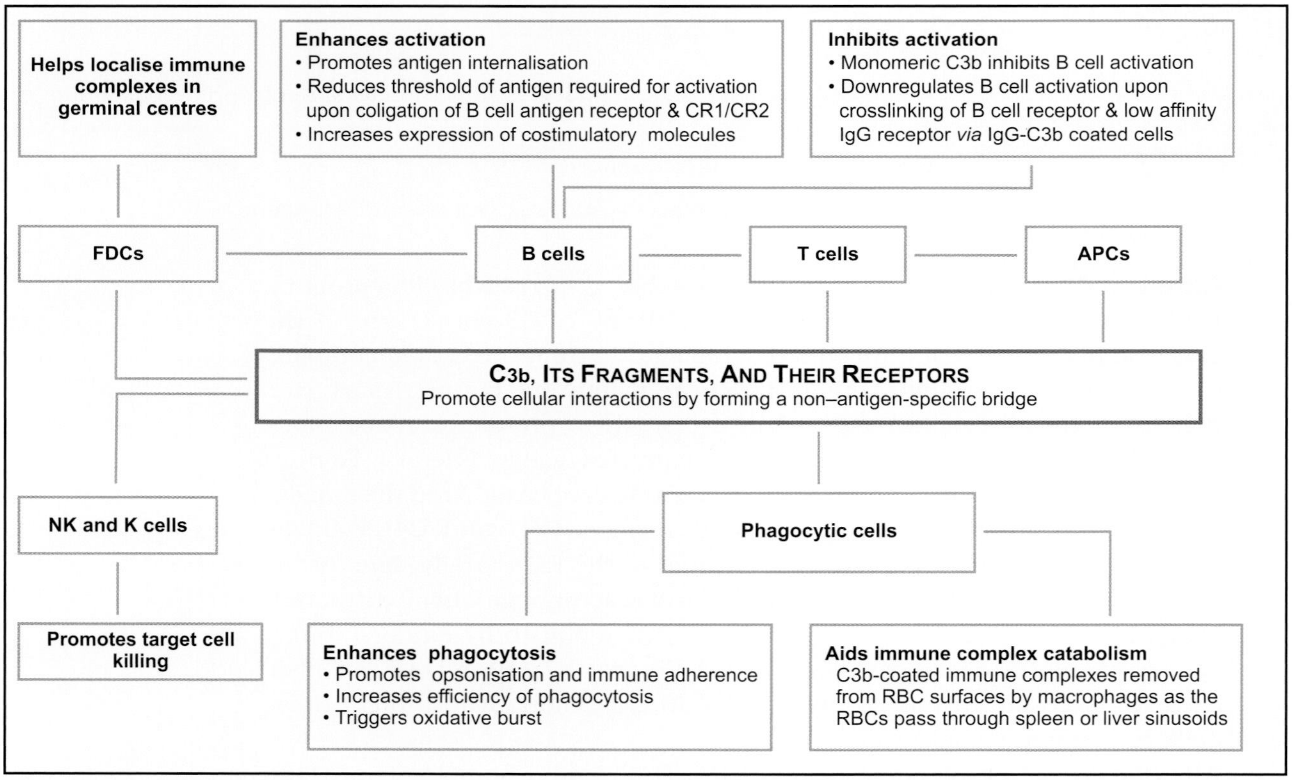

Mind Map 3.2 Interaction of C3b, C3b fragments and their receptors (CR1, CR2, CR3, CR4)

- CD59, a protein anchored in the cell membrane**,** can bind C8 and prevent C9 polymerisation.
- Another membrane-anchored protein, HRF (**H**omologous **R**estriction **F**actor), can prevent channel formation and C9 polymerisation.

3.9 Complement Receptors

Effector molecules combine with specific receptors on the target cell surface and exert their action. These receptors are involved in ligand recognition, signal transduction, and induction of cellular responses. Receptors that bind split-products of complement components are expressed on a variety of cell types. CRs exist in the intracellular pool and cycle to the cell surface. Expression of some of these receptors increases with cell activation or with stimulatory signals. The variety of cells that express these receptors can help us gauge, to some extent, the wide range of effects that complement components can induce (Table 3.3).

4 Entities of Adaptive Immune Response: Immunogens

Are there any queers in the theatre tonight?
Get them up against the wall!
There's one in the spotlight, he don't look right to me,
Get him up against the wall!
That one looks Jewish!
And that one's a coon!
Who let all of this riff-raff into the room?
There's one smoking a joint,
And another with spots!
If I had my way,
I'd have all of you shot!

— Pink Floyd, *In the Flesh*

4.1 Introduction

A perceived threat to the well-being of an immunocompetent host results in a specific immune response — *the adaptive immune response* — intended to eliminate that threat. In this chapter and the next, we will examine the two entities involved in this adaptive immune response, namely, the entity that triggers the response and the organs and tissues that are the site of the response.

People involved in software development are known to have borrowed biological terms, with the result that words, such as *virus* and *worms*, have become commonplace in information technology jargon. Interestingly, even the way that computer security specialists define and detect intrusions is eerily similar to the way our immune systems seem to define and detect potential threats. Computer security systems look upon the act of an intrusion as someone attempting to break into or misuse a system. The exact understanding of what constitutes *someone, break into,* and *misuse* is left to each organisation's management. Two major models are used in detecting intrusions. The *anomaly detection model* detects intrusions by looking for activity different from a user's or system's normal behaviour, and the *misuse detection model* detects intrusions by looking for activity that corresponds to known intrusion techniques (signatures) or system vulnerabilities.

Abbreviations

BcR:	B cell antigen Receptor
CTL:	Cytotoxic T Lymphocyte
ISCOM:	Immune Stimulatory Complex
MDP:	Muramyl-Di-Peptide
MW:	Molecular Weight
SRBC:	Sheep RBC
TcR:	T cell antigen Receptor
TD antigens:	T-Dependent antigens
TI antigens:	T-Independent antigens
TSST-1:	Toxic Shock Syndrome Toxin-1

4.2 Immunogens and Antigens

Immunogens are substances capable of stimulating the immune system (eliciting a B or T cell response) **and reacting with the product(s) of such stimulation** (antibodies and/or cells expressing specific receptors). The term *immunogen* is often confused with *antigen*[1]. Specific reactivity is the only criterion for defining an antigen, and the definition does not include the capability for immune stimulation. For example, a molecule will be termed an antigen if it reacts specifically and observably with an Ig molecule without activating a B cell to produce Ig molecules; to be termed an immunogen, the molecule must be capable of doing both. Put differently, all immunogens are antigens, but not all antigens are immunogens. The terms antigen and immunogen have been used interchangeably, creating ambiguity. Regrettably, this continues, and except for in academic discussion, the two terms remain synonymous.

As we explained in chapter 1, specific immune responses are mediated by T and/or B lymphocytes. T cells themselves are of two types — αβ and γδ — depending upon their antigen receptors. The manner in which these cells are stimulated differs.

- To stimulate αβ T cells, the immunogen has to be processed and presented to T lymphocytes in the context of MHC molecules. αβ **T c**ell antigen **R**eceptors (TcRs) cannot bind or recognise the antigen in its native forms. αβ TcRs only recognise antigen-derived peptides (or lipids) that have been loaded on appropriate MHC or CD1 molecules.
- γδ T cells are not constrained by the need for MHC molecules in antigen recognition. γδ TcRs recognise antigen in its native (unprocessed and free) forms.
- **B c**ell antigen **R**eceptors (BcRs), being membrane forms of the Ig molecule, recognise antigen in its native form (see chapter 6). However, this recognition alone is insufficient for B cell activation. B cells can be activated only when they receive T cell help in conjunction with antigen recognition[2].

[1] Both the terms were perhaps derived similarly. Antigen = **anti**body **gen**erator; immunogen = **immuno**globulin **gen**erator. Whatever the source of these words, these molecules give rise to both B and T cell responses, i.e., they are **immune** response **gen**erators.

[2] The exceptions are T-Independent antigens (these should really be called T-Independent immunogens, but old nomenclature is hard to get rid of), which can stimulate B cells independent of T cell help (see section 4.3). However, these antigens do not elicit a sustained, robust, humoral, and cell-mediated immune response.

Table 4.1 Noncovalent Forces Involved in Antigen–Antibody Interaction

Noncovalent forces	Nature	Example
Electrostatic force	The result of the attraction between oppositely charged ionic groups on protein side chains, such as R-NH_3^+ and R-COO^-	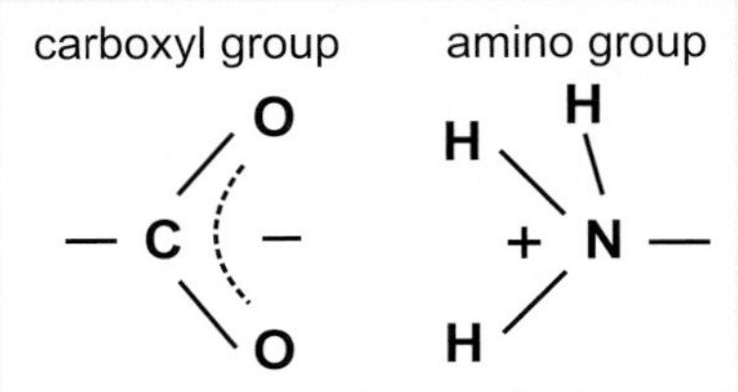
Hydrogen bond	A reversible bond formed between hydrogen attached to strongly electronegative atoms (N, O); it results in the hydrogen ion/atom acquiring a partial positive charge and a pair of negatively charged electron pairs on the electronegative atom May be formed between charges located on different molecules or on different parts of one large molecule; in proteins, the normal hydrogen bond donors are O-H and N-H, and the lone pair acceptors are =O, –O–, and –N	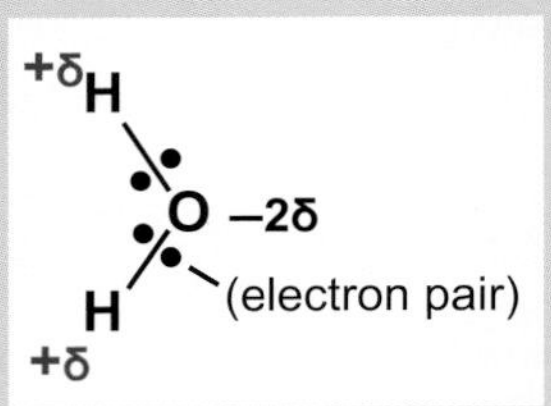 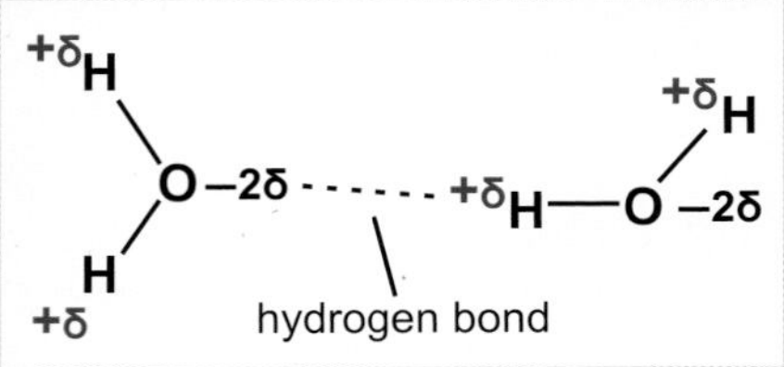
Salt bridges	Ionic interactions in proteins; these occur between carboxyl groups and amino groups of lysines and arginines and often occur whenever arginine or lysine residues are buried in the interiors of proteins	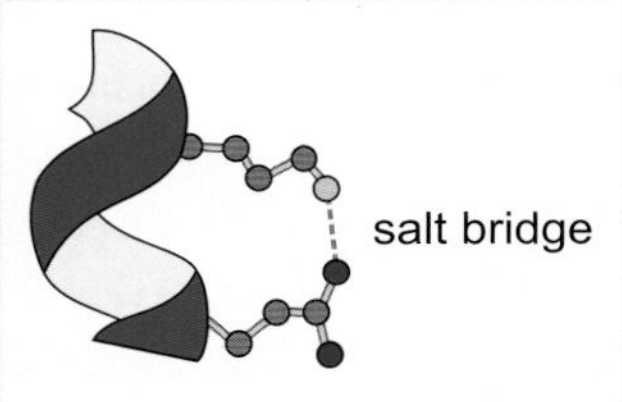
Van der Waal's forces	The result of attractions between induced oscillating dipoles in two electron clouds Induced oscillating dipoles are caused by the 'sloshing' of electrons in a molecule; such sloshing causes the molecule to temporarily become slightly negative at one end and slightly positive at the other end; a molecule approaching this temporarily polar molecule will be induced to become polar — when the original dipole reverses charges, the charges on the induced dipole will also change	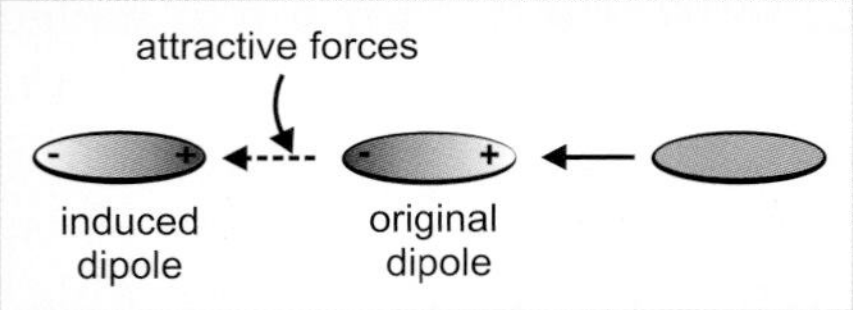
Hydrophobic forces	They are a result of a combination of Van der Waal's forces and the tendency of hydrophobic groups to pack together to exclude water Strength of the hydrophobic interaction is proportional to the surface area hidden from water	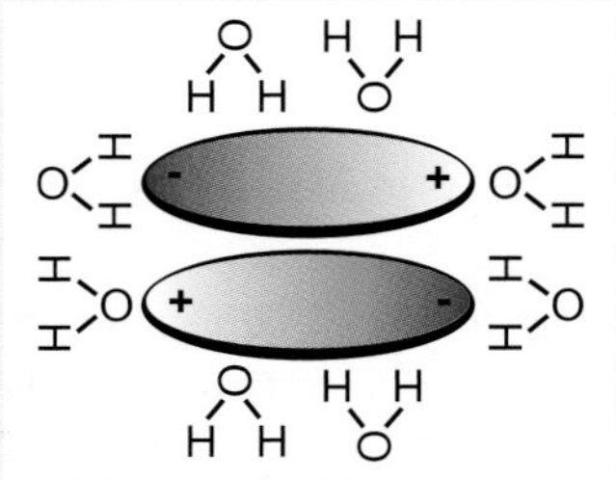

To elicit a sustained immune response, the immunogen should ideally stimulate both T and B cells. The portion of the immunogen that binds specifically with membrane receptors on T or B cells is called its *antigenic determinant* or *epitope*. An epitope can be defined as a discrete site on a macromolecule which is recognised by and binds to a lymphocyte receptor. An immunogen may have more than one such epitope; the total number of epitopes on a given immunogen is its valency. Some epitopes are called *immunodominant* because the immune response seems to be preferentially (or disproportionately) directed against them rather than other (subdominant) epitopes. Although TcRs and BcRs belong to the same structural family (the Ig superfamily discussed in chapter 6), their epitopes differ.

4.2.1 B Cell Epitope

The BcR is an Ig molecule expressed on B cell surfaces. As Igs recognise and bind soluble antigens, B cells have similar recognition capabilities. Major characteristics of the B cell epitope are listed below.

- **It forms a binary complex with the BcR**. Because the BcR is an Ig molecule, the interaction between the BcR and its epitope is an antigen–antibody interaction — one epitope reacts with one antigen-binding site of the BcR (called the *paratope*).
- **The B cell epitope is generally hydrophilic in nature**. When assuming a three-dimensional conformation, hydrophobic amino acids of proteins get buried deeply in the structure, away from the water molecules of the surrounding medium. The hydrophilic portions of the molecule are exposed to water molecules. As B cell epitopes are located on the molecule's surface, they tend to be hydrophilic.
- **It is conformational**. The antigen–antibody reaction can be viewed as a lock and key arrangement (the epitope is the key to the antibody's lock), or more appropriately, it is like two pieces of a jigsaw puzzle (Figure 4.1). For a proper antigen–antibody fit, the molecules must have complementary shapes. Antigen–antibody binding takes place by the formation of multiple noncovalent bonds, such as Van der Waal's bonds, H-bonds, and hydrophobic interactions (Table 4.1). Because of the weak nature of these bonds, a large number of them are needed to hold interactants together. Water molecules may help by filling gaps and increasing binding. Proteins and polysaccharides have a peculiar three-dimensional structure; the surface of these molecules is not smooth but consists of projections and contusions. These conformational structures behave as B cell epitopes. Thus, the B cell epitope is often found on bends on the molecule or in areas of high segmental mobility[3]. Even a single change in the amino acid

[3] Proteins are not static molecules. They undergo a range of motions from simple side-chain rotations to entire domain movements, all of which can play important functional roles. A large-scale molecular motion occurs when two parts of the molecule move rigidly relative to each other, and it can be traced to a small segment of the molecule that acts as a hinge. B cell epitopes are often located in such areas of segmental flexibility.

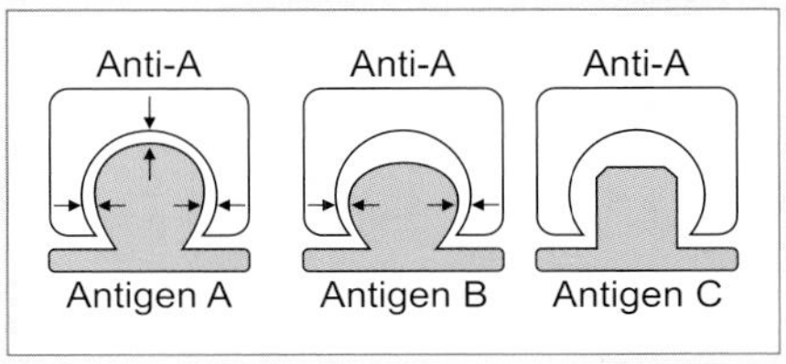

Figure 4.1 Conformations of the B cell epitope and the antigen-binding pocket of the antibody molecule (paratope) must be complimentary, like the fitting contours of a jigsaw puzzle, for the antigen and antibody to interact. Antigen–antibody binding takes place by the formation of multiple noncovalent bonds and hydrophobic interactions. A large number of such bonds are required to hold the molecules together, and suitable atomic groups must be present on the corresponding parts of the molecules to allow simultaneous formation of these bonds (left panel). If the epitopes and paratopes are not complimentary, the intermolecular attractive forces are not strong enough to withstand the repulsive forces that will drive the reactants apart (middle panel). If the conformations of the epitopes and paratopes are completely different, repulsive forces will far exceed attractive forces, and the reactants will fail to interact (right panel).

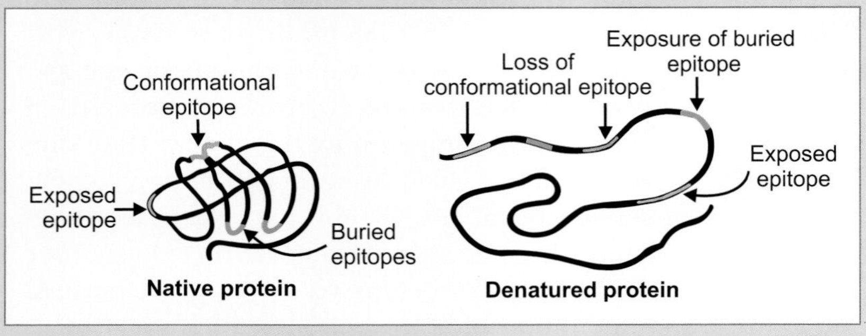

Figure 4.2 B cell epitopes can be sequential or nonsequential, but they have to be accessible. The three-dimensional conformation of the epitope is important for antigen–antibody interaction. It therefore does not matter whether the residues that form the epitope are continuous or not, as long as they are brought together by folding of the molecule (left panel). The denaturation of proteins results in linearisation of the protein molecules and destroys the nonsequential, but not the sequential, epitope. Denaturation also exposes formerly inaccessible hidden epitopes (right panel). Hence, antisera raised against the native protein may not bind the denatured form with equal strength.

sequence of a protein or carbohydrate in a glycoprotein/polysaccharide can change the three-dimensional conformation of that molecule, leading to altered antigenicity. Similarly, if a molecule unfolds or is partially denatured (e.g., by heating or chemical modification), the conformation of the molecule changes; the original epitopes may be destroyed or rendered inaccessible, and normally hidden epitopes may be exposed (Figure 4.2). The denatured protein is therefore likely to be antigenically different from the native protein and will give rise to antibodies against the newly exposed epitopes. Moreover, antibodies produced against the original molecule may not react with the denatured molecule. For the same reason, antibodies produced against a linear polypeptide will not react with a α-helical polypeptide, and *vice versa*.

Karl Landsteiner used a protein carrier coupled to various haptens[4] *via* the NH_2 group of azobenzene to demonstrate such conformational specificity of BcRs. This was a simple but elegant approach to the problem because the haptens behaved like the dominant antigenic determinants of the carrier molecule. By changing the chemical nature and/or position of the hapten, Landsteiner could alter the conformation of the carrier–hapten complex. The results of his experiments are summarised in Figure 4.3.

- **It can be sequential (continuous) or nonsequential (discontinuous)**. A stretch of amino acids on the surface of a protein that assumes a specific conformation and allows antibody binding is a sequential epitope. Because the three-dimensional conformation of the molecule is important in BcR recognition, B cell epitopes need not be restricted to continuous stretches of amino acids. Amino acids or carbohydrates located on different segments of the primary structure may contribute to epitope formation if they are brought together by molecular folding.
- **The B cell epitope is small**. The size of the B cell epitope is dictated by the size of the antigen-binding site of the antibody. The epitope cannot be larger than its binding pocket. The shape of the epitope is entirely dependent upon the protein's tertiary structure. Smaller ligands, such as oligonucleotides, peptides, and haptens consisting of 4–8 hydrophilic residues, can form compact structures, and they often bind deep within the antibody-binding pocket. With

[4] Haptens (from Greek *haptein* — to fasten) are small molecules that are too small to be immunogenic when administered by themselves but which can elicit an immune response when conjugated to a large carrier molecule. However, haptens can react specifically with the Igs formed in response to hapten–carrier stimulation. Thus, haptens are partial immunogens; they have the property of antigenicity but not immunogenicity. Generally, the carrier activates T cells, and the hapten reacts with BcR. Hence, hapten–carrier conjugates elicit predominantly anti-hapten Igs. Large proteins are excellent carriers. The large number of peptides generated from such proteins allows the activation of several TH cells, enhancing the speed and efficiency of the resultant immune response.

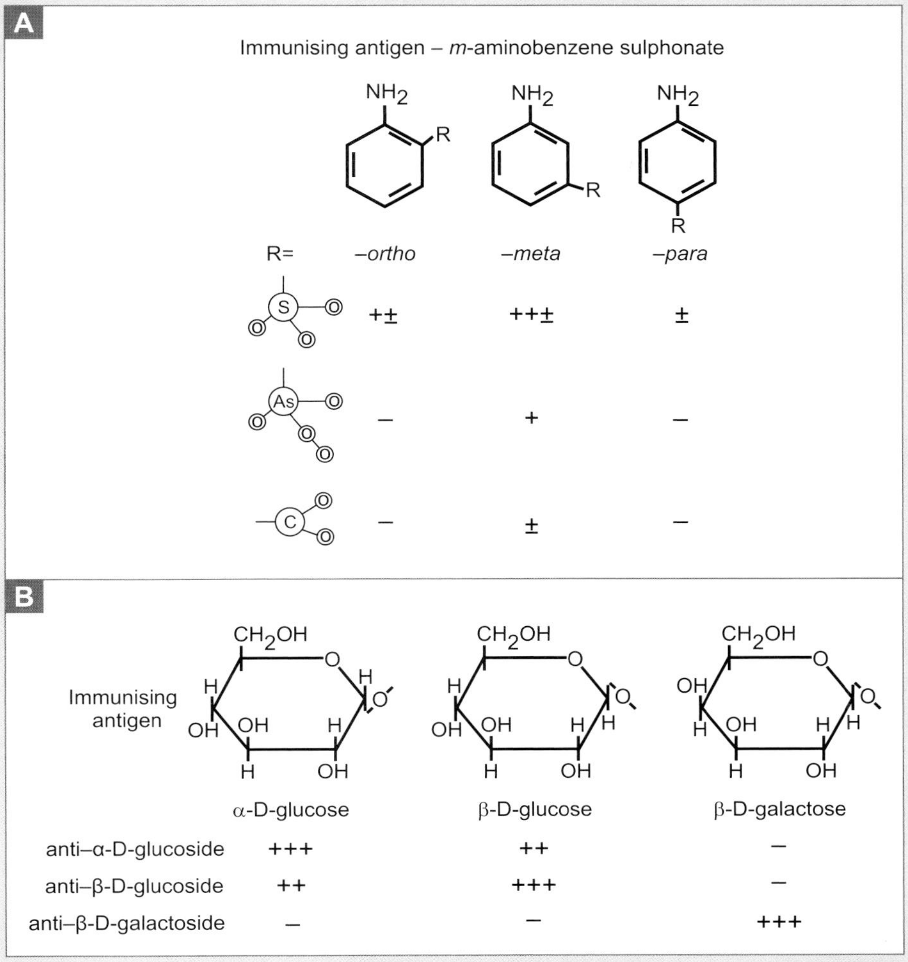

Figure 4.3 Landsteiner's experiments demonstrated the importance of three-dimensional conformation in antibody recognition and that it is the shape of the epitope, and not its chemistry, that is important in antibody recognition. Antibodies raised against a particular antigen readily bind that antigen, but subtle differences in epitope position or chemical nature result in reduced (or loss of) binding. Thus, antibodies raised against the tetrahedral *meta*-isomer of aminobenzene sulphonate react strongly with it, weakly with the *o*-isomer, and negligibly with the *p*-isomer (panel A). They also weakly recognise another tetrahedral hapten, aminobenzene arsonate, at the *m*-position but not in the *o*- or *p*-positions. Only negligible recognition is observed in the planar hapten aminobenzene carboxylate, and that too only in the *m*-position. If antigens are chemically related, their conformations may be close enough for antibodies raised against one to bind the other (e.g., α-D-glucose and β-D-glucose; panel B). However, if they are conformationally different (e.g., α-D-glucose and α-D-galactose), antibodies raised against one fail to recognise the other.

large globular proteins, the interacting face between the antibody and the antigen can be a flat or undulating surface in which the protrusions on the protein surface complement the depression in the antibody-binding site. In such cases, up to 15-20 amino acids on the protein surface may make contact with a similar number of amino acids on the antibody molecule.

- **It must be accessible**. In 1969, Sela demonstrated the importance of accessibility in immunogenicity. He used a nonimmunogenic synthetic peptide of repeated units of a single amino acid (a poly-amino acid) and attached an aromatic amino acid, tyrosine, at the branched chains. He found that when tyrosine was at the outer end of the branched chains and accessible to BcRs, the molecule was immunogenic. However, if the tyrosine was attached to the inner ends of the poly-amino acid backbone where it was inaccessible, the molecule was nonimmunogenic. For proteins, this means that a native protein will elicit antibodies only to the exposed determinants on its cell surface. The antibodies elicited will fail to react with those epitopes that are buried deep in the molecule (Figure 4.2).

4.2.2 The αβ T Cell Epitope

About 95% circulating T cells express the αβ TcRs. Unless otherwise specified, the term T cell is used to mean αβ T cells throughout this book.

- **The αβ T cell epitope is recognised only in the context of an MHC molecule**. The T cell epitope has to be loaded on an MHC molecule (either class I, class II, or CD1[5]) for TcR recognition and binding. Consequently,
  - αβ T cells do not recognise either native or soluble antigens;
  - only proteinic (and lipid) antigens can stimulate T cells — they do not recognise polysaccharides or nucleic acids; and
  - the three-dimensional conformation of the protein is not important in T cell recognition, and a similar T cell response will be generated whether a native or denatured protein is used in immunisation.
- **It forms trimolecular complexes with TcR and MHC molecules**. Two interaction sites on the peptide are involved in antigen recognition — the site that binds to the TcR (i.e., the epitope) and the site that reacts with the MHC molecule (called *agretope*; Figure 4.4)
- **The αβ T cell epitope is generated by APCs**. APCs internalise and digest native proteins. Peptides generated in this process are loaded onto the MHC molecules and exported to the cell surface; the MHC:peptide complex binds TcR. Small linear peptides may be loaded onto MHC molecules at or near the cell surface, but such loading is inefficient.
  - Viable, functional APCs are generally required for T cell epitope generation.
  - The epitopes are generally internal, that is, they need not be accessible.
  - The haplotype of the MHC dictates the kind of epitope presented. MHC molecules are highly polymorphic, and every individual expresses a unique set of MHC molecules (see chapter 7). Of the number of peptides generated from an antigen during processing, only those peptides that can bind to the particular MHC molecules expressed by the individual will be presented to T cells.

[5] Until recently, T cells were thought to recognize only peptides loaded on MHC molecules. Some conventional T cells and NKT cells have now been established to have antigen receptors that can recognize lipids or glycolipids loaded onto CD1 molecules. These CD1 molecules are structurally similar to MHC class I molecules. Although this discussion focuses on proteinic antigens, it applies equally to lipid antigens.

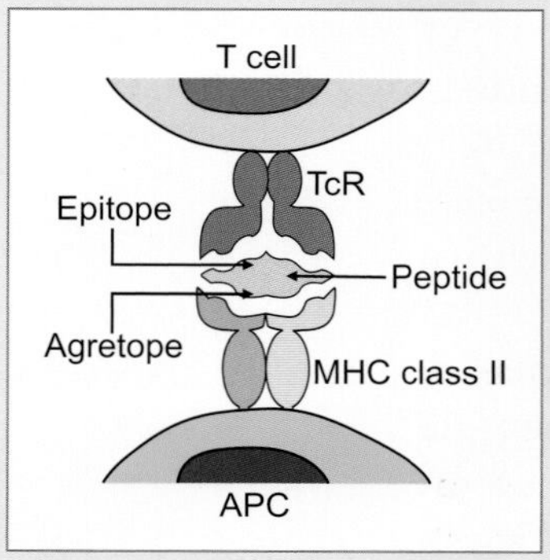

Figure 4.4 The αβ T cell epitope forms a trimolecular complex with TcR and MHC. The αβ T cell epitope is recognised only in the context of MHC molecules. APCs internalise and degrade protein antigens. Peptides derived from these antigens are loaded on MHC molecules. αβ TcR recognises this peptide:MHC complex. The TcR must recognise the MHC and the peptide. The site on the peptide that is recognised by MHC is called the agretope; the site that binds TcR is the epitope.

B Cell Epitope

- forms a binary complex with BcR,
- is conformational,
- can be continuous or discontinuous,
- consists of small 4-8 hydrophilic residues that are accessible to BcR, and
- can be derived from proteins, carbohydrates, nucleic acids, etc.

αβ T Cell Epitope

- forms a trinary complex with TcR and MHC molecules,
- has to be processed and loaded onto MHC molecules by APCs,
- is sequential,
- is derived from proteins or lipids, and
- consists of 8-20 amphipathic residues.

γδ T Cell Epitope

- can be proteinic or nonproteinic in nature,
- forms a binary complex with γδ TcR, and
- is not processed and loaded onto MHC; γδ TcR recognises native molecules independent of MHC.

- **It is sequential (continuous)**. Given that peptides generated by digestion of the native antigen are loaded onto MHC molecules, the epitope consists of a primary sequence of amino acids in the protein.
- **The αβ T cell epitope is between 8-20 amino acids in length**. The MHC molecule that binds the peptide dictates the length of the T cell epitope. MHC class I molecules have grooves that are closed at both the ends. They can bind peptides 8-11 amino acids in length. The binding grooves of class II molecules are open at both ends and can bind longer peptides. Generally, class II peptides are 13-20 amino acids long.
- **It is generally amphipathic**, that is, it contains both a hydrophilic and hydrophobic region. The peptide-binding grooves on MHC molecules have hydrophobic regions. To fit into such hydrophobic grooves, the T cell epitope needs hydrophobic amino acids. Hydrophilic regions are needed for interaction with TcR.

4.2.3 The γδ T Cell Epitope

γδ T cells comprise less than 5% of total T cells. Relatively less is known about these T cells, their signalling pathways, and their role in health and disease. Their unique features are listed below:

- The γδ T cell epitope is similar to the B cell epitope. It is recognised independent of MHC molecules.
- It need not undergo processing and presentation by APCs.
- It may be either proteinic or nonproteinic. Intact proteins or glycoproteins (e.g., glycoproteins from the herpes simplex virus) or small nonproteinic molecules, such as phosphates (e.g., pyrophosphomonoesters from *M. tuberculosis*) or amine-containing molecules (e.g., alkylamines from several natural sources), are recognised by the γδ TcRs (Figure 4.5).

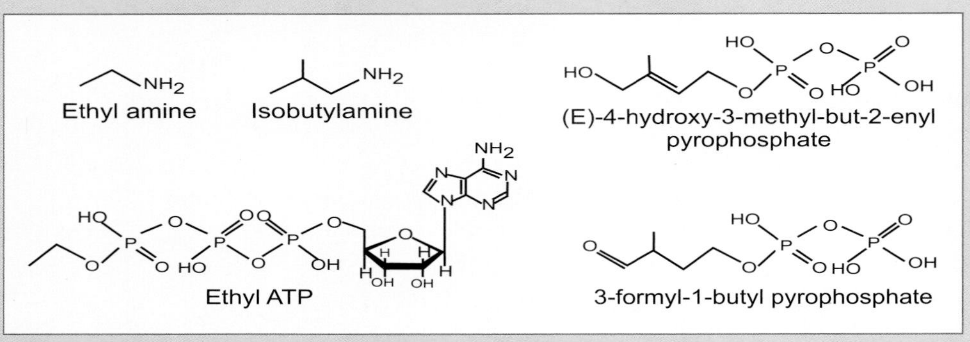

Figure 4.5 Proteinic or nonproteinic epitopes are recognised by γδ TcRs. The γδ TcR epitope is recognised independent of MHC molecules. The diverse molecules recognised by γδ TcRs include intact proteins, small molecules, such as like ethylamine or isobutyl amine, ethyl ATP, and pyrophosphates.

4.3 Types of Antigens

- **Heterophile antigens**: Antigens or epitopes shared by unrelated (widely divergent) species are called heterophile antigens. They were described by Forsmann, who found that antiserum raised by injecting guinea pig tissues (such as liver, kidney, and brain) in rabbits was capable of reacting with **S**heep **R**ed **B**lood **C**ells (SRBC), that is, antibodies produced against the guinea pig tissue recognised epitopes on SRBC. The antigen was termed *Forsmann antigen*, and the resultant antibodies were called *heterophile antibodies*. Human B group RBCs and *E. coli* also share heterophile epitopes. The Epstein-Barr virus and SRBC share antigenic epitopes as well; antibodies in the sera of patients suffering from infectious mononucleosis (caused by the virus) lyse SRBC and bovine RBC in the presence of complement. Sera of patients suffering from some rickettsial infections have antibodies that cross-react with certain *Proteus* strains. An old serological test, the Weil-Felix test used in the diagnosis of rickettsial fevers, exploited this phenomenon.
- **Isophile antigens**: Antigens found in some individuals of a species that are capable of eliciting an immune response in other, genetically distinct members of the same species are termed *isophile antigens*. Blood group antigens are the best-known example of isophile antigens (see chapter 17). Different alleles of the same genes code for these proteins. Thus, isophile antigens reflect allotypic variation in the genetic makeup of an individual.
- **Sequestered antigens**: *Sequestered* means secluded, isolated, or set apart. Certain tissue antigens, such as eye lens proteins or sperm antigens, are sequestered from circulation and are not easily accessible to cells of the immune system. Hence, sequestered antigens can elicit an immune response if introduced at other sites or if exposed to the cells of the immune system following trauma or infection. For example, the barrier membrane that protects sperm-forming tissue is damaged in mumps, resulting in a testicular inflammation (orchitis) that may cause an immune response against sperm-forming tissue. Therefore, mumps in adult males may lead to impotency. Similarly, trauma to the eye can allow the immune system access to lens proteins, causing an autoimmune inflammatory reaction (see the sidetrack *Privileged Being, Privileged Seeing* in chapter 12). The term *sequestered antigen* is also used to describe antigens (or parts thereof) that are retained on FDCs for a long time. Such sequestered antigens have a role in affinity maturation (see chapter 8).

Figure 4.6 Superantigens bind antigen receptors outside the antigen-binding pocket and activate lymphocytes independent of antigen specificity. T cell superantigens, such as staphylococcal enterotoxins, bind to T cells having a particular variable region in the α or β chain of the TcR. They do not undergo processing by APCs (left panel). B cell superantigens bind to all Ig molecules having a particular VH antigen-binding region and can stimulate B cells nonspecifically (right panel).

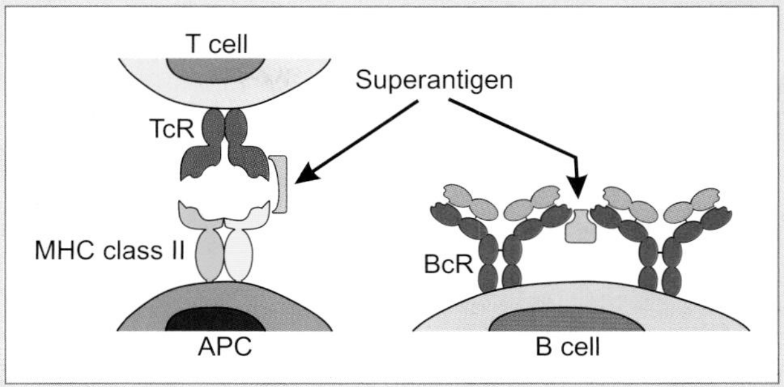

- **Superantigens**: Characterised by their ability to bind antigen receptors outside the antigen-binding site, superantigens can stimulate >5% of the naïve lymphocyte pool. Conventional antigens normally stimulate less than 0.01% of naïve lymphocytes. Both T and B cells can be activated by superantigens.
  - **T cell superantigens**: Certain bacterial toxins are powerful T cell mitogens[6] that stimulate proliferation of 2-20% of T lymphocytes. These include bacterial superantigens, such as staphylococcal enterotoxins (A through I except F), toxins produced by *S. pyogenes* (A–C and F), and certain mycoplasmal and viral antigens (mammary tumour virus, Epstein-Barr virus, etc.). Microbial T cell superantigens are medium-sized proteins (22-29 KD) that are resistant to proteases, endure temperatures of 60°C and higher, and survive in the 2.5-11 pH range. They can induce biological effects at femtomolar concentrations[7]. They probably evolved as mechanisms of exploiting MHC:TcR interaction to the selective advantage of pathogens. The main characteristics of T cell superantigens are:
    - They do not undergo processing by APCs prior to MHC-binding, unlike T cell epitopes.
    - Although originally shown to bind to T cells having a particular variable region β chain (Vβ) of the TcR, recent research has shown that superantigens can bind certain Vα regions as well.
    - Superantigens glue MHC and TcR antigen-nonspecifically. Thus, TcR, MHC, and superantigens form a trimolecular complex, although both the MHC peptide-binding groove and the TcR antigen-binding groove are not involved (Figure 4.6).
    - Superantigens cause massive immune activation. As superantigens bind both MHC and TcR, they end up stimulating APCs and T cells irrespective of antigen specificity. As a result of this antigen-nonspecific stimulation, the immune system goes into high alert, where monocytes, B cells, T cells, and NK cells are in activated mode. Superantigen stimulation leads to massive production and release of cytokines by T cells (IL-2, TNF-α/β, IFN-γ), monocytes, and NK cells (IL-1, IL-6, TNF-α). This enormous release of cytokines is responsible for clinical manifestations, such as hypotension (rapid fall in blood pressure), shock, nausea, and vomiting. The released cytokines recruit other effector cells, resulting in a runaway immune response that is injurious and potentially fatal.
    - They can induce anergy[8], apoptosis, and even active suppression of T cells. In addition, superantigens can mediate the killing of APCs. They can thus suppress adaptive immune responses.

[6] Mitogens are substances that cause cells to divide (induce mitosis).

[7] 1 femtomole = 10^{-15} M

[8] *Anergy* literally means a lack of energy. Anergic cells do not respond to signalling through their antigen receptor.

- **B cell superantigens**: Naturally occurring proteins that bind to antibodies *via* their antigen-binding site, irrespective of their antigen specificity, can cause the nonspecific activation of B cells. These include Staphylococcal protein A, gp120 from certain HIV isolates, and protein L from *Peptostreptococcus magus*. Intriguingly, a human gut-associated sialoprotein, termed protein Fv, also has B cell superantigen activity and may influence B cell selection in gut-associated lymphoid tissue. These superantigens bind to all Ig molecules having a particular V_H region (antigen-binding region). They have also been implicated in certain autoimmune disorders. The features of B cell superantigens are:
  - They bind Ig molecules at a site close to, but other than, the antigen-binding site. Therefore, the binding of the superantigen to an Ig molecule does not interfere with its antigen-specific binding.
  - They are functionally multivalent. They can bind two or more Fab regions, allowing them to crosslink BcRs and activate complement.
  - B cell superantigens cannot stimulate B cells by themselves. B cells generally require two signals for stimulation — the first is for the crosslinking of the BcRs by the antigen, and the second is the one delivered by either T cells or accessory cells. B cell stimulation by superantigens follows this inherent B cell physiology, and a second signal is generally required for complete activation and proliferation.
  - Superantigen activation causes the proliferation and eventual apoptosis of B cells.

Super Shocker: Toxic Shock Syndrome and Superantigens

Toxic shock syndrome is a spectrum of symptoms (high fever, hypotension, generalised skin rash, dysfunction of multiple organs associated with effects such as chills, headache, vomiting, diarrhoea, muscle pain, and hallucinations) first described in menstruating women using superabsorbent tampons. Investigations revealed that the causative agent was an exotoxin produced by *S. aureus*, named TSST-1 (**T**oxic **S**hock **S**yndrome **T**oxin-**1**). Superabsorbent tampons afforded this pathogen an ideal environment to grow and secrete the toxin. Guidelines for tampon composition and use (replacement every 4-6 hours) led to a dramatic decline in incidence. The notion of TSS being a female disorder was soon dispelled when it was found to be prevalent in the general populace as well. Anyone suffering from an infection by *S. aureus*, such as a burn victim or patient with postoperative infections, is potentially likely to suffer TSS.

The list of diseases caused by superantigens has been growing steadily since the staphylococcal toxin was first identified. Superantigens are implicated in food poisoning, necrotising fasciitis (rapidly spreading soft tissue infection), scarlet fever, rheumatoid fever, Kawasaki's disease (a leading acquired heart disease in children), diabetes mellitus, psoriasis, SLE, multiple sclerosis, and sudden unexpected nocturnal death syndrome, amongst others.

Superantigens are now being explored for disease treatment.

- The T cell proliferating abilities of superantigens can be used in cancer immunotherapy. Fusion proteins consisting of superantigens fused with Fab fragments of antibodies against a particular peptide:MHC complex expressed by tumours allow T cells to target these tumours.
- The ability of superantigens to cause anergy or the deletion of a particular subset of T cells or to induce active immunosuppression is now being investigated to eliminate/silence self-reactive T cell populations.

- B cell superantigens can bind soluble Ig molecules because soluble Igs and BcRs have the same structure. Igs can therefore compete with BcRs for sites on the superantigen.
- Igs formed later in the immune response bind superantigens with lowered affinity. Normally, antibodies formed later in the immune response show an increased affinity for the challenging antigen because of a change in the antigen-binding region of the Ig molecule. Termed *affinity maturation*, this phenomenon results from mutations in the DNA region encoding the antigen-combining site of the Ig molecule (see chapter 6). Thanks to this change in the antigen-binding site, antibodies produced later in the immune response fit the antigen better than those produced in the initial stages. For superantigens, such mutations result in a decreased rather than increased binding.

❒ **T-dependent and T-independent antigens**: As explained, antigens can be divided into two subtypes, depending upon their B cell activation pattern.

- **T-Dependent (TD) antigens**: These antigens are unable to stimulate B cell proliferation on their own. Cell to cell contact (called *cognate help*) between T and B cells is needed for B cell responses to TD antigens. Thus, two signals are required for B cell stimulation by TD antigens — the first delivered by the antigen *via* the BcR and the second by cognate interaction between T and B cells.

  Most proteinic antigens are TD antigens. Conventional B cells ($CD5^-$ B2 B cells), respond to TD antigens. T cell-associated phenomena, such as germinal centre formation and affinity maturation, are only observed for TD antigens. Mice lacking T cells (e.g., SCID mice, nude mice, or neonatally thymectomised mice described in chapter 5) are incapable of producing antibodies to TD antigens. Immune responsiveness to TD antigens is determined by the T cell repertoire of an animal, and if there is a hole in this repertoire (i.e., if T cells capable of responding to a particular MHC:peptide complex are missing), the animal becomes unresponsive to that antigen.
- **T-Independent (TI) antigens**: Physical contact between B and T cells is not required for B cells to respond to TI antigens. The B cell response occurs in the absence of the second signal, although soluble cytokines secreted by APCs and/or T cells can enhance this response. TI antigens are normally polysaccharide antigens with repetitive subunits and multiple antigenic

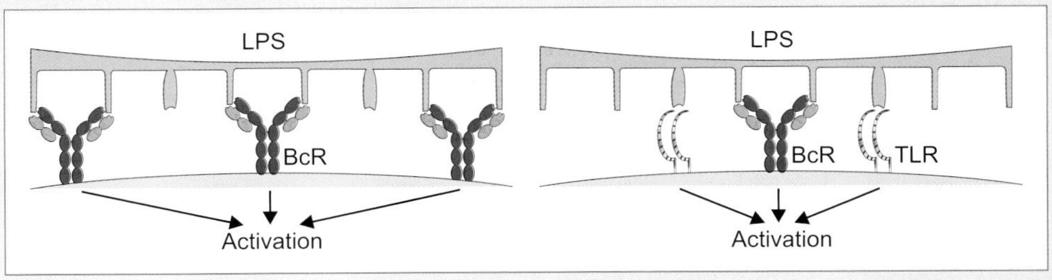

Figure 4.7 The engagement of multiple repetitive epitopes on TI antigens by BcRs generates a signal large enough to activate B cells in the absence of T cell help. TI type I antigens, such as LPS, can stimulate B cells by antigen-specific (left panel) and antigen nonspecific mechanisms (right panel). Type I antigens can therefore behave like B cell mitogens and cause their polyclonal activation.

determinants. Examples of naturally occurring TI antigens include the envelope of the Epstein-Barr virus, LPS, and flagellin. Examples of synthetic TI antigens are dextran sulphate, Ficoll, and polyvinyl pyrrolidone. The B1 subset of B cells ($CD5^+$) usually responds to TI antigens. Secondary responses to TI antigens do not essentially differ from primary responses.

- TI antigens do not lead to germinal centre formation.
- They fail to cause affinity maturation. The overall affinity of antibodies to TI antigens remains low even after repeated immunisations.

TI antigens are further divided into type 1 and type 2.

- **TI type 1** antigens are truly T-independent; response to these antigens does not decrease in nude mice. They are large polysaccharide antigens that are mitogenic to B cells and can stimulate neonatal as well as mature B cells by antigen-nonspecific mechanisms, such as *via* TLRs. At high concentrations, they behave as polyclonal B cell activators (i.e., activate B cells in an antigen-nonspecific manner). If, however, the TI type I antigen is specifically recognised by the BcR, B cell activation is antigen-specific (Figure 4.7). LPS is the prototypical type 1 antigen.

Defending the Defenceless: New Approaches to Fighting Childhood Infections

Children under the age of four are susceptible to a number of invasive bacterial infections. Many of these infectious agents (such as *H. influenzae* type b, *N. meningitides*, and *S. pneumoniae*) cause severe infections because children are unable to mount effective immune responses to these capsulated organisms. The capsules allow the organisms to escape phagocytosis, and therefore T cell involvement, in the ensuing response. Being carbohydrates, capsules are poor immunogens. Although purified capsular polysaccharides can be used in vaccines in immunocompetent individuals, their TI nature precludes use in infants and children. Different strategies are therefore used to convert the TI response to a TD type.

- Conjugating purified capsular polysaccharides or oligosaccharides to a protein carrier renders the carbohydrate immunogenic even in very young children. Four vaccines containing such conjugates have been approved for clinical use in the US. They consist of a conjugate of the capsular antigen of *H. influenzae* type b attached to a protein carrier. In trials the vaccine resulted in >97% reduction in disease. Efforts are now underway to develop conjugate vaccines against *Salmonella typhi, Shigella* spp., and type B streptococcus.
- The use of peptides that mimic the carbohydrate's antigenic structure can result in a TD immune response. Such antigens have the advantage of being stable, easy to produce, and of a defined chemical nature. Unlike in conjugated vaccines, this approach has not reached the clinical trial stage.
  - Anti-idiotypic antibodies can be used to define specific peptides that mimic carbohydrate antigens. Such peptide mimics have been used successfully in some animal models of viral, bacterial, and parasitic infections.
  - Peptide mimics of carbohydrate antigens can also be identified by screening a phage-display library with anti-polysaccharide monoclonal antibodies.
  - Nucleic acid vaccine technology and peptide mimicry can be combined in a novel approach that involves the cloning of an oligonucleotide encoding a peptide mimic into a eukaryotic expression plasmid. Such peptide mimic DNA vaccines have been developed against the capsular polysaccharide *S. pneumoniae* serotype 4. They have been shown to confer protection in laboratory animals.

- **TI type 2 antigens** are highly repetitious molecules; their multiple, identical epitopes can extensively crosslink BcRs, thereby stimulating B cells. They are not B cell mitogens and do not polyclonally activate B cells. Although type 2 antigens do not need the direct involvement of TH cells, some cytokine production by T cells (e.g., IL-2, IL-5) is required for efficient B cell stimulation. Type 2 antigens are incapable of eliciting an immune response in neonatal mice or T cell deficient nude mice. They can activate complement and stimulate only mature B cells. Bacterial flagellin and the synthetic polysaccharide TNP-Ficoll are examples of TI type 2 antigens. The exact mechanism by which type 2 antigens activate B cells is not clearly understood. Immunocompromised individuals (such as HIV seropositives), people older than 65, and children under two years of age seem unable to respond to these antigens.

The TI antigen pathway seems to provide a quick protective response against common viral, bacterial, and fungal pathogens. Many pathogens have extracellular polysaccharides that allow them to resist ingestion by phagocytes and escape destruction. Escaping phagocytic digestion also helps them avoid T cell activation, which can only occur when APCs display peptides of pathogenic origin in the context of MHC molecules. By enabling antibody production to polysaccharide antigens, the TI pathway allows for humoral immune responses to many bacterial pathogens that would otherwise have escaped immune destruction. This pathway may be viewed as a first line of adaptive response — a stopgap arrangement until the conventional adaptive immune response is generated.

4.4 Factors Influencing Immunogenicity

Immunogenicity is not intrinsic to a molecule. It is determined by both the nature of the molecule and the animal host. Substances that are immunogenic in one animal need not be so in another. **Properties of the immunogen** that influence its immunogenicity include:

❐ **Foreignness**

According to the Self–Nonself Theory of immune recognition, an immunogen must be recognised as nonself (foreign) to elicit an immune response. Molecules present in the tissues or tissue fluids of an animal fail to elicit an immune response in that animal[9]. Thus, bovine serum albumin can elicit an immune response in mice, whereas mouse serum albumin cannot. This failure to mount an immune response to self-antigens is because of the elimination of self-reacting clones during the process of T cell maturation in the thymus. The presence of regulatory T cells that suppress immune responses in an antigen-specific manner strengthens this curb on the response to self-antigens (regulatory T or Treg cells are described in chapter 9). For the same reason, ubiquitous proteins that have a highly conserved structure across a variety of species (e.g., cytochrome *c* and dextran) are not immunogenic. Bovine cytochrome *c* is too similar to murine cytochrome *c* to be recognised as foreign in mice. T cells that can react with self-antigens not expressed in the thymus during the process of maturation escape elimination (proteins from eye lens, testes). Consequently, these antigens elicit an immune response. The Self–Nonself Theory of immune recognition is being challenged as too simplistic (see chapter 1). Nonetheless, it

[9] It needs no stretch of imagination to realize that the mounting of an immune response against self-antigens will prove disastrous for an animal. Diseases that result from an immune response to self-antigens are called *autoimmune diseases* (chapter 15).

is far easier to induce an immune response to foreign antigens than to self-antigens.

- **Degradability**

  The processing and presentation of an antigen in the context of MHC molecules is indispensable to the development of a robust humoral and/or cell-mediated immune response. Macromolecules, such as polystyrene, that cannot be degraded for presentation are poor immunogens. Polymers of D-amino acids, which are stereoisomers of the naturally occurring L-isomers, are also poor immunogens, as APC enzymes are incapable of degrading the D-isomer. On the other hand, substances that are rapidly broken down by plasma enzymes or lysosomal enzymes are weakly- or non-immunogenic. Such highly degradable proteins get digested to fragments that are too small to load onto MHC molecules in the highly proteolytic lysosomal environment of APCs.

- **Molecular size and complexity**

  Molecular shape is unimportant in determining a substance's immunogenicity. Proteins and polypeptides with rod-like, globular, or random-coil configuration can all be potent immunogens. Molecular size, though, does influence immunogenicity. Generally, substances with a **M**olecular **W**eight (MW) of 6 KD and above tend to be better immunogens than those with low MWs. Even the polymerisation of a low-MW substance can increase its immunogenicity — bovine insulin (MW 6 KD) is a poor immunogen, but if polymerised by chemical methods or aggregated by heat, it is immunogenic. Similarly, intact flagella are more immunogenic than polymerised flagellin, which is in turn, is more immunogenic than flagellin itself. Some molecules may bind tissue proteins and become immunogenic. For example, picryl chloride, formaldehyde, or drugs, such as penicillin and sulphonamides (all with MW <1 KD), can elicit an immune response. These molecules behave like haptens; they bind to tissue proteins and act as antigenic determinants of the tissue-protein–drug complex.

- **Complexity**

  Obviously, a complex molecule has a larger variety of potential determinants than a simple molecule. The larger the variety of determinants, the greater the chance that one or more of these will preferentially stimulate cells of the immune system. Increasing complexity therefore generally increases immunogenicity. This could be one of the reasons for an increase in immunogenicity with an increase in MW or size. Crosslinking of surface antigen receptors is necessary to stimulate B cells (see section 6.2.3). A complex molecule with multiple epitopes is more likely to stimulate B cells. Structural stability or rigidity seems to be another factor that can influence immunogenicity. Molecules, such as gelatine, that lack structural stability are poor immunogens. A linear homopolymer (a linear polypeptide of an amino acid) is usually nonimmunogenic, but a branched poly-amino acid structure tends to be immunogenic. The presence of aromatic amino acids in a molecule enhances its immunogenicity, probably because these amino acids lend rigidity to the molecule. The physical form of the antigen is also important. In general, particulate antigens are more immunogenic than soluble ones, and denatured antigens are more immunogenic than native forms. Particulate antigens, probably because they are more irritable, cause an inflammatory response and are better immunogens[10].

As it takes two for this immune-response tango, the induction of an immune response depends not only on the immunogenicity of the molecule but also on the nature of the animal and the method of antigen administration.

[10] See the Danger Hypothesis outlined in the sidetrack *Us and Them* in chapter 1.

- **Genetics**

  An immunogen may elicit a prolonged response in an individual of one species but fail to do so in another of the same species. Many studies have clearly established that many aspects of the immune response are genetically determined. Immune responsiveness is partially under the control of MHC genes located on chromosome 6 in humans and chromosome 17 in mice. These genes and their impacts on immune responsiveness are discussed in chapter 7. For obvious reasons, genes that encode BcRs, TcRs, and other proteins involved in immune regulation also influence an animal's ability to respond to different immunogens.

- **Age**

  As a rule, inducing a robust immune response in very young animals is difficult. If an animal encounters an immunogen before it is immunologically mature (i.e., neonatally), the T cells capable of reacting to that antigen are eliminated. The animal then becomes tolerant to that particular antigen. The capacity to respond to immune stimuli is also hampered by increasing age. A variety of changes occur with aging, including the appearance of new differentiation antigens, modification of cell membranes, and quantitative differences in surface Ig expression on B lymphocytes. Age may also lead to impaired B lymphocyte, T lymphocyte, and macrophage functions. Consequently, the appearance of autoantibodies, the production of paraproteins[11], and immune deficiencies can occur with increasing age. It is therefore essential to consider an animal's age when interpreting immunological experiments.

- **Antigen administration**

  The dose of the immunogen, its route of administration, and the presence of adjuvants all influence the outcome of antigen administration.

  - **Dose of the immunogen**

    Every immunogen has an optimal immunogenic dose. Amounts much larger than the optimal dose generally lead to tolerance (i.e., failure to elicit an immune response), termed the high-zone tolerance; much smaller doses may lead to low-zone tolerance. This is especially true of TI antigens. Generally, a single dose of an immunogen is less effective in eliciting an immune response than repeated administrations. The dose of the antigen can also influence the type of immune response. Small repeated antigen doses have been found to induce IgE antibody responses in experimental animals; larger doses result in an IgG antibody response.

  - **Route of administration**

    For deliberate immunisation of experimental animals, an immunogen is usually injected in the body by specific routes. The most common routes of administration are intradermal or subcutaneous, intramuscular, intravenous, and intraperitoneal. Regardless of the route, most antigens eventually reach the nearest draining lymph node, and once there, activate antigen-specific T and/or B lymphocytes. When administered intravenously, the antigen reaches the spleen, which acts as a filter for blood and traps all antigens entering the blood. As a rule, an immunogen will elicit a better response if introduced parenterally[12] into the body. This is because the immunogen is more likely to be digested before it reaches an APC if it is introduced by the digestive route. The same holds true for immunisations through epithelial tissue. The route of administration also influences the

[11] A paraprotein is an abnormal concentration of a single Ig or Ig light chain found in urine and/or blood. It is produced by plasma cells formed by the clonal proliferation of a single B cell. Also called Bence Jones proteins, paraproteins allow the elucidation of Ig structure. See the sidetrack *Unwelcome and Unwanted B* in chapter 8.

[12] Parenterally: *para* — around, *enterally* — gut; i.e., by routes other than the digestive tract.

type of immune response elicited. Minute doses repeatedly inhaled through respiratory mucosa or given subcutaneously tend to result in an IgE antibody response, but mucousal administration of large doses of the antigen elicits an IgA antibody response. Also, soluble antigens tend to be immunogenic when injected in tissues but tolerogenic when administered intraperitoneally.

- **Use of adjuvants**

  Adjuvants (from Latin *adjuvare* — to aid) are agents that are administered along with the antigen and which can enhance humoral or cellular immune response by nonspecific mechanisms. Thus, a substance that can accelerate, prolong, and/or enhance the quality of the immune response to an immunogen can be labelled an adjuvant. A highly heterogeneous group of molecules or compounds have such immunostimulatory properties. The mode by which these molecules stimulate the immune system varies widely. Because adjuvants have diverse mechanisms of action, they must be chosen for use with a particular immunogen based upon the route of administration and desired immune response. The exact mechanism of adjuvant action is unclear, although six broad modes can be delineated.

  - **Enhanced translocation of the immunogen from the site of administration to the local draining lymph node**: Immune responses are not initiated at the site of immunogen injection. An immunogen can trigger an immune response only when it reaches the local lymph node. Adjuvants may promote such translocation by protecting the immunogen from rapid, nonspecific elimination. The immunogenicity of peptide antigens that would normally be cleared rapidly from sites of injection can be improved with the use of particulate adjuvants that associate with or otherwise hold the antigen.
  - **Improved antigen delivery to, or antigen processing and presentation by, APCs**: By promoting phagocytosis, adjuvants assist in the translocation of an immunogen besides increasing its presentation to T cells.

    Adjuvants, such as purified saponins, ISCOMs (**I**mmune **S**timulatory **Com**plexes; see the sidetrack *Spiking the cocktail*), and liposomes, have been shown to improve the induction of CTL (**C**ytotoxic **T** **L**ymphocyte) responses by delivering the antigen directly to the cytosol for presentation by class I molecules.

    Particulate antigens, such as liposomes, ISCOMs, polymer microspheres, and MF 59, can be used to target antigen to phagocytic cells. Macrophages and DCs that have internalised antigens move rapidly to draining lymph nodes, where the phagocytosed antigen is processed and presented.

    Some adjuvants enhance cross-presentation by DCs, increasing CTL responses to the antigen (see section 7.2.3).
  - **Increased biological or immunological half-life of the antigen**: Many adjuvants are good adsorbing agents. They cause the slow and prolonged release of immunogen; they have a depot effect. The adjuvant activity of aluminium salts, Freund's adjuvant, and oil-based adjuvants is thought to be attributable, at least partially, to this depot effect.
  - **Engagement of PRRs**: Many cells of the innate immune system express receptors that specifically recognise patterns or motifs on pathogenic organisms (section 2.2). Engagement of these PRRs is known to enhance immune responses.

Microbial components or their derivatives, such as pertussis toxin, LPS (or its active derivative monophosphoryl lipid A), mycobacterial cell wall-derived muramyl dipeptide, and CpG rich DNA, engage TLRs and activate cells of the innate immune system.

The fusing of antigen to the complement component C3d can result in antigen-specific antibody responses because C3d acts as a molecular adjuvant of innate immunity.

The cage-like structures of ISCOMs may mimic PAMPs and stimulate immune responses.

- **Induction of local reaction at the site of injection**: Signals from damaged or stressed cells induce immune responses. Many common adjuvants, such as aluminium salts, $Ca_3(PO_4)_2$, and mineral oils, can cause a local inflammatory reaction. Nevertheless, the efficacy of the adjuvants must be weighed against the severity of local reactions. For this reason, Freund's adjuvant (complete or incomplete) is unacceptable for use in humans.
- **Induction of costimulatory molecules or immunomodulatory cytokines**: Costimulatory molecules that deliver the second signal for T cell activation hold adjuvant potential. As explained in section 9.3, two signals are required for naïve-T cell activation. The engagement of TcRs by MHC:peptide complexes delivers the first signal. Binding of costimulatory molecules (such as CD80/CD86) on APCs to their ligands on T cells delivers the second, activating T cells. DNA encoding antigen–CD86 fusion proteins deliver both the signals required for T cell activation. Administration of these with the antigen enhances T cell responses.

  Adjuvants, such as the Freund's adjuvant, cause a strong local inflammatory reaction and induce the expression of costimulatory molecules and the production of pro-inflammatory cytokines. This recruits macrophages and DCs to the site of antigen injection.

  Adjuvants can selectively modulate cytokine responses, inducing the desired TH1 or TH2 type response[13]. Bacterial toxins, such as the cholera and pertussis toxins, preferentially drive TH2 responses and enhance IgA and IgE antibody production. Conversely, monophosphoryl lipid A induces IFN-γ, and hence, a TH1 response. Cytokines may also be used as adjuvants. For instance, IL-12 selectively induces a TH1 response when administered with antigens. IFN-γ and GM-CSF are two other cytokines being evaluated for adjuvant activities.

Choosing the appropriate adjuvant is crucial for inducing the appropriate immune response (high antibody/CTL response) and key to the immunised animal's welfare. Characteristics of the antigen (size, net charge, and presence/absence of polar groups) and the species of animal to be immunised affect the selection of the adjuvant. Perhaps the most desirable feature of an adjuvant is its ability to specifically enhance an immune response to the antigen that it is administered with. An adjuvant with broad, nonspecific effects is more likely to induce adverse immunological effects. The benefits of incorporating adjuvants into vaccine formulations must be weighed against risks of adverse reactions. Adjuvant-induced local adverse reactions include inflammation at the site of injection, induction of granulomas, or abscess formation. Systemic reactions include malaise, fever, adjuvant arthritis, and anaphylactic shock. The risks of adverse reactions

[13] TH1 and TH2 are two types of T helper cells. They differ in the cytokines they produce, the Ig isotypes they promote, and the types of immune responses they drive. They are described in chapter 9.

to adjuvants and concerns about their safety have restricted their development since the introduction of alum salts in the early 1900s. The acceptability of side-effects often depends upon intended use. For standard prophylactic use in healthy individuals, only adjuvants with minimum side-effects are acceptable (hence the popularity of rather weak alum salts). For adjuvants that are designed to be used in life-threatening conditions (e.g., cancer), acceptable side-reaction levels are higher. Other issues important to adjuvant design and usage include biodegradability, stability, ease and cost of manufacture, and applicability to multiple immunogens. Admittedly, in spite of knowledge advances, adjuvant selection remains largely empirical. Many experimental adjuvants currently undergoing clinical trials are highly immunostimulatory. Future adjuvants are likely to exploit more site-specific delivery systems to better target antigens achieving the two-fold objective of optimal immunostimulation with minimal side-effects.

Spiking the Cocktail: Adjuvants 101

Freund's Adjuvant: Complete Freund's adjuvant is an extremely effective adjuvant. A water-in-oil emulsion containing suspended live or dead mycobacteria, it stimulates almost all immune responses. An intense inflammatory response develops at the site of its injection. The adjuvant activity of mycobacteria is largely due to a complex glycolipid — **M**uramyl-**d**i-**P**eptide (MDP). The complete adjuvant causes a reaction that is too strong to be of use in humans. **Freund's incomplete adjuvant,** consisting of the water-in-oil emulsion but not the mycobacterial suspension, is less inflammatory. Even then, its use is restricted to animals. Derivatives of MDP, when administered through saline, induce mainly humoral immunity. Water-in-oil emulsions of MDP can induce humoral and cell-mediated immune responses and may be used in humans.

- **Mineral salts**: These were the first set of adjuvants licensed for use in humans. Aluminium-containing compounds are popular adjuvants ($Al(OH)_3$, $Al_2(SO4)_3$, alum, etc.), especially for inducing humoral responses against proteinic antigens. Calcium phosphate nanoparticles have been used in a variety of vaccines (diphtheria, pertussis, tetanus, polio). These particles have been found to induce predominantly TH1 responses.
- **Immunostimulatory adjuvants**: Adjuvants that exert their effect primarily at the cytokine level are called immunostimulatory adjuvants.
  - **LPS**
    Although a powerful immunostimulant, LPS is not suitable for clinical applications because of its toxicity. **Monophosphoryl lipid A** is a less toxic derivative of LPS that retains adjuvant activity. It promotes IFN-γ release in clinical trials. It does not promote a potent antibody response if used alone.
  - **Saponins**
    Triterpenoid glycosides derived from *Quillaia saponaria* have long been used in veterinary medicine. Because of their ability to interact with cholesterol, saponins intercalate in cell membranes and cause pore formation. The antigen easily penetrates cells through these pores and gets carried to the regional lymph node. The immunostimulatory fraction of saponin — Quil-A — is unsuitable for clinical use because of its toxicity. QS21, a pure fraction of Quil-A, has low toxicity and induces TH1 type responses (induction of IFN-γ, IL-2, CTLs, and IgG1 antibodies). It has been used in clinical trials for multiple immunogens (cancer, HIV-1, influenza, malaria, and hepatitis B).
  - **CpG DNA**
    The investigation of the tumour-reducing effects of the BCG vaccine led to the discovery of the immunostimulatory capabilities of bacterial DNA. This immunostimulatory effect was later linked

to the presence of unmethylated CpG dinucleotides (see chapter 2). These unmethylated CpG dinucleotides are a common motif in bacterial DNA but are rare in vertebrate DNA. Unmethylated CpGs can be recognised by TLRs in the context of selective flanking sequences. CpGs are found to release pro-inflammatory cytokines (TNF-α, IL-1, IL-6, and IL-12), and they are potent inducers of TH1 responses. Several preclinical and clinical vaccine trials using CpG DNA as an adjuvant are currently underway.

- **Cytokines**
  Cytokines, such as IL-1, IL-2, IL-12, IFN-γ, and GM-CSF, are being evaluated as adjuvants. However, all show dose-related toxicity. A high manufacture cost, stability problems, and a short *in vivo* shelf-life make their use in routine vaccination unlikely. Nonetheless, they hold immense promise for cancer immunotherapy.

- **Particulate adjuvants**: These adjuvants have a diameter of 10-100 nm, comparable to that of microbes, and are easily taken up by DCs and macrophages.
  - **MF59**
    This oil-in-water emulsion of biodegradable oil (squalene) was found to increase the influenza vaccine's immunogenicity. It targets the antigen to macrophages and DCs and induces a potent antibody response. It has been licensed for use in humans in a variety of vaccines (HIV, influenza, hepatitis B, and herpes simplex).
  - **Liposomes**
    Liposomes are particles made of concentric lipid membranes containing phospholipids and other lipids in a bilayer configuration. Aqueous compartments separate these lipid membranes. They are often used with other immunostimulators, such as monophosphoryl Lipid A. A liposomal hepatitis B vaccine is in the final stages of commercial development. The development of polymerised liposomes is also being investigated. Modified liposomal structures called chochleates are being similarly evaluated for the mucosal delivery of vaccines. A virosomal preparation consisting of liposomes and inactivated hepatitis A virus is now approved for clinical use.
  - **I**mmuno**s**timulating **com**plexe**s or ISCOMs**
    Formed by the the incorporation of Quil A into lipid particles comprising cholesterol, phospholipids, and cell-membrane antigens, ISCOMs have relatively low toxicity and target the antigen directly to APCs. They induce TH1 type responses. An influenza ISCOM vaccine is nearing the end of a successful clinical trial.
  - **Polymeric microparticles**
    Biodegradable polyester microparticles made of poly(lactide-co-glycolides) are being investigated for targeting antigens to APCs. The safety of poly(lactide-co-glycolides) is well established — it has been used as suture material for a long time. Polymeric microparticles are used to encapsulate the antigen; they induce prolonged immunity by allowing the antigen's controlled release. They induce potent mucosal and systemic immunity when administered orally.
- **Mucosal adjuvants** are used for mucosal antigen delivery and induce robust IgA antibody responses.
  - **Bacterial toxins**, including the cholera toxin and the enterotoxin of *E. coli,* are the most potent mucosal adjuvants available. Sadly, their toxicity limits usage. Recombinant mutated toxins are being studied as possible adjuvants. The mutated cholera toxin is reportedly being investigated as an adjuvant to be used in topical applications.
  - The development of **transgenic plants** expressing antigens and adjuvants is the most recent advance in the field of mucosal adjuvants. Their acceptance and applicability remains questionable.

4.5 Immunogenicity of Natural Molecules

Proteins, polysaccharides, lipids, and nucleic acids differ in their capacity to stimulate an immune response, proteins being the most efficient and nucleic acids the least.

- Many naturally occurring globular proteins are highly immunogenic. A globular protein is a chain of amino acids folded and refolded to yield a tight tertiary structure, or if associated with other chains, a quaternary structure. Determinants on the surfaces of the proteins, especially more exposed areas, such as the loops and cones of the folded chains, are important in determining the antigenic specificity of the antibody response. The whole protein may be visualised as a carrier–hapten complex — the larger globular part (comprising the internal surface nondistinctive peptides) functions as the carrier, and antigenic determinants act as haptens. The presence of lipid or polysaccharide molecules on these proteins enhances their immunogenicity. As a rule, the density of determinants is in the range of one determinant per five KD MW. A single amino acid in a highly exposed position may, however, contribute largely to the entire molecule's specificity. Given that a large globular protein will have many determinants, antisera raised against the same protein often differ in their specificities as they react with different determinants of the protein. Unlike B cell responses, T cell responses to proteinic antigens are determined by the primary amino acid sequences of the proteins. T cell clones raised against a protein will be specific for particular peptides of that protein loaded onto a specific MHC molecule. Both the protein and the MHC (its class and haplotype) in whose context the protein is being recognised need to be stated for clone identification. Thus, a T cell clone raised against human serum albumin in mice may be referred to as anti-HSA 64-72 (sequence of amino acids in the protein that constitutes the peptide recognised) IA^k (MHC class II haplotype, in whose context the peptide is recognised).

- Polysaccharides are serologically important cell components. Many polysaccharides are weakly- or non-immunogenic by themselves, but they dominate the serologic specificity of the proteins they are attached to. By contrast, ubiquitous polysaccharides, such as starches, dextrans, and glycogens, are largely nonimmunogenic. Pentoses and hexoses — either unsubstituted or substituted — are most frequently encountered in one or more antigenically dominant positions in immunogenic polysaccharides. The typing of RBCs into blood groups is on the basis of polysaccharides. Putative blood group antigens can also be found in bacteria and in various tissues and secretions of humans and other animals (see chapter 17). The highly heat-stable, microcapsular Vi antigen found in many pathogens is also a polysaccharide. Cell-surface polysaccharide antigens are often conserved in bacteria. The presence of such polysaccharide antigens can hence be used in taxonomic classification. This grouping of microbes according to their antigens is called *serotyping*. Serotyping is an important distinguishing character in many capsulated bacteria, such as streptococci, pneumococci, and salmonella.
- Lipids are triesters of organic acids with various alcohols, nitrogenous bases, and radicals, such as PO_4^{3-} and SO_4^{2-}. Lipids are largely nonimmunogenic. When conjugated with proteins or polysaccharides, lipids can be immunogenic. Lipids are loaded onto CD1, a MHC-like molecule, and presented to lipid-specific T cells and NKT cells (a subset of NK cells that express both NK and T cell markers). These cells are discussed in chapter 9.
- Nucleic acids are poor immunogens. They have to be complexed with larger molecules, or altered, to be immunogenic. The autoimmune disease SLE is characterised by antibodies against nuclear proteins and DNA.

5 Entities of the Adaptive Immune Response: The Lymphoid System

Win or lose, sink or swim
One thing is certain — we'll never give in
Side by side, hand in hand
We all stand together
Play the game, fight the fight
But what's the point on a beautiful night?
Arm in arm, hand in hand
We all stand together

— Paul McCartney, *We All Stand Together*

Abbreviations

BALT:	Bronchus-Associated Lymphoid Tissue
FDC:	Follicular Dendritic Cell
GALT:	Gut-Associated Lymphoid Tissue
HEV:	High Endothelial Venule
IEL:	Intraepithelial Lymphocyte
MALT:	Mucosa-Associated Lymphoid Tissue
M cells:	Membranous cells
MZ:	Marginal Zone
PALS:	Periarteriolar Lymphoid Sheath
SALT:	Skin-Associated Lymphoid Tissue
SCID:	Severe Combined Immuno-deficiency
SED:	Subepithelial Dome
sIgA:	Secretory IgA
VIP:	Vasoactive Intestinal Peptide

5.1 Introduction

Cells of the adaptive immune system are organised into tissues and organs which are collectively referred to as *the lymphoid system* (Figure 5.1). This system comprises cellular components responsible for antigen-specific host defence. Its components include lymphocytes, epithelial cells, and stromal cells[1], arranged either in discretely encapsulated organs (thymus, spleen, etc.), or in accumulations of diffuse lymphoid tissue (e.g., Peyer's patches). **Organs and tissues of the lymphoid system are connected to each other by the lymphatic system**. The lymph flowing through these lymphatic vessels is plasma derived. Plasma tends to seep through thin capillary walls into surrounding tissues because of the high pressure under which blood circulates. A major portion of this plasma re-enters blood through capillary membranes. The remainder, called *lymph*, flows into a separate system of tiny capillaries — the lymphatic capillaries. These vessels have valves that only allow unidirectional flow, and this forces the lymph to flow through the lymph nodes. Lymph vessels channel the constant stream of lymph through successively larger branches into the thoracic duct, where it enters blood circulation, ensuring that the levels of fluid in the circulatory system remain constant. Mature lymphocytes continuously circulate between various lymphoid organs and other tissues through the lymph and the blood stream, increasing their chances of encountering immunogens. Approximately 1-2% of the available lymphocyte pool circulates every hour. Most lymphocytes leave blood circulation through a specialised section of the postcapillary venule (i.e., a very tiny vein) known as the **H**igh **E**ndothelial **V**enule (HEV[2]), although some lymphocytes may also leave blood circulation through nonspecialised venules. In HEVs, lymphocytes react with high endothelial cells. The constricted space in these venules forces lymphocytes and DCs to come into close contact, enhancing chances of antigen recognition[3]. All lymphocytes eventually re-enter the blood stream through the lymphatic system.

It is not difficult to see that the anatomical features that foster cellular interaction also greatly increase the efficiency of the adaptive immune system. These cellular interactions do not occur at the site of the entry of immunogens or perceived threats, but they occur in specialised compartments. Immunogens or pathogens are transported by the lymph to tissues of the lymphoid system. Those deposited in peripheral tissues are carried through lymphatic channels and trapped in the lymph nodes directly downstream of sites of infection. Pathogens entering the blood are trapped in the spleen; those infecting the mucosal surfaces accumulate in the tonsils or Peyer's patches. Thus, *in vivo*, antigens are presented in a complex environment in which movements of antigens, of cells that present antigen, and of responding cells (T and B) are subject to anatomical constraints. The architecture of the whole lymphatic system and the lymphoid organs is designed to maximise chances of trapping immunogens entering the system. Specialised sub-compartments appear to facilitate cell-to-cell contact and recognition and provide the most favourable milieu for signalling and induction mechanisms. The juxtaposition of cells further increases response efficiency; cells that must interact for the generation of an immune response are in close proximity.

5.2 Primary Lymphoid Organs

The primary or central lymphoid organs are major sites of lymphopoiesis, that is, the generation of lymphocytes. Lymphoid stem cells divide and differentiate

[1] The connective tissue, nerves, and vessels that form the frame-like support of an organ or a part thereof are known as the stroma. Stromal cells are the cells that form this supporting framework.

[2] The name *HEV* alludes to the fact that the endothelial lining of these venules consists of cells having a plump cuboid morphology rather than the flat morphology observed in other blood vessels. The *high* in the name thus refers to these cells' thickness.

[3] Ever entered a crowded station at rush hour?

They Have Feelings Too: Lymphoid Organs as Neuroendocrine Organs

Although we think of the thymus as a purely immunological organ, it is also rich in nerve fibres. Noradrenergic and peptidergic fibres enter the thymus with nerve bundles and plexuses around blood vessels, penetrate the cortex from subcapsular plexuses, and branch out among lymphocytes in the thymic cortex. The vasculature and lymphatic tissue of the outer and deep cortex are enervated by these fibres. Neuropeptides are present in thymic neurons. *In vitro* studies indicate that T cell responses can be affected by neurosecretory products, and anatomical studies clearly show the association of nerve fibres with aggregates of lymphoid cells *in vivo*. Other organs of the immune system, including the bone marrow; lymph nodes; spleen; and lymphoid tissues, such as Peyer's patches and tonsils, show extensive enervation and penetration by noradrenergic nerve fibres. Little is known about the function of this enervation outside its assumed role in controlling blood flow. Many cells of the lymphoid system, including T cells and DCs, express receptors for hormones and neuromediators, such as **V**asoactive **I**ntestinal **P**eptide (VIP)[4], somatostatin, calcitonin generated peptide, and substance P. These neuropeptides are released from unmyelinated nerve endings in lymphoid organs. Reciprocally, neural cells express receptors for cytokines, and lymphoid cells can secrete neuropeptides. Thus, both neural cells and immune cells can influence each other in paracrine and autocrine manners. The central nervous system and the immune system have been suggested to have bidirectional circuits. Thymosins, cytokines, complement components, enkephalins, adrenocorticotropic hormone, and thyroid-stimulating hormone present in the thymus produce neuroendocrine effects on both systems. The precise relation of the anatomy and physiology of neuroimmune functions in T cell development and differentiation remains unresolved. Recent research further stresses the relation between the immune and neuroendocrine systems — thymectomy in mice not only reduces the immune response but also deteriorates learning performance. The body cannot be viewed as a series of isolated systems; it is best understood holistically.

into lymphocytes in the primary lymphoid organs. The rate of cell division and differentiation is stupendous. Around 10^8 lymphocytes are formed daily in the bone marrow through cell division. **Newly formed lymphocytes undergo a process of proliferation and maturation in the primary lymphoid organs before migrating to the secondary lymphoid organs**. Two primary lymphoid organs have been recognised in mammals — the thymus and the bone marrow. Recent evidence suggests that in humans, the intestinal epithelium is also a primary lymphoid organ[5]. Both primary and secondary lymphoid organs are often referred to as *lymphoepithelial* because they consist of a mass of lymphocytes and a smaller number of epithelial cells.

5.2.1 The Thymus

In mammals, the thymus is located in the thorax, which overlies the heart and the major blood vessels. Immature lymphocytes produced by the yolk sac or liver in early stages of development or by bone marrow stem cells later in life migrate to the thymus for maturation. These lymphocytes that mature in the thymus are termed *T lymphocytes* or *T cells*[6]. Unlike other lymphoid organs, the thymus is not involved in lymphocyte circulation and does not receive lymph from other tissues.

The thymus is a bilobed, encapsulated organ. Each lobe is organised into lobules or follicles. Strands of connective tissue, called *trabeculae*, separate the lobules. These lobules are organised structures consisting of an outer cortex and inner medulla. A three-dimensional network of stromal cells, consisting of epithelial cells,

[4] Recent studies indicate that in lymphoid organs, lymphocytes are the major source of VIP. VIP has potent anti-inflammatory effects and promotes T cell differentiation to T_H2 type while inhibiting T_H1 pathways (section 9.5.1) — this neuro-peptide seems to behave suspiciously like a cytokine.

[5] It represents a major site of T cell lymphopoiesis and houses a substantial fraction of T cells in the body.

[6] All types of T cells, whether $\alpha\beta$ T cells, $\gamma\delta$ T cells, or NKT cells, mature in the thymus. The process of maturation of peptide-specific $\alpha\beta$ T cells is described here; other types of T cells undergo maturation in an essentially similar manner.

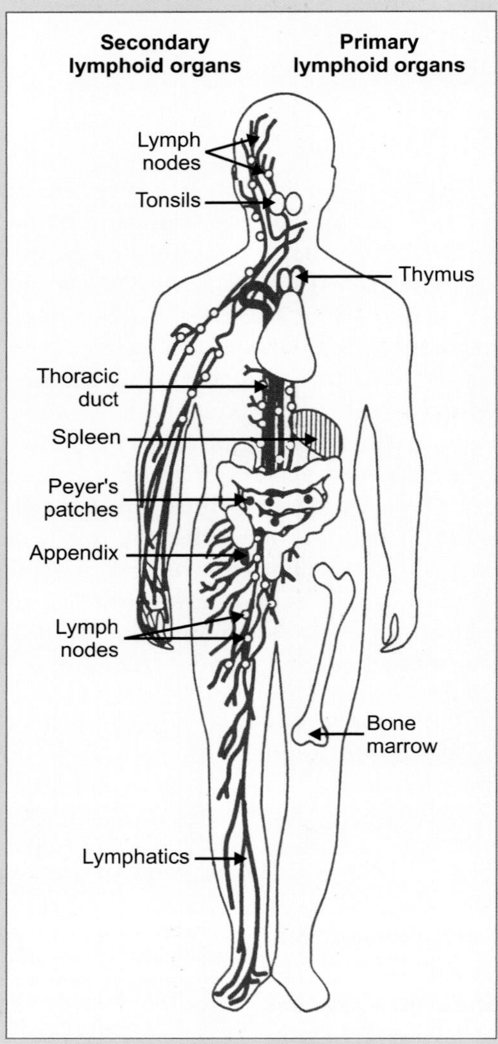

Figure 5.1 The lymphoid system consists of lymphocytes, epithelial cells, and stromal cells arranged either in discretely encapsulated organs or in accumulations. The thymus and the bone marrow are the primary human lymphoid organs. These are the sites of lymphopoiesis and lymphocyte maturation. The lymphocytes that have matured in the primary organs get lodged in the secondary lymphoid organs (lymph nodes, spleen, and MALT — the tonsils, intestinal epithelium, Peyer's patches, appendix, urinogenital epithelium, skin, and bronchial lining), where they undergo further differentiation upon antigenic challenge. The lymph nodes are found at the junctions of lymphatic vessels traversing the body. They drain and filter lymph from tissue spaces as well as from internal organs. The lymph eventually collects in the thoracic duct and enters the blood stream (courtesy the Department of Immunology, Erasmus University, Rotterdam, The Netherlands).

interdigitating DCs, and macrophages, crisscrosses the cortex and the medulla. The cortex is densely packed with immature lymphocytes (also known as thymocytes) with a few scattered macrophages. The medulla appears epithelial because of a relative scarcity of lymphocytes. The lymphocytes found in the medulla are mature thymocytes. Most macrophages are found at the corticomedullary junction and in the medulla. Most of these thymic macrophages originate in the bone marrow and migrate to the developing thymus. Like these macrophages, the DCs found in the medulla originate in the bone marrow (Figure 5.2).

A layer of epithelial cells (the *subcapsular thymic epithelial cells)* lines the internal surface of the thymus. Both the cortex and medulla also have their own layers of epithelial cells. They show heterogeneity in their ultrastructural morphology and origin. The cortical epithelial cells are stellate cells with long cytoplasmic extensions. Medullary epithelial cells are spindle shaped. Some epithelial cells in the outer cortex are referred to as *thymic nurse cells,* as each nurse cell seems to almost completely surround many thymocytes and nurse them. These nurse cells play a role in T cell maturation. They act as scavengers, internalising and destroying cells that are destined to die. They are also involved in positive selection (section 9.2 and Figure 9.1).

The thymic epithelium, like the skin, undergoes progressive keratinisation and maturation. The subcapsular thymic epithelium contains less mature keratin than the specific agglomerations of epithelial cells found in the medulla — called *Hassal's corpuscles* — where the most mature keratin is found. Hassal's corpuscles are filled with whorls of keratinised epithelium, leukocytes, and cell debris. The epithelium of the thymus is postulated to undergo continuous turnover, like the epithelium of the skin. Hassal's corpuscles are thought to be the internal sites for the endocytosis, degradation, and carrying away of exfoliated epithelium by leukocytes.

Chemotactic peptides secreted by thymic epithelial cells and/or mononuclear cells are believed to initiate the immigration of circulating lymphocytes into the developing thymus. The thymus is populated with lymphoid cells in three discrete 24-36 hour waves separated by refractory periods. Temporally expressed receptors or periodic releases of chemotactic factors by thymic epithelial cells are believed to regulate this population of the thymus. Epithelial cells also have an important role in T lymphocyte maturation. It is now recognised that thymic epithelial cells have immunomodulating and neuroendocrine functions that affect the functioning of T cells inside and outside the thymus. They produce soluble thymic hormones, such as thymulin, thymosins, thymopoietin, and the thymic humoral factor, as well as cytokines such as IL-3 and IL-7, that regulate the proliferation, differentiation, and maturation of T cells. Granular deposits of IL-1 are found to be present in thymic cortical epithelial and mononuclear cells. Thymic epithelial cells express MHC molecules in three basic patterns. The subcapsular epithelial layer

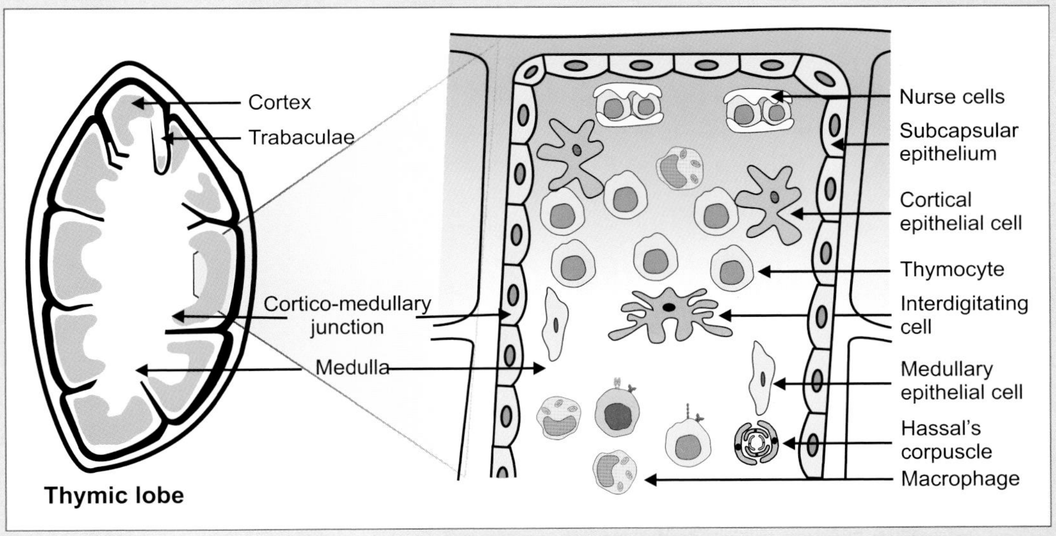

Figure 5.2 The thymus is a bilobed primary lymphoid organ where T lymphocytes mature. Each thymic lobe is organised into lobules that are separated by strands of connective tissue called trabaculae (left panel). The lobules consist of an outer cortex and an inner medulla, and they are lined by subcapsular thymic epithelial cells (right panel). Both the cortex and medulla have their own layer of epithelial cells, called the cortical epithelial cells and medullary epithelial cells, respectively. Some cortical epithelial cells seem to completely surround many thymocytes and are known as nurse cells. The cortex is densely packed with immature lymphocytes. It also contains a scattering of macrophages. Interdigitating cells found at the corticomedullary junction are thought to be important in thymic maturation. The medulla shows the presence of Hassal's corpuscles, stellate medullary epithelial cells, macrophages, and a relatively few thymocytes that survive the maturation process.

is MHC class II negative. The remaining cortex, including thymic nurse cells, is strongly MHC class II positive. The inner membranes of thymic nurse cells also strongly express MHC class I molecules. The medulla is strongly MHC class I and II positive. All cells in the thymus express MHC class I molecules, but the highest expression is on nonepithelial cells.

In the thymus, T cells learn to distinguish between self- and nonself antigens. This process is referred to as *thymic education*. T cells undergo both positive and negative selection — positive selection for those T cells capable of recognising antigens bound to self-MHC molecules and negative selection to ensure the deletion of self-reacting clones. Both these processes are discussed in chapter 9. The earliest cells to enter the thymus lodge in the subcapsular region of the cortex and form a distinctive population. They undergo a phase of intense proliferation and give rise to large, self-renewing lymphoblasts. The stream of newly produced thymic lymphocytes goes from the cortex to the medulla. As the thymic lymphocytes go deeper into the cortex, they are subjected to selection processes. In the medullary region, interdigitating cells derived from the bone marrow lie superimposed on the epithelial cells. Both these cells are rich in MHC class I and II molecules. They are thought to be important in the twin processes of self–nonself education and acquisition of MHC restriction (i.e., the capacity to recognise antigen only in the context of self-MHC molecules). The relative contribution of the epithelial cells *vis á vis* other stromal cells in these processes is not known. Although MHC class II expressing epithelial cells are thought to be important in thymic education, nonepithelial DCs in the thymus are thought to control the development of $CD4^+$ T_H cells — depletion of these DCs leads to a loss of peripheral T_H cells.

T cell precursors that arrive from the bone marrow to the thymus reside there for a week before entering a phase of intense proliferation. An estimated 5×10^7 thymocytes are generated daily in young adult mice. This number corresponds to about a quarter of the total thymocytes in the young thymus. About 95–98% of these cells undergo apoptosis in the thymus and never leave it, presumably because they fail the twin tests of self–nonself discrimination and autologous MHC restriction. Daily, only about 10^6 cells survive the selection processes to leave the thymus. Macrophages in both the cortex and the medullary regions are important in the digestion of those T lymphocytes that are dead or destined to die. Immature lymphocytes that enter the thymus at the cortex journey towards the medulla and receive maturation and survival signals *en route.* Only mature T cells coexpressing TcRs for nonself proteins and the appropriate MHC coreceptor (CD4 if the TcR recognises class II molecules, CD8 if it recognises class I) finally survive to reach the medulla. From there, these T cells eventually leave the thymus. Mature T cells that leave the thymus are

- ❐ capable of recognising antigen only in the context of self-MHC molecules,
- ❐ take part in cellular interactions,
- ❐ possess homing receptors, and
- ❐ are equipped with all the properties necessary to perform the immune functions of T lymphocytes.

The thymus is relatively large at birth. Once peripheral tissues are populated with diversified T cells, its major function is over. The thymus was believed to atrophy at puberty. It is now established that this is not entirely correct. Reduction in true thymic tissue starts as early as at one year of age and continues at the rate of 3% per year until middle age. The adult thymus is nonetheless functional, and it continues to produce precursors of T lymphocytes at a slow rate.

5.2.2 The Bone Marrow

Some lymphocytes were discovered to mature in the Bursa of Fabricus (an organ found only in birds and reptiles) in birds. These cells were termed *B lymphocytes* or *B cells*. In mammals, B cells mature in the foetal liver and adult bone marrow[7]. Besides being a site of B cell maturation, the bone marrow is also an important site of antibody production — unlike the thymus, the bone marrow acts as an important secondary lymphoid organ.

The bone marrow is physiologically the most important site of haematopoiesis. In adults, pleuripotent stem cells in the bone marrow give rise to platelets, erythrocytes, and lymphocytic cells (T, B, and NK cells). Bone marrow stem cells also give rise to myeloid lineage cells (neutrophils, basophils, eosinophils, macrophages, and DCs) (Figure 5.3). The aggregate volume and weight of the bone marrow surpasses that of the liver. The bone marrow has a highly organised anatomical structure, comprising vascular and nonvascular components. It is divided into wedge-shaped haematopoietic compartments filled with proliferating and differentiating blood cells in connective tissue matrices bordered by venous sinuses. It has a closed circulation in which arterioles flow out into venous sinuses, finally emptying into a large central sinus that is connected with the efferent (*efferent*: conducting outwards) venous system. Haematopoietic cells move centripetally as they differentiate and mature from the bone marrow periphery (i.e., near the surrounding bone) inward towards extravascular tissue spaces. Therefore, cells near the bone are immature, whereas those near the central sinus are mature. As with the thymus, bone marrow stromal cells are essential for the regulation of haematopoietic cell development. The microenvironment of the bone marrow consists of a unique endothelium and connective tissue stroma combined with

[7] A fortuitous occurrence; otherwise, we would have to come up with yet another name for these lymphocytes ☺. That being said, the bone marrow does not perform the function of a primary lymphoid organ in all mammals. In ungulates, such as cattle and sheep, B cell maturation occurs in the Peyer's patches.

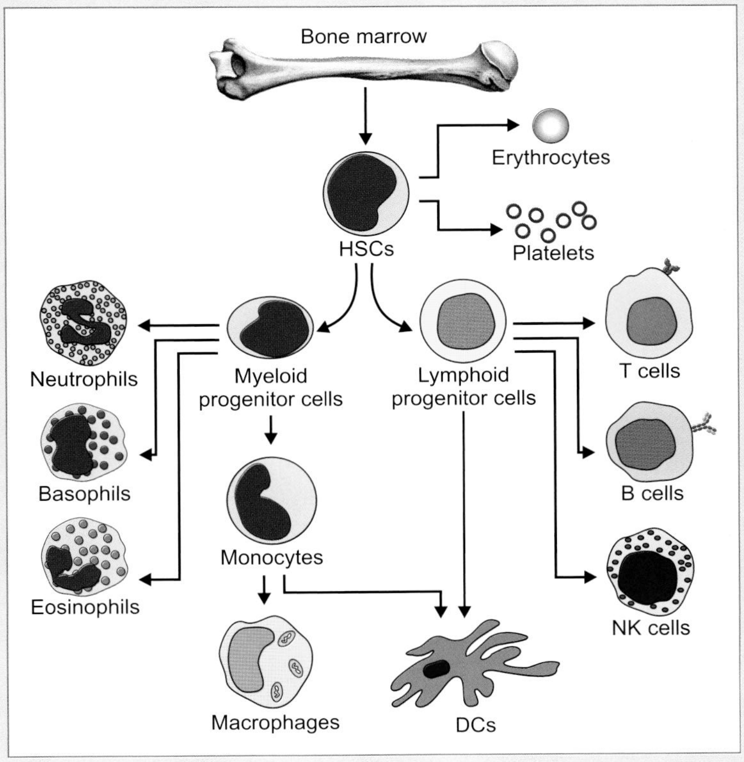

Figure 5.3 The bone marrow is the most important site of haematopoiesis. Pleuripotent haematopoietic stem cells (HSCs) in the bone marrow give rise to erythrocytes, platelets, lymphoid progenitor cells, and myeloid progenitor cells. Lymphoid progenitor cells further differentiate to T lymphocytes, B lymphocytes, and NK cells. Lymphoid progenitor cells also give rise to plasmacytoid DCs. Myeloid progenitor cells give rise to the myeloid lineage cells (neutrophils, eosinophils, and basophils). They also give rise to monocytes that can further differentiate to macrophages or DCs, depending upon their exposure to different cytokine milieus.

The Thymus

- It is a bilobed organ which is the site of T lymphocyte proliferation and maturation.
- In the thymus, T cells acquire appropriate antigen and homing receptors.
- T cells undergo positive and negative selection in the thymus.
  - Positive selection ensures the recognition of antigen bound to self-MHC molecules.
  - Negative selection ensures the deletion of self-reacting clones.
- The thymus consists of two regions — the cortex and medulla — each having its own epithelial layer of cells.
- The epithelial cells have several functions.
  - Chemotaxins secreted by thymic epithelial cells are responsible for the movement of lymphocytes to the developing thymus.
  - The epithelial cells have immunomodulatory and neuroendocrine functions.
  - Hormones, cytokines, and other signals emanating from thymic epithelial cells are responsible for T cell development and maturation.

Bone Marrow

- It is the most important site of haematopoiesis.
- It is a site of B cell maturation and selection.
- Long-lived plasma cells are lodged here; it is the major site of antibody production.

locally deposited cytokines that regulate haematopoietic stem cell compartmentalisation, proliferation, and differentiation. The signals and factors involved in this development are not yet fully characterised. It is known that the stroma assists in haematopoiesis through the glycosaminoglycan-rich extracellular matrix that binds and distributes growth factors, such as GM-CSF. Stromal cells also secrete IL-7, a growth factor essential for lymphocyte development.

Emerging lymphocytes are subjected to various influences in the microenvironment of the bone marrow to become mature B cells capable of Ig production. Like T cells, B cells are also subjected to a selection process that eliminates self-reactive clones. Those pre-B cells that do not form functional Ig molecules or that possess Ig molecules that recognise and bind self-antigens undergo apoptosis in the bone marrow. B cells that survive this selection process leave the bone marrow through efferent blood vessels.

5.3 Secondary Lymphoid Organs

Lymphocytes that have matured in primary organs are lodged in the secondary or peripheral lymphoepithelial organs. These organs create an environment in which lymphocytes can interact with each other and the antigen**. They also act as the centre through which the immune response**, **once generated**, **is disseminated through the body**. Secondary lymphoid organs are sites of further maturation and differentiation of lymphocytes. Peripheral lymphoid organs have a typical structure, with defined T and B cell compartments. The peripheral lymphoid organs and their characteristics are as follows.

- **Lymph nodes** are distributed throughout the body. These are encapsulated organs found at the junctions of lymphatic vessels. They act as filters for the lymph and trap antigens and cells containing antigens that flow into them *via* afferent (*afferent*: conducting inwards) lymphatics. They also provide sites for the activation of naïve lymphocytes and clonal expansion of activated lymphocytes.

- The **spleen** acts as a filter that traps bloodborne organisms or antigens. Like the lymph nodes, the spleen is a major site of the activation and clonal expansion of activated lymphocytes.
- **Epithelium-associated lymphoid tissue** comprises nonencapsulated lymphoid tissue found in a variety of organs. These tissues can be broadly divided into two:
  - **M**ucosa-**A**ssociated **L**ymphoid **T**issue (**MALT**), found in submucosal areas of the gastrointestinal, respiratory, and urinogenital tracts, such as tonsils, adenoids, and the lining of the bronchi, and
  - **S**kin-**A**ssociated **L**ymphoid **T**issue (**SALT**).

5.3.1 Lymph Nodes

Human lymph nodes are encapsulated, kidney-shaped structures between 1-25 mm in diameter. They are found at the junctions of lymphatic vessels (Figure 5.1). The capsule covering the lymph node is a collagenous structure that penetrates it. The radial trabeculae formed by the capsule and reticulin fibres form the supporting structure for various cellular components within a lymph node. Just beneath the capsule is the subcapsular sinus lined with phagocytic cells. Lymph passes from the surrounding capsule into this sinus by way of afferent lymphatic vessels. The lymph leaves the node by the efferent lymphatic vessels (Figure 5.4). Blood vessels enter and leave the node through the hilus. Lymphocytes travel to lymph nodes through the blood stream and enter it across the HEV; HEV is an important site of APC-lymphocyte interaction and antigen-dependent activation. The lymph nodes are fed by two vascular systems:

- the lymphatic system, which delivers antigens and antigen-transporting cells from peripheral tissues to the nodes and returns fluid and cells to circulation, and
- the blood vasculature, which brings circulating lymphocytes into the system.

Histologically, a lymph node can be divided into two zones — the cortex and the medulla (Figure 5.4). The cortex consists of an outer cortex and inner paracortex. The outer cortex lies just under the subcapsular sinus and consists of a macrophage-rich zone and B cell follicles. It is often referred to as the *B cell area*. The cells of the cortex are localised to discrete follicles. Around each follicle is a condensation of reticular cells. These are termed *primary follicles*. Primary follicles are present in lymph nodes as early as the second trimester of human foetal life. They consist of a network of FDCs (**F**ollicular **D**endritic **C**ells) and recirculating small B lymphocytes. Primary follicles are found in the spleens of germ-free, as well as normal, animals. This implies that they are formed independent of pathogenic challenge. They seem to provide the microenvironment that is essential for B cell survival, although the nature of the signals from the primary follicles that allow for this survival is not clear. Naïve B cells emerging from the bone marrow have to lodge in the primary follicles briefly before entering circulation. Failure to enter the primary follicles results in B cell death within days. Primary follicles develop into secondary follicles upon antigenic stimulation[8]. These secondary follicles contain many FDCs and macrophages, as well as a fine network of interdigitating cells that are rich in MHC class II molecules. **B cells that have encountered antigen in the lymph nodes**

[8] Antigenic stimulation also increases extravasation of lymphocytes into the lymph nodes near the site of stimulation. With the proliferation of lymphocytes, this gives rise to what is popularly called *swollen glands*.

migrate from the primary to the secondary follicles and differentiate to plasmablasts and memory cells. **The plasmablasts give rise to antibody-secreting plasma cells**. These plasma cells then migrate to the medullary cords, from where a majority leave for the bone marrow. Unstimulated B cells migrate through the primary follicle and leave the lymph node by means of efferent lymphatics.

The paracortex, consisting mostly of T cells, lies interior to the outer cortex. It is a high-traffic zone where migrant or recirculating T and B lymphocytes enter from the blood. Lymphocytes in the paracortex are directed towards specific B or T cell microenvironments by fibroblastic reticular cell corridors, where the lymphocytes encounter APCs. The paracortex is populated with T cells and many APCs — the interdigitating cells — rich in MHC class II molecules. If T lymphocytes do not get signals through the engagement of their TcRs and costimulatory molecules, they crawl out of the lymph node through the lymphatic channels. For a naïve lymphocyte that does not encounter its cognate antigen, such a journey is thought to take less than 24 hours.

The medulla lies to the inner side of the paracortex and consists of medullary strands separated by medullary sinuses and interconnected medullary cords. Scavenger phagocytic cells are arranged along the medullary strands. During the passage of lymph across lymph nodes, phagocytic cells trap any particulate antigens that may be present. Antigens that escape phagocytosis within one lymph node face phagocytes in other lymph nodes which the lymph passes through before entering major efferent collecting ducts. Antigens that succeed in eluding lymph node entrapment are ultimately captured by blood monocytes or macrophages in the spleen, liver, or bone marrow.

Challenge by a TD antigen results in the appearance of germinal centres consisting of rapidly multiplying lymphoblasts in the secondary follicles. These germinal centres are discrete lymphoid compartments where B cells (that are stimulated by the twin stimuli of antigen and T cell help) divide, switch the isotype of Ig expressed, and differentiate. B cell follicles are found in all peripheral lymphatic

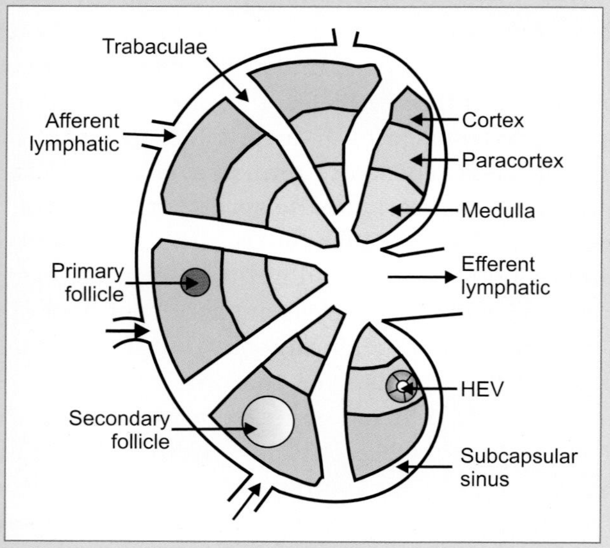

Figure 5.4 Lymph nodes are encapsulated structures that provide sites for the activation and clonal expansion of lymphocytes. Lymph passes from the surrounding capsule into the subcapsular sinus through the afferent lymphatic vessels. It leaves the node by way of the efferent lymphatic vessels. Lymphocytes travel to the lymph nodes through the blood stream and enter it across the HEVs. The lymph node can be divided into three regions — the cortex, the paracortex, and the medulla. The cortex lies beneath the subcapsular sinus and is rich in B lymphocytes (B cell area; depicted in blue colour). The cells of the cortex are localised to discrete follicles termed the primary follicles. They provide the microenvironment essential for B cell survival. The cortex also shows the presence of secondary follicles. These are formed upon antigenic stimulation and are the sites of antigen-stimulated B cell proliferation and differentiation. The paracortex lies interior to the outer cortex and is the T cell area (depicted in pink colour). The medulla lies to the inner side of the paracortex and is rich in scavenger macrophages.

tissues, including the foci of chronic inflammation. A germinal centre develops inside a follicle as proliferating B cells displace cortical reticular fibres into a basket-like enclosure that separates central lymphoblasts from the peripheral mantle of small B cells. Lymphoblasts, FDCs, and macrophages reside in this enclosure. Proliferating B cells in germinal centres have a defined nuclear shape (handy for identification). The sizes and numbers of germinal centres are related to the intensity of antigenic stimulation. Germinal centres display a well-defined architecture consisting of a dark zone, light zone, and mantle (Figure 8.2). Germinal centres last around three weeks after antigen administration and are the sites of antigen-dependent B cell proliferation, selection and differentiation, and affinity maturation (see below).

FDCs in germinal centres bind and retain unchanged antigen for long periods of time. During the process of antigen-dependent proliferation, B cells undergo extensive mutation in the genetic region that codes for the antigen-combining site of the Ig molecule. As a result, they give rise to clones of cells that express Igs with an altered antigen-combining site. Antigen-based selection allows the survival of B cells with a better fitting antigen-combining region. Antibodies produced by such cells react much more strongly with antigen, and antibodies produced later in the immune response have increased affinity for the antigen. This phenomenon is termed *affinity maturation*. FDCs are thought to help in the selection of clones that have increased affinity for the antigen. Macrophages in the follicles phagocytose those newly emerging B cells that are destined to die because of reduced complementarity of fit (and hence, lowered affinity).

5.3.2 The Spleen

A fist-sized organ that lies at the upper left of the abdomen, behind the stomach and close to the diaphragm, the spleen contains up to 25% of the body's mature lymphocytes. It is also the body's largest blood filter, instrumental in retrieving iron from old erythrocytes and in removing pathogens and cellular debris from the blood. Innate and adaptive immune systems are uniquely intertwined in this organ. The spleen is surrounded by a dense fibrous collagenous capsule with muscular trabeculae that penetrate the organ and subdivide the spleen into lobules. As with other encapsulated lymphoid organs, these intrusions, along with the reticular framework, support the variety of cells found in the spleen. The spleen consists of three types of tissues.

- **The red pulp** makes up the bulk of the spleen. It contains erythrocyte-rich blood in the cords of the reticulum and contains erythrocytes, lymphocytes, macrophages, granulocytes, and plasma cells. The red pulp is the site of blood

> The immune system's response to an antigenic challenge provides a good lesson in business management. Don't believe us? Read this description of how 'BEANS factory', a provider of ecommerce solutions, deals with customer requirements.
>
> "Our dynamic, highly responsive and flexible organisational structure is centred on a customer-oriented, market, and innovation-driven focus. Our strategic business units are structured such that different teams of specialists with the necessary knowledge and skills are assembled to address specific market, project, or client needs. These project teams are created, managed, and then dissolved over the project cycle." (http://www.beansfactory.com, accessed on July 10, 2004)
>
> We rest our case.

filtration and erythrocyte disposal. Aged or damaged erythrocytes are removed from circulation in the red pulp. The red pulp is also where plasma cells and plasmablasts lodge after differentiation of B cells in the white pulp.

- **The white pulp** is formed of cylindrical collections of lymphocytes around the arteries. It represents the organised lymphoid compartment in which regulated activation, maturation, and differentiation of antigen-dependent B and T cells occurs.
- **A border region called the Marginal Zone** (MZ) separates the red and white pulps. A principal function of the MZ is antigen trapping, and it seems to be important in generating rapid humoral responses to bloodborne antigens. The MZ contains a heterogeneous assortment of mononuclear cells with specialised functions. It is home to a subset of B cells called MZ B cells (section 8.3) that respond to predominantly TI type 2 antigens. Macrophages and DCs present in the MZ also appear to be important in MZ B cell generation, maintenance, and functioning. The spleen is structured such that most of the blood which flows to it passes through the MZ and directly along the white pulp, leading to efficient monitoring of the blood by the immune system.

The spleen, unlike the lymph nodes, has a single vascular supply. Immune cells and antigens enter the tissue through blood flowing in from the splenic artery, which is the spleen's only point of entry. The splenic artery branches several times before entering the splenic hilus. After branching into trabecular arteries, it further branches into central arterioles that penetrate the white-pulp nodules. Ultimately,

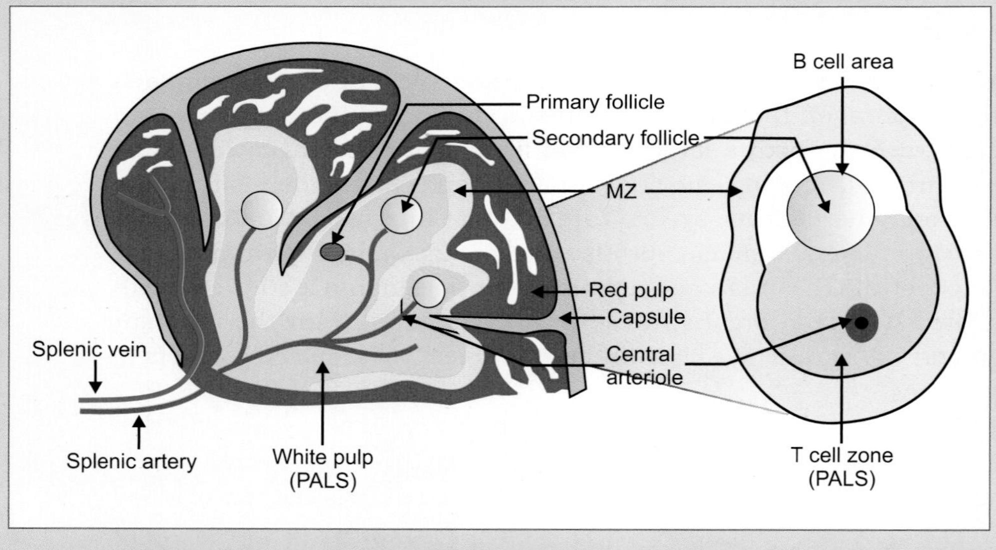

Figure 5.5 The white pulp of the spleen is the centre through which immune responses to bloodborne antigens are initiated and disseminated. A schematic cross-section of the spleen is shown in the left panel. The right panel is an enlargement of a small section of the white pulp. The bulk of the spleen consists of the red pulp. Blood carrying lymphocytes enter the spleen at the hilus through the splenic artery. After branching into trabecular arteries, the splenic artery further branches into central arterioles. Cells (and antigens) then pass into a marginal sinus and empty into the trabecular vein that in turn empties into the splenic vein. Cylindrical collections of lymphocytes around the trabecular arterioles constitute the white pulp. A border region known as the MZ separates the red and white pulps. The bulk of the lymphoid tissue of the white pulp is arranged around the central arterioles and consists of a coaxial layer rich in T cells, called the PALS, flanked by the B cell area containing B2 B cells. Primary and secondary follicles are found in this area. The MZ that separates the white and red pulps contains a special subset of B cells that is similar in characteristics to B1 B cells.

these vessels terminate in small arterioles which empty into the reticulum of the red-pulp cords or white-pulp sinuses. Majority of the lymphoid tissue of the white pulp is arranged in sheaths around the central arterioles which are called the **P**eri**a**rteriolar **L**ymphoid **S**heath (PALS). The organisation of the white pulp closely resembles that of the lymph nodes. It consists of T and B cell compartments arranged around branching arterial vessels. The PALS consists of a coaxial layer rich in T cells with abundant interdigitating DCs that are thought to act as APCs early in the immune response. In the white pulp, B cells are organised into two compartments. The first consists of predominantly naïve cells and includes cells from the MZ that lies adjacent to the marginal sinus. The second compartment is composed of cells associated with follicles which are primary follicles in their resting states (Figure 5.5). Secondary follicles develop upon antigenic stimulation. Both primary and secondary follicles are similar in composition and structure to those found in the lymph nodes (see section 5.3.1).

Lymphocyte traffic in the spleen moves across the blood:tissue interface into sites of antigen presentation in the MZ and PALS. **The spleen is the primary site for the initiation of immune responses to bloodborne antigens and pathogens. It is also a partner in every immune response in the body**. After 48-100 hours of antigen priming, antigen-laden mononuclear cells and lymphoblasts are released into efferent lymph from other lymphatic tissues. These cells lodge in the spleen and set up satellite zones of T and B cell proliferation. As mentioned, during active immune responses, lymphoblastic B cells committed to plasma cell differentiation lodge in red-pulp cords and sinuses, where they mature and secrete antibodies.

5.3.3 Epithelium-associated Lymphoid Tissue

Nonencapsulated lymphoid tissue is found at two major sites in the body — tissues associated with the mucosa or those associated with the skin.

i. MALT: Mucous membranes that cover the digestive, respiratory, and urogenital

The Lymph Nodes

- These are encapsulated bean-shaped organs found at the junctions of lymphatic vessels.
- Afferent and efferent lymphatics maintain a continuous, active flow of lymphocytes to these organs.
- Lymph nodes filter and trap antigens or antigen-containing cells in lymph.
- They are the sites of the activation and clonal expansion of antigen-activated B and T lymphocytes.
- They consist of a cortex (B cell area), paracortex, and medulla (together, T cell area).
  - Cells of the cortex are arranged in discrete areas called primary follicles.
  - Antigen stimulation with TD antigen gives rise to secondary follicles — the site of memory B development, isotype switching, and affinity maturation.

The Spleen

- It is an encapsulated organ just behind the stomach. It consists of a red pulp and white pulp separated by a marginal zone.
  - Erythrocytes are disposed off in the red pulp.
  - The white pulp represents the organised lymphoid compartment. It contains PALS, which has primary and secondary germinal centres.
  - The marginal zone is involved in antigen trapping. MZ B cells respond to TI antigens.

tracts are spread over an area of approximately 400 m^2. They are the major portals of immunogen entry. Being the primary sites for immunogen encounter, mucosal epithelia need to be well defended. The number of plasma cells in MALT exceeds that in the lymph nodes, spleen, and bone marrow combined. **MALT can be either well organised, as in the appendix and Peyer's patches, or barely organised, as in the lamina propria of the intestinal epithelium**. Organised lymphoid tissues of the respiratory and gastrointestinal tracts contain the largest numbers of lymphocytes and are most fully characterised. Afferent lymphatics carry the immunogens that gain entry at the mucosal surface to draining lymph nodes, where specialised APCs present them to the immune system. Some important MALTs include:

- Nasopharyngeal-associated lymphoid tissue which includes lingual tonsils at the base of the tongue, palatine tonsils at the sides of the back of the mouth, tubal tonsils at the pharyngeal openings of the eustachian tubes, and nasopharyngeal tonsils (adenoids) in the roof of the nasopharynx.
- Bronchus-associated lymphoid tissue, which in humans, is normally found only in the lungs of children and adolescents and may be found in adult lungs in certain pathological states.

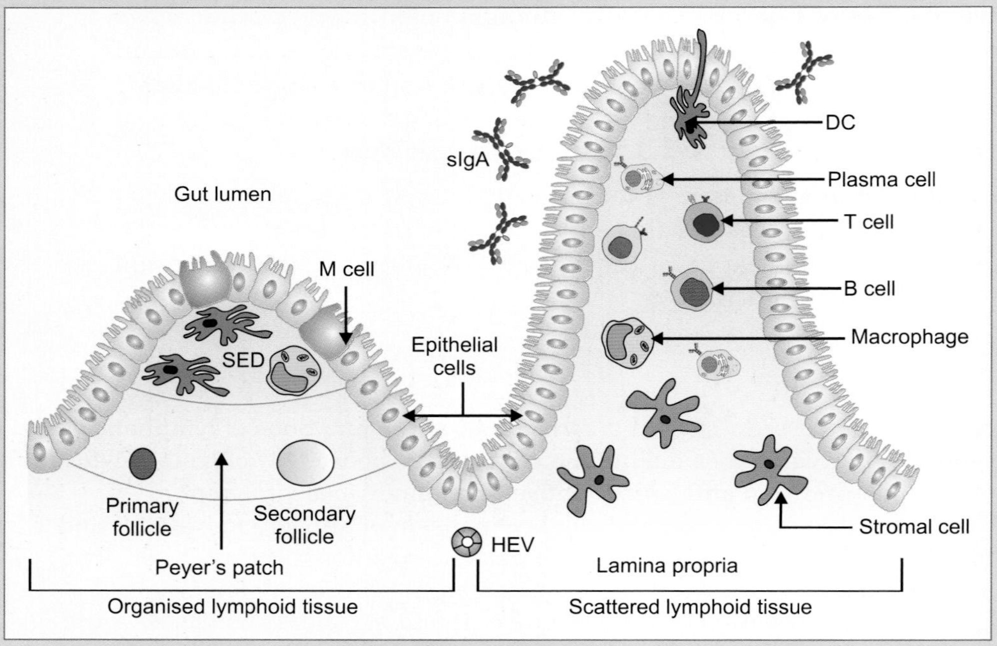

Figure 5.6 GALT is the collective name given to multiple types of lymphoid tissues found at various sites in the gastrointestinal tract. The lamina propria lies just under the intestinal epithelium and contains loose clusters of lymphoid cells consisting of B cells, plasma cells, T cells, and macrophages interspersed with stromal cells. The diffuse tissue of the lamina propria is rich in IgA-secreting plasma cells. These are the effector sites of the immune response generated in the organised lymphoid tissue such as Peyer's patches, mesenteric lymph nodes, and lymphoid follicles. The submucosal layer of the lamina propria contains aggregations of organised lymphoid tissue, called Peyer's patches. The Peyer's patches have a unique SED specialised to sample environmental antigens. M cells are interspersed in the cuboidal absorptive epithelial cells of the dome. These M cells, along with DCs, are important in sampling antigens in the gut lumen. Located beneath the SED are numerous primary and secondary follicles surrounded by a mantle of B cells. The interfollicular area is rich in T cells. The B cells undergo isotype switching and affinity maturation in the secondary follicles.

- **G**ut-**A**ssociated **L**ymphoid **T**issue (GALT) which includes the lymphoid tissue of the epithelium of the gastrointestinal tract.The outer mucosal epithelium contains dispersed **I**ntra**e**pithelial **L**ymphocytes (IELs) with unknown functions. The lamina propria under the epithelium has large number of B cells, T cells, plasma cells, and macrophages, all arranged in follicles. The submucosal layer contains the Peyer's patches, which are nodules of 30-40 lymphoid follicles.

The mucosal immune system is the body's most dispersed, diverse, and complicated lymphocytic system. The microenvironment in the mucosal barriers has a marked influence on the immune response. Like the thymus, the mucosal immune system plays a role in generating antigen-reactive lymphoid cells that mature into effector cells. The mucosal lymphatic tissues are also responsible for inducing tolerance to antigens that are commonly experienced in the enteric canal. Mucosal tolerance is manifested by antigen-specific suppression of delayed cutaneous hypersensitivity (see chapter 16) and reduced IgG antibody formation. Thus, MALT has to achieve a fine balance between opposing immune functions — amplifying the development of committed B cells to IgA-secreting plasma cells in response to environmental antigens and inducing systemic tolerance to others.

A great deal remains to be learnt about the mucosal immune system, but GALT is its best studied component. The intestinal epithelium has a dual role — it has to allow nutrient absorption while disallowing pathogen entry. The cells of the epithelium have *tight junctions*, which are junctions that do not allow the passage of molecules of more than two KD. The outer mucosal epithelial layer contains IELs. IELs are a large heterogeneous population of immune cells[9] in the intestinal epithelium consisting mostly of T lymphocytes. One type expressing the αβ TCRs

[9] As lymphocytes are dispersed within the epithelium and not in epithelial cells, *intra-epithelial* lymphocytes could be misleading.

Well Throat: Tonsils

The tonsils are lymphoepithelial structures found at the openings of respiratory and digestive tracts. The Swiss scientist Waldeyer first noted the importance of this ring of lymphoid tissues in the pharynx (hence the name, *Waldeyer's ring*). The ring comprises the tonsils and the subepithelial lymphoid tissues in the mucosa of the pharynx. Tonsils are secondary lymphoid organs containing aggregations of lymphoid cells located in the lamina propria of the pharyngeal wall. The subepithelial lymphoid compartment of the tonsils consists of secondary follicles (B cell areas) surrounded by interfollicular T cell areas. The epithelium contains cells ultrastructurally similar to M cells which are thought to act as a portal to antigens. The mucosa of the pharynx has a complex secretory immune system. B cells stimulated by the antigen in the mucosa of the pharynx migrate to the nearest lymph nodes for differentiation to Ig-producing plasma cells. Most Igs are IgA polymers that are exported by serous salivary cells through the action of a special protein complex that is formed between the J chain of the Ig molecule and a receptor protein, pIgR (section 10.7). Key to successful protection of the mucosa is the ability of IgA polymers to prevent the adherence of bacteria and viruses to the pharyngeal epithelium.

Tonsils had been suspected of causing or exacerbating various diseases (rheumatoid arthritis, myocarditis, glomerulonephritis, and gout, to name just a few), and the Greeks were reported to have performed tonsillectomies as early as 3000 BC. Tonsillectomies have now declined thanks to an appreciation of the tonsils' immune function. Nonetheless, viruses (e.g., Epstein-Barr and measles viruses), as well as post-viral bacterial infections, may use tonsils as routes of entry. Recent evidence suggests that the tonsils could also serve as sites of HIV entry and replication.

'M' for the Gut: M cells

M cells are specialized cells found in the follicle-associated epithelium of intestinal Peyer's patches, in the appendix, and in mucosal-associated lymphoid tissue of the lungs. They originate from epithelial cell precursors that are induced to differentiate into actively pinocytosing cells containing few lysosomes and low levels of phosphatases. In the gut, M cells transport substances from the lumen of the intestine, across the epithelial barrier and to underlying immune cells where processing and initiation of immune responses occur. M cells serve as a channel for the entry of molecules and particulate matter — including viruses and bacteria — into the underlying mucosal follicular region. In the case of bacteria, adhesiveness seems to be a major determinant of transport. M cells do not take up normal intestinal flora as efficiently as intestinal pathogens; commensal organisms seem to be less adhesive to the M cells than noncommensals. Once taken up by M cells, bacteria are transferred to follicular macrophages. Not having the capacity to resist lysosomal action in these macrophages, the transported bacteria are killed after transfer, and they prime the system to elicit a noninflammatory immune response. There is also evidence to suggest that transcytosis of bacteria results in LPS-induced IL-1 release, aiding the proliferation of lymphocytes. Uptake by M cells provides an easy entrance into host tissues and this property is exploited by pathogens such as, poliovirus, *Salmonella typhimurium*, and *Shigella flexneri*. These pathogens subvert the immune mechanisms and survive and proliferate in the mucosal tissue. Given the unique features of M cells and their specialized ability to transcytose numerous microorganisms and particulates, M cells are now being investigated for targeted delivery of antigens.

and the CD4 or CD8 coreceptors seem to have originated from the thymus. The other group expresses CD8$\alpha\alpha$ homodimers (instead of the CD8$\alpha\beta$ heterodimers found on conventional T cells) and may express $\alpha\beta$ or $\gamma\delta$ TcRs. Some IELs are CD4$^-$, CD8$^-$ cells that express $\alpha\beta$ TcRs and NK cell markers. These NKT lymphocytes are thought to be of extrathymic origin, as they are found in neonatally thymectomised or nude mice. Their precise ontogeny (i.e., origin and development) is under investigation. IL-7, secreted by epithelial enterocytes, is thought to be important for their development. Recent work in the murine model suggests that some NKT cells may develop in the crypt of the lamina propria.

The lamina propria lies just under the epithelial layer and contains loose clusters of lymphoid cells consisting of B cells, plasma cells, TH cells, and macrophages. Progenitors of T cells are also found in the crypts of the lamina propria of both large and small intestinal villi. The submucosal layer of the lamina propria contains aggregations of lymphoid follicles known as *Peyer's patches* (Figure 5.6). These patches are found primarily in the distal ileum of the small intestine. Peyer's patches have a unique **S**ub**e**pithelial **D**ome (SED) that is specialised to sample environmental antigens. The dome epithelium covering is composed of cuboidal absorptive epithelial cells interrupted by delicate membranous cells that have luminal microfolds on their surfaces. These cells, called *M cells*[10], endocytose and transport various materials without lysosomal degradation[11]. Peyer's patches contain lymphoid compartments that are analogous to the deep cortex and follicles of lymph nodes. Each Peyer's patch contains multiple individual B cell follicles located beneath the dome epithelium and separated by interfollicular areas that are rich in T cells. A mantle of B cells surrounds each follicular germinal centre. Antigens are deposited into small lymphocytes, small mononuclear phagocytes, or DCs immediately beneath the M cells and above the B cell mantle of germinal centres. Minute quantities of

[10] The *M* in the name is for membranous or microfold — not for James Bond's boss ☺.

[11] M cells were originally thought to be the only cells involved in transport of antigens from the intestinal lumen. Recently, DCs have been shown to be involved in antigen sampling, too. They extrude in and out of the epithelial tight junctions to capture and transport a sampling of antigens present in the intestinal lumen.

intact antigens and products of digestion are transported to the lamina propria by ordinary absorptive epithelial cells found anywhere in the small bowel. Peyer's patches facilitate the generation of an immune response within the mucosa. B cell precursors and memory cells are stimulated by antigen in Peyer's patches. These then migrate to the mesenteric lymph nodes, where they undergo further proliferation and maturation. Mature lymphocytes then enter systemic circulation, migrate throughout the MALT, and finally travel to the gut by passing through HEVs, maximising their chances of antigen encounter. Current research points to the existence of two types of Peyer's patches, which behave like primary and secondary lymphoid organs in humans. **Peyer's patches are important sites of IgA antibody secretion**.

ii. SALT: The skin, being exposed to the external environment, is under continuous assault by microorganisms. The impervious epidermis is important in keeping most invaders out. Its own lymphoid system also helps control breaching of the outer layer. Keratinocytes, specialised cells found in the epidermis, secrete cytokines that can induce inflammatory responses. These cells also present antigens to T_H cells. Langerhans cells, a type of DC found interspersed in this epidermal matrix, are perhaps the most important component of SALT. They are specialised APCs that reside in the epidermis and constantly monitor the epidermal microenvironment by internalising microorganisms (or immunogens) that penetrate the epidermal layer. Following microbial uptake, Langerhans cells migrate to regional lymph nodes. Here, they differentiate into interdigitating DCs that are potent activators of naïve T cells and initiate an immune response. The epidermis also contains intraepidermal lymphocytes that are similar to the IELs of MALT. The dermal layer underlying the epidermis also contains scattered DCs, T cells, macrophages, and mast cells. Thus, immune responses initiated in the SALT are effective in maintaining cutaneous immunity and are normally sufficient to prevent the systemic entry of invaders.

Epithelium-associated Lymphoid Tissue

- It comprises nonencapsulated lymphoid tissue in skin and mucosa.
- MALT — consisting of nasal-associated, and gut-associated lymphoid tissue — is the most diverse and dispersed of the lymphocytic systems.
  - IELs in GALT consist of αβ T cells, γδ T cells, and NKT cells.
  - IELs are thought to be important in innate immune responses.
  - In humans, GALT acts as a primary lymphoid organ — crypts in the lamina propria are the site of T cell lymphopoiesis.
  - Peyer's patches, found mainly in the distal ileum of the small intestine, have a unique epithelium. M cells and DCs in the epithelium sample environmental antigens.
  - Peyer's patches also contain primary and secondary follicles; B cells in these follicles are responsible for antigen-specific mucosal defence and produce mainly IgA.
- SALT is involved in the antigen-specific defence of external skin surfaces.
  - Keratinocytes and Langerhans cells are specialised APCs of the epidermal layer; intra-epidermal layers contain B and T cells.
  - Langerhans cells monitor the epidermal environment for antigens.
  - The dermal layer also contains DCs, macrophages, and mast cells capable of antigen presentation.

Lessons in Wanting: Immunodeficient Mouse Models

Single-gene mouse mutants have proven to be useful experimental tools in immunological research. Two key single-gene naturally occurring mutations are the **nu**de (*nu*) and **S**evere **C**ombined **I**mmuno**d**eficiency (*SCID*) mutations.

- *nu* mutation: Mice having this mutation are called *nude mice* (no funny ideas, please) because the mutation results in hairlessness. Later, the thymus was found not to develop normally in homozygous nude mice, that is, these mice are athymic. As in mice whose thymus is removed neonatally, T cell areas of lymphoid organs are heavily depleted in nude mice. Although they lack T cells, nude mice have a normal complement of B cells. This mouse model is therefore a unique tool for the study of the role of the thymus in lymphocyte differentiation and B cell functioning. Reconstituting these mice with populations of cells derived from normal mice allows investigations into the role of selective immune cell populations. As the NK cells and macrophages of these mice are found to be more potent than those of normal mice, they can be used for research on NK cells as well. The nude mouse model is also the first animal model to be widely used in studying factors regulating tumour growth and metastases.
- *SCID* mutation: SCID mice lack a key enzyme (DNA-dependent Protein Kinase) needed for the development of a normal immune system (see chapter 6). Therefore, SCID mice lack both T and B cells. The SCID mouse model was the first known animal model for human SCID, a congenital syndrome that is usually fatal. As they lack a functional adaptive immune system, SCID mice are excellent models for studying the relationship between immune defects and lymphoid system cancers. SCID mice have great potential in reconstitution studies because they lack normal B and T cell compartments. The successful implantation of functional human foetal haematolymphoid organs (including the thymus and lymph node tissues) into SCID mice has led to the development of the SCID-hu chimeric model that can be used in cancer research, chemotherapy, and radiation therapy.

6 Molecules of Adaptive Immune Recognition: Lymphocyte Receptors

One way or another I'm gonna find ya
I'm gonna getcha getcha getcha getcha
One way or another I'm gonna win ya
I'm gonna getcha getcha getcha getcha
One way or another I'm gonna see ya
I'm gonna meetcha meetcha meetcha meetcha
One day, maybe next week
I'm gonna meetcha, I'm gonna meetcha, I'll meetcha
I will drive past your house
And if the lights are all down
I'll see who's around

One way or another I'm gonna find ya
I'm gonna getcha getcha getcha getcha

— Blondie, *One Way or Another*

Abbreviations

AID:	Activation-Induced cytosine Deaminase
AP-1:	Activating Protein-1
BLNK:	B cell Linker Protein
Btk:	Bruton's Tyrosine Kinase
C: region	Constant region
CDR:	Complementarity-Determining Region
CSR:	Class Switch Recombination
DAG:	Diacylglycerol
DNA: -PK	DNA-dependent Protein Kinase
ER:	Endoplasmic Reticulum
HMG: proteins	High-Mobility Group proteins
IP_3:	Inositol-(1,4,5)-trisphosphate
IκB:	Inhibitors of NFκB
LAT:	Linker of Activated T cells
CTLA: -4	Cytotoxic T Lymphocyte-associated Antigen-4
MAP: kinase	Mitogen-Activated Protein kinase
mIg:	membrane-bound Immunoglobulin
NFAT:	Nuclear Factor of Activated T cells
NFκB:	Nuclear Factor κB
NHEJ: pathway	Non-Homologous DNA End-Joining pathway
PI3: -kinase	Phosphatidyl-inositol-3-kinase
PIP_2:	Phosphatidyl-inositol-(4,5)-bisphosphate
PKC:	Protein Kinase C
PLC:	Phospholipase C
PTK:	Protein Tyrosine Kinase

6.1 Introduction

A *receptor* is a general term used for a molecule that receives signals from its ligand, that is, the molecule it binds. Receptors expressed on membranes generally have three domains:

- ❒ an extracellular domain that (specifically) binds the ligand,
- ❒ a transmembrane domain that spans the plasma membrane, and
- ❒ a cytoplasmic domain that participates in signal transduction; if the receptor has very short cytoplasmic domains (as happens with BcRs and TcRs), it is associated with accessory signal transducers.

Thus, receptors themselves, or molecules associated with them, link the cell's exterior to its interior. The binding of a ligand to a receptor is usually communicated to the cell's interior by a conformational change in the receptor. The most common result of this conformational change is either the opening of ion channels in the cell membrane or a change in the receptor's cytoplasmic domain. The opening of ion channels changes the intracellular concentration of ions such as Ca^{2+} and this signal produces an intracellular response. On the other hand, the conformational change of the receptor's cytoplasmic domain enables it to associate with and activate enzymes and/or signalling proteins (see *General Concepts in Signal Transduction* in section 6.2).

The recognition of an antigen by receptors causes the receptors to cluster, triggers a signalling cascade, and ultimately results in the transcription of multiple genes. Although both BcRs and TcRs are involved in antigen recognition, they recognise antigens in fundamentally different ways. Antibodies and BcRs recognise antigens in their native states. Whereas αβ TcRs recognise only fragments of the antigens loaded on self-MHC (or CD1) molecules, γδ TcRs recognise antigen directly, without processing and loading. This multiplicity in antigen recognition aids defence of the body.

- ❒ **Ig molecules, the main units of BcR complexes, serve two purposes in the immune system**. Circulating Igs help in the defence against microorganisms and their toxic products. Igs trigger a host of processes, such as phagocytosis and complement activation, which kill invading pathogens. They effectively neutralise toxins by helping in their removal or by blocking their activity. As part of the BcR complex, antibodies also help activate the process that results in the secretion of copies of themselves. These antibodies secreted by plasma cells remain in circulation even after the removal of the challenger, prolonging conferred protection.
- ❒ **TcRs help in the detection of microorganisms that can live or hide inside cells**. Such organisms take refuge within the host cell, use it as a means of transport, or replicate within the cell while evading the immune system. αβ T cells detect intracellular parasites by probing cell surfaces for pathogen-derived peptides loaded onto MHC (or CD1) molecules[1]. γδ TcRs recognise molecules of microbial origin as well as self-molecules expressed as a consequence of infection.

Both BcRs and TcRs are multiprotein complexes consisting of an antigen-recognition unit and associated proteins involved in signal transduction. The antigen recognising units have similar structures and belong to the Ig superfamily (see the sidetrack *Molecular Mafias*). However, the receptors differ greatly in terms of the epitopes they recognise and the cascade of events (and hence, the functions)

[1] Unless otherwise specified, this chapter describes processes and signalling pathways of peptide-specific αβ TcR.

they trigger. Clustering (also known as crosslinking) of receptors is required to generate the initial signal. For BcRs, this clustering is easily achieved by the binding of multivalent antigen to multiple BcRs on the B lymphocyte. A similar mechanism operates in γδ T cells as well. Such clustering is more complex for αβ TcRs. It is probable that multiple TcRs on the T cell engage multiple MHC:peptide/CD1:lipid complexes on APCs or target cells. The signal generated by the crosslinking of BcRs or TcRs is converted to a form recognised by the intracellular machinery, initiating a series of reactions that ultimately change gene expression in the nucleus. We describe the receptors, some of the downstream signalling pathways, and the genetic mechanisms involved in these receptors' assembly in this chapter. A cautionary note — numerous new terms are introduced here; it is important to understand the concepts without getting bogged by the terminology.

RAG:	Recombinase Activating Gene
RSS:	Recombination Signal Sequence
S: region	Switch region
SH2: domain	Src-Homology domain 2
TdT:	Terminal deoxynucleotidyl Transferase
V: region	Variable region
ZAP: -70	70 KD ζ-Associated Protein

6.2 B Cell Antigen Receptor

The immune system must recognise and deal with myriad antigens. Encoding and expressing a radically different protein for every possible antigen the system may or may not encounter, however, would be a waste of energy and DNA. In the face of this dilemma, evolutionary forces yielded an ingenious solution — antigen receptors having a highly variable structure at the antigen-binding end but an invariant structure at the cytoplasmic end (distal from the antigen-binding site). This allows the same basic structure to be used for binding different antigens. As a further refinement, the signalling function is bifurcated from the receptor function, allowing the use of pre-existing cellular machinery and making synthesis of new molecules unnecessary.

Antigen receptors and signalling molecules comprising BcR (and TcR) complexes are not randomly distributed in the plasma membrane. They are found clustered in special structures termed *lipid rafts*[2]. These are specialised lipid-rich areas, or microdomains, found in the outer leaflet of the plasma membrane. The precise nature and organisation of lipid rafts is unknown; they are enriched in sphingolipids, cholesterol, glycosylphosphatidylinositol-linked proteins, and acylated signalling molecules. Certain cellular proteins can associate with lipid rafts because of their structural features, but others are excluded from them. This selective transport of certain cytosolic proteins to lipid rafts while excluding others has promoted the idea that lipid rafts act as assembly points and platforms that facilitate interaction of particular signalling components. They are thought to facilitate immunoreceptor signalling by producing a physical environment rich in kinases, adaptor molecules, and intracellular effector molecules while excluding negative regulators such as CD22 (section 6.2.2).

[2] The plasma membrane was for long incorrectly viewed as a featureless ocean of lipids randomly dotted with islands of proteins. The fluid plasma membrane has distinct detergent insoluble regions or domains called lipid rafts. The plasma membrane is now viewed as dotted with lipid rafts containing specialised subsets of proteins to which new proteins can be recruited upon receptor activation.

6.2.1 Structure of the B Cell Antigen-Recognition Unit

The antigen-recognition unit of the BcR complex is a **m**embrane-bound **Ig** (mIg) molecule associated with numerous signal transduction accessory molecules. The antigen-binding portion of the mIg molecule provides specific epitope recognition; accessory molecules allow the membrane-anchored portion to associate with cytosolic enzymes needed for intracellular signalling. A detailed description of different types of Igs is given in chapter 10. In this section, only features important to antigen recognition are reviewed.

- The prototypical Ig molecule has a quaternary structure consisting of four polypeptide chains:

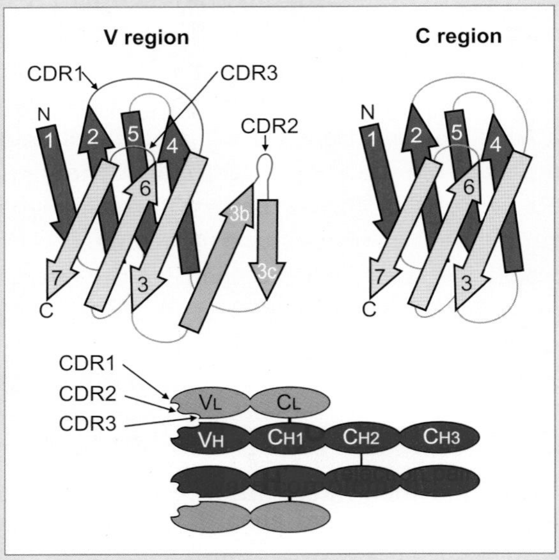

Figure 6.1 The prototypical Ig molecule consists of four polypeptide chains — two heavy (H) and two light (L) — and is bilaterally symmetrical. Each polypeptide chain consists of globular domains folded into antiparallel β-pleated sheets that are connected to the neighbouring globular domains by amino acid sequences that do not participate in the β-sheet structure (top panels). The H chains consist of four such domains, and the L chains consist of two domains. Globular domains of the adjacent L–H or H–H chains interact in the quaternary structure of the complex to form domains with discrete functions (bottom panel). The amino acid sequences in the first N-terminal domains of the H and L chains vary greatly and are called V regions (VH and VL respectively). The remaining domains have little sequence variation and are called C regions. The L chain has one constant domain (CL), and the H chain has three (CH1, CH2, CH3). The C domains consist of seven β-strands, packed in two sheets of four and three strands each (top right panel). The two sheets are joined by disulphide linkages (not shown here), giving a stable hydrophobic core. The structure of the V region is similar to that of the C region except that it has two additional strands (top left panel). Sequence variability is not uniformly distributed in the V region but is clustered in three distinct regions known as CDRs. Peptide loops that connect the β-strands have maximum variability. The less variable regions occur in the β-pleated sheets that provide the basic framework of the Ig-fold. The CDRs of both chains constitute the antigen-binding region of the antibody, and the typical Ig molecule has two antigen-binding sites.

  - two 50 KD H chains, of around 440 amino acids each, and
  - two 25 KD L chains of 220 amino acids each.

- The secondary structure of the Ig molecule is formed by the folding of polypeptide chains into antiparallel β-pleated sheets. Each polypeptide of the Ig molecule is folded into compact globular domains that are connected to neighbouring globular domains by amino acid sequences that do not participate in the β-pleated sheet structure. This kind of polypeptide folding is a universal structural feature found in a multitude of proteins, and it is called *the Ig-fold* or domain (see the sidetrack *Packed and Folded*).
- Ig H chains consist of four Ig domains; L chains consist of two domains each.
- Globular domains of the adjacent L–H or H–H chains interact in the quaternary structure of the complex to form discrete functional domains (Figure 6.1).
  - Globular domains at the amino-terminus of the H and L chain are noncovalently associated through an extensive hydrophobic interface. This association forms the antigen-binding pocket of the Ig molecule. Because the Ig molecule consists of two H and two L chains, it has two antigen-binding sites. The sequence of amino acids in these globular domains at the amino-terminus, for both the H and L chains, is highly variable. These domains are therefore termed *variable* domains or regions (VH and VL, respectively).
  - Globular domains at the COOH-terminus of the L chains are associated with the second globular domain of the H chain. The remaining two globular domains of the two H chains are held together by noncovalent hydrophobic interactions to yield a Y-shaped molecule. Apart from the VH and VL domains, the sequence of amino acids in the remaining H and L chains has limited variation. These domains are therefore termed the *constant* domains or regions (CH and CL).

Packed and Folded: The Ig Domain

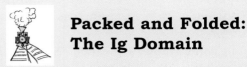

As the first protein structures were solved, proteins were established to have structurally different lobes. In 1973, Wetlaufer assigned the word *domain* to these compactly folded structures. Such domains form an important level in the hierarchical organisation of the three-dimensional structures of globular proteins, and they are important in protein folding and biological functions. They may be considered connected units that are independent in terms of their structure, function, and folding behaviour.

Each Ig domain consists of approximately 110 amino acids, and each fold consists of a sandwich of two β-pleated sheets. Each sheet consists of antiparallel β-strands connected by loops of varying lengths. The β-strands are stabilised by hydrogen bonds that connect the amino group of one strand with the carboxyl groups of the adjacent strand. These β-strands are characterised by alternating hydrophobic and hydrophilic amino acids whose side chains are perpendicular to the plane of the β-sheet (Figure 6.S1). The side chains of the hydrophobic amino acids are oriented inwards between the two β-pleated sheets. The side chains of the hydrophilic amino acids are pointed outward, away from the two opposing β-pleated sheets. The sandwich of the two β-pleated sheets is stabilised by hydrophobic interactions between the two β-pleated sheets and the presence of a disulfide bond, which forms a loop of about 60 amino acids.

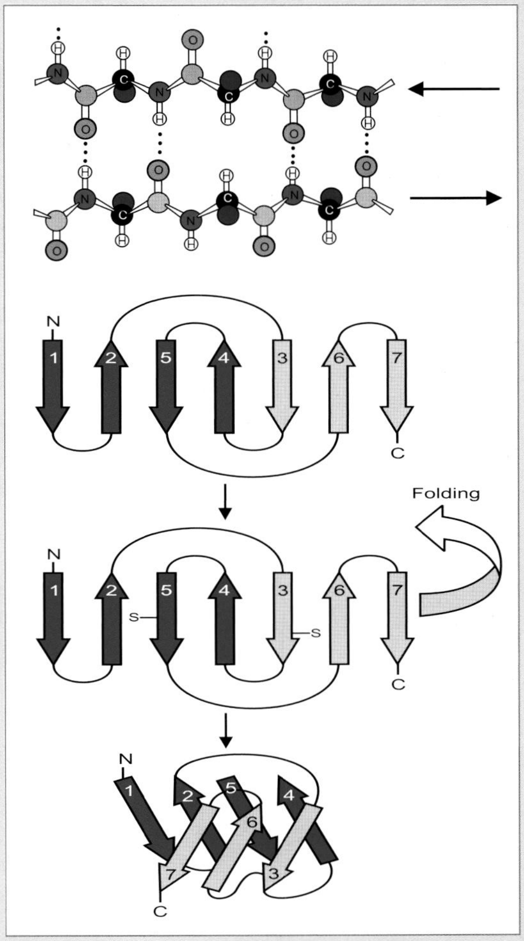

Figure 6.S1 The Ig domain consists of a sandwich of two β-pleated sheets stabilised by hydrophobic interactions and a disulphide bond.

- C domains have seven strands — four making one sheet and three making the other. The two sheets are held close together and joined by disulphide linkages emanating from strands 2 and 6, giving a stable structure and a hydrophobic core.
- The V region structure is slightly different from the C region structure in that it has two additional strands.
  - In the V_H and V_L domains, variability is not uniformly distributed; it is clustered in the three hypervariable regions or **C**omplementarity-**D**etermining **R**egions (**CDR**s; as they form the surface that matches the topology and physicochemical character of the antigen). The less variable regions in the β-pleated sheets are called *framework regions*. They provide the basic framework of the Ig- fold.
  - Each V region has three CDRs. The peptide loop that links the β-strands 2 and 3 constitutes CDR1 (amino acids 24-34). The loop connecting strands 3b and 3c represents CDR2 (amino acids 50-56), the loop linking strands 6 and 7 forms CDR3 (amino acids 89-97).

- CDRs are brought together to form a continuous surface that constitutes the antigen-binding site of the antibody, that is, amino acids in the CDR regions take part in antigen binding.

Each B cell expresses about 10^4-10^5 antigen receptors per cell. Naïve cells express either mIgD or mIgM molecules, or both. Memory cells express mIg molecules of a switched isotype (mIgG/mIgA/mIgE) alone or with mIgM. The mIg molecule is identical to secreted Ig except in two ways.

1. It has a stretch of amino acids at the –COOH termini of the H chains needed for anchoring the molecule to the cell membrane.
   - As the molecule is anchored in the cell membrane and traverses it, the stretch of amino acids consists of a hydrophobic region sandwiched between the hydrophilic portions.
   - The hydrophobic portion is the transmembrane segment of the mIg molecule. It forms a stretch of α-helix within the membrane. This transmembrane segment is the most evolutionarily conserved part of the mIgM H chain. It is identical in species as apart as mice and humans, suggesting a critical function.
   - The DNA that encodes this transmembrane part lies beyond the last CH domain on the C region gene. The differential transcription of this gene produces the two forms of Igs (section 8.6).
2. mIg molecules exist only as basic four-chain units. Unlike the secreted forms of IgA or IgM antibodies, they do not form polymers.

General Concepts in Signal Transduction

Signal transduction can be defined as the process of converting a signal from one form to another. In lymphocytes, a signal is generated by the binding of a ligand to its receptor, followed by receptor clustering. This original clustering signal is transduced into chemical signals in the cytoplasm by the activation of receptor-associated protein kinases.

Kinases are essentially phosphorylating enzymes. In receptor-mediated activation, they add a phosphate group to suitable amino acid residues on other proteins (or enzymes). Phosphorylation by protein kinases is a common general mechanism for the regulation of biochemical activity in cells. Phosphorylated membrane proteins bind other signalling proteins that are normally free in the cytoplasm, transferring the signal from the membrane to the cytosol. Such binding also increases the local concentration of the signalling protein and its ability to be phosphorylated.

- Phosphorylation can occur at tyrosine, serine or threonine, and histidine residues; separate classes of kinases phosphorylate different amino acids.
- Phosphorylation may activate or inactivate enzymes; enzymes may become active when phosphorylated and may be inactive when dephosphorylated or *vice versa*. Some enzymes have two phosphorylation sites — one inhibitory, the other activatory. The enzyme is inactive when phosphorylated at the inhibitory site and active when this site is dephosphorylated. Enzymes involved in dephosphorylation are called *phosphatases*. Phosphatases can remove an inhibitory phosphate group and allow the enzyme to be activated, or they can deactivate an activated protein.
- Phosphorylation creates a binding site on the enzyme; many receptor-associated kinases are associated with the inner surface of the cell membrane and cannot interact with targets that are freely moving in the cytosol. Phosphorylation of such membrane-associated enzymes creates a binding site for these target proteins. Removal of the phosphate group by phosphatases reverses this signalling event.

Phosphatases have an important role in signalling. They help in inactivating (bringing to ground level) the signal generated after the ligand is removed from the receptor. Such removal of phosphate groups adds a time factor to the signalling mechanism — the individual signal is active only for a given period. Phosphatases thus limit the length of time that the signalling event occurs, preventing a runaway cellular response. CD45, associated with both BcRs and TcRs, is an example of a transmembrane tyrosine phosphatase. Kinase/phosphatase activities ensure that the lymphocyte is rapidly activated upon encountering a homologous antigen and equally rapidly inactivated once the antigen is removed.

Adaptor proteins are small molecules that connect different receptors to common intracellular signalling components. These proteins do not have kinase activity; instead, they recruit other molecules to the activated receptor. They often contain several SH2 domains (**S**rc-**H**omology domain 2) flanked by SH3 domains. They bind to the phosphotyrosine residue by their SH2 domain and to other proteins through the SH3 domains. Thus, adaptor proteins position the proteins bound to their SH3 domain at or near the receptor-associated tyrosine kinases.

Small G proteins are a class of GTP-binding proteins that exist in two states, depending upon whether they are binding GTP or GDP. The GTP-bound form is active. Removal of the phosphate yields the inactive, GDP form. As these G proteins have phosphatase activity, they are normally inactive. Adapter proteins bind the G proteins to membrane receptors. Receptor-activated factors then displace the GDP and allow GTP to bind and activate the G protein. In turn, activated G proteins stimulate a cascade of kinases — e.g., MAP kinases — that phosphorylate transcription factors in the nucleus. Ras is a well-known example of a G protein.

Transcription factors bind to the promoter region of DNA and promote DNA polymerase binding and mRNA synthesis.

Signal transductions in the immune system follow the same general pathway, whether it is the antigen receptors, costimulatory molecules, activating receptors on NK cells, or FcRs that are engaged. The details (the kind of enzymes and adaptor molecules, genes transcribed) differ. The coligation of activating receptors is necessary to start the signalling cascade. The general pathway of antigen receptor activation is outlined below.

- Signal transducing units of activating receptor complexes are associated with intracellular protein kinases localised at the cell membrane's inner surface.
- Ligation of activating receptors (represented by a single Ig-fold in the figure) brings protein kinases together, allowing them to act on each other and on the tyrosines on the cytoplasmic tails of the signal transducing units, initiating the signalling process. Receptor clustering results in the phosphorylation of tyrosines in the ITAMs of its signal transducing unit by receptor-associated kinases.
- Phosphorylation of the tyrosines allows cytosolic **P**rotein **T**yrosine **K**inases (PTKs) to dock into the activated (i.e., phosphorylated) protein; adaptor proteins may help in the process by bringing target cytosolic molecules to the cell membrane.
- The cytosolic kinases then get phosphorylated by other PTKs.
- Small G proteins are also activated.
- This chain reaction results in the recruitment and activation of a key enzyme that can cleave the membrane-associated PIP_2 into two potent messenger molecules — DAG and IP_3. As one molecule of the enzyme can produce many molecules of these messengers, this represents a major amplification step in receptor-mediated signalling.
- Together these proteins activate a cascade of protein kinases that result in the phosphorylation and activation of multiple transcription factors.
- The transcription factors translocate to the nucleus, resulting in the transcription of new genes.

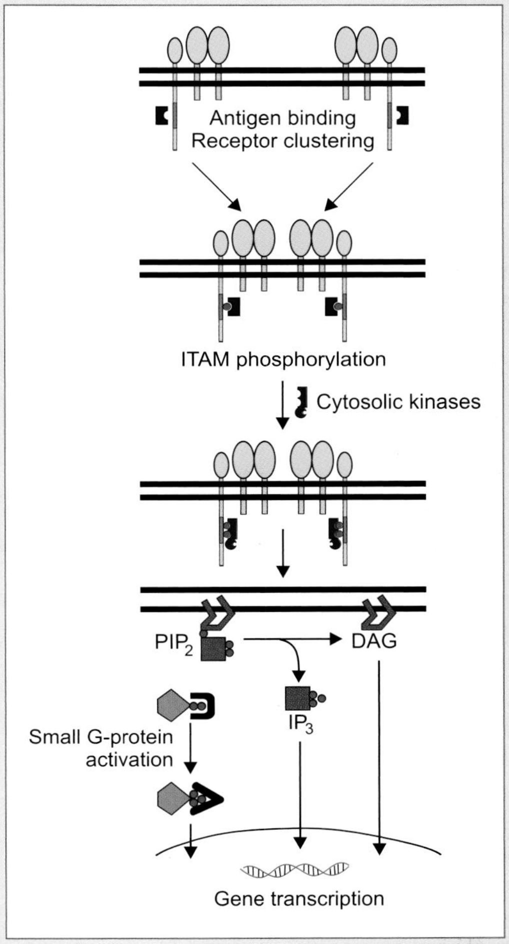

Figure 6.S2 The general pathway of signal transduction that follows clustering of antigen receptors is similar in B and T cells.

6.2.2 The BcR Complex

The BcR complex is a multiunit structure of one molecule of mIg associated with an Ig-α-Ig-β heterodimer and other molecules (CD19, CD21, CD22, CD45, CD72, and CD81) involved in the signalling pathways (Figure 6.2). Of these, CD19, CD21, CD45, and CD81 are positive regulators of BcR function whereas CD22 and CD72 are negative regulators.

- **Ig-α (**CD79a**) and Ig-β** (CD79b) belong to the Ig superfamily. A disulphide bond links the two molecules to form a heterodimer. One molecule of the heterodimer is associated with one molecule of mIg. CD79a and CD79b show a sequence homology with T cell receptor–CD3 complex components (section 6.3.1), suggesting a common origin. Ig-β is unique to B cells and it is expressed at all stages of B cell development. By contrast, Ig-α is not expressed at the plasma-cell stage.

  The heterodimer associates with mIg molecules in the **E**ndoplasmic **R**eticulum (**ER**). Quality control mechanisms in the ER ensure that only mIg molecules associated with Ig-α and Ig-β are exported to the cell surface. Unassociated mIg molecules are retained in the ER. Both Ig-α and Ig-β molecules undergo conformational changes when surface mIgs aggregate because of the binding of multivalent antigen. Both these proteins have long cytoplasmic tails containing two ITAMs each. These ITAMs get phosphorylated when mIgs are crosslinked by the antigen. The generated signal is transmitted to the B cell nucleus through an intracellular signal transduction pathway.
- **B cell coreceptor complex** consisting of CD19, CD21 (CR2), and CD81 (TAPA-1) is physically associated with the BcR and is a positive regulator of BcR function.
  - CD19, a member of the Ig superfamily, is the earliest cell-surface molecule related to B lineage differentiation; all cells of this lineage express CD19.

Molecular Mafias: The Ig Superfamily

Proteins having the same overall architecture can be clustered into superfamilies. Proteins in a given superfamily are encoded by genes derived from a common primordial gene that encoded the basic domain structure. These genes may have evolved independently and do not necessarily share genetic-linkage functions.

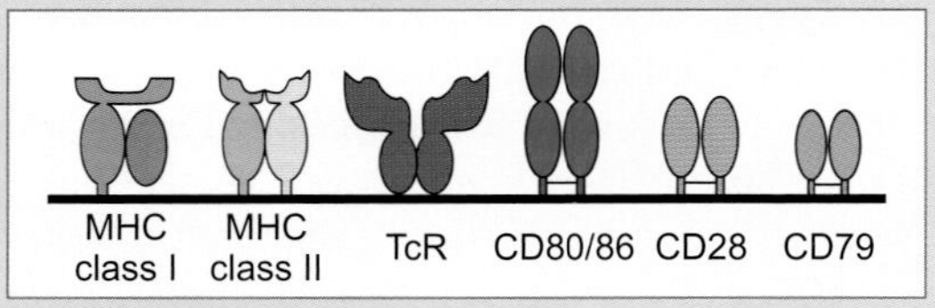

Figure 6.S3 Many proteins involved in immune interactions belong to the Ig superfamily.

The Ig superfamily is an ever-increasing group currently comprising about 80 members. Its membrane-bound proteins are found in species as diverse as insects, worms, and humans. The common factor in these molecules is the Ig-like fold consisting of an antiparallel β-sheet sandwich formed from nine β-strands held together by a disulphide bond. This architecture is thought to confer protease resistance. Human proteins belonging to this group include TcR and its accessory proteins (CD2, CD3, CD4, and CD8); MHC (-like) molecules (class I, class II, CD1, and Qa1); BcR and its accessory molecules (Ig-α, Ig-β, CD19, and CD22); FcRs; a number of cytokine receptors; costimulatory molecules (CD28, CTLA-4 (CD152), their ligands B7-1 (CD80), and B7-2 (CD86)); and adhesion molecules — PECAM-1 (CD31), ICAM-1 (CD54), ICAM-2 (CD102), and ICAM-3 (CD50). Most of these proteins do not bind antigens. The Ig-fold is thought to facilitate interactions between membrane proteins; such interactions can occur between the faces of the β-pleated sheet of both homologous and nonhomologous Ig domains.

Figure 6.2 The B cell antigen receptor is a multiprotein complex consisting of an Ig molecule and other associated molecules. The antigen-recognition unit of the complex, mIg, has a short cytoplasmic tail unsuitable for signal transduction. The Ig-α:Ig-β (CD79) heterodimer associated with the mIg molecule is the signal transducing element. The BcR complex also has a molecule of CD19 covalently associated with CD81 and CD21. This CD19:CD21:CD81 complex and CD45 are both positive regulators of B cell function. Negative regulators associated with the complex include CD22, CD72, and FcγRIIB1.

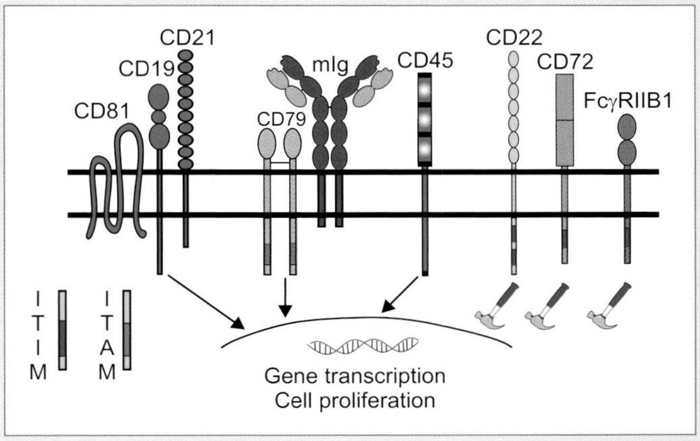

CD19 is phosphorylated upon BcR ligation. The coligation of CD19 with BcR decreases the threshold of antigenic stimulation by at least two orders of magnitude. It also regulates the phosphorylation and activation of multiple downstream adaptor proteins and enzymes — including Btk (**B**ruton's **T**yrosine **K**inase) and MAP-kinases (**M**itogen-**A**ctivated **P**rotein-kinases). Genetically engineered mice deficient in CD19 show an impaired antibody response to antigens, pointing to its importance in BcR signalling.

- CD21 is a receptor for the complement component C3d (see chapter 3). BcR binds surface antigen of C3d-coated microbes; C3d binds CD21. This coligation of the CD19–CD21 complex results in an enhanced B cell response (Figure 6.3).
- CD81 is a member of the tetraspanin family[3] of integral membrane proteins widely expressed in the body. It has gained notoriety as a receptor for the hepatitis C virus envelope protein. Both B and T cells express CD81. In B cells, it is physically and functionally associated with CD19, along with two other members of the tetraspanin family. Although its actual role in signal transduction is unclear, CD81 may partially relay or amplify the signal initiated by CD19.

- **CD22**, a sialoadhesin that binds to sialic acids, is constitutively associated with BcR. Its cytoplasmic tail carries two ITIMs. By recruiting phosphatases and inhibiting the tyrosine phosphorylation of multiple cellular proteins, CD22 limits BcR signalling. Its inhibitory functions are dominant over the costimulatory functions of the B cell coreceptor complex. Because CD22 ligands are expressed on the endothelial cells of murine bone marrow (but not other tissues), CD22 has been implicated in the homing of recirculating mature B cells to the bone marrow.
- **CD72** is a type II membrane protein[4] that has a C-type lectin domain in the extracellular region and an ITIM in its cytoplasmic tail. It is also a negative regulator of BcR functioning.
- **FcγRIIB1**, an isoform of FcγRIIB (CD32), is expressed on B cells and downmodulates BcR signalling only when it is coligated to BcR by antigen complexed with IgG antibodies. The single ITIM in its cytoplasmic tail gets phosphorylated by crosslinking with cell membrane-anchored Ig and suppresses the phosphorylation of CD19. The lipid phosphatase SHIP is essential for this inhibitory function.

[3] Members of the tetraspanin family have four transmembrane domains.

[4] Type I membrane proteins are anchored in the cell membrane at the amino-terminus (e.g., the invariant chain of MHC class II molecules and CD72), whereas type II membrane proteins are anchored through the carboxy-terminus (e.g., MHC class I and II molecules and Ig).

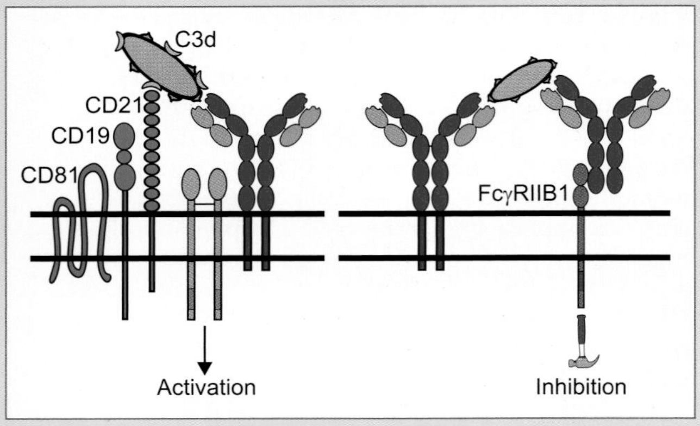

Figure 6.3 Coligation of CD21 with BcR can positively regulate its function, whereas its coligation with FcγRIIB1 negatively regulates its function. CD21, a receptor for cleavage products of the complement component C3, is associated with CD19 and CD81. Coligation of CD21 and BcR by C3d-coated antigen results in enhanced B cell response and reduction in the concentration of antigen required for B cell activation. Conversely, the coligation of the FcγRIIB1 with BcR by the antigen–IgG complex inhibits B cell responses.

- **CD45**, known as the leukocyte common antigen, is a membrane-associated protein tyrosine phosphatase expressed on all nucleated haematopoietic cells. Its ability to dephosphorylate Src family kinases gives it a crucial role in antigen receptor signalling. The B220 isoform of CD45 is a very early marker for B cell lineage. CD45 is thought to promote BcR signal transduction by constitutively maintaining Src family kinases in a partially active state.

6.2.3 Simplified Outline of BcR Signalling

BcR and TcR signalling pathways are essentially similar. Ligation of receptors results in the phosphorylation of ITAMs on accessory molecules and leads to the recruitment of a key enzyme that phosphorylates **P**hospho**l**ipase **Cγ** (PLCγ). As a consequence, potent messengers **D**i**a**cyl**g**lycerol (DAG) and **I**nositol-(1,4,5)-tris**p**hosphate (IP_3) are formed, ultimately activating multiple transcription factors (Figure 6.4). The shared nature of the enzymes and molecules involved in these downstream signalling pathways make them attractive targets for controlling undesirable immune responses (Table 6.1).

- **mIg clustering leads to the phosphorylation of ITAMs on CD79a and CD79b** by several Src family **P**rotein **T**yrosine **K**inases (PTKs), including Lyn, Blk, Fyn, and Lck.
  - Phosphorylation appears to occur at the membrane-proximal tyrosine.
  - Tyrosine phosphorylation of CD79 further facilitates the recruitment of additional molecules of these PTKs.
  - The recruitment of PTKs amplifies CD79a and CD79b phosphorylation.
  - CD45 may help in the process by dephosphorylating Fyn (see section 6.3.2).
- Only a subset of CD79a and CD79b has both these molecules with their ITAM tyrosines phosphorylated. These phosphorylated molecules serve as docking sites for the SH2 domains of the cytosolic PTK Syk. **Syk recruitment** by BcR is a key event that triggers multiple downstream signalling events and **results in the phosphorylation of PLCγ**.
  - The docking of Syk allows it to get activated through tyrosine phosphorylation by Src family PTKs. Lyn is important in this activation — Lyn-deficient B cells fail to induce Syk tyrosine phosphorylation.
  - Phosphorylated Syk phosphorylates BLNK (**B** cell **Link**er protein), an adaptor protein needed for transmitting the signal generated at the cell membrane to downstream targets.

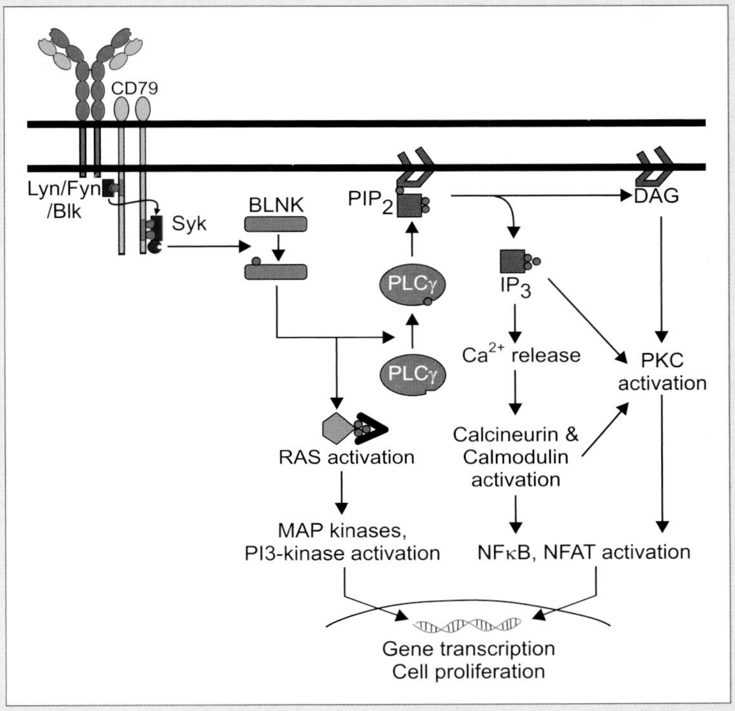

Figure 6.4 Crosslinking of the antigen-recognition units of the BcR complexes by the antigen activates a cascade of downstream signalling events that result in B cell proliferation. Heterodimers of Ig-α:Ig-β (CD79) are the signal transducing elements of BcRs. mIg clustering results in a change in the conformation of CD79. The resulting signal activates receptor-associated tyrosine kinases such as Lyn, Fyn, and Blk. They activate and phosphorylate ITAMs present in the cytoplasmic tails of each other and the CD79 heterodimer. CD79, which has both its ITAMs phosphorylated, serves as a docking site for Syk. Phosphorylated Syk phosphorylates the adaptor protein BLNK. BLNK recruits and phosphorylates the key enzyme of the cascade — PLCγ. This enzyme acts on membrane-associated PIP_2 and splits it into two potent messengers — DAG and IP_3. DAG remains associated with the membrane; IP_3 is released in the cytosol. IP_3 causes the release of intracellular Ca^{2+} stores. These events activate Ca^{2+}-binding calcineurin and calmodulin. Together with DAG, these molecules activate PKC and induce the transcription and nuclear localisation of the transcription factors NFAT and NFκB. BLNK phosphorylation also activates Ras. Ras activation triggers PI3-kinase and MAP kinases and results in B cell proliferation.

- Multiple cell surface receptors get rapidly tyrosine phosphorylated. These include CD19 and CD22 (both of which have SH2 docking sites), and the PLCγ that are recruited by BLNK.
  - PLCγ is phosphorylated by the PTKs Syk and Btk.
  - Phosphorylation of CD22 is a feedback loop providing for downregulation of the escalating response. The ligation of CD22 and FcγRIIB1 is especially efficient in inhibiting BcR signalling.

❒ Tyrosine phosphorylated **PLCγ hydrolyses P**hosphatidyl**i**nositol-(4,5)-bis**p**hosphate (**PIP_2**) **to yield** two potent messengers — **DAG and IP_3**. These products of PIP_2 cleavage activate **P**rotein **K**inase **C** (**PKC**).

- DAG remains associated with the inner surface of the plasma membrane. It is an activator of several members of the serine/threonine protein kinase family — PKC is the most important amongst these.
- IP_3, in contrast, interacts with receptors on the ER, resulting in the release of intracellular Ca^{2+} stores. IP_3 also opens Ca^{2+} channels in the plasma membrane, leading to an influx of Ca^{2+} into the cell. These events result in increased intracellular Ca^{2+} levels and the translocation of transcription factors.
  - The massive increase in intracellular Ca^{2+} results in the activation of the Ca^{2+}-binding proteins calmodulin and calcineurin. Ca^{2+} released by IP_3 further activates PKC.
  - Calmodulin and calcineurin induce nuclear localisation of the **N**uclear **F**actor of **A**ctivated **T** cells (NFAT) and NFκB (**N**uclear **F**actor **κB**) proteins.

Table 6.1 Drugs Targeting Lymphocyte Signalling Pathways

Drug	Target	Effect
Cyclosporin A	Calcineurin phosphatase	Inhibits T cell activation and cytokine synthesis *via* TcR engagement but cannot block T cell activation by exogenous cytokines
Tacrolimus (FK506)	Calcineurin phosphatase	Inhibits B and T cell activation
Sirolimus (Rapamycin)	Signals delivered to T cells *via* IL-2, IL-4, and IL-6 p70^{s6} kinase involved in the synthesis of proteins required for cell-cycle progression	Inhibits the proliferation of T cells by arresting them in the G1 phase

- **PKC activation results in the activation of** multiple transcription factors, including **NFκB**.
- BcR stimulation also recruits adaptor proteins to the receptor site. These proteins induce activation of the small G protein Ras.
- **Ras activation ultimately activates P**hosphatidyl**i**nositol-**3**-kinase (PI3-**kinase**) **and MAP kinases**. Several transcription factors, including AP-1 (**A**ctivating **P**rotein-**1**), which regulate the expression of numerous genes involved in cell growth, are activated because of PI3-kinase and MAP kinases' activation.
- The end result is B cell proliferation and differentiation.

6.3 T Cell Antigen Receptor Complex

TcR complex, like BcR complex, consists of immune-recognition receptors and associated signal transduction molecules (Figure 6.5). The antigen receptors of circulating mature T lymphocytes comprise two highly variable glycoprotein heterodimers (either αβ or γδ pair), noncovalently associated with other invariant chains collectively called the *CD3 complex*. Also found to be associated are CD4/CD8 and CD45 molecules. The αβ and γδ heterodimers are responsible for antigen recognition and confer specificity. CD4 and CD8 are the coreceptors that interact with MHC molecules; the CD3 complex and CD45 are involved in signal transduction.

The αβ TcR complex is expressed by over 95% human peripheral blood T cells. The heterodimer consists of one chain each of an acidic α chain of 39–46 KD and a more basic 40-44 KD β chain held together by disulphide linkages. The γδ T cells

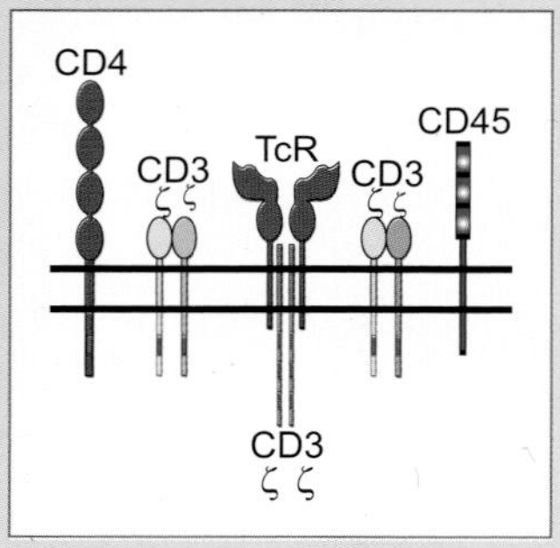

Figure 6.5 The TcR receptor complex consists of an antigen-recognition unit, the coreceptors CD4/CD8, and associated molecules, such as CD3 and CD45. The CD3 complex associated with the TcR is involved in signal transduction. It is composed of five chains (γ, δ, ε, η, and ζ) organised to form three dimers — γε, δε, and ζζ. Of these, the γ, δ, and ε chains contain an extracellular Ig-fold and a long cytoplasmic tail with one ITAM each. The ζ chain has a short extracellular fold and three ITAMs in its cytoplasmic chain. The complex also contains the phosphatase CD45 and coreceptors CD4 or CD8. CD4, the coreceptor for MHC class II molecules (shown here), and CD8, the coreceptor for MHC class I molecules (not shown here), are found on different subset of mature T cells. Both these molecules have one ITAM each in their cytoplasmic tail. The coreceptors must bind to the same MHC molecule as the TcR for optimal signal transduction.

B Cell Antigen Receptor

- BcR is a multimolecular complex consisting of an antigen-recognition unit (mIg) with associated molecules.
- BcR recognises antigen in its native form.
- mIg, the antigen-recognition unit of BcR, is similar in structure to secreted Ig except that mIg contains a stretch of hydrophobic amino acids that anchors it in the cytoplasmic membrane.
  - mIg consists of two H and two L chains; each H chain associates with one L chain, yielding a molecule that is a homodimer of a heterodimer $(H–L)_2$.
  - Each chain consists of a 110 amino acid V region and 110 amino acids (L chain) or 330 amino acids (H chain) C region.
  - Globular domains present at the amino-terminus of the H and L chains are held together by hydrophobic interactions and form the antigen-binding pocket.
- Other molecules associated with BcR include:
  - Ig-α and Ig-β heterodimers, which are responsible for transduction of the signal generated by mIg ligation.
  - CD19–CD21–CD81 complex, a positive regulator of BcR signalling — CD19–BcR coligation decreases the threshold of antigenic stimulation, CD19–CD21 coligation results in enhanced B cell responses.
  - CD22, which limits BcR signalling by recruiting phosphatases to the signalling complex.
  - CD72, a negative regulator of B cell signalling.
  - FcγRIIB1, a negative regulator of BcR signalling — crosslinking of BcR with this molecule downmodulates BcR signalling.

have a more restricted distribution, but structurally, the γδ receptor is essentially similar to the αβ receptor.

- The TcR heterodimer is structurally homologous to the Ig molecule in that it also consists of distinct domains — the V and C regions. The V region is distal to the cell membrane and is responsible for antigen binding; the C region is proximal to the cell membrane.
- The V domain of TcR comprises CDR1, 2, and 3, corresponding to those found in the Ig molecule.
  - The CDR1 and CDR2 regions of the αβ TcR interact with the α helices of MHC molecules.
  - The highly polymorphic region corresponding to CDR3 interacts with antigenic peptides bound in the MHC cleft.
- Proximal to the membrane, each TcR chain has a short connecting sequence containing a cysteine residue that is involved in the formation of a disulphide bond between the chains.
- Transmembrane domains of TcR subunits have a stretch of positively charged amino acids that allow them to interact with the CD3 complex and a short cytosolic tail of 15-20 amino acids.
- The CD3 associated with TcR, like CD79 of the BcR, is involved in signal transduction. CD3 expression is necessary for αβ or γδ heterodimer expression. Antibodies to the CD3 complex can block T cell function. The CD3 complex is composed of five chains, known as γ, δ, ε, η, and ζ.
  - The three chains are organised to form three dimers — a heterodimer of gamma and epsilon chains (γε), a heterodimer of delta and epsilon chains (δε), and a homodimer of zeta chains (ζζ).
  - Differential splicing of the zeta chain gives the eta (sometimes called nu) chain and the corresponding ζη dimer. About 90% of TcR complexes express ζζ heterodimers and 10% express ζη heterodimers.

Seeing Red (or Green): ITIMs and ITAMs

An **ITAM** (**I**mmmunoreceptor **T**yrosine-based **A**ctivating **M**otif) is composed of two tyrosine residues separated by around 13 amino acids. The canonical ITAM has the general sequence of Yxx(L/I)x (6-8)Yxx(L/I), where each letter represents an amino acid (Y = tyrosine, L = leucine, I = isoleucine, and x = any amino acid). Although these motifs were originally discovered in Ig-α and Ig-β, they are now known to be present on other molecules of the immune system. ITAMs are found on the cytoplasmic tails of numerous receptors, including FcγRIII (found on NK cells, macrophages, and neutrophils), FcεRI on mast cells and basophils, and activatory receptors on NK cells. The phosphorylation of tyrosines in the ITAMs by protein kinases allows them to recruit other enzymes/molecules involved in signal transduction. The Syk family of tyrosine kinases (Syk and ZAP-70), involved in BcR and TcR signal transduction respectively, have two SH2 domains that can bind two phosphotyrosines. These two tyrosines have to be spaced precisely for proper binding. Thus, the presence of ITAMs in the cytoplasmic tail of a protein implies the involvement of the Src family and Syk family kinases in signal transduction.

ITIM (**I**mmunoreceptor **T**yrosine-based **I**nhibitory **M**otif) has a large hydrophobic residue (such as isoleucine or valine) one or two residues upstream of a tyrosine, followed by leucine two amino acids later ([I/V]xYxxL). ITIMs are also not exclusive to TcRs and BcRs. They are found on a number of inhibitory receptors of the immune system — CD22, FcγRIIB, CTLA-4 — as well as on a number of inhibitory receptors on NK cells. Tyrosine-phosphorylated ITIMs recruit **SH**2-containing **i**nhibitory **p**hosphatases (called SHP-1 and SHIP) that are essential for the negative regulation of cell activity. These inhibitory phosphatases carry a SH2 domain that preferentially binds the phosphorylated tyrosines in the ITIM. SHP-1 removes phosphate groups added by tyrosine kinases, and SHIP is thought to inhibit the activation of PLCγ and the production of DAG and IP_3.

- γ, δ, and ε chains are members of the Ig superfamily, and they consist of an extracellular Ig-fold, a transmembrane domain, and a long cytoplasmic tail.
- ζ (and therefore η) chains do not have the Ig-fold. They have a very short extracellular domain (only nine amino acids) and a long cytoplasmic domain (ζ has113 amino acids; η has 155).
- The transmembrane domains of all CD3 chains are negatively charged and as a result, interact with positively charged TcR transmembrane domains.
- All CD3 chains have ITAMs — one each in γ, δ, and ε chains and three each in ζ and η chains.

- Coreceptors CD4 and CD8 are transmembrane glycoproteins found on different subsets of T cells. These molecules are important in antigen recognition by αβ T cells. They are coreceptors for MHC molecules. The coreceptors must bind to the same MHC molecule as the TcR for optimal signal transduction. Antibodies to CD4/CD8 inhibit T cell activation.
  - Both CD4 and CD8 belong to the Ig superfamily.
  - CD4 is a transmembrane monomeric glycoprotein expressed on T_H and T_{reg} cells. CD4 binds MHC class II molecules. It contains four extracellular Ig domains, a transmembrane domain, and a cytoplasmic tail with serine residues that can be phosphorylated by serine/threonine kinases. It is internally associated with the Src family kinase Lck.
  - CD8 is a dimer found on CTLs that binds MHC class I molecules. It consists of αβ or αα subunits. Each subunit consists of a single Ig-fold, a transmembrane region, and a cytoplasmic tail that can be phosphorylated at several sites.

The SAARC Family

The Src family of PTKs, pronounced *Sark* (or SAARC, as in South Asian Association of Regional Cooperation), is a group of closely related proteins that regulate crucial cellular processes in response to the activation of transmembrane receptors. The first member was discovered as the oncogene v-src, responsible for the tumour causing ability of Rous sarcoma virus. Incidentally, v-src was established to be a modified form of c-src — a normal cellular gene that the Rous sarcoma virus had picked up. Src kinases are common components of signalling pathways involved in the control of a number of cellular processes in vertebrates — proliferation, differentiation, adhesion, migration, and survival, to name just a few. They are constitutively anchored to the inner leaflet of the plasma membrane and tend to be concentrated in lipid rafts.

Src PTKs having a key role in signal transduction in lymphocytes include Lck, Lyn, Blk, and Fyn.

- Lck is associated with the cytoplasmic domain of CD4 and α chain of CD8 in memory T cells, but it is distributed throughout the cytosol in naïve T cells.
- Fyn associates with ζ and ε CD3 chains.
- Blk is a B cell-specific kinase. In B cells, Lck, Blk, and Fyn seem to be functionally redundant.
- Lyn is the most abundant kinase in B cells.

Src PTKs are proteins of 52–62 KD composed of distinct functional regions. Of special interest in lymphocyte signalling are the two small modular units known as the SH2 (**S**rc-**H**omology **2**) and SH3 (**S**rc-**H**omology **3**) domains. The SH2 domain binds short amino acid sequences containing phosphotyrosine. The ligand-binding surface of the SH2 domain has two pockets — one contacts the phosphotyrosine and the other contacts the +3 amino acid residue that follows the phosphotyrosine. Src kinases have a preference for leucine at this position (hence, the importance of L/I in YxxL/I in the ITAM motif). By contrast, the SH3 domain recognises left-handed polyproline helices. The SH2 and SH3 domains are important in the control of Src PTKs activity. In the absence of input signals, Src proteins are kept in an inactive state. The SH2 and SH3 domains are unexposed, and they cooperate to turn off catalytic machinery. When the kinase is turned on, the SH2 and SH3 domains are exposed, and they target the protein kinase to appropriate sites in the cell.

The enzymatic activity of Src family kinases is itself regulated by phosphorylation. Phosphorylation at one site activates the enzyme, whereas phosphorylation at another site inhibits it. The kinases are kept in an inactive state by the action of the kinase Csk, which phosphorylates the inhibitory tyrosine at the carboxy-terminus. In the resting state, constitutively expressed Csk keeps enzymes inactive. Upon activation, CD45 dephosphorylates this site. Thus, the balance between Csk and CD45 activity regulates the action of Src kinases in lymphocytes.

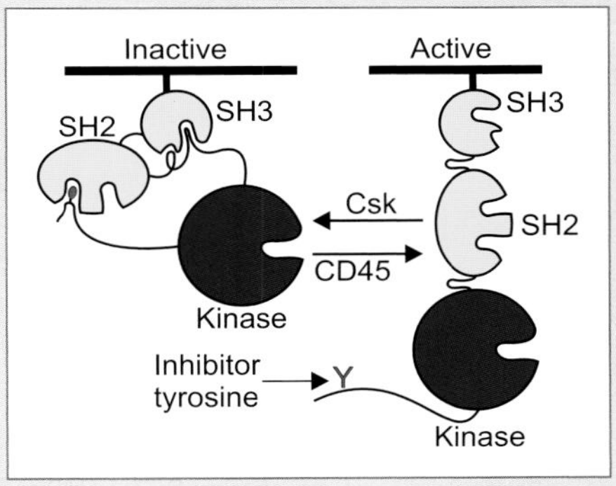

Figure 6.S4 Src kinase activity is regulated by phosphorylation and dephosphorylation.

- CD45, a phosphatase associated with TcRs and BcRs, dephosphorylates Src family kinases. It is especially important in T cell activation, as disruption of CD45 signalling affects tyrosine phosphorylation, inositol phosphate generation, and Ca^{2+} mobilisation

6.3.1 Simplified Outline of TcR Signalling

As for BcRs, the clustering of TcRs leads to the nuclear localisation of transcription factors and ultimately results in gene transcription, which allows the proliferation and differentiation of cells (Figure 6.6).

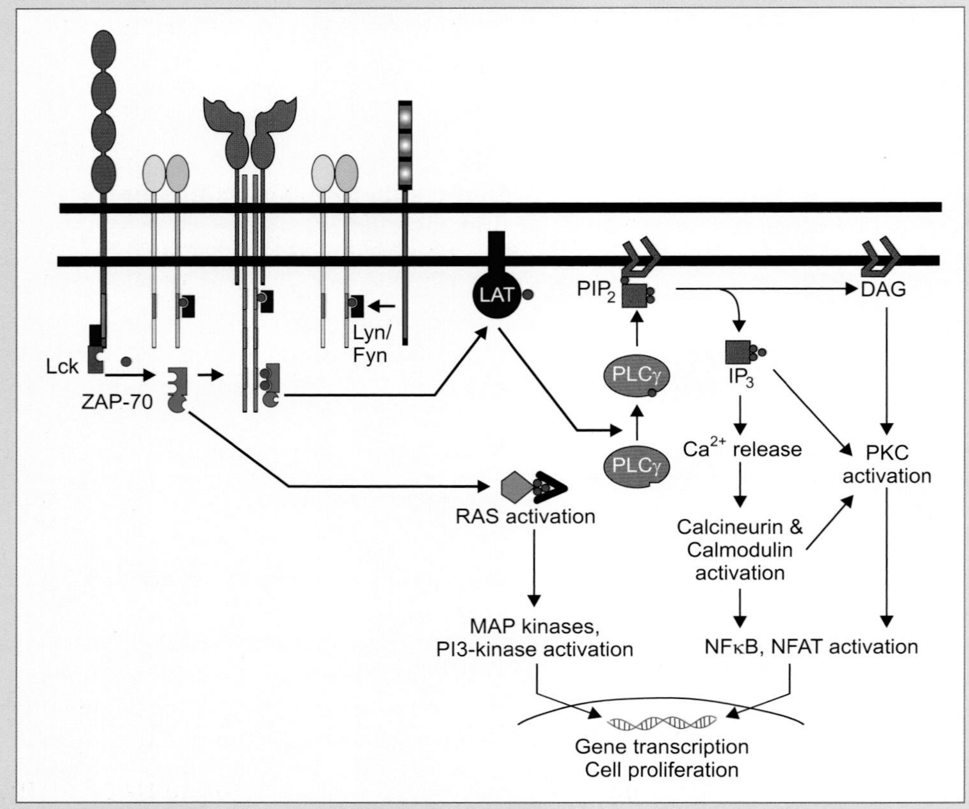

Figure 6.6 Coligation of TcRs by peptide:MHC complexes along with CD4/CD8 engagement triggers downstream events that result in T cell proliferation. Antigen recognition by TcRs results in a change in CD3 conformation. CD3-associated protein kinases, such as Lyn and Fyn, are normally in an inactive state and need to be activated by CD45. CD45 removes their inhibitory phosphate and activates them. Activated Lyn and Fyn phosphorylate the ITAMS on CD3 dimers. The engagement of CD4 (or CD8) by MHC molecules activates the coreceptor-associated tyrosine kinase Lck. Activated Lck is brought into the vicinity of CD3ζ-associated ZAP-70, which it then phosphorylates. ZAP-70 activates the Ras pathway, which in turn activates MAP kinases and PI3-kinase. ZAP-70 also phosphorylates the adaptor protein LAT. LAT recruits PLCγ and other kinases to the cell membrane. Activated PLCγ cleaves PIP_2 to IP_3 and DAG. The downstream events of PIP_2 cleavage are similar in B and T cells and result in new gene transcription and T cell proliferation.

- The clustering of the TcRs and coreceptors activates PTKs associated with the cytoplasmic domains of CD3 and coreceptor proteins; **CD3 and Fyn**, **associated with it**, **initiate signalling**.
- Lck, a Src family kinase bound to the cytoplasmic tail of CD4 (or CD8), is brought into the proximity of the ITAMs of CD3 ζ chain, and this results in the phosphorylation of the Syk family kinase ZAP-70 (70 KD **ζ-A**ssociated **P**rotein).
  - Both Fyn and Lck are regulated by tyrosine phosphorylation at two sites. Phosphorylation at one site induces activation, whereas it has an inhibitory effect at the other site (called the regulatory site). The enzymes are normally inactive because of phosphorylation at the regulatory site. CD45 helps Lck and Fyn activation by removing the regulatory phosphate.
  - Lck and Fyn phosphorylate tyrosines in the ITAMs of the CD3 complex. The tyrosine phosphorylated ITAMs in the ζ chain become specific docking sites for PTK ZAP-70. The docked ZAP-70 remains inactive until CD4 or

Major Steps in B Cell and T Cell Signalling Pathways that Result in their Proliferation

	B CELL	T CELL
Ligation of Antigen Receptor	Phosphorylation of CD79a and b by tyrosine kinases (Lyn, Blk, Fyn, and Lck)	Phosphorylation of CD3 ζ chain by Lck (associated with CD4 or CD8) and Lyn/Fyn (associated with CD3)
	↓	↓
	CD79 molecules, with both their sites phosphorylated, serve as docking sites for Syk	Phosphorylated CD3ζ becomes a docking site for the Syk family kinase ZAP-70
	↓	↓
PLCγ Phosphorylation	Syk is phosphorylated by Src family kinases (e.g., Lyn)	ZAP-70 is phosphorylated by Lck or Fyn
	↓	↓
	Syk phosphorylates BLNK	ZAP-70 phosphorylates LAT
	↓	↓
	BLNK recruits PLCγ	LAT recruits PLCγ
	↓	↓
	Syk and Btk phosphorylates PLCγ	ZAP-70 phosphorylates PLCγ
	↓	↓
Activation of PKC and Translocation of Transcription Factors	Phosphorylated PLCγ hydrolyses PIP_2 to DAG and IP_3	
	DAG activates many kinases, including PKC	Increased intracellular levels of Ca^{2+}, caused by IP3, result in the activation of calmodulin and calcineurin and the further activation of PKC
		↓
		Calmodulin and calcineurin induce the nuclear translocation of transcription factors (NFAT, NFκB)
	↓	↓
Activation of PI3-Kinase, Map Kinases, and Gene Transcription	Adaptor proteins activate the Ras pathway	ZAP-70 activates the Ras pathway
	↓	↓
	Ras activation results in the activation of PI3-kinase, MAP kinases, and AP-1	
	↓	↓
	Together, these lead to transcription of genes involved in cell proliferation and differentiation	

Turn On: NFκB Proteins

The NFκB family of transcription factors has a central role in coordinating the expression of several genes involved in the control of innate and adaptive immune responses. The family comprises five members — RelA, RelB, cREL, NFκB1, and NFκB2. The primary activated form of NFκB is a heterodimer of RelA and NFκB1. The NFκB family of proteins is normally present in the cytoplasm in association with a family of inhibitory proteins, IκBs (**I**nhibitors of NF**κB**). Upon activation, IκBs get phosphorylated and subsequently degraded. The degradation of IκBs allows NFκB proteins to translocate to the nucleus and bind their cognate DNA-binding sites. NFκB proteins can thus regulate the transcription of a large number of genes, including those of antimicrobial peptides, cytokines, chemokines, stress-response proteins, and antiapoptotic proteins.

NFκB proteins get rapidly activated in response to a variety of stimuli, including stress signals, the engagement of PRRs (especially TLRs) by PAMPs, and pro-inflammatory cytokines, such as IL-1 and TNF-α, produced by activated macrophages and monocytes. IL-1 and TNF-α produced by APCs in particular can induce NFκB phosphorylation and enhance T cell activation. TcR clustering alone is not enough to activate NFκB or the transcription factor AP-1. It results in a deficiency in IL-2 production. However, engagement of the costimulatory molecule CD28 in conjunction with TcR clustering activates these proteins and their nuclear translocation, resulting in the IL-2 production necessary for T cell proliferation. NFκB is important in numerous B cell processes, such as germinal centre formation and isotype switching. It is also involved in B cell maturation, as mice lacking NFκB1 and NFκB2 show a complete absence of B cell maturation. NFκB function is thus essential for lymphocyte activation and survival and for normal immune responses. The constitutive activation of NFκB pathways, on the other hand, is associated with diseases such as inflammatory bowel disease, multiple sclerosis, and asthma.

CD8 engagement results in Lck activation. **Lck or Fyn then phosphorylate** (and therefore, activate) **bound ZAP-70**.

- ZAP-70 activation is crucial in T cell activation. It initiates three separate cascades, each of which ultimately activates transcription factors.
  1. **ZAP-70 phosphorylates the adaptor protein LAT** (**L**inker of **A**ctivated **T** cells).
     - LAT, located in the plasma membrane, is thought to link early tyrosine phosphorylation events to distal portions of the signalling pathway. In the absence of LAT, TcR engagement does not lead to gene transcription.
     - Tyrosine phosphorylated LAT recruits PLCγ, PI3-kinase, and other molecules to the membrane.
  2. ZAP-70, like Syk, activates PLCγ to cleave PIP_2 into IP3 and DAG (section 6.2.3).
     - The cleaving of PIP_2 ultimately leads to the activation of the transcription factors NFAT and NFκB.
     - Both factors enter the nucleus and activate the transcription of several genes, including that of IL-2.
  3. ZAP-70 also activates the Ras pathway.
     - Ras activation leads to activation of PI3-kinase and MAP kinases.
     - The transcription factor AP-1 is also activated.

> New mRNA is synthesised as a consequence, and this results in the proliferation and differentiation of the T cell.

Agonists and Antagonists

Changing some TcR contact residues of a peptide can alter the signalling events associated with TcR clustering.

- Peptides that activate T cells are known as *agonist peptides.*
- Structurally related peptides may prevent T cells from responding to agonist peptides instead of activating them; alternatively, they may deliver negative signals to T cells. Such peptides are termed *antagonist peptides.*
- Some peptides lead to only partial T cell activation and are called *partial agonists* or *altered peptide ligands.*

Altered peptide ligands seem to inhibit TcR signalling through the altered phosphorylation of CD3ε and ζ chains. TcR ligation in the absence of costimulation is also thought to generate such incomplete phosphorylation. Pathogens, especially viruses, may persist and survive through this mechanism of regulating T cell activation. For example, in HIV infections, mutant viruses that circumvent CTL killing by partially activating cells arise in later stages of infection. The malarial parasite *P. falciparum* also evades the immune system by this mechanism. The reduced activation induced by altered peptide ligands is now being explored for therapeutic modulation of T cell function in diseases characterised by unwanted T cell activation (e.g., autoimmune disorders or allergies), as well as for disorders of suboptimal T cell activation (e.g., cancers).

6.4 Costimulators and B and T Cell Signal Transduction

Two simultaneous signals are required for both B and T cell activation. BcR clustering by the antigen delivers the first signal for B cell activation. However, most antigens lack the multiple epitopes required for sufficient BcR clustering (except of course, the TI antigens). A second signal delivered by cognate (physical) interaction with TH cells is therefore needed to respond to such (TD) antigens (see section 8.4). A large number of molecules are implicated in this cognate interaction. The most important interaction is the engagement of CD40 on B cells by its ligand CD154 (also known as CD40L) on T cells. CD40–CD154 interaction enhances NFκB protein transcription.

With T cells, both primary and secondary signals are delivered by the same APC. TcR clustering by the MHC:peptide complexes deliver the primary signal. The second signal is delivered by the engagement of costimulatory molecules. CD28 is the prototypical costimulatory molecule expressed by naïve and antigen-primed T cells. The ligands for CD28 are CD80 and CD86. The molecular mechanism by which CD28 crosslinking affects T cell activation is not clearly understood. The cytoplasmic tail of CD28 lacks enzymatic activity but has several tyrosines that can be phosphorylated by Src family PTKs. Phosphorylation of the cytoplasmic tyrosines is thought to lead to PI3-kinase binding and result in NFκB activation. Thus, CD28 is believed to amplify TcR signal by augmenting transcriptional activity and increasing mRNA stability. CD28 ligation is also thought to recruit lipid rafts to the site of T cell–APC interaction, causing a local increase in lipid-associated enzymes and adaptor proteins. It also promotes T cell survival by inducing the upregulation of Bcl-XL (see the sidetrack *Stimulating Company*, chapter 9). LFA-1, another costimulatory molecule, is an integrin family molecule that interacts with its ligand ICAM-1 on the APC. This interaction helps in cytoskeletal reorganisation during T cell activation and amplifies TcR signals by recruiting Src family kinases

T Cell Antigen Receptor

- The T cell antigen receptor complex consists of an antigen-recognition unit and associated molecules.
- The antigen-recognition unit is a heterodimer of αβ or γδ chains held together by disulphide linkages.
  - Each of these chains has a membrane-distal V region and a membrane-proximal C region.
  - Both Vα and Vβ contribute to the antigen-binding pocket.
  - Cytoplasmic segments of the heterodimer contain hydrophilic amino acid sequences that allow them to associate with CD3.
- αβ TcR recognises digested fragments of the antigen loaded onto MHC class I, MHC class II, or CD1 molecules.
- CD3, associated with the TcR antigen-recognition unit, is a complex of five chains (γ, δ, ε, ζ, and η) that yield a set of three heterodimers (γε, δε and ζζ or ζη), and it is involved in signal transduction.
- TcR is also associated with the coreceptors CD4 or CD8. CD4 is a monomeric glycoprotein; CD8 is a dimer.

and PI3-kinase to the site of interaction and promoting activation of transcription factors.

CTLA-4 (**C**ytotoxic **T L**ymphocyte-associated **A**ntigen-**4**) shows a 30% homology with CD28 and binds to the same ligands as CD28 (CD80/86). However, it is a negative regulator of T cell activation. CTLA-4 ligation reduces TcR-dependent activation of MAP kinases and transcription factors NFAT, NFκB, and AP-1. Ligation of CTLA-4 during TcR stimulation reduces cytokine production and causes cell-cycle arrest (section 9.1). The molecular mechanisms behind these effects are still being investigated.

6.5 Generation of BcR and TcR Diversity

The specific immune recognition mediated by BcRs and TcRs defends the body against forever evolving pathogens and transformed cells. This implies that the immune system must have the wherewithal to specifically recognise and react to practically infinite molecules. The different Ig molecules that the individual's immune system produces constitutes its antibody-specificity repertoire, or B cell repertoire; the diversity of epitopes recognised by T cells is the T cell repertoire. This section describes the molecular and genetic mechanisms involved in generating this repertoire diversity. It requires some knowledge of the principles of protein synthesis and genetics. Admittedly, this is a difficult section to follow; some may find the details too extensive. To facilitate comprehension, the essentials of the processes are summarised in the main bullets and details appear as sub-bullets.

Faced with the dilemma of generating a virtually unlimited repertoire from a limited number of genes, nature devised an elegant solution, with some unusual mechanisms to achieve diversity (Figure 6.7).

- BcR and TcR genes are assembled from separate (randomly chosen) gene segments, much as different structures can be assembled from the same Lego pieces.
- The antigen-binding sites of BcR or TcR are not preformed in the cell but require assembly. Variable (V) regions of the Ig H chain (V_H) and TcR β and δ chains (Vβ and Vδ, respectively) are assembled from three segments — *V*[5] (coding for the first ~97 amino acids), *D* (coding for the next 2-3 amino acids), and *J* (coding for the remaining 10 or 11 amino acids). Similarly, Ig L chain (V_L) and TcR α and γ

[5] To avoid confusion between the V region of the antigen receptor and the V gene segment that codes for only a part of the assembled V region, gene segments are represented by *V*, *D*, and *J* in this chapter.

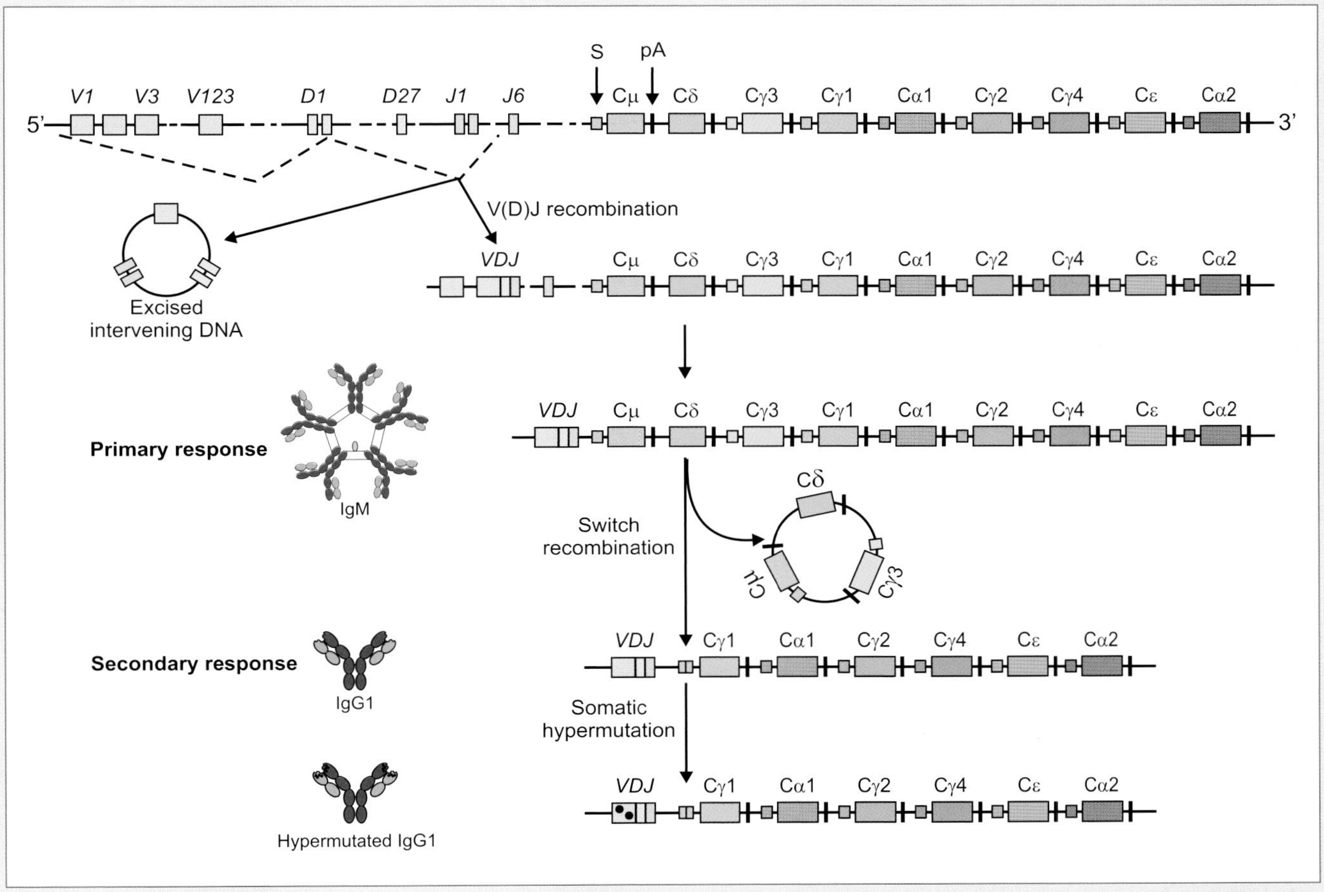

Figure 6.7 Multiple mechanisms are involved in generating Ig diversity. The germline locus of human Ig H chain on chromosome 14 contains several *V*, *D*, *J* subexons and multiple C region gene segments encoding the different isotypes (schematic organisation not to scale, shown in the top panel). Each C region is flanked by a switch region (S) except for *Cδ* on 5' end and a polyadenylation site (pA) that allows termination of transcription on the 3' end. For simplicity, exon details within each C region and pseudogenes are not shown. The Ig V region is assembled from separate segments in a process called *V(D)J* recombination. The random selection of *V*, (*D*), and *J* regions, addition or deletion during recombination, and random pairing of H and L chains generates diversity and gives rise to the large and varied repertoire of the primary Ig response. Following antigenic challenge, Ig H CSR allows the fully assembled and expressed VH region gene to be coupled to a new CH region (here, Cγ1), increasing the functional diversity of the Igs produced. Locus-specific somatic hypermutation (depicted by black dots) further alters the affinity of the antibody for the antigen.

chains (Vα and Vγ respectively) are put together from the *V* segment (encoding the first 97 amino acids) and the *J* segment (encoding the remaining 13 amino acids). As the germline genome contains several *V*, *D*, and *J* segments, there are multiple choices for each segment (Table 6.2). The random selection of *V*, (*D*), and *J* regions during recombination generates the antigen receptors, and the resultant diversity is termed *combinatorial diversity* (Table 6.3).

- The joining of the *V*(*D*)*J* segments is imprecise, allowing for the addition or deletion of nucleotides, further increasing the potential diversity of the repertoire. This diversity is generated through alterations at *V–D*, *D–J*, or *V–J* junctions and is called *junctional diversity*.
- The assembled V region couples with the C region[6] to yield the complete Ig L, Ig H, and TcR α, γ, β, or δ genes.

[6] In the case of Ig L and TcR (α, β, γ or δ) chains, the C region consists of one Ig fold ~110 amino acids. In the case of Ig H chains, it consists of three Ig folds ~330 amino acids.

Table 6.2 Elements of Human *V(D)J* Recombination

	Ig H chain	Ig L chains κ	λ	TcR chains α	β	γ	δ
Chromosome	14	2	22	14	7	7	14[$]
V region subexons	3 (*V–D–J*)	2 (*V–J*)	2 (*V–J*)	2 (*V–J*)	3 (*V–D–J*)	2 (*V–J*)	3 (*V–D–J*)
Number of subexons*	*V* – 48 *D* – 23 *J* – 6	*V* – 41 *J* – 5	*V* – 34 *J* – 5	*V* ~55 *J* – 61	*V* – 67 *D* – 2 *J* –14	*V* – 14 *J* – 5	*V* – 4 *D* – 3 *J* – 3
Minimum recombination events required for V region assembly	3	2	2	2	3	2	3
Order of recombination events	*D* joins *J*; *D–J* joins *V*	*V* joins *J*	*V* joins *J*	*V* joins *J*	*D* joins *J*; *D–J* joins *V*	*V* joins *J*	*V* joins *D*; *V–D* joins *J*

* These numbers are derived from the exhaustive cloning and sequencing of DNA from one individual. As a result of polymorphism, the number will not be the same for all individuals. Also, the numbers do not include pseudogenes, which are essentially mutated and nonfunctional versions of a gene sequence.

$ The δ gene cluster lies within the α gene cluster, and hence, rearrangement of the α chain genes inactivates genes encoding the δ chain. The exact number of Vδ genes is unclear.

- The functional heterodimeric B cell antigen receptor is formed by the random pairing of a L chain with a H chain. Similarly, the random pairing of α (or γ) with the β (or δ) chains yields αβ (or γδ) TcRs. This further increases diversity and gives rise to the large and varied repertoire of the primary immune response. This primary B (or T) cell repertoire is achieved through combinatorial and junctional diversity. It does not require exposure to antigens, that is, it is shaped during the process of B (or T) cell maturation. In this way, the immune system is prepared to recognise and deal with antigens it has not yet encountered.
- Two additional processes increase only the B cell repertoire following antigenic challenge — **C**lass **S**witch **R**ecombination (CSR) and somatic hypermutation.
  - Ig H CSR allows the fully assembled and expressed V_H region gene to be coupled to a new C_H region so that the antigenic specificity of the antibody remains unchanged while effector functions vary.
  - Locus-specific somatic hypermutation in V_H and V_L regions alters the affinity of the antibody for the antigen following antigen challenge.

6.5.1 *V(D)J* Recombination

As explained above and summarised in Table 6.2, V region exons of BcRs and TcRs are assembled from subexons. Separate chromosomes encode the BcR (and Ig) H and L chains. Similarly, gene segments encoding TcR α, β, and α chains also lie on different chromosomes. The TcR δ chain locus lies within the α chain locus. Hence, rearranging the α locus results in the deletion of the entire δ locus. The respective C regions are encoded by separate exons downstream of the V gene segments. DNA recombination events that occur at the three Ig (H, Lκ, and Lλ) and four TcR (α, β, γ, and δ) loci to yield productive V genes are similar. The Ig H, TcRβ,

Table 6.3 Combinatorial Diversity

BcR V region subexons	Number	Combinations generated
*V*κ *J*κ	41 5	205 κ chains
*V*λ *J*λ	34 5	170 λ chains
*V*H *D*H *J*H	48 23 6	6.624×10^3 H chains
Random association of H and L chains		$6624 \times (205+170) = 2.5 \times 10^6$

or TcRγ chains rearrange before the Ig L, TcRα, or δ chains, and the recombination process assembles the complete (and unique) VH/Vβ/Vδ exon from linear gene arrays of *V*, *D*, and *J* gene segments. Assembly is tightly regulated and occurs in a preferred temporal order in the case of Ig H and TcRβ chains — *D* joins *J* and the combined *D–J* is joined to the *V* subexon. However, in the case of TcRδ, *V–D* joining precedes *V–D–J* joining. In the case of the light chains (whether Ig L, TcRα, or TcRγ), the question of order does not arise — *V* joins *J*. The main features of *V*(*D*)*J* recombination are given below.

- **V(D)J recombination occurs at conserved noncoding Recombination Signal Sequences (RSSs) that lie adjacent to each V, D, and J segment.**
  - An RSS consists of a palindromic[7] heptamer (seven DNA basepairs) and an A/T rich nonamer (nine DNA basepairs) separated by intervening spacers of either 12 basepairs (one turn of the DNA helix) or 23 basepairs (two turns of the helix).
  - The length of the spacer is important in determining RSS functionality; efficient recombination can occur only between RSS with 12- and 23-basepairs (termed *the 12/23 rule*) spacers. Thus, for the Ig κ locus, all *V* segments are attached to 12-spacer RSSs and all *J* segments are attached to 23-spacer RSSs, ensuring that *V–J* joining is much more efficient than *V–V* or *J–J* joining. Whenever the V region is assembled from three subexons (i.e., H/β/δ), both *V* and *J* segments must join *D*, so the *D* segments are flanked by RSSs of appropriate spacer lengths on each side (Figure 6.8). For Ig H chains, both *V* and *J* segments have 23-spacer RSSs, and the *D* segment is flanked by 12-spacer RSS on both sides.
  - Distances between the RSSs does not seem to affect the efficiency of *V*(*D*)*J* recombination. For example, human Ig κ and Ig H loci have *V* subexons extended over 2300 kilobases, and distal *V* segments are still used with reasonable frequency. How RSS pairs locate each other over such large distances is not clear. However, recombination within a single chromosome is strongly preferred. Thus, recombination between κ and λ loci (lying on different chromosomes) is 1/1000 as frequent as *V–J* recombination within the same loci.
  - RSSs are generally arranged such on the antigen receptor loci that the joined coding segments remain in the chromosome and the junction of the RSSs (known as the *signal joint*) is excised on a circular DNA that is eventually lost from the cells.

[7] A palindrome is a word, verse, sentence, or numerical sequence that reads the same backwards and forwards. A classic example — Madam I'm Adam.

Figure 6.8 VDJ recombination occurs at RSSs adjacent to each *V*, *D*, and *J* segment. Each coding region RSS contains moderately well-conserved heptamer and nonamer sequences (represented here by consensus sequences) separated by an intervening stretch of 12 or 23 basepair, nonconserved DNA, called the spacer (upper panel). The bottom panel shows the arrangement of 12 (open triangle) or 23 (closed triangle) basepair spacers in the three Ig (H, κ, λ) and four TcR (β, δ, α, γ) loci (adapted from Gellert, *Annual Review of Biochemistry*, 2002, 17:101).

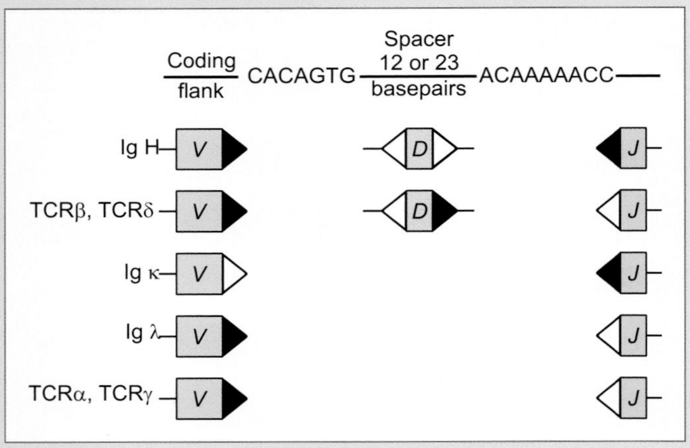

- ***V*(*D*)*J* recombination has two distinct stages — DNA breakage and strand break repair.**
  - In the first stage, the lymphoid-specific proteins RAG-1 and RAG-2 (**R**ecombinase **A**ctivating **G**enes **1** & **2**) recognise the RSSs, ensure their correct 12/23 pairing, and break the DNA between each heptamer and the neighbouring coding sequence.
  - The second stage of the process is the joining phase and has many aspects in common with general DNA double-strand break repair. Ubiquitously expressed NHEJ proteins process and link the ends into coding joints and signal joints. At the end of the process, two (for Ig L, TcRα, TcRγ) or three (for Ig H, TcRβ, TcRδ) separate segments of DNA recombine to yield a single V region.
- ***RAG-1* and *RAG-2* are highly conserved genes that are expressed in lymphoid cells undergoing *V*(*D*)*J* recombination, and they are indispensable for this process**[8].
  - All known RAG activities require the presence of both proteins. By themselves, they seem to preferentially bind RSS with 12-spacers. However, in the presence of HMG-1 and -2 (**H**igh-**M**obility **G**roup **P**roteins; one of a group of nonspecific, DNA-binding and -bending proteins), the RAG proteins bind coding sequences flanked by either 12- or 23-spacers.
  - The complex consisting of RAG-1, RAG-2, HMG-1, and 12- and 23-spacer DNA is highly stable. It is also resistant to other nonspecific DNA present in the vicinity.
- **RAG (1&2) proteins bind DNA and introduce DNA double-strand breaks between the target RSS heptamer and the flanking sequence of *V*, *D*, or *J* coding segments**, yielding two types of termini — a blunt 5'-phosphorylated signal end and a hairpin coding end that retains the full coding sequence. The four RAG-liberated DNA ends remain associated with RAG proteins in a stable, postcleavage synaptic complex.
  - The cleavage occurs in two steps and seems to require Mg^{2+}; first a nick is introduced at the 5' end of the signal heptamer of the two participating coding sequences, leaving a 5'-phosphoryl group on each RSS and a 3'-OH group on the coding end (Figure 6.9).
  - The second step is a transesterification reaction catalysed by RAG proteins; the 3'-OH group of the coding strand invades and joins the phosphoryl

[8] The indispensability of RAG proteins in the process of antigen-receptor generation can be judged from the fact that RAG knock out mice lack mature functional T and B cells.

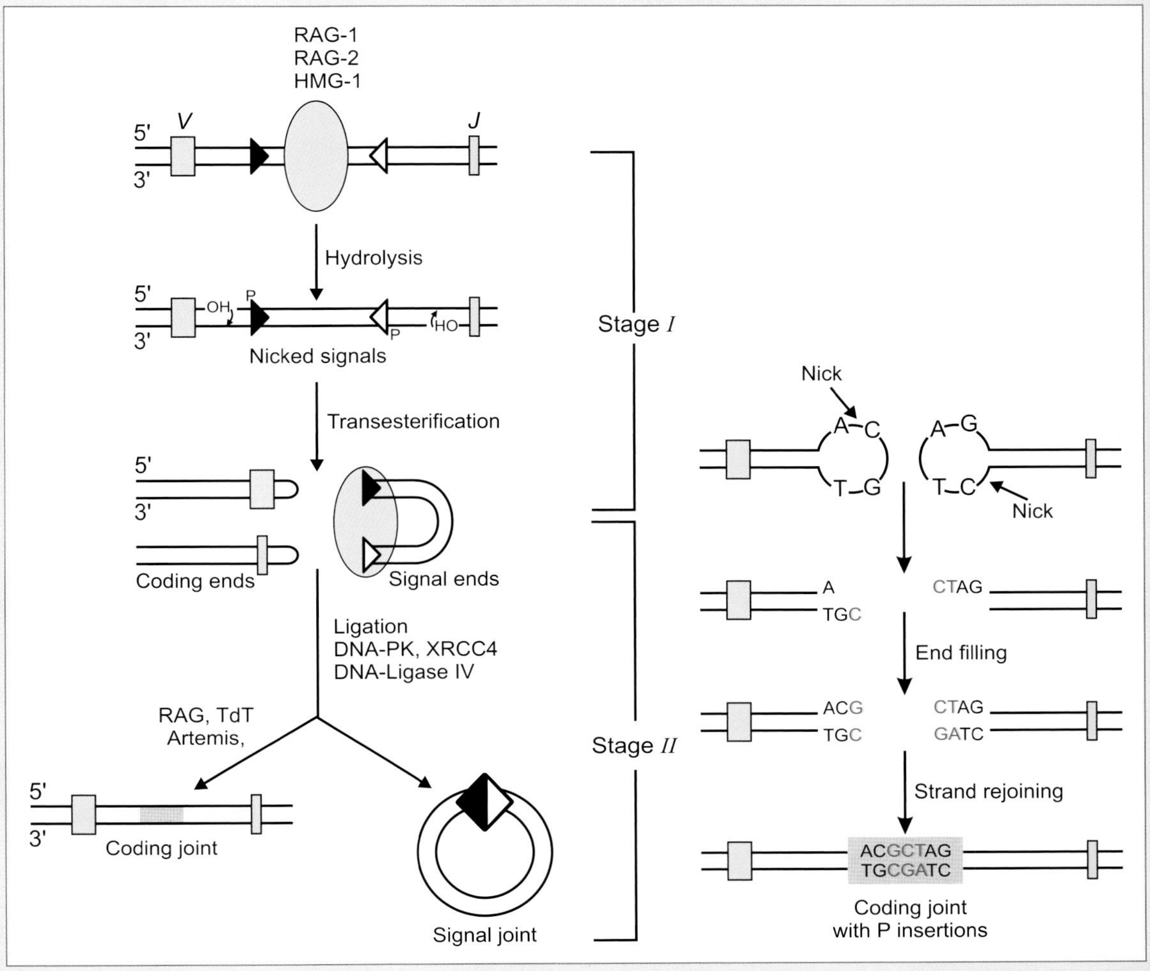

Figure 6.9 *V(D)J* recombination occurs in two stages — DNA breakage catalysed by RAG proteins and DNA double-strand break repair catalysed by NHEJ proteins. RAG-1 and RAG-2 recognise the RSSs and ensure their correct 12/23 pairing (open and closed triangle). They bind to the DNA and introduce double-strand breaks between the target RSS heptamer and the flanking sequence of the coding segments to yield a 5'-phosphoryl group on each signal end and a 3'-OH group on the coding end. Transesterification catalysed by RAG proteins results in the 3'-OH group of the coding strand invading and joining the phosphoryl group at the same nucleotide position on the opposite strand. This generates a hairpin coding end and a blunt signal end. The two signal ends are joined to yield a signal joint catalysed by NHEJ enzymes such as DNA-PK, DNA ligase IV, and XRCC4. The coding ends undergo nucleotide additions or deletions (grey area) before ligation, and they require RAG proteins, TdT, and Artemis in addition to the NHEJ machinery. The left panel depicts the process of P insertions in the coding joints. For simplicity, the two coding ends are shown adjacent to each other. The hairpin coding ends are opened by introducing nicks that are a few bases off each other, leaving self-complimentary single-strand extensions. End-filling using the overhang as template results in P insertions (grey box).

group at the same nucleotide position on the opposite strand, yielding the DNA hairpin coding end and a blunt, 5'-phosphorylated signal end.

- **The two blunt signal ends are joined to yield a signal joint on a circular DNA**; this signal joint has no further role and is eventually lost[9].
  - Signal joints are relatively simple, usually precise end-to-end fusions of the two heptamer sequences.

[9] Thus, *V(D)J* arrangement results in *loss* of genetic information. Incidentally, this is a unique process involving the destruction of genetic information rather than the regulation of gene expression.

Understanding the Terminology

Exons and introns are coding sequences in genes. Introns are intervening, noncoding nucleotide sequences. Both exons and introns are transcribed into RNA. Primary RNA transcripts are converted to mature mRNA molecules by excision of introns and splicing together of exons.

Promoters are relatively short nucleotide sequences extending up to 200 basepairs upstream from the transcription initiation site (i.e., 5' region of the DNA) that promote initiation of RNA transcription in a specific direction. Promoter regions facilitate the binding of RNA polymerase to DNA and orient it for proper transcription of the gene.

Enhancers are nucleotide sequences, situated some distance upstream or downstream from a gene, that activate transcription from the promoter sequence in an orientation-independent manner. Enhancers activate nearby promoters, probably by binding a regulatory protein that can also bind the promoter and RNA polymerase.

Silencers are nucleotide sequences that downregulate transcription in both directions over a distance.

Recombination is the reciprocal exchange of genetic material between DNA fragments. It involves the breaking and rejoining of DNA pieces to generate new DNA pieces and can occur between two different double-stranded DNA molecules or between two parts of the same DNA molecule.

- Homologous recombination occurs between DNA strands that have long stretches of homology. Double-stranded breaks can be repaired if a chromosome or chromatid that is homologous to the broken DNA is available in the cell.
- **N**on-**H**omologous DNA **E**nd-**J**oining (NHEJ) is a pathway that rejoins DNA strand breaks without relying on marked homology. The main known pathway uses the Ku protein-binding complex and is regulated by the DNA-dependent protein kinase. This pathway is often used in mammalian cells to repair strand breaks caused by DNA-damaging agents.
- Site-specific recombination occurs at specific sequences of DNA because of targeting by a specific enzyme called recombinase. *V(D)J* recombination is an example of site-specific recombination that occurs between DNA regions of a single chromosome.

A **mutation** is a change in DNA sequence. Transition is the swapping of one pyrimidine base (cytosine, thymine, or uracil) for another pyrimidine base or of one purine base (adenine and guanine) for another purine base. Transversions occur when one pyrimidine base is swapped for a purine base or *vice versa*.

Error-prone DNA polymerases are DNA polymerases that copy templates inaccurately and introduce mutations in the process. Examples include POLζ, POLν, POLμ, and POLι. Some of these are candidates for enzymes that introduce base-changes during somatic hypermutation.

  - DNA-ligase IV and XRCC4, along with the heterodimeric subunit of the **DNA**-dependent **p**rotein **k**inase (DNA-PK),[10] are required for this ligation.
- Joining coding ends is a more complex process. **The coding ends undergo nucleotide addition or deletion** before ligation.
  - The hairpin ends produced by RAG cleavage must be reopened before the end-joining pathway can process and join them.
  - Hairpin coding ends are opened at the apex or points nearby; RAG proteins along with the recently identified protein Artemis seem to be involved in this opening, although other DNA-repair proteins may also be involved.
  - The opened coding ends undergo nucleotide excision or nucleotide addition; the junctional sequences lie within the antigen-binding site (in the CDRs), so alterations in coding joints are responsible for increasing the diversity of antigen receptors beyond that generated by combinatorial joining of gene segments.
  - The mechanism of nucleotide deletion at the coding junctions is not clear; it is generally believed that exonucleases may be involved.

[10] DNA-PK is a multimeric protein consisting of a catalytic unit (DNA-PKcs) and a heteromeric binding subunit Ku (consisting of Ku70 and Ku80). As explained in chapter 5, the loss of DNA-PK results in SCID because developing lymphocytes fail to rearrange their antigen receptors.

- Two types of nucleotide additions are observed in the coding joints: nontemplated and templated.
  - Nontemplated addition may result in an addition of up to 15 nucleotides at the coding joint. This template-independent addition is due to the enzyme **T**erminal **d**eoxynucleotidyl **T**ransferase (TdT) that is normally expressed only in early lymphoid cells where *V(D)J* recombination is active. TdT adds deoxynucleotides without a template to the ends of DNA chains.
  - Templated nucleotide additions occur because of the off-centre nicking of hairpin DNA intermediates, which results in a self-complementary overhang. These nucleotide additions are termed *P* (for **P**alindromic) *insertions*.
  - The catalytic subunit of DNA-PK along with DNA-ligase IV and XRCC4 seems to be important in the formation of the coding joint.

- All V regions have a weak promoter. Proper ligation places the weak promoter adjacent to the assembled V region into close proximity of enhancers present downstream of the J region or upstream of the C region. These enhancers activate transcription from the particular promoter, allowing the cell to make complete Ig (H or L) or TcR (α, β, γ, δ) chains.

Regulation of *V(D)J* recombination: As explained, lymphocyte-specific RAG protein expression limits *V(D)J* recombination activity to nonproliferating stages of developing lymphocytes. The randomness and imprecision of *V(D)J* recombination results in only about one in three *V(D)J* rearrangements being in frame, and therefore, productive. Those cells that make nonproductive rearrangements go on to rearrange their second allele so that most differentiating lymphocytes eventually achieve productive rearrangements. V–J joining affords even more chances of corrections, as recombination may be tried again on the same allele by the use of a *V* region upstream and a *J* region downstream of the erroneous junction. If a lymphocyte fails to rearrange its receptor productively, it is pushed into apoptosis. Upon productive rearrangement, the newly synthesised Ig H (or TcRβ) chain associates with a surrogate L (or pre-TcRα) chain to form a pre-receptor complex. The expression of these surrogate pre-receptors results in the cessation of further rearrangements, ensuring allelic exclusion. It also activates the rearrangement of the Ig L or TcRα genes. Allelic exclusion makes certain that only a single allele at a particular locus is expressed in a single B or T cell; as the second allele does not undergo productive rearrangement, it is automatically excluded from expression. Allelic exclusion is an actively regulated process that occurs at the progenitor to precursor transition through feedback control of the *V*– to –*DJ* joining step, possibly by the surrogate light chain. The exact mechanism that enforces allelic exclusion is not clear. If the receptor generated after gene rearrangement recognises self-antigens, the lymphocytes get one more chance at gene rearrangement in the process of receptor editing. Receptor editing is possible only because of the continued expression of RAG proteins in immature lymphocytes. Receptor editing allows secondary rearrangement of the antigen receptor locus and avoids clonal deletion of the newly formed immature B or T cells (Figure 6.10). It is relatively easy to understand such rearrangements at the Ig Lκ and TcRα loci; the genomic organisations of these loci, consisting of only *V* and *J* subexons, permit successive rearrangements. However, the mechanism of receptor editing of the Ig H/TcRβ is not clear.

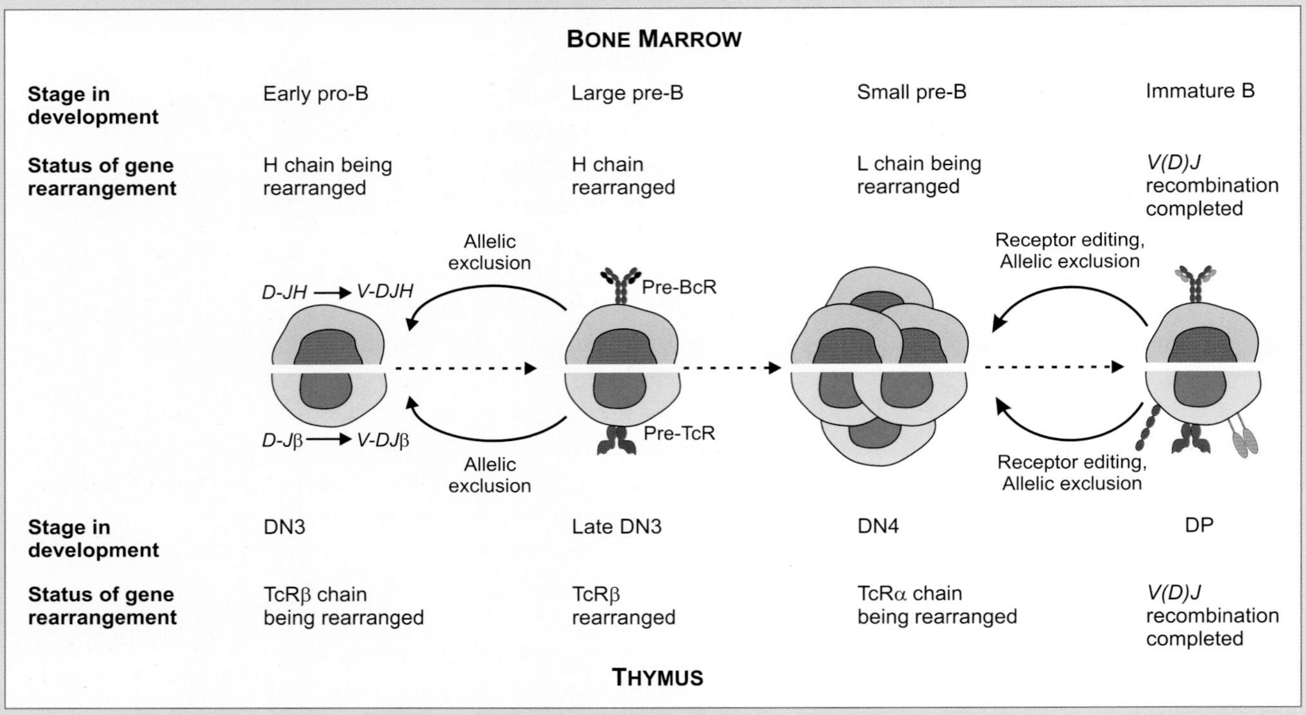

Figure 6.10 Developing B and T lymphocytes undergo similar ordered *V(D)J* gene rearrangements; feedback mechanisms link appropriate antigen receptor expression to further development. V region genes of the H chain and TcRβ chain are the first to undergo recombination in the developing lymphocytes. The *D–J* rearrangement occurs first, and the assembled *DJ* is joined to the *V* subexon. Productive rearrangement results in gene expression. The newly synthesised Ig H or TcRβ chain associates with its surrogate partner (indicated by blue and red symbols, respectively) and is expressed at the cell surface. Pre-receptor expression is a major checkpoint in the life of the developing lymphocyte. Signalling through the pre-receptors ensures allelic exclusion, and it allows clonal expansion and further progression on the developmental pathway. However, TcRγ and TcRδ gene segments rearrange without any strict order and without the clonal expansion and selection observed after Ig H and TcRα rearrangements. Ig L chain or TcRα chain rearrangement is also activated. A productive rearrangement causes the cessation of the rearrangement of these chains and enforces allelic exclusion. Cells that fail to undergo productive rearrangement may undergo receptor editing. Only those cells that have undergone productive rearrangement express antigen receptors on their cell surfaces and become immature (B or T) lymphocytes.

6.5.2 CSR and Somatic Hypermutation

Following productive rearrangement, lymphocytes undergo selection in the primary lymphoid organs before leaving. T cells do not undergo any further recombination events at their antigen receptor locus; they (and their progeny) are stuck with the TcRs assembled in the primary lymphoid organs. B lymphocytes, on the other hand, are subject to further recombination events. This happens in the later phases of the primary immune response to TD antigens, when proliferating B cells differentiate to memory cells and plasma cells. B cells undergo a second wave of genetic alterations — CSR and somatic hypermutation at the Ig gene loci — at this stage. **CSR allows the switching of the CH region expressed** from Cμ to Cγ, Cα, or Cε, resulting in the expression and secretion of IgG, IgA, and IgE antibodies, respectively, without changing antigen specificity[11]. The different Ig isotypes use the same set of V genes and only the C regions of H chains are shuffled so that the Ig effector functions vary but their antigenic specificity remains unchanged. In contrast, **somatic hypermutation involves a change in the**

[11] This is similar to stretching your wardrobe by pairing one pair of trousers with different tops to create multiple outfits. Languages use a similar trick. For example, in English, prefixes to the word *logy* (meaning *study of*) are switched to describe various fields of study — zoology, biology, sociology, etc.

V(D)J Recombination

- *V(D)J* recombination is the process by which antigen receptors of B and T lymphocytes are assembled.
- The antigen-binding regions of the receptors are assembled from separate subexons.
  - Ig V_L region and TcR α, γ chains are encoded by two subexons, *V* and *J*.
  - Three separate subexons (*V*, *D*, and *J*) code for Ig V_H region and TcR β, δ chains.
  - Assembling occurs in a temporal fashion; generally *D* joins to *J* and *DJ* is joined to *V*.
- *V(D)J* recombination occurs at conserved noncoding RSSs that lie adjacent to each *V*, *D*, and *J* segment.
  - Each RSS consists of a palindromic heptamer and A-T-rich nonamer separated by a spacer of 12- or 23-basepairs.
  - The nonamer and heptamer come together during recombination.
  - Efficient recombination occurs only between RSSs with 12- and 23-basepair spacers.
- RAG-1 and RAG-2 proteins are involved in the recombination process.
  - They recognise the RSSs, ensure their correct 12/23 pairing, and introduce a double-strand break between the target heptamer and flanking coding sequence.
  - The breakage yields two types of termini — a blunt 5'-phosphorylated signal end and a hairpin coding end that retains the full coding sequence.
  - The two blunt signal ends are joined to yield a signal joint on a circular DNA.
  - The two coding joints are ligated by nucleotide modification (i.e., either addition or deletion).
- Those cells that successfully recombine their Ig H or TcR β, δ chains express these chains as a part of the pre-receptor complex and start rearranging Ig L or TcR α, γ chains.
- Cells that make nonproductive rearrangements go on to rearrange their second allele; if they fail to rearrange the second allele they apoptose.

nucleotide sequences of the genetic loci of the antigen-binding pocket **of the antibody V regions**, and it results in increased affinity for the antigen. Although CSR may be observed in B1 or MZ cells responding to TI antigens, somatic hypermutation is generally not observed for TI antigens.

Mature B cells that have completed the functional *V(D)J* recombination of both H and L chain genes express IgM molecules at their cell surfaces and get negatively selected for self-antigens. Those that do not express BcRs capable of reacting with self-antigens survive this selection, express mIgD molecules, and migrate to the secondary lymphoid organs (e.g., spleen and lymph nodes), where they compete for survival signals. Those B cells that receive these survival signals enter the pool of circulating lymphocytes and remain quiescent until they encounter TD antigens. When these B cells encounter a TD antigen recognised by their BcRs and receive appropriate signals from T cells, they proliferate vigorously in the lymphoid follicles in special microenvironments known as germinal centres. As described in chapter 8, these specialised structures in B cell areas provide an appropriate niche for proliferation, BcR alteration, affinity-based selection, and differentiation of activated B cells. Nevertheless, it should be noted that the principle of affinity-based selection operates right from the earliest stages of the immune response. Thus, of all the B lymphocytes available in circulation during contact with antigens, only those cells with an affinity high enough to permit the molecules to remain bound together until internalisation are recruited into the immune response.

i. CSR: The murine Ig H locus consists of eight C_H genes located downstream of the *V(D)J* locus. The $C\mu$ gene is located at the V_H proximal end of the C_H gene cluster. $C\alpha$ is at the distal end (Figure 6.11). Each C_H gene (except that of $C\delta$) is

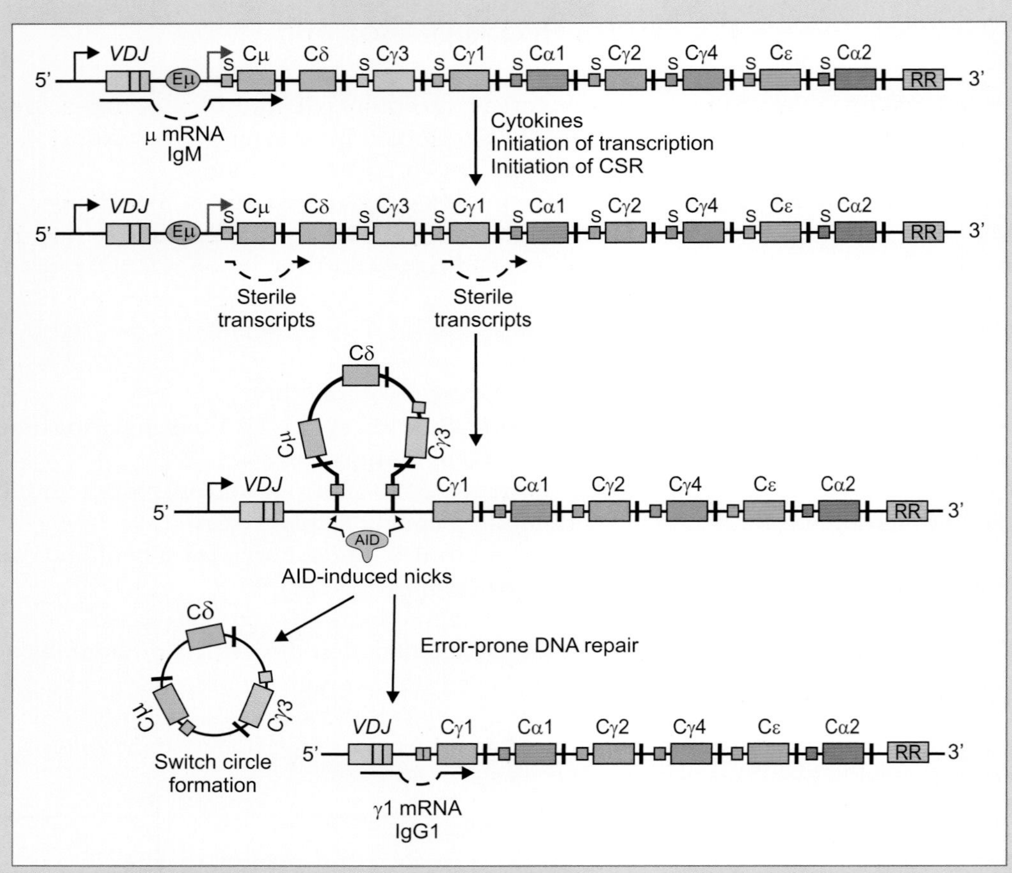

Figure 6.11 DNA recombination in switch regions of the Ig *CH* genes allows isotype switching in antigen-activated B cells. The figure depicts a simplified schematic of the germline murine Ig H chain on chromosome 12. Each *CH* gene (except that of *Cδ*) is flanked at its 5' by a Switch (S) region that is the site of CSR. Transcription is initiated from a promoter 5' of the I exon (depicted by the bent black arrow), runs through the S region, and undergoes polyadenylation downstream of the CH exon. In cells that have not switched their isotype, this yields a μ chain transcript. Regulatory region (RR), downstream of the last *CH* gene (*Cα*) and the intronic enhancer (Eμ, purple bent arrow) are responsive to cytokines and hence influence class-switching. CSR is preceded by the transcription of the two S regions and is influenced by cytokines, such as IL-4 or IFNγ. Deamination by AID starts the process by introducing nicks in the two S regions, resulting in the looping out and deletion of intervening DNA segments as circular DNA. Enzymes of the NHEJ pathway and error-prone repair enzymes then ligate the DNA so that the rearranged V region is juxtaposed to a new CH region (here Cγ1).

flanked at its 5' by the **S**witch (S) region, a 1-10 kilo basepair region, composed of tandem (i.e., one behind the other), repetitive sequences with many palindromes. Although their exact primary sequences are not similar, all S regions contain G-rich pentameric sequences that are major repeat units of Sμ. **CSR occurs at these S regions**. Because the *Cδ* gene is not flanked by an S region, an isotype switch to IgD cannot occur. Instead, the entire *VDJCμCδ* region is transcribed into a long primary RNA transcript and differentially spliced to generate *VDJCμ* (yielding IgM) or *VDJCδ* (yielding IgD). The formation of this long primary transcript is feasible because of the proximity of the *Cμ* and *Cδ* genes (only 5 kilo basepairs apart). This differential splicing results in naïve B lymphocytes expressing both mIgM and mIgD molecules on their cell surface. Upon activation, naïve B cells initially produce the μ heavy chain and, hence, IgM antibodies[12]. Activation also starts CSR, which

[12] Although most B cells initially express IgM isotype and then switch isotypes in the microenvironment of germinal centres, CSR can also occur outside germinal centres; in the case of the IgA isotype, it can occur without the prior expression of mIgM molecules.

Class Switch Recombination

- CSR allows B cells to produce Igs with the same antigenic specificity but different isotypes.
- Isotype switching is feasible because the V and C regions of Ig molecules are encoded by different genes.
  - The murine (and human) Ig H locus consists of eight *CH* genes downstream of the *V(D)J* locus.
  - The *Cμ* gene is at the VH proximal end of the *CH* gene cluster; *Cα* is at the distal end.
- S regions flank each *CH* gene (except of *Cδ*) at its 5' region.
- CSR occurs within two S regions, resulting in the looping out and deletion of intervening DNA segments as circular DNA.
- Transcription through the S region plays a primary role in targeting CSR; CSR is preceded by transcription of the two S regions undergoing CSR.
- AID and NHEJ machinery are both necessary for CSR.
- Cytokines can influence CSR outcome.

occurs within two S regions, resulting in the looping out and deletion of intervening DNA segments as circular DNA. The *V(D)J* exon is juxtaposed to a downstream *CH* gene, allowing the generation of a different isotype. As shown in Figure 6.11, CSR between the Sμ and Sγ1 region 5' to the *CH* gene brings the *Cγ1* gene adjacent to the VH exon, resulting in isotype switching to IgG1. Subsequent switching to other isotypes may occur at the recombinant switch region. Thus, CSR could result in direct isotype switching from IgM to IgE or sequential isotype switching from IgM to IgG1 to IgE. CSR has been shown to generate two products — the rearranged chromosome and extrachromosomal circles containing the deleted intervening regions.

On the basis of current knowledge, the molecular mechanism of CSR can be divided into four steps.

1. **Selection of target S region**: A dedicated enhancer and a dedicated promoter upstream of each S region together activate transcription in response to extracellular signals delivered by T cells and cytokines that regulate production of specific classes of Ig molecules.
   - Germline CH genes are organised into germline transcription units in which transcription initiates from a promoter 5' of the I exon,[13] runs through the S region, and undergoes polyadenylation downstream of the CH exon (Figure 6.11). A 3' enhancer complex located 40 kilo basepairs downstream of the last *CH* gene (*Cα* in the case of the murine Ig H locus) can affect germline transcription and, therefore, CSR. RNA splicing generates a processed germline transcript by fusing the I exon to the CH exons and deleting the intervening S region-derived sequences. Gene-targeting studies demonstrate the necessity of the promoter integrity of the I exon for efficient CSR. Transcription through the S region plays a primary role in targeting CSR; CSR is preceded by transcription of the two S regions, starting from the I promoter located 5' to each S region.
   - Cytokines can promote switching to a particular isotype. Thus, IL-4 is known to promote isotype switching to IgG1 and IgE in mice (IgG4 and IgE in humans). IFN-γ has been shown to promote isotype switching to IgG2b/IgG2a in mice and IgG1 in humans, and TGF-β induces germline transcripts and subsequent switching to IgA.

[13] The Ig H *μ* gene, like other CSR-capable *CH* genes, has Ig H intronic enhancer — a two kilo basepairs DNA sequence that lies between the V region and the Sμ region. Immediately downstream of the I enhancer lies the I exon that serves as a promoter of the *Cμ* gene and regulates the transcription of the Ig H gene. This noncoding I exon is spliced on to the first exon of the CH region being transcribed. Disruption of I promoter or I exons prevents CSR.

Affinity Maturation

- Affinity maturation is the increase in affinity of antibodies for their homologous antigen with respect to time. Thus, antibodies produced later in the immune response have a much greater affinity for the antigen than those produced earlier.
- Observed only for TD antigens, it occurs in germinal centres and is the result of hypermutation in the rearranged V gene.
- Hypermutation occurs during a small window in the proliferative stage of B cells in the germinal centre.
  - It occurs stepwise, with brief bursts of high mutation rates interspersed with mutation-free growth.
  - Hypermutation is nonrandom; the RGYW motif is the preferred target of the hypermutation machinery.
  - Most accumulated mutations are point mutations that alter antibody CDRs.
  - AID, an RNA-editing enzyme specific to germinal centre B cells, is vital to affinity maturation.

- Cytokines influence CSR by regulating the expression of germline transcripts. Segments of DNA located 5' (or upstream) of I exons contain promoters/enhancers that regulate transcription of germline transcripts. These promoters have elements responsive to cytokine-induced transcription factors, and hence, cytokines can influence switching to a particular isotype by promoting the transcription of a particular S region.
- Cytokines may influence CSR to a particular *CH* gene by inducing its transcription and influencing interaction with the downstream enhancer elements. Four enhancer elements downstream of *Cα* are thought to influence the germline transcription of distal genes.
- Thus, cytokines accompanied by appropriate costimulatory signals (e.g., CD40–CD154 interaction) induce the production of sterile transcripts[14] from promoters that are upstream of the targeted switch regions.

2. **Formation of G loops in the transcribed S region**: As explained, S regions are G-rich on the nontemplate strand and must be transcribed in *cis* and the correct orientation to support recombination.
   - Recent research has revealed that CSR (and somatic hypermutation) are triggered by the targeted deamination of deoxycytidine residues by AID (**A**ctivation-**I**nduced cytosine **D**eaminase), an enzyme specifically expressed in activated B lymphocytes. AID-deficient mice and humans lack CSR and somatic hypermutation. Some hyper-IgM syndrome patients with impaired CSR have mutations in the human gene encoding AID.
   - S regions carry two characteristic sequence motifs — G-rich repeats, and hotspot motifs for AID deamination.
   - S region transcripts form characteristic structures that electron microscopic images have identified as extended G-loops in which the G-rich RNA transcript pairs stably with the C-rich DNA template strand and displaces the G-rich DNA strand to a single strand.
   - It is thought that AID interacts specifically with and deaminates the G-rich strand of G-loops.
3. **Introduction of double-stranded breaks**: Two double-stranded cleavages are known to occur in CSR. The exact mechanism of how and where these nicks are made is unclear. It is hypothesised that after DNA deamination by AID at S

[14] Sterile transcripts are driven from I promoters located upstream of all S regions. They are believed not to encode proteins. Instead, they are thought to be spliced to form mature sterile transcripts that contain the C region exons and sequences upstream of the S region.

regions, uridine removal occurs because of the action of the uracil DNA glycosylase, leaving a nicked site where endonucleases generate a single-strand break in the DNA followed by double-strand break.

4. **Repair and ligation**: The constituents of the NHEJ pathway, namely, Ku complex and DNA-PKcs, are involved in CSR but may not be the only mechanisms of DNA repair. Error-prone DNA polymerases may also play a role in the repair process; they are thought to be responsible for CSR's high frequency of mutations. The repair and ligation rearranges the chromosome, where the *V(D)J* exon is juxtaposed to a downstream *CH* gene, and an extrachromosomal circle carrying the intervening segment is formed.

ii. Somatic hypermutation: BcR diversity in uncommitted B cells arises through combinatorial usage of *V(D)J* gene fragments to form functional H and L chains and through the junctional diversity generated by the imprecision of the joining process. **Antigen-stimulated cells show further diversification because of somatic point mutations in their V regions**. The frequency of such mutations is extraordinarily high, with a mutation rate of 1×10^{-3} per basepair per generation. As this rate is about 10^6 times higher than spontaneous mutations, the process is referred to as *hypermutation*. This increase in affinity observed in the latter phases of the primary response and especially in the secondary response is termed *affinity maturation*, and it can be directly correlated to a steady increase in the number of mutations. Hypermutation seems to occur specifically on the rearranged V gene, independent of whether this gene is translated into a functional H or L chain or whether it is on the nonactive chromosome. Even an unrelated sequence introduced on the V region undergoes hypermutation.

Linking Back: Theories of Antibody Diversity

Two theories were proposed for antibody diversity. The first, the *Germline Hypothesis*, proposed that all genes needed for generating the antibody repertoire are present in the fertilised ovum (i.e., the germline) and, therefore, in every cell. These genes were postulated to arise during evolution through conventional mechanisms for gene duplication, mutation, and selection. Each antibody-generating cell was thought to express only one set of V genes from the whole complement of V genes present in its germline. Thus, the germline hypothesis claimed that selective gene expression determined antibody specificity of lymphocytes.

The second theory postulated that relatively few genes for the V locus were inherited. The V regions of cells destined to be lymphocytes were proposed to mutate faster than the remaining DNA (i.e., the V region became a mutational hotspot) during development. This was called the *Somatic Mutation Theory*, as nongerm cells were thought to generate diversity. A small number of germline V genes were thought to become diversified through mutations in somatic cells, yielding many clones of immunologically competent cells. Of the numerous clones thus generated, only nonself-reactive clones were postulated to survive, while self-reactive clones were deleted. The germline theory implies that antibody diversity was generated in a species over evolutionary time, whereas the somatic variation theory implies that antibody diversity is generated within individuals during their lifetime.

As happens often in science, both theories proved partially correct. The germline does contain a large repertoire of antibody genes. However, this repertoire is insufficient to encode for the actual repertoire of antibodies that an individual is capable of producing. The antibody repertoire is increased through combinatorial and junctional diversity in the developing lymphocytes. Only after antigenic challenge is the antibody diversity further increased by somatic mutations.

To understand how hypermutation increases affinity, we need to understand antigen–antibody interactions at the molecular level. Complexes between proteinic antigens and homologous antibodies involve contacts between multiple amino acid residues on both molecules. Multiple noncovalent bonds formed between these residues determine the strength of the reaction (Table 4.1). Bond formation is critically dependent on the distance between interacting groups. The closer the groups, the stronger the forces between them, and the higher the affinity of the antibody for that antigen. For example, in an antibody–hen egg lysozyme complex, 17 residues of the antibody interact with 16 residues of lysozyme. A change in a single amino acid at or around these residues may permit additional salt links, hydrogen bonds, and so forth, resulting in firmer binding. A maturation system based on random mutation is obviously a game with few winners and many losers. Most point mutations do not result in increased affinity for the antigen; they may result in either a lowered affinity for the antigen, early termination of the protein chain, or recognition of self-antigens. Such B cells are the 'losers' eliminated by apoptosis. Only a few B cells that show increased affinity for the challenging antigen are the 'winners' and are permitted to expand and differentiate. Selection thus has a major role in affinity maturation. A short burst of somatic hypermutation, followed by selection and clonal expansion, forms the basis of affinity maturation.

Although hypermutation has been recognised since the early 1960s, the underlying process is only recently being understood.

- Hypermutation is nonrandom, with distinct areas showing a high frequency of mutations (these areas are called *hotspots*). **Mutations are mostly confined to a region spanning about two kilo basepairs downstream of the Ig promoter region**.
  - Majority mutations are point mutations, with transitions more frequent than transversions and with A nucleotides in the coding strand replaced twice as frequently as T nucleotides.
  - Certain nucleotide motifs are the preferred targets of hypermutation machinery. These are the RGYW (where R is A or G, Y is C or T, W is A or T)

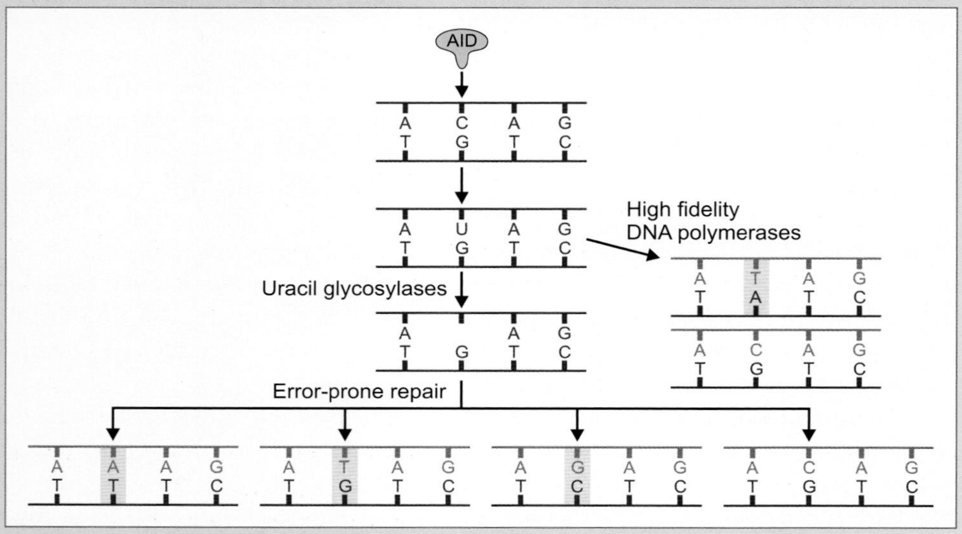

Figure 6.12 AID, found in germinal centre B cells, starts the process of hypermutation by the deamination of DNA. AID is believed to convert C in the rearranged V regions of Ig chains to U. High-fidelity DNA polymerases read U as T and pair it with A, resulting in a mutation (grey box). The removal of U by uracil glycolases leaves a gap at the site, and low-fidelity polymerases fill the gaps more or less at random. For simplicity, only the mutated strand is shown in low-fidelity DNA repair (adapted from Gearhart, *Nature*, 2002, 419:29).

Figure 6.13 B cells activated by TD antigens undergo diversification of their antibody repertoire as a result of somatic hypermutation in their V region genes. The figure depicts the proposed model of hypermutation. The uppermost panel is a schematic of Ig H chain genes of a B cell that has undergone VDJ recombination. Also shown are the promoter and enhancer regions (bent arrows) that seem to permit hypermutation at particular loci and the preferred targets (RGYW nucleotide motif) of the hypermutation machinery. AID introduces a DNA lesion at the target motif, resulting in a double-stranded break. A repair complex is formed at the site and leaves a gap in one strand that is filled in by error-prone DNA polymerases. This results in a mutated gene (grey box) upon strand replication. For simplicity, only the mutated strand is shown in the figure.

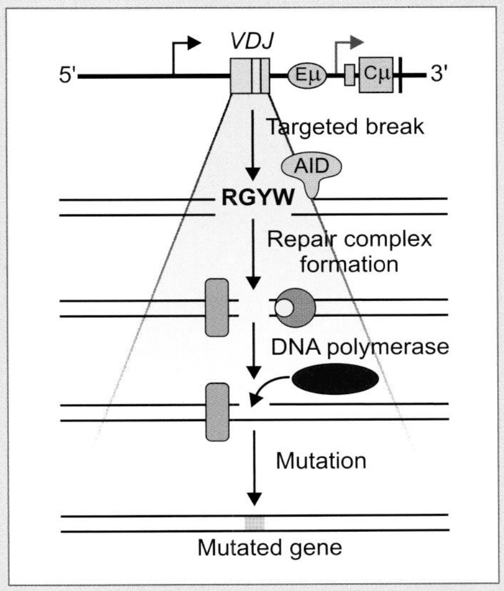

or its complement, WRCY motifs. They were identified by establishing and analysing a large database of somatically mutated Ig genes. The surrounding sequences seem important in deciding if the RGYW motif becomes a substrate of the mutation machinery. These mutational properties probably reflect the specificities of the mutator enzyme(s).

- **Somatic hypermutation is dependent on transcription**. Hypermutation of the targeted region appears to be directed by transcription-related elements, including the promoter and enhancer regions. Experimental data suggest that the transcriptional promoter determines the precise region that will mutate, and specific enhancers allow mutations at particular loci.
- **AID is crucial for somatic hypermutation**.
  - Hypermutation is postulated to start because of the deamination of DNA by AID; such deamination converts the C in DNA to U.
  - If mismatches are not repaired before DNA replication, U is read as A, resulting in the addition of T, that is, a transition of the deaminated strand.
  - Alternatively, uracil glycosylases may remove U before replication, leaving a gap which may be filled during replication or through DNA mismatch repair pathways (Figures 6.12, 6.13). Mutations have been postulated to occur when error-prone DNA polymerases try to correct the error.
  - Although many recent experiments support the role of AID in hypermutation, many questions remain unanswered. It is unclear how AID targets the rearranged V genes, and how and why error-prone polymerases are involved in the DNA repair. DNA repair is usually carried out by the more accurate DNA polymerase β. Thus, factors that subvert DNA polymerase β usage and promote error-prone polymerase usage need to be identified.

To summarise, *V(D)J* recombination with hypermutation and CSR allows the use of a small number of genes to generate an open-ended repertoire. Each individual inherits sketchy information — a broad game plan — which is refined as the need arises. The sketchy information is in the form of relatively few *V(D)J* genes from which an individual produces a comparatively large number of receptors by *V(D)J* recombination. The generated receptors recognise epitopes with moderate to low affinity. If, in its lifetime, a particular B lymphocyte bearing a certain receptor does not come across its antigen, it does not get activated or improved upon. However, upon antigen contact, somatic hypermutation refines the receptor, increasing its affinity for that antigen. CSR allows the same antigen to be attacked by different mechanisms as it endows a new effector function upon the Ig molecule without altering its antigenic specificity. Although TcRs and BcRs are generated by similar mechanisms, TcRs do not undergo somatic hypermutation. Further refinement of the T cell repertoire could be potentially deleterious because of the high risk of self-reactivity and may therefore not have evolved for TcRs.

Molecules of Adaptive Immune Recognition: Antigen-presenting Molecules and Antigen Presentation

7

I know I left too much mess
And destruction to come back again
And I caused nothing but trouble
I understand if you can't talk to me again
And if you live by the rules of "it's over"
Then I'm sure that that makes sense

Well I will go down with this ship
And I won't put my hands up and surrender
There will be no white flag above my door

— Dido, *White Flag*

Abbreviations

β_2-m:	β_2-microglobulin
CLIP:	Class II-associated Invariant chain Peptide
HSP:	Heat Shock Protein

7.1 Introduction

The binding of B or T cell antigen receptors to their respective ligands activates the adaptive immune system. The preceding chapters have shown that B cell recognition is rather simple, as BcRs recognise native ligands. The story gets more complicated with αβ T cells. The αβ TcR ligand has to be processed and loaded onto specific molecules that are expressed on APC surfaces. These molecules on which antigen fragments are loaded are MHC or MHC-like (CD1) molecules[1]. One can think of MHC molecules as flagpoles; the displayed flag will depend upon the availability and wherewithal of the cell. MHC molecules are unique in that they bind a variety of peptides — different flags can be flown from the same MHC poles. TcRs are the T cells' flag-recognition systems. MHC molecules can be loaded with fragments of any proteins that the cells have internalised or synthesised and expressed on cell surfaces. Molecules loaded with self-peptides act as white flags, sparing the cell from the adaptive immune system's ferocity. The presence of fragments derived from infecting agents or from proteins synthesised during infection or transformation serves the opposite purpose. These fragments behave like red flags, signalling potential danger and targeting the cell for destruction.

7.2 MHC Class I Molecules

Located on chromosome 6 in humans and chromosome 17 in mice, the MHC region codes for polypeptides of three classes. Two of these are highly polymorphic[2] peptide receptors involved in T cell antigen recognition; the third class of molecules includes some complement components, cytokines and enzymes that are important in innate immunity, and proteins involved in growth and development.

Except for a few cell types, such as neurons, almost all **nucleated cells of the body express MHC class I molecules**[3]. The degree of expression differs for different cell types. Lymphocytes show the highest level of expression. Expression is very low in fibroblasts, neural cells, and muscle cells. Pro-inflammatory stimuli, such as IFNs (including α, β, and γ), TNF-α, and LPS, upregulate the expression of these molecules on most cell types. Murine class I molecules are termed H2-K, -L, or -D, with the haplotype indicated by a letter in superscript (e.g., H2-K^d, read as H2-K of d); human equivalents are HLA-A, -B, or -C, with the haplotype indicated by a number (e.g., HLA-A1).

7.2.1 MHC Class I Structure

MHC class I molecules are transmembrane glycoproteins that have an Ig-fold in their membrane-proximal domain; the membrane-distal domains form a cleft or groove for peptide binding. Class I molecules are composed of a glycosylated 45 KD polypeptide chain termed the α (or heavy) chain that is noncovalently associated with a nonglycosylated peptide — β_2-microglobulin (β_2-m). β_2-m is a 12 KD peptide sometimes referred to as the light chain of class I proteins. Both heavy and light chains belong to the Ig superfamily. Calling β_2-m *light chain* is misleading — it is not encoded by MHC genes but by genes located on chromosome 15 in humans and chromosome 2 in mice. It is a soluble protein that can be found by itself in serum or urine. Although β_2-m can be synthesised independent of the class I molecule, the reverse does not hold true. β_2-m is necessary for the processing and expression of class I molecules, and individuals with a congenital defect in β_2-m production fail to express class I molecules.

[1] *Major histocompatibility* complex seems a strange name for molecules involved in antigen presentation to T cells. The name was coined long before the function of these molecules was established. The *major* and *histocompatibility* in the name came from the observation that this genetic region appeared to affect transplant (allograft) rejection; *complex* referred to the fact that the region consisted of numerous loci closely linked to each other involved in different functions. Because these molecules gave rise to antibodies, they were called MHC *antigens*. In mice, the MHC complex is called the H-2 complex (**H**istocompatibility antigen-**2**) because it represents the second antigen originally defined by Gorer, a British scientist who studied murine MHC. The human MHC is called the HLA (**Hu**man **L**eukocyte **A**ntigen).

[2] In genetics, *poly*, Greek for many, and *morph*, Greek for shape, together mean variation at a single genetic locus and variation within a species. Individual variant genes are called *alleles*. Each set of alleles is known as a haplotype or allotype.

[3] Human erythrocytes, being non-nucleated, do not express class I molecules.

Ii:	Invariant chain
Ir: genes	Immune Response genes
LMP:	Low Molecular weight Peptide
MECL: -1	Multicatalytic Endopeptidase Complex Like-1
TAP:	Transporter associated with Antigen-Processing

The α chain of the class I molecule is a type II membrane protein anchored in the cell membrane at the -COOH terminus, and it has a short cytoplasmic tail. The extracellular NH_2-terminus of the class I molecule has three globular domains, termed α1, α2, and α3, each about 90-92 amino acids in length. The α3 domain is closely associated with β_2-m and has a site for binding the CD8 coreceptor. β_2-m is not anchored in the cell membrane but held in position solely by its interaction with the α chain. The α1 and α2 domains, consisting of two α-helices resting on a sheet of eight β-strands, form the peptide-binding groove. The α1 domain is highly polymorphic, and along with α2, is responsible for the wide variations observed in peptide binding by class I molecules (Figure 7.1).

Both MHC class I and class II molecules differ from other peptide-binding proteins in two major respects.

- Each MHC molecule can bind multiple peptides (although not at the same time). An individual can be infected by a wide variety of pathogens whose proteins differ widely in amino acid sequences. In order to activate T cells, MHC molecules, whether class I or II, must stably bind different peptides.
- The binding of the peptide is essential for the correct folding and stabilisation of MHC molecules. The MHC–peptide complex is extremely stable, and the peptide copurifies with the MHC molecule in experiments designed to isolate MHC molecules from cell lysates.

The peptide lies in an elongated conformation along the peptide-binding groove, with both amino- and carboxy-termini tightly fixed in this groove. The peptide-binding groove contains pockets that accommodate particular peptide side chains — termed *anchor residues* — that anchor the peptide in the groove. These pockets vary in depth and chemical nature between allelic variants and thus determine the set of peptides that can be bound by a particular class I allele. Six pockets have been identified in the class I groove (P1 through P6). P1 and P6 react with the N- and C-termini of the peptide respectively, and P2-P4 react with the peptide side chains. For human class I molecules, the C-terminal anchor residue of the peptide needs to be either hydrophobic or basic[4]. The terminal anchor residues of the peptide form multiple hydrogen bonds and salt bridges with conserved amino acid residues in the two pockets the peptide-binding groove ends. These interactions are essential for stable association between the peptide and the class I molecule, and they constrain the length of bound peptides to about 8-10 residues.

[4] Murine class I molecules are generally more hydrophobic than human class I molecules and can bind only peptides having a hydrophobic residue at the C-terminus.

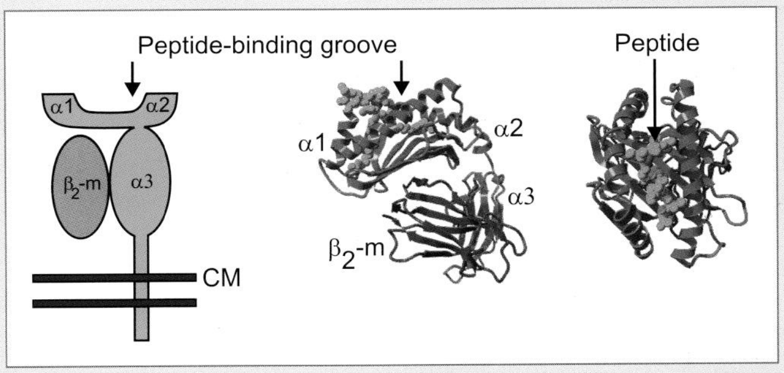

Figure 7.1 The peptide-binding groove of MHC class I molecules is present at the membrane-distal region of the α chain that is noncovalently associated with β_2-m. The extracellular region of the class I molecule has three globular domains termed α1, α2, and α3. The α3 domain is closely associated with β_2-m and also has a site for the binding of the CD8 coreceptor. β_2-m is not anchored in the cell membrane; it is held in position solely because of its interaction with the α chain. The left panel is a schematic depiction of the molecule. The centre panel is a ribbon diagram of the molecule. A space-filling model of the top view of the peptide-binding groove (that is looking down into the groove) is depicted in the right panel. The floor of the groove is formed by β-pleated sheets, and the walls are formed by α-helices. The α1 and α2 domains contribute to the formation of the peptide-binding groove. These domains also have sites that contact TcRs. The peptide lies in an elongated conformation along the peptide-binding groove with both amino- and carboxy-termini tightly fixed in this groove.

7.2.2 MHC Class I Synthesis and Assembly

Both the light and heavy chains of class I molecules are synthesised in the ER. A newly synthesised heavy chain binds to a number of ER-resident chaperone proteins, beginning with calnexin, during the assembly process. The interaction with calnexin is thought to facilitate folding of the nascent heavy chain and promote assembly with β_2-m. The Ig-binding protein BiP can substitute for calnexin. Erp57 (a thiol reductase) also associates with this complex. Once the heavy chain β_2-m heterodimer is formed, it dissociates from calnexin and associates with calreticulin. The whole complex then interacts with TAP (**T**ransporter associated with **A**ntigen-**P**rocessing; see below). This interaction promotes the loading of peptides onto the class I molecule. A TAP-associated transmembrane glycoprotein known as tapasin[5] stabilises this process. Tapasin promotes the stability and peptide transporting activity of TAP and holds the class I molecule in its peptide-receptive conformation. Once loading is accomplished, class I molecules are released from TAP. The peptide:MHC class I complex is then transported through the *trans*-golgi network to the cell surface, where it undergoes periodic recycling between the endosomes and the cell surface (Figure 7.2). Eventually, the molecules are internalised and

[5] Except for tapasin, all other chaperones (calnexin, calreticulin, and BiP) involved in MHC class I assembly are *housekeeping proteins*, which participate in the folding of a variety of multimeric proteins in the ER. The only known function of tapasin is in MHC class I peptide-loading. It is encoded by the MHC, and its expression is induced by IFNs.

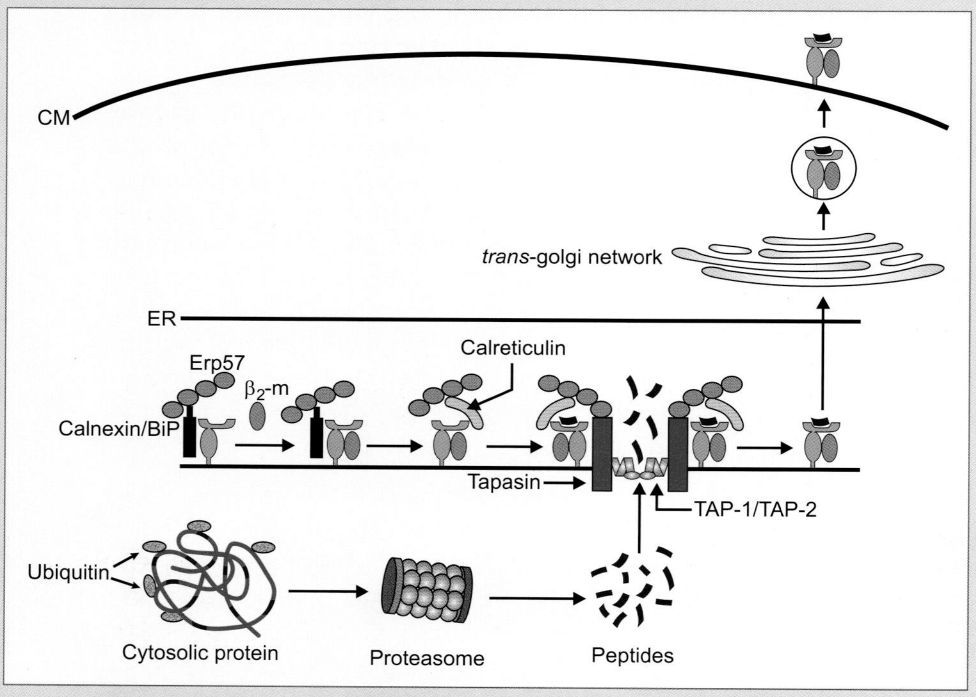

Figure 7.2 MHC class I molecules are synthesised and loaded with peptides in the ER. A number of chaperones are involved in MHC class I synthesis. Calnexin or BiP associate with the newly synthesised α chain of the class I molecule. Another chaperone — Erp57 — also associates with this complex. These chaperones help in the folding of the α chain and promote its association with β_2-m. The $\alpha:\beta_2$-m heterodimer dissociates from calnexin and associates with calreticulin. Class I molecules present peptides derived from cytosolic proteins. In the cytosol, poly-ubiquinated proteins are degraded by the proteasome to yield oligopeptides that may be further trimmed by cytosolic endopeptidases. The peptides are translocated across the ER membrane and into the ER lumen by the heterodimeric protein TAP, which consists of two subunits — TAP-1 and TAP-2. TAP proteins are thought to form a pore into the ER membrane that allows peptide translocation. Tapasin is also involved in the translocation process. Once loaded, class I molecules are transported across the *trans*-golgi network to the cell surface.

Figure 7.3 A majority of the peptides loaded on MHC class I molecules is generated by proteasomes. The proteasome is the main proteolytic system of eukaryotes, responsible for degradation of most cytosolic and nuclear proteins. It consists of a barrel-shaped core 20S subunit. This core subunit is composed of heptameric rings of α and β subunits held together to form a hollow cylinder. The association of this core subunit with two 19S regulators at either end yields the 26S proteasome involved in ATP-dependent protein degradation. IFN-γ induces the formation of additional proteins that can associate with the proteasome. Prominent amongst these is the 11S regulator. The 20S subunit can associate with either the 11S regulator alone or simultaneously with one 11S and one 19S subunit. These novel forms of proteasomes are called immunoproteasomes. The incorporation of other IFN-γ induced proteins, such as LMP-2, LMP-7, and MECL-1 (not shown here), in the immunoproteasome results in the formation of peptides suitable for MHC class I loading.

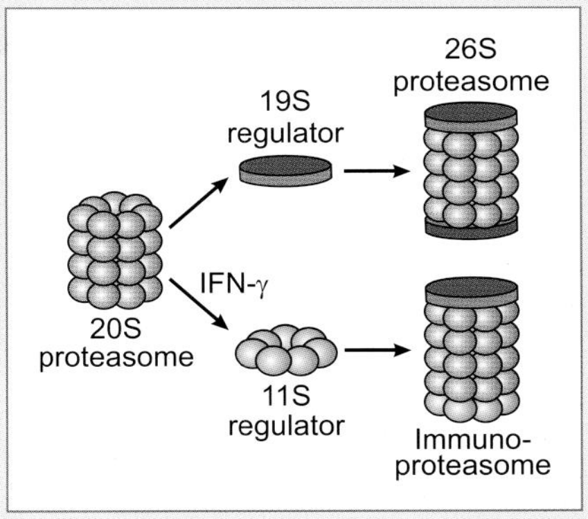

degraded. Under physiological conditions, binding of a peptide is essential for the stability and transport of class I molecules. Heavy chains that do not get associated with β_2-m are not loaded with the peptide (i.e., are empty). Chaperones retain them in the ER. These incompletely assembled units are translocated to the cytosol and degraded by the proteasome.

7.2.3 MHC Class I Antigen Processing and Loading

MHC class I molecules were thought to only present peptides derived from proteins in the cytosol and endogenous biosynthetic pathways. It is now clear that they can also present exogenous antigens, and the process is termed *cross-presentation* (see below). Two proteolytic processes are important in the generation of endogenous peptides.

- **Proteasomal degradation**: The degradation of proteins by the 26S proteasome is the first step in antigen processing.
  - The proteasome is the main proteolytic system in eukaryotic cells. To ensure that the proteasome degrades the correct protein, it is tagged by ubiquitination. In this process, multiple molecules of ubiquitin are covalently attached to the ε-amino group of lysine residues in the protein.
  - The proteasome consists of two regulators (19S and 11S) that attach like caps to a core 20S proteolytic unit (Figure 7.3). The polyubiquitin chain is recognised by its 19S subunit. The association of the 20S proteasome with two 19S regulators at either end yields the 26S proteasome involved in ATP-dependent degradation of ubiquitin-conjugated proteins. The core 20S structure is a barrel-shaped multicatalytic complex. It is composed of two heptameric outer rings of structural α subunits and two heptameric inner rings of catalytic β subunits, which together form a hollow cylinder. The 20S proteasome is involved in the degradation of unfolded proteins and polypeptides and can function independently of, or in association with, regulatory subunits. The immune system has evolved mechanisms to modify this pathway to enhance the efficiency of peptide generation by proteasomes.
  - IFN stimulation[6] induces the 11S regulator and three additional subunits of the proteasome. Two of these, LMP-2 (**L**ow **M**olecular weight **P**eptide-**2**) and LMP-7, are encoded by MHC genes. MECL-1 (**M**ulticatalytic **E**ndopeptidase **C**omplex **L**ike-**1**) is non-MHC encoded. Together, these

[6] This is an example of the links between innate and adaptive immunity. As a result of the innate immune response, cells at the site of viral invasion secrete IFNs (type I and type II); IFN stimulation leads to the upregulation of MHC class I and/or MHC class II expression. Induction of the immuno-proteasome further ensures that viral proteins are efficiently loaded onto class I molecules, triggering an adaptive immune response.

When Harry Connects With Sally: MHC, Mate Selection, and Neuronal Connections

MHC genes play a central role in immune recognition. Growing evidence indicates that MHC molecules also perform crucial roles outside the immune system.

- MHC genes are a source of specific odours that influence individual recognition, mating preferences, nesting behaviour, and selective blocking of pregnancy in animals. Experiments in rodents establish that mice prefer mates that are as genetically different from their MHC haplotyes as possible. Soluble MHC in animal urine and sweat seems to influence rodents' choice of mates. The involvement of MHC class I genes in the generation of strain-specific urinary odours (odourtypes) in mice is supported by several experiments. In the murine and human genomes, a large cluster of odorant receptor genes is tightly linked to the MHC. There is a high degree of linkage disequilibrium between the two types of loci in humans, suggesting the existence of a functional connection between the MHC and MHC-linked odourant receptor loci. Peptides that specifically bind MHC class I molecules activate chemosensory receptors in rodents' nasal cavities. Independent studies have clearly established the chemosignal role of MHC-bound peptides.
- The MHC seems to influence not only mate selection but also reproductive behaviour. *In vitro* fertilization experiments in animals illustrate that the chances of bringing a pregnancy to full term are maximised when the foetus and mother have different MHC alleles. This is hypothesised to be a way of ensuring that MHC diversity is maintained in a population, maximising the chances of survival of the species, because the MHC is directly linked to host defence. Thus, the increased chance of favouring of heterozygotes and rare alleles likely explains the connection between MHC and the mate selection.
- The role of the MHC is difficult to study in human mate selection, as MHC loci are the most polymorphic loci in the human genome. Recent experiments in humans seem to indicate that the same compulsions observed in other animals are operative in *Homo sapiens sapiens*. Individuals describe body odours as pleasant when they are from people who have few HLA alleles matching their own. Research suggests that at least in women, mate preference seems to be dictated by the paternal MHC, and women prefer mates with MHC haplotypes different from their own. There is some evidence to suggest that oral contraceptives may reverse this choice. If true, this would imply that the women using oral contraceptives might not make the most immunologically appropriate choice!
- MHC class I proteins are expressed by normal, uninfected neurons throughout life. In adults, neuronal MHC class I expression is primarily dendritic and intracellular (unlike that observed for cells of the immune system). This MHC class I expression is regulated by the naturally occurring electrical activity that sculpts developing projections in the neurons. This expression is also linked to normal and pathological changes in neuronal activity. In mice that lack functional MHC I proteins, neural connections between the eye and brain do not develop properly. Growing evidence shows that MHC class I molecules may act as a brake on synaptic plasticity, helping prune unnecessary connections during development and preventing the formation of inappropriate synapses that interrupt normal function in mature brains. These molecules are also required for responses to injury in the central nervous system. MHC class I molecules are crucial for translating neuronal activity into changes in synaptic strength and neuronal connectivity *in vivo*.

additional subunits alter the catalytic activity of the proteasome to generate peptides having the basic or hydrophobic C-terminal residues required for efficient binding to class I molecules. LMP-2 and LMP-7 containing proteasomes are labelled *immunoproteasomes* because they generate a different spectrum of peptides than the constitutive proteasome.

- Most peptides resulting from proteasomal degradation are too short to be loaded onto class I molecules and are rapidly destroyed by endopeptidases and exopeptidases; the amino acids are recycled for protein synthesis.

- **Peptide trimming**: Proteasome-generated oligopeptides are often longer than those required for loading onto class I molecules. Aminopeptidases in the cytosol or ER trim these oligopeptides to antigenic peptides of the length (8-10 amino acids long) needed to bind to MHC class I molecules. Thus, the C-terminus of the antigenic peptide is determined by the proteasome; the N-terminus is often the result of peptide trimming.

Peptides produced in the cytosol have to be translocated to the ER to enable their association with class I molecules. A heterodimer called TAP, consisting of two transporter proteins (TAP-1 and TAP-2), is involved in this translocation. TAP is a member of the ATP-binding cassette family of transport proteins. The generated peptide is first bound to TAP and then translocated to the ER. Both TAP-1 and -2 cooperate in this translocation. They are thought to jointly form a pore in the ER membrane through which the peptide is translocated from the cytosol to the ER lumen. Peptides having a hydrophobic or basic C-terminus are preferentially translocated by TAP. Initial binding of the peptide is ATP independent, but the pumping of peptides across the ER membrane is an ATP-consuming process. Peptides that are not bound to class I molecules are exported from the ER to the cytosol and eventually degraded by resident peptidases.

Given that almost all proteins residing in the cytosol and ER are synthesised by an APC, MHC class I molecules display a sampling of the genes expressed by that cell in peptide form to the immune system. In most cases, these peptides are derived from autologous proteins and ignored by the immune system because of self-tolerance. However, if MHC class I:peptide complexes are recognised as foreign, the immune system is triggered, activating CTLs and killing the offending cell. **MHC class I molecules are thus important in presenting intracellular parasite-derived or tumour-derived peptides**. Because many infected/transformed cells are deficient in MHC class I expression, absence of MHC class I molecules results in NK cell-mediated lysis of these cells. This is because the engagement of inhibitory receptors on an NK cell by peptide:MHC class I complexes on the target cell delivers a negative signal to the NK cell (see section 2.3.2). Surface expression of class I molecules is therefore needed to spare the cell from NK cell-mediated lysis. Through these checks and balances, the immune system maximises the chances of eliminating potentially infected or transformed cells. MHC class I molecules are, in this respect, double-edged swords. On one hand, the recognition of a particular MHC:peptide complexes by TcRs on activated CTLs results in the lysis of APCs bearing that MHC:peptide complex; on the other hand, a complete lack of class I molecules makes the cell susceptible to NK cell-mediated lysis.

Naïve $CD8^+$ T cells recognise antigen loaded onto MHC class I molecules. If MHC class I molecules were to present only endogenous antigens, only infected APCs (e.g., DCs or macrophages) would be able to activate naïve $CD8^+$ T cells. A mechanism for presenting exogenous antigen by the MHC class I pathway is therefore needed to trigger CTL activation in the absence of APC infection. This is achieved by cross-presentation. **Antigen processing of exogenous antigens in the MHC class I pathway is termed *cross-presentation***; *cross-priming* is the priming of CTLs by exogenous antigens. Although both DCs and macrophages are efficient scavengers, macrophages are poor at stimulating naïve T cells. DCs, however, are the body's most efficient APCs, and current experimental data suggests that different subsets of DCs are capable of cross-presentation under different conditions.

DCs internalise apoptotic and necrotic cells and cross-present the antigens derived from such cells. Because apoptosis is part of normal cell turnover, a tolerogenic response is hypothesised to be initiated when macrophages or immature DCs internalise apoptotic cells and cross-present the antigen to naïve T cells. Conversely, if immature DCs are exposed to inflammatory stimuli (such as LPS, cytokines, dsRNA, or CpG DNA) it will result in DC maturation and immune activation. Peptides from the apoptosed cells will be loaded onto MHC class I molecules because of cross-priming. Cells undergoing necrotic cell death as a result

Class Interrupted: Viral Interference in MHC Class I Antigen Presentation

MHC class I molecules present virus-derived peptides to naïve $CD8^+$ T cells. The activation of naïve $CD8^+$ T cells results in the generation of an antiviral CTL response and the recruitment of other components of the immune system to this response. Hence, the downregulation of MHC class I expression can help a virus escape recognition and targeting by the immune system. Nevertheless, complete shutdown of class I expression is disadvantageous to the virus, as it leaves the infected cell susceptible to NK cells. It requires some juggling on the virus' part to successfully evade immune detection. From the viewpoint of the virus, it should ideally downregulate, without completely shutting down, MHC class I expression. Viruses have evolved ingenious ways of reducing MHC class I expression and antigen presentation.

- Kaposi's sarcoma virus (a gamma herpes virus) encodes two proteins, K3 and K5, that increase the endocytosis of class I molecules from the cell surface, decreasing surface expression of these molecules and preventing recognition by CTLs.
- The Nef protein, encoded by HIV, relocates cell-surface MHC class I molecules to the *trans*-golgi network, downregulating their surface expression.
- Vpu, expressed by HIV-1, induces the degradation of newly synthesised class I molecules.
- The human cytomegalovirus encodes a virtual catalogue of proteins that interfere with the surface expression and peptide-loading of MHC class I molecules.
  - Two proteins encoded by the cytomegalovirus, US2 and US11, induce the translocation of newly synthesised MHC class I molecules to the cytosol for degradation. At the same time, NK cell activation is avoided, as US2 and US11 downregulate only some MHC class I alleles. The alleles that are allowed to be expressed at the cell surface deliver inhibitory signals to NK cells (see below).
  - The encoding of UL40 brilliantly illustrates how a chance mutation bestows a survival advantage. A peptide generated by the processing of UL40 is similar to the HLA-E ligand (a nonamer derived from HLA-C signal sequence) and helps stabilise and express HLA-E at the infected cell's surface. The HLA-E:UL40-derived peptide complex delivers an inhibitory signal to NK cells and protects the virus from NK cell-mediated lysis.
  - UL18, a virus-encoded MHC class I homologue, is thought to help prevent NK cell-mediated lysis.
  - US6 inhibits the TAP-mediated peptide-loading of MHC class I molecules.
  - The glycoprotein US3 retains MHC class I molecules in the ER.
- Adenovirus types 2 and 5 express a protein that binds to MHC class I molecules, retaining them in the ER/*cis*-golgi.
- Adenovirus type 12 represses the transcription of LMP, TAP, and MHC class I molecules.
- The herpes simplex virus blocks the transport of MHC class I molecules from the ER to the cell surface. The immediate early protein ICP47 that is encoded by the virus binds TAP, inhibiting the binding and translocation of peptides from the cytosol to the ER lumen — interference with peptide-loading causes the molecules to be retained in the ER.
- The Epstein-Barr virus encoded nuclear antigen-1 is resistant to proteasomal degradation and escapes antigen processing.

of infection or trauma can also provide both the antigen and the stimulus necessary for immunogenic cross-priming. The end result will be an immunogenic CTL response against the infected non-APCs. Cross-priming is thought to be especially important in immunity to viral infections that are localised to peripheral, nonlymphoid compartments (e.g., human papilloma virus infection, in which the infection is confined to epithelial cells of the skin). Cross-priming may also be vital in generating immunity to viruses that infect professional APCs and inhibit or interfere with MHC class I antigen processing and presentation. Under these conditions, uninfected DCs can internalise infected DCs and induce immunity by priming naïve $CD8^+$ T cells through cross-presentation. Although their exact mechanisms are unclear, three general pathways of cross-presentation are recognised.

- **Direct translocation** of pathogen-derived antigenic material in the cytosol of host APCs by the pathogen itself, or by specialised mechanisms of transport, allows processing of the protein by the normal cytosolic machinery of the MHC class I pathway. This mode of gaining access to the cytosol is observed for certain viruses and bacteria, such as *Listeria monocytogenes*.
- **Direct endosomal loading of preformed (recycling) MHC class I molecules** with peptide determinants that are generated in the endosomal compartments is the second mode of cross-priming. Alternatively, the peptide antigen may be exocytosed (regurgitated) from the endosomal compartment onto the cell surface for association with preformed MHC class I molecules.
- **The diversion of exogenous proteins** from endosomal compartments or from extracellular fluid into the cytosol for processing by the conventional MHC class I pathway may also result in cross-priming.

MHC Class I Molecules

- Class I molecules are expressed by all nucleated cells of the body.
- They are transmembrane glycoproteins consisting of an α chain that is noncovalently associated with β_2-m.
  - The α chain is anchored in the cell membrane by its -COOH terminus; the extracellular NH_2-terminus consists of three globular domains (α1, α2, α3).
  - The peptide-binding groove is formed by the α1 and α2 domains and is closed at both ends.
  - The terminal anchor residues of the peptide form multiple hydrogen bonds and salt bridges with amino acid residues in the two pockets at the ends of the peptide-binding grooves.
  - The α3 domain associates with β_2-m and has a site for CD8 binding.
- MHC class I molecules predominantly present peptides derived from endogenous antigens and are important in presenting parasite-derived or tumour-derived peptides; under certain conditions, class I molecules can be loaded with peptides derived from exogenous antigens; this phenomenon is called cross-presentation.
  - Both the carboxy- and amino-termini of the peptide are tightly fixed in the peptide-binding groove.
  - Because of the closed peptide-binding groove, class I molecules bind only peptides 8-10 amino acids in length.
  - The proteasome in the cytosol generates peptides for loading.
  - TAP, a heterodimer of two transporter proteins, translocates peptides from the cytosol across the ER membrane and into the ER lumen. Tapasin stabilises the process and holds the class I molecule in a peptide-receptive conformation.
  - The peptide-loaded class I molecule is eventually transported to the cell surface.

7.3 MHC Class II Molecules

MHC class II molecules are vital to the functioning of the immune system. **TH cells must recognise peptide:MHC class II complexes on APCs to initiate or cooperate in an immune response**. MHC class II molecules have a limited tissue distribution; their expression is essentially restricted to cells of the immune system. They are found on professional APCs (DCs, B cells, macrophages and related cells), as well as on thymic epithelial cells. The degree of expression varies not only according to cell type but also according to the maturation and activation status of the cell. Immature DCs do not stably express class II molecules on their cell surfaces, but mature DCs express very high levels of these molecules. Similarly, only mature B cells constitutively express MHC class II molecules. Human, but not murine, activated T cells also express MHC class II molecules. The expression of class II molecules is also upregulated in cells of the monocyte/macrophage lineage upon activation. IFN-γ can induce the expression of these molecules in many cell types that do not normally express them. Murine class II molecules are referred to as IA or IE, with the haplotype written as a superscript (e.g., IA^d). Human class II molecules are HLA-DR or DP or DQ, followed by a number that defines the haplotype.

7.3.1 MHC Class II Structure

Class II molecules are heterodimers composed of two glycosylated subunits, a 33-35 KD α chain, and a 25-29 KD β chain held together by noncovalent bonds. Both chains are anchored in the cytoplasm at the -COOH termini. They have a short cytoplasmic domain of 10-15 amino acids and an extracellular domain of 90-100 amino acids. Both the α and β chains consist of two domains each (α1 and α2; β1 and β2). The β2 domain contains the CD4-binding site. **The membrane-distal domains of both chains, that is α1 and β1 domains, form the peptide-binding groove**. The floor of this groove consists of β-pleated sheets; α-helical regions form the walls. The peptide-binding groove is open at both ends. An MHC class II ligand can therefore vary in length from 12-25 residues. **The peptide lies in an extended conformation along the class II peptide-binding groove**. The peptide is held in the groove by multiple hydrogen bonds formed between the backbone of the peptide and the amino acids lining the binding groove. This is in contrast to class I molecules, where hydrogen bonds are clustered at terminal residues. **Both peptide ends hang out of the groove**; this is in contrast to the MHC class I molecule, where peptide ends are deeply buried within the molecule (Figure 7.4). As in class I molecules, the peptide-binding grooves of class II molecules

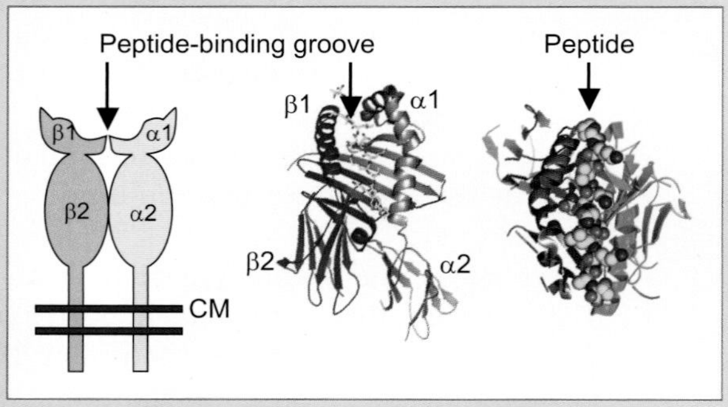

Figure 7.4 MHC class II molecules are heterodimeric proteins consisting of an α and a β chain held together by noncovalent interactions. Both chains contribute to the peptide-binding groove, which lies at the membrane-distal region. Each chain consists of two domains (α1 and α2; β1 and β2). The β2 domain has the CD4-binding site. The left panel is a schematic depiction of the molecule; the centre panel is a ribbon diagram of the molecule. A space-filling model of the top view of the peptide-binding groove is depicted in the right panel. Both α1 and β1 domains contribute equally to the peptide-binding groove. The floor of the groove is formed by β-pleated sheets, and the walls are formed by α-helices. The peptide-binding groove is open at both ends, and the peptide lies in an extended conformation along it, with both the peptide ends hanging out of the groove.

also contain pockets to accommodate particular peptide side chains (anchor residues). The side chains of the amino acids in the centre of the group protrude out of the groove and make contact with TcR residues. Those of the remaining amino acids point into the groove and are accommodated in the pockets. The anchor residues determine the set of peptides that can be loaded onto the molecule. The highly polymorphic β1 domain is the major determinant of the binding specificity of MHC class II haplotypes. Of the nine pockets identified, P1, P3, and P7 seem particularly important for binding.

The interactions of MHC class I and class II molecules with their peptides is governed by similar principles. The product of a particular allele of MHC class I or class II molecules is capable of binding any one of a large number (thousands) of peptides. These peptides differ in their sequences but share two or three amino acid residues (called *motifs*) that fit into anchoring pockets on MHC molecules. Thus, peptides binding to different alleles can be distinguished by their motifs. Viewed from the top, the peptide-binding groove presents a rather flat surface, with the peptide in the middle bordered by α-helices of the MHC molecule. The TcR interacts with this surface. The receptor is positioned diagonally over the surface, with CDR1 and CDR2 loops of the TcRα chain over the N-terminal of the peptide and the CDR1 and CDR2 loops of the TcRβ chain looming over the C-terminal of the peptide. The less variable CDR1 and CDR2 loops of the TcR interact primarily with relatively conserved α-helices of the MHC; the more variable CDR3 of the TcR touches most variable parts of the peptide (Figure 7.5).

7.3.2 MHC Class II Synthesis and Assembly

Like all proteins, MHC class II molecules are synthesised in the ER with their chaperone molecules. The chaperone molecule is called the **I**nvar**i**ant chain[7] or Ii. A trimer of Ii forms a scaffold on which newly synthesised αβ chains are added. An absence of Ii leads to misfolding and the aggregation of class II molecules. A nonamer consisting of three subunits each of Ii, α, and β subunits is formed after this process. The insertion of the CLIP (**Cl**ass II-associated **I**nvariant chain **P**eptide) region of Ii into the peptide-binding groove of the class II molecule stabilises Ii–class II interaction. CLIP protects the peptide-binding groove and prevents premature loading by peptides present in the ER.

[7] Unlike MHC class II molecules, the sequence of amino acids in this chaperone does not vary for different haplotypes and is therefore termed *invariant*. It is a non-MHC protein that is encoded by genes on chromosome 5 in humans and chromosome 18 in mice. Originally thought to be involved only in MHC class II synthesis, Ii is now known to have multiple functions, including roles in B cell maturation and in the chaperoning of the protease cathepsin L and the MHC-like CD1 molecule.

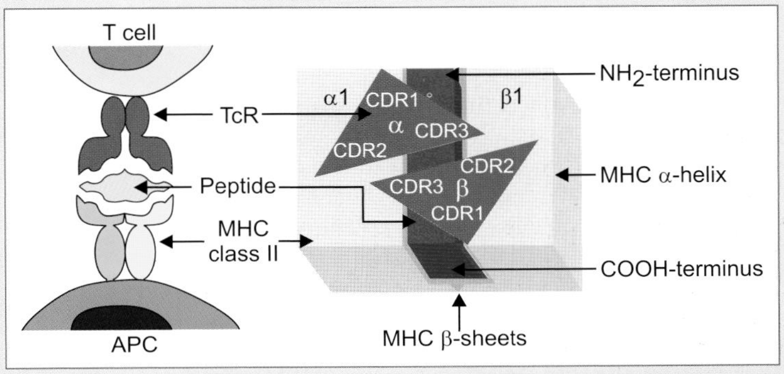

Figure 7.5 TcRs physically interact with MHC:peptide complexes on the APCs. The left panel is a schematic side-view of the trimolecular complex formed between the TcR, MHC class II molecules, and the peptide present in the peptide-binding groove. Viewed from the top, the β-sheets of the MHC class II molecules (right panel) form the floor of the peptide-binding groove. The peptide in the groove forms a rather flat surface in the middle and is bordered by the α helices of the MHC molecule. The TcR sits diagonally across this surface, with the CDR1 and CDR2 loops of the TcRα chain positioned above the amino-terminus of the peptide, with the CDR1 and CDR2 loops of the β chain looming over the carboxy-terminus of the peptide. The less variable CDR1 and CDR2 loops of the TcR make contact with the relatively less variable α-helices of the MHC molecules. The most variable CDR3 region of the TcR makes contact with the most variable parts of the peptide. TcR:MHC class I:peptide contact occurs on similar lines (adapted from Klein & Sato, *New England Journal of Medicine*, 2000, 343:702).

The newly synthesised $(\alpha\beta Ii)_3$ nonamer is exported across the *trans*-golgi network to endosomal peptide-loading compartments. Cytoplasmic motifs present in the tails of Ii and β chains of class II molecules are thought to be responsible for this targeting of class II molecules to endosomal compartments. Once the nonamer reaches the endosomes, the Ii trimer scaffold is cleaved to yield three αβIi units. Ii then undergoes sequential proteolysis from the N- and C-termini, leaving CLIP in the peptide-binding groove until it is displaced by the peptide ligand. The removal of CLIP and loading of the peptide onto the MHC class II molecule is facilitated by two other molecules — HLA-DM and HLA-DO (H2-M and H2-O in mice). The peptide-loaded class II molecules are then exported to the cell surface (Figure 7.6). The cell surface peptide:class II complexes are fairly stable (average half-life of around 48 hours), allowing ample opportunity for a T cell encounter.

7.3.3 MHC Class II Antigen Processing and Loading

The peptides presented by MHC class II molecules are generated mostly by the degradation of proteins that access the endocytic pathway. Lysosomal proteases and other hydrolases degrade the proteins and generate peptides that can be loaded onto MHC class II molecules. Thus, they present exogenous antigens endocytosed by APCs. As endosomes/lysosomes are also sites of endogenous protein degradation, class II molecules display a fair sampling of the cells' own proteins as well.

Exogenous antigens enter the endocytic pathway by a variety of mechanisms — pinocytosis (DCs), phagocytosis (macrophages, DCs), and receptor-mediated endocytosis (B cells, DCs, and macrophages). The importance of the different routes of internalisation differs in various APCs. As a result of internalisation, antigens are enclosed in endocytic vesicles and transported along the endosomal–lysosomal

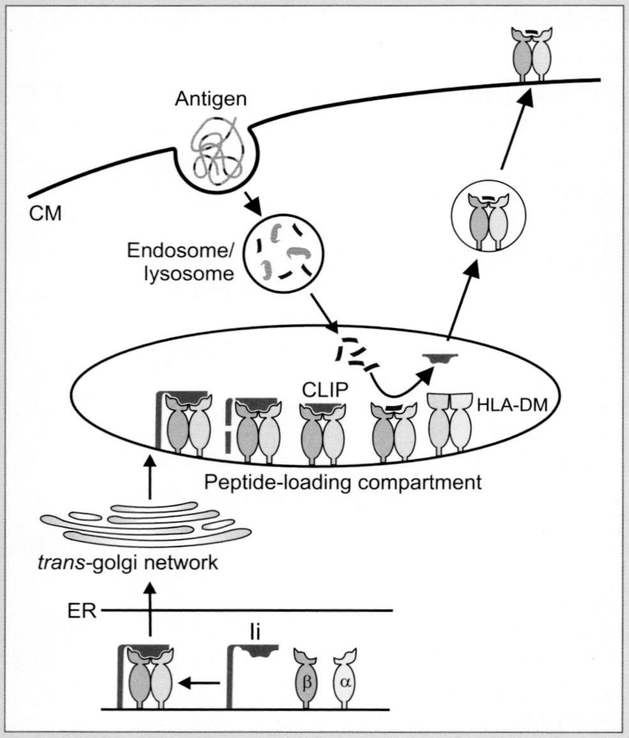

Figure 7.6 MHC class II molecules are synthesised in the ER and loaded with peptides in the lysosomal peptide-loading compartments. The α and β chains of the molecule are synthesised with a chaperone molecule, Ii. A part of Ii (termed CLIP) is inserted in the peptide-binding groove of the αβ heterodimer. This interaction stabilises the complex and prevents the premature loading of class II molecules. Nonamers consisting of three molecules of α:β:Ii are formed in the ER and transported across the *trans*-golgi network to the peptide-loading compartments. Ii undergoes sequential proteolysis in these compartments. CLIP remains in the groove until it is displaced by peptides generated in the endosomal or lysosomal compartments. This peptide-loading is facilitated by HLA-DM and -DO (not shown here). Any proteins (whether self or foreign) that are internalised by the cell enter the endosomal/lysosomal pathway. Because of the increasing proteolytic environment in these compartments, the proteins undergo degradation. Peptides of appropriate lengths and motifs are loaded onto MHC class II molecules. The loaded MHC:peptide complexes are then exported to the cell surface. The recognition of these complexes by $CD4^+$ T cells triggers an immune response.

Determinants of Survival: Immune Responsiveness and MHC

McDevitt and Chinitz were the first to observe that the capacity to respond to several antigens is controlled by a gene cluster that they termed the ***Immune response (Ir)*** genes. Later, *Ir* genes were found to encode MHC class II molecules. All immune response genes, however, are not MHC linked. MHC class II genes control immunological responsiveness essentially through their effect on T cell functioning. The failure to respond to a particular antigen may be linked to a number of reasons:

- **A failure of association** — an immune response is triggered when naïve T cells recognise a peptide in the context of MHC class II molecules. The failure of a particular MHC class II allotype to interact or associate with a given peptide will result in failure to activate antipeptide T cells, and the individual will be termed a nonresponder for the antigen that gave rise to that peptide.
- **A hole in the T cell repertoire** — this is caused when T cells recognising a particular peptide:MHC complex are either absent or unable to respond to it. Mechanisms that lead to holes in the T cell repertoire include the following:
  - An absence of genes that code for a particular TcR or set of TcRs
  - Deletion, anergisation, or strict regulatory control of T cells reacting to a particular MHC:peptide configuration. Autoreactive T cells are deleted or anergised during thymic selection; if the association/interaction of a given epitope with a MHC class II molecule results in a conformation similar to a self-antigen, T cells responding to such complexes will either be absent or unable to respond because of the action of T_{reg} cells
  - Failure to positively select particular T cell clones during thymic education

pathway. As the antigen is transported from the endosome to the lysosome, it is subjected to an increasingly acidic environment that is rich in proteases and hydrolases. Thus, antigens are exposed to conditions that are increasingly denaturing and proteolytic during transport, and the antigen is reduced to peptides of varying lengths by the time it reaches lysosomal peptide-loading compartments. **The formation of MHC class II:peptide complexes therefore occurs as a result of the intersection of two endocytic pathways — the one transporting the exogenous antigen along the endocytic route and the other exporting class II molecules from the ER to the cell surface**.

In the peptide-loading compartments, peptides encounter CLIP-loaded MHC class II molecules. The loading of the peptide on class II molecules entails the exchange of CLIP for the peptide ligand. The MHC-encoded, nonpolymorphic HLA-DM facilitates the loading process. HLA-DM is thought to facilitate peptide-loading by physically associating with the MHC class II molecule and holding it in an open conformation conducive to the dissociation of CLIP and subsequent re-association with the peptide ligand. HLA-DM is said to act like a peptide editor because the spectrum of peptides presented by an APC is influenced by the presence or absence of HLA-DM. HLA-DO, another nonclassical MHC molecule, modulates HLA-DM functioning.

7.4 Organisation of MHC Genes

The MHC gene complex contains many individual genes. Although the complex performs similar functions in different species, the detailed arrangement of the genes differs amongst species. Originally, different genetic loci of the MHC were identified by functional and serological analysis. The result has been a rather

MHC Class II Molecules

- MHC class II molecules are expressed mainly by cells of the immune system.
- They are heterodimers of an α chain and a β chain held together by noncovalent bonds.
  - Both chains consist of two domains each (α1 and α2; β1 and β2) and are anchored in the cell membrane by their -COOH terminus.
  - The binding site for CD4 lies in the β2 domain.
  - The peptide-binding groove is formed by the α1 and β1 domain.
  - The binding groove is open at both ends; both ends of the peptide hang outside the groove.
  - The peptide is held in the groove by multiple hydrogen bonds.
- MHC class II molecules are synthesised in the rough ER along with Ii.
  - Ii promotes proper folding of the molecule.
  - It protects the peptide-binding groove from premature loading.
  - In the peptide-loading compartments, Ii undergoes sequential proteolysis, leaving CLIP behind.
  - Peptide loading occurs by CLIP displacement.
  - Peptide-loaded class II molecules are then exported to the cell surface.
- Class II molecules can be loaded by endogenous or exogenous peptides.
  - Antigens are degraded in the endocytic lysosomal compartments.
  - Peptides generated by these degradative processes are then loaded onto the class II molecules.
  - HLA-DM (H2-M) and HLA-DO (H2-O) facilitate the loading process and influence the spectrum of loaded peptides.

complicated mess as far as the names of the various genes and gene products is concerned (it was really not just to test students' patience ☺). Identification by serology has also led to the gene products being referred to as *antigens*. Recently established genetic maps of the human and murine MHC have helped identify the genes that encode particular polypeptides. Only genes with known functions will be described here using the most accepted nomenclature to minimise confusion (so we hope). Many other genes with possible functions in immunity also map to this DNA region but will not be discussed. This is because the proteins that some genes code for have been identified, but their exact functions are not known; in the case of other genes, the encoded proteins have not been fully characterised.

Figure 7.7 shows the genetic organisation of the human and murine MHC which are essentially similar. The entire complex extends over 4×10^6 basepairs. Separate regions of the complex code for MHC class I and class II molecules; several genes within these regions encode each chain.

- **In humans, three genes — *HLA-A, -B, -C* — encode MHC class I heavy chains**; the HLA class I loci are highly polymorphic and more than 80 alleles for HLA-A, 180 for HLA-B, and 40 for HLA-C have been identified. Both alleles of each locus are expressed; therefore, an individual can express up to six different MHC class I molecules. **Murine MHC class I genes are known as *H2-K*, and *H2-D*.** In humans, the MHC class I region is located on the same stretch of the chromosome. The murine class I region seems to be translocated when compared to the human MHC class I; it is split in two in the mouse, with the class II and class III regions located between the two class I regions.
- **Three pairs of MHC class II α and β chains are found in humans — HLA-DR, -DP, and -DQ.** Some individuals contain up to three β chain genes in their HLA-DR cluster (represented by a single β gene in the figure). These DRβ gene products can also associate with the DRα chain, yielding four sets of DR

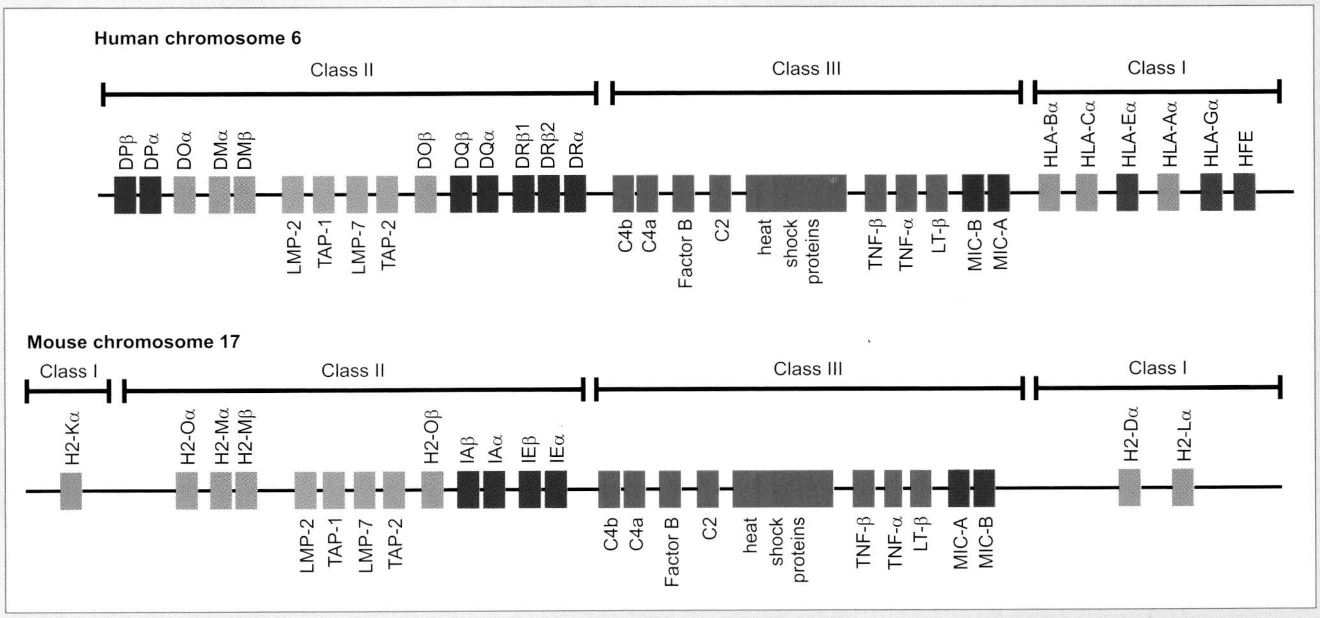

Figure 7.7 Genes encoding MHC molecules are located on chromosome 6 in humans and chromosome 17 in mice. Conventionally, the complex is divided into class I, II, and III regions. A schematic representation of the organisation of human (top panel) and mice (bottom panel) MHC genes is depicted here and does not reflect actual distances between the genes. In humans, the class II region codes for the α and β chains of the MHC class II molecules. This region also codes for molecules involved in loading of class II molecules (HLA-DM and HLA-DO) and molecules involved in the processing and loading of class I molecules (LMP-2, LMP-7, TAP-1, and TAP-2). Adjacent to this region are the genes encoding MHC class III molecules. These molecules are not involved in antigen processing and presentation and include components of the complement pathway (C4a, C4b, Factor B, and C2), a number of HSPs (here, depicted by a single, large box), cytokines (TNF-α, -β, LT-β), and the MHC-like molecules MIC-A and MIC-B. The α chain of MHC class I molecules is encoded by the class I region. It also encodes for haemochromatosis protein (HFE), involved in iron metabolism, and the nonclassical MHC class I molecules HLA-E and HLA-G. The organisation of the murine MHC genome is essentially similar to the human genome, with the class II region encoding for the α and β chains of murine class II molecules (IA and IE) and molecules involved in loading of class I and class II molecules. The murine class I region, unlike its human counterpart, is split into two, coding for H2-K and H2-D. In some haplotypes, a third MHC class I molecule (H2-L) is encoded by genes lying adjacent to H2-D.

molecules. **Two sets of genes encode murine class II molecules — *H2-A* (also called IA) and *H2-E* (also called IE).**

- *HLA-DM* and *-DO* (and their murine equivalents *H2-M* and *H2-O*), involved in peptide-loading onto MHC class II molecules, are also located in the class II region.
- *LMP* genes in the MHC class II region code for two subunits of proteasome involved in the cytosolic breakdown of proteins. Genes encoding TAP molecules involved in class I loading also lie in this region, in close association with the *LMP* genes.
- The MHC class III region encodes a variety of proteins (see the sidetrack *A Class Apart*).

Regulation of MHC expression: A 5' promoter flanks MHC class I and class II genes. Specific transcription factors can therefore regulate MHC gene expression. Defects in MHC class II transcription factors can result in bare lymphocyte syndrome, described in chapter 8 (the sidetrack *Left Defenceless*).

A Class Apart: MHC Class III Molecules

The MHC represents about 0.1% of the human genome and is traditionally divided into class I, II, and III regions. It is one of the most gene-dense regions of the genome. MHC class I and II regions encode the proteins involved in antigen presentation. Although the entire class III region has now been sequenced, the functions of all the proteins encoded by this region are far from understood. Products of the MHC class III region are not involved in antigen presentation. Some proteins encoded by the telomeric end of the class III region appear to be involved in inflammatory responses and are often dubbed MHC class IV molecules. These include members of the TNF family and HSP70 (**H**eat **S**hock **P**rotein 70). Complement components and other molecules involved in growth and differentiation remain in class III. Products of class III and IV genes seem to be associated with a number of immune and nonimmune diseases. The more important proteins with known functions include the following —

- Components of the complement cascade: C2, C4a, and C4b of the classical pathway and Factor B of the alternative pathway are encoded by the MHC class III region.
- Several products critical to growth, development, and differentiation:
  - NOTCH4, a transmembrane receptor that determines cell fate and differentiation through cell-to-cell interaction, is encoded by this region. Murine *NOTCH4* is a proto-oncogene.
  - Tenascin-XB, an extracellular matrix protein involved in connective tissue cell migration and muscle morphogenesis during embryonic development, is clustered close to the *NOTCH4* gene.
  - PBX2, a transcription factor probably involved in regulating the expansion of haematopoietic precursors, is also encoded by the class III region.
- Enzymes related to the metabolism of lipids or steroids:
  - Lysophosphatidic acyl transferase-α is required for the acylation of glycerols for lipid biosynthesis.
  - PPT2 is a palmitoyl protein thioesterase required for the hydrolysis of lipid thioesters from lipoproteins.
  - CYP21B (21-hydroxylase) is an enzyme required for the hydroxylation of steroids important in the biosynthesis of glucocorticoids and mineral corticoids.
- The receptor for advanced glycation end-products: This receptor binds advanced glycation end-products that accumulate with aging; it is thought to play an important role in the chronic inflammation that contributes to complications in diabetes, inflammatory bowel disease, and Alzheimer's disease.
- Two proteins distantly related to class I molecules — MIC-A and MIC-B: They activate NK cells, $\gamma\delta$ T cells, and CD8$^+$ T cells that carry the NKG2D receptor.
- Three related cytokines — TNF-α, lymphotoxin-α (also called TNF-β), and LT-β (TNF-χ):
  - TNF-α is a pro-inflammatory cytokine important in innate immune responses. A lack of TNF-α leads to an absence of splenic primary B cell follicles.
  - TNF-β has a similar mode of action as TNF-α but with a limited tissue distribution. Deletion of the TNF-β gene results in a specific absence of lymph nodes, Peyer's patches, and splenic germinal centres in mice, indicating that TNF-α and -β are involved in the development of secondary lymphoid organs.
  - LT-β is a membrane-bound protein that forms a heterodimer with TNF-β and can induce activation of the transcription factor NFκB.
- HSP70: HSPs are a family of chaperone molecules induced by environmental stress, such as oxidative injury. HSP70 is a member of the HSP family that is thought to be involved in the processing and presentation of bacterial and tumour antigens.

As explained, most nucleated cells express MHC I molecules, whereas professional APCs express class II molecules. However, all professional APCs do not express class II molecules constitutively. For example, only activated macrophages express class II molecules. By contrast, resting B cells express class

II molecules, and activation further upregulates this expression. Cytokines, especially pro-inflammatory cytokines, upregulate MHC expression. IFNs (whether α, β, or γ) seem to induce MHC class I expression on many types cells. IFN-γ in particular can induce or upregulate MHC class II expression on non–antigen-presenting cells, such as intestinal epithelial cells, keratinocytes, vascular endothelium, and pancreatic β cells. A prolonged inflammatory response can therefore result in improper antigen presentation by tissue cells and precipitate autoimmune responses (see chapter 15).

7.5 Polygenism and Polymorphism in MHC Molecules

The immune system's ability to respond to numerous foreign immunogens can be attributed to MHC molecules' ability to bind a range of peptides. This ability is due to two characteristics of classical MHC molecules — polygenism and polymorphism. In contrast, nonclassical MHC molecules show little polymorphic variation. These include the molecules HLA-DM and HLA-DO involved in MHC class II peptide loading, HLA-E, which presents peptides from the signal sequences of other HLA-molecules to NK receptors, and CD1.

Classical MHC genes are polygenic; several sets of MHC class I and class II genes encoding proteins with different ranges of peptide-binding specificities are found in the same individual. An individual inherits one set of genes from each parent; hence, one MHC haplotype set will be inherited from both the parents by the offspring. As these genes are polymorphic (see below), most individuals are likely to be heterozygous at these loci. MHC alleles are codominant, that is, products of both alleles are expressed, and both present antigens to T cells. This means that the potential number of distinct MHC molecules expressed by each cell of an individual is doubled. With three sets of MHC class I genes and three sets of class II genes on each allele, an individual human can typically express six class I and class II molecules each on his/her cells. Because MHC class II molecules consist of two chains (α and β) and because α and β chains from different chromosomes may combine to give functional molecules — two α and two β chains giving rise to four different products — the number of class II molecules expressed by an individual may be even greater.

MHC genes are the most polymorphic genes known. There is great variation in individual genes and their products within a single species. Maximum polymorphism is observed for residues that line the peptide-binding groove of the molecules and thus directly influences the binding specificity. **MHC polymorphism directly affects T cells' antigen recognition**. Generally speaking, T cells can only be activated by APCs that have the same MHC allele. If APCs of one mouse haplotype (say $H2\text{-}K^b$) infected with a particular virus are mixed with the CTLs of another strain ($H2\text{-}K^k$) induced by the same virus, the $H2\text{-}K^b$ APCs will not be killed. The CTLs of the $H2\text{-}K^k$ haplotype type will not recognise peptide:MHC complexes presented by the APCs of $H2\text{-}K^b$ haplotype. This is the histocompatibility (or MHC) restriction of cellular interaction. The restriction implies that T cells are programmed (or restricted) to respond to cells bearing only a particular histotope[8] (MHC molecule). Histocompatibility restriction develops along with antigen specificity during the thymic maturation of T cells (see chapter 9). Thus, **TcR specificity is defined both by the peptide and the MHC molecule on which the peptide is loaded.** Direct contact of the TcR with polymorphic residues on the MHC molecule affects

[8] Although it is well established that T cells are MHC restricted, between 1-7% T cells can recognise nonself MHC (allogeneic) molecules, that is, between 1-7% T cells are alloreactive. The exact mechanism of this alloreactivity is not clear. TcRs recognise peptides loaded on self-MHC molecules. However, if the configuration of a peptide:nonself MHC complex is similar to that recognised by the TcR, alloreactivity results. Some TcRs also respond to distinctive features of the nonself-MHC, independent of the bound peptide. Alloreactivity is of special significance in transplantations and graft rejections.

antigen recognition. MHC restriction in antigen recognition reflects the combined effects of differences in peptide-binding and of direct contact between the MHC molecule and the TcR.

It is important to understand the need for MHC diversity and the advantages this diversity confers upon the species. Although MHC genes do not directly encode TcRs, through their association with antigens, they are part of the T cell antigen recognition process. Polygenism and polymorphism in MHC proteins expands the range of antigens to which the immune system can respond. Polymorphism and codominance double the number of different MHC molecules expressed by an individual. They increase diversity in MHC molecules and, consequently, the peptides presented to the T cell. The population as a whole benefits from such diversity because a pathogen that evades the immune system of one individual may not be able to evade that of another. In the large evolutionary scheme of things, individuals are expendable — the species must survive[9]. The discovery of superantigens and the emergence of the AIDS pandemic help us appreciate the dynamic nature of the host-parasite relationship. Whenever man comes up with a better mouse trap, nature corresponds with a better mouse — every strategic evolutionary change in the host results in a corresponding, survival-bestowing mutation in the parasite.

If indeed MHC diversity bestows an evolutionary advantage, then why limit MHC loci? Each addition of a distinct MHC molecule to the MHC repertoire would increase the peptide:MHC conformations recognised as self, and this in turn would necessitate the silencing of a greater number of T cell clones in order to maintain self-tolerance. As it is, less than 5% lymphocytes entering the thymus survive the selection process. If the number of MHC loci were increased, even less T cell clones would be selected. Given that between 100-200 identical MHC:peptide complexes (representing about 1% of a particular MHC haplotype) are required to activate a T cell, an increase in MHC molecules will mean an effective decrease in the concentration of a single type of MHC molecule that is expressed on a cell surface (provided the total number of MHC molecules expressed by the cell remains constant). As only a portion of MHC molecules are loaded with peptides, this means that the number of one particular peptide:MHC complexes on APCs would be insufficient to activate T cells. By having multiple MHC loci but limiting their numbers, the system achieves a balance between the advantages of diversity and the disadvantage caused by the increased presentation of self-peptides.

7.6 CD1

Until recently, T cells were thought to recognise protein-derived antigens that were presented in the context of MHC class I and class II molecules exclusively. Research has now established that antigen presentation is not restricted to proteins and that the immune system has evolved the capability of presenting lipid antigens to T cells. CD1 molecules are involved in this presentation. CD1 molecules show little polymorphic variation and hence are considered nonclassical MHC molecules.

CD1 molecules are β_2-m associated glycoproteins that are structurally related to MHC class I molecules (Figure 7.8). Their primary function is the presentation of lipid antigens. Most lipids presented by CD1, such as diacylglycerols, sphingolipids, polyisoprenoids, or mycolates, have two hydrophobic tails and a relatively small, polar head group (Figure 7.9). The CD1 groove contains two large,

[9] Communism, anyone?

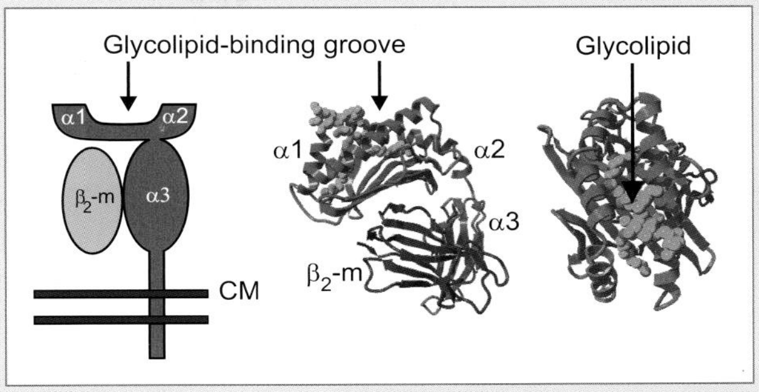

Figure 7.8 CD1 molecules are structurally similar to MHC class I molecules and present lipid antigens to αβ T cells or NKT cells. CD1 molecules consist of an α chain that is noncovalently associated with β_2-m. The left panel is a schematic depiction of the molecule. The centre panel is a ribbon diagram of the molecule. The α1 and α2 domains form the antigen-binding groove. The α3 region anchors the molecule in the cell membrane. The antigen-binding groove is lined with nonpolar amino acids, allowing interaction with aliphatic chains of the antigen. This allows the polar region of the molecule to make contact with TcR residues. A space-filling model of the top view of the binding groove is depicted in the right panel. Two to four hydrophobic pockets in the groove accommodate the hydrophobic side chains (adapted from Barral & Brenner, *Nature Reviews Immunology*, 2007, 7:929).

deep pockets, and the inner surface of the groove is lined almost exclusively with nonpolar amino acids. This lining provides a hydrophobic surface for interaction with the aliphatic hydrocarbon chains of the amphipathic glycolipids presented. The insertion of aliphatic hydrocarbon chains of antigen into the hydrophobic CD1 groove positions the rigid and hydrophilic elements of the antigen on the α-helical surface of the groove so they are available to interact with TcR. Thus, the lipid moieties serve to anchor the antigens within the groove formed by the α1-α2 domains of CD1 proteins, and they mainly contact CD1 proteins rather than TCRs. The polar head group protrudes from the groove, is solvent exposed, and is held in place by a hydrogen-bonding network. This polar head group makes direct contact with the TcR.

CD1 molecules are expressed by APCs, such as monocytes, DCs, and B cells. Five CD1 isoforms (CD1a to CD1e) encoded by genes located on chromosome 1 have been identified in humans; only one isoform (CD1d) is found in mice. The different CD1 isoforms seem to localise to different endosomal/lysosomal compartments. In humans, only the first three isoforms (CD1a, 1b, 1c) are known to function in T cell responses to naturally occurring self- and foreign antigens. CD1d presents antigen to NKT cells. Recent evidence suggests that CD1 isoforms are involved in immunity to intra-cellular parasites, such as *Leishmania* spp. and *Plasmodium* spp., as well as in autoimmune diseases. Most experimental studies are carried out in mice who express only CD1d, so these studies are restricted to NKT cells. Humans, however, have lipid-specific T cells in addition to NKT cells, and these occur at a frequency comparable to peptide-specific T cells. Thus, CD1 proteins, like classical MHC molecules, are thought to have a dual role in immune responses:

- **They present exogenous antigens to T cells** to activate host defence against infections; such presentation may be especially important for immune responses to pathogens, such as mycobacteria, that have glycolipid-rich cell walls. *M. tuberculosis* downregulates CD1 expression as part of its immune evasion strategy (see chapter 13).
- **CD1 molecules present self-sphingolipids**, such as gangliosides and ceramides, and have a role in tumour surveillance, autoimmunity, and immunoregulatory interactions with other cells.

The exact mechanism of synthesis and assembly of these molecules and the antigen processing and presentation of their ligands is still being elucidated. The current model suggests that although they are structurally similar to MHC class I

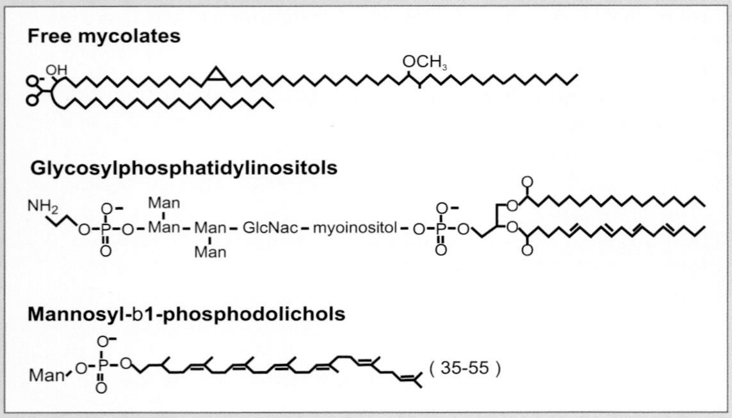

Figure 7.9 A variety of lipids, phospholipids, and glycolipids can be loaded onto CD1 molecules. The antigens presented include those derived from microbial origins (mycolates) and self-protein residues (phosphatidylinositols, phosphodolichols).

molecules, CD1 molecules are functionally similar to MHC class II molecules in that CD1 proteins survey the endocytic pathway to intersect and bind lipid antigens. Newly synthesised CD1 molecules are loaded with a self-lipid in the ER that protects its groove. Different isoforms of CD1 seem to be loaded by different pathways. Ii has been shown to act as a molecular chaperone for the murine CD1d molecule (and its human equivalent) during its synthesis in the ER. The newly synthesised CD1d molecules are transported through the golgi and *trans*-golgi networks to cell surface. The cell-surface CD1 molecules are then internalised and sorted to early/ late endosomes or lysosomes for loading with fragments of exogenous lipids that have been internalised and degraded to appropriately sized fragments. Functional equivalents of HLA-DM/DO have (as yet) not been identified in the CD1 pathway. However, cross-priming and cross-presentation is known to occur by the uptake of apoptotic cells and the release of exosomes by living, infected cells.

7.7 Antigen-Presenting Cells

Lymphocytes by themselves cannot initiate a robust adaptive immune response. Antigen capture and T cell activation are integral to triggering the adaptive immune response. Accessory cells (or APCs) intimately involved with the lymphoid system are essential in the development and expression of immune responses. **Physical interaction between APCs and T cells is required to initiate an immune response**. Many types of cells can process and/or present antigen to T cells (DCs, macrophages, B cells, FDCs, vascular endothelial cells, epithelial cells, eosinophils, mast cells). However, antigen presentation is an important function of only the first three cell types.

7.7.1 Antigen Processing and Presentation

Antigen presentation by specialised accessory cells is the first step in triggering an immune response. Recognition of the antigen by T cells is largely determined by how the antigen is processed by APCs. A complex series of steps generates peptides (or glycolipids) from the antigen. The type of MHC (CD1) molecules that bind these peptides (or glycolipids), in turn, determine the subset of T cells stimulated — MHC class I molecules stimulate **C**ytotoxic **T** **L**ymphocytes (CTLs), MHC class II molecules stimulate T_H cells, and glycolipid-loaded CD1 molecules activate lipid-specific αβ T cells or the more innate-like NKT cells. Of these, T_H cells that respond

CD1 Molecules

- CD1 molecules are expressed by thymocytes and professional APCs, such as monocytes, DCs, and B cells.
- They are similar to MHC class I molecules in structure; they consist of an α chain noncovalently associated with one molecule of β_2-m.
- Five isoforms of the molecule are identified in humans; only one is found in mice.
- Different isoforms of CD1 seem to be loaded by different pathways.
- CD1 molecules present lipid antigens to lipid-specific T cells and NKT cells.
  - The CD1 antigen-binding groove contains two large, deep pockets.
  - The inner surface of the groove is lined almost exclusively with nonpolar amino acids.
  - Lipids, such as diacylglycerols, sphingolipids, polyisoprenoids, and mycolates, can be loaded in this hydrophobic groove.

to peptide antigens seem to be indispensable for generating an immune response — only these cells can co-opt other cells types in developing the response. The expression and loading of MHC class II molecules is necessary for a full-blown immune response. Once the immune system is activated, APCs must be removed from circulation to avoid overstimulation of the system. Hence, the APCs themselves eventually become the target of cytotoxic cells, such as NK cells, K cells, and CTLs. Such removal also helps in the destruction of any intracellular pathogens that may have survived in the APCs' phagosomes. The major steps in antigen presentation are given below.

- **Internalisation**: The antigen must be taken up by APCs for successful presentation.
  - Soluble antigen can be taken in by pinocytosis[10].
  - Particulate antigens can be internalised by various means. The most obvious route is phagocytosis. PRRs help in the initial attachment of the particulate antigen to the phagocyte cell membrane. Apolipoprotein E, secreted by APCs, was found to mediate the uptake of lipids or glycolipids dissolved in serum. Digested fragments derived from phagocytosed cells are then displayed in the context of MHC or CD1 molecules on the surface of scavenging cells. Scavenger cells are involved in clearing cellular debris — whether self or foreign. Therefore, they display a sampling of both self- and nonself antigens, with other mechanisms (requirement for costimulatory molecules, presence of T_{reg} cells, elimination of self-reactive clones, etc.) in place to ensure the absence of autoimmune responses.
  - Nonphagocytic or poorly phagocytic cells express a variety of receptors that can bind the antigen and aid internalisation. These include PRRs (such as mannose receptors, TLRs, and LBP) that are described in chapter 2. Ligands for these receptors are expressed only on microorganisms (i.e., nonself cells). This ensures that the emanating response is not self-deleterious. Lipid antigen uptake is facilitated by low-density lipoprotein receptors, C-type lectins that bind mannose residues on glycolipids, and scavenger receptors that bind modified forms of low-density lipoproteins and apoptotic cells.
  - Many accessory cells also express FcRs. Antigen–antibody complexes bind to the surface of the accessory cell *via* FcR. Such binding activates the cell membrane, allowing internalisation of bound complexes. This method of

[10] Similar to drinking, pinocytosis is the ingestion of dissolved materials by endocytosis. The cytoplasmic membrane invaginates and pinches off, placing small droplets of fluid in a pinocytic vesicle. The liquid contents of the vesicle are then slowly transferred to the cytosol.

internalisation requires the presence of specific antibodies in the system, and hence, it is of particular use in secondary responses. The antigen internalised through the FcR is presented more efficiently than by other routes, aiding in quicker responses.

  - B cells can take up and internalise soluble antigen *via* their antigen receptors. B cells are unique in this respect — unlike other accessory cells, they behave as antigen-specific APCs. They are especially important in presenting soluble antigen, as neither macrophages nor DCs can efficiently take up soluble antigens.

- **Antigen processing**: The internalised antigen is transported across a series of endocytic compartments rich in degradative enzymes (proteases, glycosidase, and glycolytic enzymes). Lipids/sphingolipids are often present in membranes, and these have to be extracted for loading. A class of proteins called *saposins* — nonenzymatic proteins that localise to lysosomes — facilitate the hydrolysis of different glycosphingolipids by glycolytic enzymes in these compartments. The lysosomal compartments gradually decrease in pH, and consequently, they have increased degradative capability. Ultimately, this produces fragments suitable for loading onto MHC or CD1 molecules.
  - Peptides of a size suitable for binding to MHC molecules (about 8-20 amino acid long) are produced by these degradative pathways and can even be secreted from the cell. The generated peptides are then loaded onto MHC molecules. Evidence suggests that some proteins may not undergo internalisation; instead, they are degraded by surface proteases and loaded onto MHC molecules at or near the cell surface.
  - The exchange of endogenous lipids, loaded onto the CD1 by fragments derived from the internalised lipid antigens, takes place in endocytic compartments, such as lysosomes. Our understanding of lipid trafficking and degradation is far from complete. It is known that different lipids are distributed differently among compartments, and different compartments of the endocytic pathway have distinct lipid compositions. Different isoforms of CD1 are therefore differentially distributed in these compartments.
- **Antigen presentation**: *Professional* antigen presentation consists of two separate interactions.

The Day After: Fates of Interacting TcRs and MHC Molecules

Receptor internalisation is often accompanied by receptor-mediated ligand internalisation. Upon physiological stimulation, receptors of the tyrosine kinase family, such as BcRs, complement receptors, and FcRs, are rapidly internalised. TcRs also belong to this family of receptors. They are rapidly downmodulated following stimulation by a variety of ligands (peptide:MHC complexes, superantigens, anti-TcR antibodies). The internalised TcRs are then degraded along with some of the molecules involved in signal transduction. In most cases, ligands are soluble molecules that bind to their receptors with high affinities and are co-internalised along with the receptors. It is more difficult to establish the fate of membrane-bound ligands, such as MHC:peptide complexes. Membrane-bound MHC:peptide complexes are extracted from APCs by T cells. In addition, other membrane-bound ligands, such as costimulatory molecules (CD80/86) and adhesion molecules, that take part in T cell-APC interaction are also acquired by T cells. The exact method of extraction and the fate of the acquired molecules have not been determined.

- In the case of protein antigens, recognition of the MHC:peptide complex by TcRs constitutes the first signal delivered to T cells. CD4/CD8, coreceptors for MHC molecules, are also engaged in this primary interaction. CD1 binds CD4/CD8 poorly; these coreceptors do not play a significant role in the activation of CD1-restricted T cells.
- The second signal is delivered by the engagement of costimulatory molecules. A number of costimulatory and adhesion molecules (CD80/CD86[11], CD54[12]) are expressed by APCs. The binding of costimulatory molecules to their ligands (CD28[13] and CD11a/CD18, respectively) expressed on T cells along with TcR engagement leads to naïve T cell activation. Only professional APCs, such as DCs, macrophages, and B cells, express these costimulatory molecules, and hence, only these cells can activate naïve T cells. Other cells, such as endothelial cells and fibroblasts considered nonprofessional APCs, do not normally express these molecules[14]. There is a virtual cross-talk between the professional APCs and T cells; the interaction of TcRs with the MHC:peptide complexes also upregulates the expression of existing costimulatory molecules or induces expression of new molecules. Little is known about the signalling pathways that control the expression of these adhesion molecules by professional APCs. It is now believed that even with professional APCs, PRR engagement is required to induce the expression of costimulatory molecules. Such a system allows the induction of costimulatory activity by common microbial constituents, enabling the immune system to distinguish between antigens borne by infectious agents and those associated with innocuous proteins, including self-proteins. This idea is supported by the observation that foreign proteins often cannot evoke an immune response on their own unless they are mixed with microbial constituents, such as LPS, that induce costimulatory activity. Such microbial constituents are therefore often coadministered with the antigen in vaccines.

Engagement of TcRs in the absence of costimulatory signals (as happens in the case of nonprofessional APCs) leads to anergy in naïve T cells. Such T cells cannot be further stimulated by APCs. Thus, whether a naïve T cell proliferates in response to antigen recognition is determined by the costimulatory signal; absence of this signal is likely to render the T cell permanently incapable of responding to antigenic stimulation. Delivering only signal two (i.e., ligation of CD28) in the absence of antigenic stimulation, on the other hand, does not have a deleterious effect on T cells. Thus, all peptides — self- or nonself — can, in principle, be presented in the context of MHC molecules. Nonprofessional antigen presentation, absence of self-reactive clones[15], and the action of T_{reg} cells prevents a potentially deleterious immune response to self-antigens.

7.7.2 APC Functions

Accessory cells have multiple functions.

- **Capturing of antigen**: APCs, such as DCs and macrophages, function as sentinels. They act as sensors, continuously circulating between the lymphoid system and blood, phagocytosing dead or apoptosing cells. Spillage of intracellular contents of either infected or dying cells can prove injurious; by removing potentially infected or normal dying cells, APCs contain damage to the system.

[11] Old name — B7.1 and B7.2

[12] Also known as ICAM-1, CD54 binds CD11a/CD18 (old name LFA-1)

[13] CTLA-4, another molecule expressed on T cells, is also a ligand for CD80/86. Instead of stimulating T cells, the engagement of these molecules makes the cells unresponsive to the MHC:peptide signal (chapter 9).

[14] Except when exposed to inflammatory stimuli like infection or exposure to IFN-γ

[15] Strictly speaking, this is not true. Although negative selection in the thymus weeds out self-reactive clones, the process is not foolproof, and self-reactive clones do exist in the periphery. A reduced frequency (and not absence) of self-reactive clones would be a more accurate statement.

- **Migration and presentation of antigen**: APCs acquire antigens in nonlymphoid tissues and then migrate to T cell areas of the lymphoid system where they present the antigen to naïve T cells, thus increasing the chances of encountering the appropriate T cells.
- **Induction of activation and proliferation of T cells**: By providing appropriate costimulation, APCs ensure the induction of an immune response. Absence of such costimulatory signals induces unresponsiveness or tolerance in T cells.
- **Modulation of immune response**: APCs influence the immune response in a number of ways.
  - Intracellular events in APCs determine the type of peptide (or lipid fragment) generated and hence the epitope against which the T cell response is directed.
  - The class of MHC molecules on which the peptide is loaded in the APCs determines the type of the T cell subset that is activated (e.g., T_H or CTL).
  - The cytokines secreted by APCs influence the course and extent of the immune response.

7.7.3 Types of APCs

APCs can be professional or nonprofessional. Antigen presentation by professional APCs is indispensable to switching on immune responses; only these APCs will be considered here.

- **DCs**: These cells are by far the most important APCs. They comprise a large family of leukocytes with related morphology and possess the ability to activate naïve T cells. They are widely distributed in tissues and are found especially at environmental interfaces. These are perhaps the only cells with no known function other than activating and controlling T and B cells. Although they differ widely in terms of anatomic localisation, phenotype, and function, all DCs have several common features.
  - DCs originate from $CD34^+$ progenitor cells and are seeded by means of the bloodstream to the tissues, where they give rise to immature DCs (see the sidetrack *Long Arms of the Police*).
  - Immature DCs actively internalise the antigen by both receptor-mediated and nonreceptor-mediated mechanisms and degrade these internalised antigens in endocytic vesicles to produce fragments capable of binding MHC (or CD1) molecules.
  - DCs mature and migrate to lymphoid organs in response to danger, such as tissue damage, presence of pathogen-derived products, inflammatory cytokines, and chemokines.
  - Mature DCs can activate antigen-specific naïve T cells. Maturation upregulates the expression of peptide-loaded MHC molecules and costimulatory molecules, allowing DCs to present antigen to naïve T cells.
  - Mature DCs secrete cytokines that can skew the immune response to the T_{H1} or T_{H2} type (i.e., they become DC_1 or DC_2 type) in response to microenvironmental factors. Conversely, the IL-10 secreted by immature DCs can result in T_{reg} cell formation and the suppression of immune responses.

DCs are migratory cells that travel from one site to another, performing specific functions at each site. Derived from the bone marrow, DCs circulate through the blood before entering the tissue, where they become resident immature DCs that monitor their environment. Interstitial DCs and Langerhans cells are examples of such DCs. These immature DCs migrate towards inflammatory foci, where they take up and process antigens. Such immature DCs are avidly endocytic; this characteristic is downregulated upon maturation (Table 7.1). DCs then migrate to the draining lymph node and home into T cell-rich areas to initiate an immune response, maturing during this migration. As they mature, they lose their ability to respond to inflammatory stimuli and upregulate surface expression of peptide-loaded MHC class II molecules.

A particular subtype of DCs — the FDCs — has a major role in secondary humoral responses. FDCs are found in secondary follicles in B cell areas of all lymphoid tissues. These are nonphagocytic cells that can retain antigen-antibody complexes on their surfaces for prolonged periods of time. They are responsible for the increased antibody affinity for the antigen in secondary responses (see chapter 8).

Table 7.1 Characteristics of Mature and Immature DCs

Characteristic	Immature DCs	Mature DCs
Cell shape	Relatively small cells with no prominent processes	Larger cells with numerous processes (dendrites or veils)
Antigen capture	Actively endocytic and phagocytic	Poorly phagocytic
	Internalise antigens by receptor- and nonreceptor-mediated processes; active in antigen processing	May internalise antigens *via* specific receptors
	Express high levels of FcRs	FcR expression is lower; it is downregulated during maturation
Antigen presentation	Have high levels of intracellular MHC molecules but do not express them at the cell surface	Express a high density of MHC:peptide complexes on the cell surface
	Costimulatory molecules are either absent or expressed at very low levels	Variety of costimulatory molecules are expressed
	Poor in antigen presentation; may anergise rather than stimulate T cells	Efficient in antigen presentation to naïve and memory T cells
Cytokine responses	Proliferate in response to GM-CSF	Do not proliferate appreciably in response to any stimuli
	Mature in response to IFN-γ, TNF-α, and noncytokine stimuli, such as contact with LPS, CD40L, etc.	Do not differentiate further; are stable end cells
	Inhibited by IL-10	Resistant to IL-10
	Do not secrete IL-2	Can secrete IL-2 early after microbial interaction; can augment T cell proliferation following antigen presentation
	Cannot secrete cytokines that skew the T_H response; may secrete IL-10 that results in formation of T_{reg} cells	Secrete IL-10 or IL-12, skewing the response to T_{H1} or T_{H2} type, respectively

The Long Arms of the Police: Human DCs

In humans, DCs are found as $CD34^+$ precursor populations in the bone marrow and blood and as more mature forms in lymphoid and nonlymphoid tissues. $CD34^+$ progenitor cells are of two subtypes — myeloid and lymphoid progenitors. Myeloid $CD34^+$ progenitor cells give rise to $CD11c^+CD14^+$ DC precursors (monocytes) that can further differentiate to interstitial DCs, and $CD11c^+CD14^-$ precursors that give rise to Langerhans cells. The lymphoid progenitors differentiate into the $CD11c^-$ $CD14^-$ blood precursors that give rise to lymphoid DCs (or plasmocytoid DCs) (Figure 7.S1).

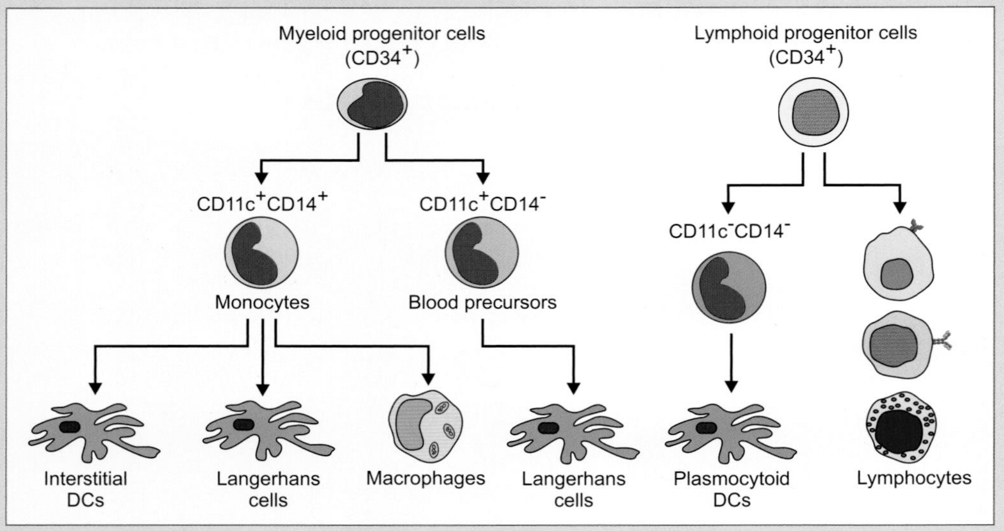

Figure 7.S1 $CD34^+$ progenitor cells can give rise to DCs, macrophages, or lymphocytes, depending upon the cytokine milieu they are exposed to (adapted from Lipscomb & Masten, *Physiological Reviews*, 2001, 82:97).

DC Subtypes: Besides being classified by degree of maturation, DCs can also be classified by location.

- **Langerhans cells** arise from the myeloid lineage and are found in nonlymphoid tissues, especially in the epidermal layer of the skin. A distinctive feature of Langerhans cells is the occurrence of Birbeck granules. These are typical structures observed using electron microscope, and they are thought to be part of the antigen-processing machinery. Langerhans cells have a varied role in the capture, migration, processing, and presentation of cutaneous pathogens/parasites. They are in an intermediate stage of maturation, wherein they are highly efficient in capturing and processing antigen but not in the presentation of the antigen to T cells. As Langerhans cells start migrating to regional lymph nodes, they mature. The processed antigen from the endocytic pathway reaches the cell surface, and the ability of Langerhans cells to present antigen to T cells improves. They are extremely efficient in sensitising hosts to low doses of antigen but appear to be damaged by high antigenic loads.
- **Interdigitating DCs** are found in the secondary lymphoid organs, such as the lymph nodes, spleen, and tonsils. These are mature DCs that capture antigen in the periphery and migrate to the secondary lymphoid organs to activate T and/or B cells. They are thus instrumental in presenting antigen to naïve T cells.
- **Interstitial DCs** are immature DCs found in almost every organ of the body — the liver, pancreas, kidney, heart, ureter, bladder, thyroid, gut, and dermis, except in parts of the brain, eye, and testes. These DCs, along with the Langerhans cells, form a sentinel network to patrol for antigens. Derived from $CD14^+$ myeloid precursors, these DCs can prime both $CD4^+$ and $CD8^+$ T cells. They are the only DCs thought to be able to induce activated B cell differentiation to Ig-secreting plasma cells in the germinal centre.
- **Plasmocytoid DCs** get their name from their resemblance to plasma cells. They are found in the T cell zones of lymphoid organs and the thymus. Their unique ability to secrete copious amounts of IFN-α/β upon viral stimulation suggests a role in defence against viral infections.
- **Thymic DCs** are important in thymic education. Unlike other DCs, these DCs present self-antigens to developing T cells to induce tolerance.

APCs

- APCs capture antigens, migrate to T cell areas, present antigens to T cells, induce activation and proliferation of T cells, and modulate immune responses.
- They internalise antigens by pinocytosis, phagocytosis, and receptor mediated endocytosis.
- They process internalised antigens, load fragments of digested antigens on MHC or CD1 molecules, and present them to T cells.
- Professional APCs, such as DCs, macrophages, and B cells, deliver two signals to T cells and activate them.
  - The first signal is antigen-specific — delivered by the recognition of peptide-loaded MHC molecules.
  - Secondary signals are antigen-nonspecific signals delivered through costimulatory and/or adhesion molecules.
- Nonprofessional APCs, such as fibroblasts and endothelial cells, normally deliver only antigen-specific signals and anergise T cells.

- **Macrophages and related cells**: By virtue of their antigen presentation capabilities, these mediators of innate immunity are also important in switching on the adaptive response. Macrophages have already been discussed in chapter 2; only their antigen presentation will be discussed here.

  Macrophages are potent APCs especially important in immune responses to intracellular pathogens, such as *Leishmania* spp. or trypanosomes. Resting macrophages have few or no MHC molecules and do not express the costimulatory molecules CD80/CD86. Upon engagement of the PAMPs expressed by microorganisms, the expression of MHC and costimulatory molecules is upregulated. Such expression of costimulatory molecules only after the ingestion of microorganisms is thought to allow the immune system to discriminate between infectious agents and innocuous self-proteins. In secondary immune responses, antibodies are already in circulation. Macrophages can bind antigen–antibody complexes *via* FcRs, leading to efficient internalisation and better antigen presentation.

  Although they are not considered professional APCs, both eosinophils and mast cells are capable of T cell stimulation. Like macrophages, resting eosinophils do not express either MHC class II molecules or costimulatory molecules. However, when exposed to IL-3 and GM-CSF, they express both these molecules and activate naïve and effector T cells. They may thus have a role in activating T cells in allergic or parasitic responses. Mast cells have also been shown to actively phagocytose Gram-negative bacteria and present immunogenic peptides to T cells, and they may have a role in the development of CTL responses to bacterial pathogens.
- **B cells**: Although not as important in antigen presentation *in vivo* as macrophages or DCs, B cells can present antigens to T cells. They are the only cells capable of presenting soluble antigens. They express MHC molecules constitutively, and costimulatory activity is induced only in the presence of microbial products, such as LPS. This ensures tolerance to self-antigens; even if resting B cells present soluble self-antigens, naïve T cells cannot be activated without costimulation.
- γδ T cells have recently been shown to be capable of antigen presentation to T cells, adding one more member to the list of known APCs.

8 B Lymphocytes

Another one bites the dust
Another one bites the dust
Another one bites the dust
Another one bites the dust
There are plenty of ways you can hurt a man
And bring him to the ground
You can beat him
You can cheat him
You can treat him bad and leave him
When he's down
But I'm ready, yes I'm ready for you
I'm standing on my own two feet
Out of the doorway the bullets rip
Repeating the sound of the beat

— Queen, *Another One Bites The Dust*

8.1 Introduction

Haematopoiesis is the formation and development of blood cells from pleuripotent (i.e., capable of giving rise to multiple cell lineages) haematopoietic stem cells (see Figure 5.3). Specific markers help distinguish different lineages. Some progenitors give rise to erythroid cells, which eventually differentiate into erythrocytes. Others give rise to myeloid lineages, which develop into phagocytes and granulocytes. Yet others develop into lymphocytes in the specialised environment of the primary lymphoid organs. Lymphocytes are essential for generating adaptive immune responses, and they are the only cells expressing receptors that recognise and bind specific antigens. This commitment to a particular antigen is pre-established. It develops in the primary lymphoid organs and once committed, the lymphocytes (and their progeny) retain this commitment for the rest of their lives. Thus the antigen-specificity of lymphocytes is determined before antigen contact. A given clone of lymphocytes, therefore, differs from every other clone of lymphocytes in the structure of its antigen receptor's combining site and hence the range of antigenic molecules it responds to. This response can be highly varied, from activation and proliferation to production of effector molecules (antibodies, cytotoxins, cytokines, or other soluble factors) and anergy.

Lymphoid stem cells give rise to lymphocytes in the microenvironment of the primary lymphoid organs (see chapter 5). The newly formed lymphocytes further mature and proliferate in these organs and give rise to lymphocytes that differ in their functional properties. Two types of lymphocytes are involved in adaptive immunity. B lymphocytes mature in the bone marrow (in mammals) or Bursa of Fabricus (in birds) and are responsible for humoral immunity. T lymphocytes mature in the thymus and are multifunctional cells involved in multiple aspects of adaptive immunity. Mature lymphocytes exiting the primary lymphoid organs are termed *naïve* (or virgin) cells because they have not yet encountered the molecules that their antigen receptors recognise. These naïve lymphocytes are exported to the periphery and localise preferentially to organised secondary lymphoid organs. Following antigenic stimulation, naïve B and T lymphocytes complete their differentiation to effector cells in the secondary lymphoid organs. B cells give rise to antibody-secreting plasma cells, and T lymphocytes give rise to T_H cells (**T H**elper cells), CTLs (**C**ytotoxic **T L**ymphocytes), or T_{reg} (**T reg**ulatory) cells.

The immune system is capable of reacting to a virtually limitless number of antigens. This ability is due to the existence of a large number of lymphocyte clones, each bearing receptors specific for a particular antigen. In this sense, lymphocytes are a heterogeneous collection of cell clones, with each clone differing in its ability to combine with and respond to different antigens. Analyses of genetic and molecular factors leading to antigen receptor diversity have led researchers to believe that the human genome has the potential to encode about 10^{18} different antigen receptors. This implies that our body can produce at least 10^{18} differing clones of lymphocytes, each capable of reacting to a different antigen. How these diverse clones arise is detailed in chapter 6.

T and B lymphocytes have different antigen receptors (see chapter 6). The membrane form of the Ig molecule is the B cell antigen receptor; the T cell antigen receptor is a structurally similar, nonsecretory glycoprotein. Lymphocyte subsets can be distinguished by their markers. Initially, these markers had peculiar names (lyt, lyb, thy, leu, etc.), usually bestowed by the (groups of) scientists who identified

Abbreviations

CD:	Cluster of Differentiation
CTL:	Cytotoxic T Lymphocyte
BAFF:	B cell Activating Factor of the TNF Family
BiP:	Binding Protein
Blimp-1:	B Lymphocyte Induced Maturation Protein-1
BLS:	Bare Lymphocyte Syndrome
Btk:	Bruton's tyrosine kinase
CAM:	Cell Adhesion Molecule
DD:	Death Domain
DISC:	Death-Inducing Signalling Complex
FADD:	Fas-Associated Death Domain
IAP:	Inhibitors of Apoptosis Protein
MZ:	Marginal Zone
RAG:	Recombination Activating Gene
RANK:	Receptor Activator of NFκB
RANK-L:	Receptor Activator of NFκB Ligand
RIP:	Receptor Interacting Protein
SCF:	Stem Cell Factor
SCID:	Severe Combined Immunodeficiency
SRP:	Signal Recognition Protein
T_H cell:	T Helper cell
T_{reg} cell:	T regulatory cell
TdT:	Terminal deoxynucleotidyl Transferase
TNFR:	TNF Receptor
TRADD:	TNFR-1-Associated Death Domain
TRAF:	TNFR Adaptor Factor
TRAIL:	TNF-Related Apoptosis Inducing Ligand
VLA-4:	Very Late Antigen-4

them. Now, an internationally accepted nomenclature identifies these markers using a single prefix, CD (**C**luster of **D**ifferentiation)[1]. In many places, we have included older names (in parentheses) for easy reference to older bibliographic material. Appendix II summarises the major CD antigens in this book.

B cells are the only Ig-producing cells. They represent between 5-15% of the circulating lymphoid pool and are found in the bone marrow, blood, lymphoid organs, and lymph. Mammalian B cells form and mature in the bone marrow, and from there, they migrate to lymph nodes through the blood stream. In the absence of antigenic stimulus, they pass through primary follicles and return to circulation by way of the lymphatic system to die a few days later. If, however, they encounter an antigen, they form a primary focus of proliferating cells and leave the primary follicle to form the germinal centre of the secondary follicle. Here, the B cells undergo proliferation and differentiation to give rise to antibody-secreting plasma cells and memory cells. The plasma cells subsequently migrate to either the medullary cords of the lymph node or the bone marrow; memory cells re-enter circulation.

Markers expressed by human B cells include the following:

- **B cell receptor complex** is the most distinctive marker of B cells. It consists of a mIgM or mIgD molecule, along with a host of other molecules that are positive and negative regulators of BcR function. These include Ig-α, Ig-β, CD19, CD21, CD22, CD35, CD45, CD72, and CD81. Section 6.2 discusses BcR receptor complex and signalling at length.
- **CD45** is of particular relevance as a B cell marker. Different isoforms of the protein are expressed on different lymphocytes or at different stages of development of the same lymphocyte — memory cells have a different isoform from naïve cells. The heavier B220 isoform is one of the earliest markers of B cell lineage.
- **CD5 and CD9** are expressed by subpopulations of B cells. CD5 is normally expressed by T cells and was historically used to identify T cells. $CD5^+$ B cells, termed B1 B cells, are involved in innate immune responses (see chapter 2).
- **MHC class II molecule** expression allows B cells to present antigen to T cells.
- **Costimulatory molecules**, such as CD80/CD86 (together referred to as B7 or B7.1/B7.2), and molecules such as CD40 and CD134 that are important to T cell–B cell interaction, are expressed by activated B cells (see section 8.4).

[1] Phew!

The Wars Within: The B cells — courtesy Niall Harris

"I want to tell you about myself. I want you to understand where I come from and what I stand for. And I want you to see, finally, why you can never win.

In the beginning, it was dark, and confusing. There were so many of us then, all crammed into such a tiny space. We clung to our surroundings as best we could, listening to the babel of voices speaking to us, and we learnt. We learnt about ourselves — where we came from, what we are, and what we are capable of.

It was a hard, dark time. As we moved through our home, staying for the most part close to the boundaries, we saw generation after generation of our kind struck down by some mysterious malaise. All around us lay the detritus of dead and dying cells. Soon — horribly soon — we too became affected. Over half of us fell then, and there was not a one of us that did not feel the changes occurring deep within our bodies, not a one of us that did not shudder at the thought of the gruesome fate that could so easily have claimed each of us. We took what solace we could from our continued existence, although it seemed meaningless then, and resolved to increase our numbers once more. Then, the second change came. By some small mercy, it was less severe than the first, and most of us survived. When finally it was all over, we found ourselves different; bigger, stronger. There was nothing for us there anymore, and so we left the comforting walls of our home behind and ventured out into the cavernous interior.

As we travelled, I looked around at my brothers, and marvelled at the variation I saw. Each of us now wore a different uniform; the same colours applied in magnificent, unique designs.

We reached our destination, the core of the marrow, and saw that others like us had come from all corners of our home. We rejoiced, for our numbers were great, and together we were strong. If we had known that our trials were not yet over, we would not have been so unwary. Before we knew it, once again our numbers were being culled — and this time the enemy was without, rather than within. Cell after cell withered and died at the touch of the deadly molecules floating amongst us. It was then that we learnt our purpose: to protect our home against invaders. Those who died were traitors to the cause, cut down because they would sabotage our home at the first opportunity. A few repented, taking on new coats. Those who would not were destroyed. We were better off without them.

And so, we left our birthplace behind and moved out into the world. We looked in awe upon the highways of this place, always busy, night or day. We marvelled at the elegantly organised structures we found. Everywhere we looked, we saw life, and activity, thousands upon thousands of cells going about their business, all playing their roles in the great machine. This is what you would destroy with barely a thought.

The greatest marvel of all, though, was the Node. It was vast, and it seemed a home away from home for us and our kinds. A centre for gossip, a haven for wanderers, and sanctuary for refugees.

The illusion was soon stripped away, though. The Node was not big enough for all who wished to use it, and cells squabbled bitterly over what little resources there were. I was one of the lucky ones; I found a pass, no doubt dropped in a brawl, which gained me admittance to the follicle, where there was food and shelter. I watched cells I had lived alongside for days wander lost and confused, unsure of where to go or what to do.

I watched them waste away and die.

And so life fell into a routine, for a time; I would leave the Node and travel the world for a while, seeing the sites, ever vigilant for incursions of your kind. Every so often I would return to my new home, checking in, as it were, and resting before my next trip. Gaining entry to the follicle was still a struggle, but it became easier with time.

Then I came here, and I sensed you. As if you could have hidden! You may have outwitted my macrophage comrades once, but they just needed to be shown where to go. And so, I returned to the Node to start the Call. I hadn't felt so alive in days! This time, I didn't go to the follicle; a T cell took me to one side and told me what I must do. He showed me the way to the germinal centre, where I increased my numbers. Some were inferior, and were culled. Most were stronger. All recognised your presence and were outraged.

I can sense you, you know. You're still there, aren't you? The last few cells, trying in vain to cling to life whilst wave upon wave of phagocytes sweep through your ranks. You might as well give up; you haven't a prayer. 'Maybe next time', you think. I wouldn't count on it; I know your face, now. And I never forget a face."

(*Reproduced with permission of the author*)

8.2 B Cell Development

B cells are formed by the multiplication and differentiation of pleuripotent stem cells derived from mesenchymal cells that migrate to the foetal yolk sac. B cell formation begins in the yolk sac, shifts to the foetal liver, and finally relocates to the bone marrow. Lymphocyte formation continues in the bone marrow throughout adult life. An estimated 2×10^7 to 5×10^7 B cells are produced daily in the bone marrow, but a large number of them die by apoptosis before reaching the peripheral lymphoid organs. A whopping 95% of the B cells that develop in the bone marrow and primary or secondary germinal centres perish there for a variety of reasons, such as faulty gene rearrangement, anti-self receptor expression, and a lack of stimulation. The total repertoire of B cell specificities approaches 10^9; the total number of B cells in adults is about 10^{11}.

All lymphoid cells arise from common lymphoid progenitor cells. These $CD34^+$ common lymphoid progenitors go through a series of steps before emerging as naïve B cells. **B cell development can be divided into four stages — pro-B cell**[2], **pre-B cell**, **immature B cell**, **and mature B cell**. These stages can be distinguished by stage-specific differentiation markers and the successive steps in the rearrangement and expression of Ig genes. The initial stages of B cell development are often referred to as part of the antigen-independent[3] phase. In the next, the antigen-dependent phase, naïve B cells undergo a process of differentiation to become plasma cells or memory cells. None of these steps or changes is spontaneous. Each step requires the interaction of extracellular signalling molecules with specific receptors on B cell surfaces. This interaction may be between the surface proteins of other cells (such as the stroma of the bone marrow) and B cells, between cytokines and their receptors expressed by B cells, or between the antigen and the mIg molecule.

[2] The pro- in the pro-B stands for progenitor, the pre- for precursor.

[3] Not to split hairs, but calling the development from lymphoid progenitor to naïve B cells *antigen-independent* is not strictly correct because self-antigens are involved in B cell selection and survival.

B Lymphocytes

- These are the only cells capable of producing Igs.
- They mature in the bone marrow in mammals or the Bursa of Fabricus in birds.
- B cells express mIg molecules on their surface; mIg is the B cell antigen receptor and is associated with a host of other molecules.
- B lymphocytes also express MHC class I and II molecules; the expression of MHC class II molecules allows them to present antigen to TH cells.
- Three major subsets of B cells are recognised on the basis of phenotypic, topographic, and functional characteristics.
  - B1 B cells are CD5-expressing, nonrecirculating, self-renewing cells found predominantly in pleural and peritoneal cavities. They respond mainly to TI antigens and have a role in innate immune responses.
  - B2 B cells are $CD5^-$, recirculating B cells found in the periphery.
    - These are conventional B cells responsible for long-lasting humoral immunity.
    - B2 cells respond to TD antigens and undergo affinity maturation and class-switch recombination.
    - Two signals are required to activate B2 cells. The crosslinking of BcRs delivers the first signal; T cells deliver the second signal.
    - The cytokines released by the TH cells are important in the proliferation and differentiation of these cells.
    - BcR ligation in the absence of TH cell-signalling anergises B2 cells.
  - MZ B cells are nonrecirculating cells located in the marginal zone of the spleen. They respond predominantly to TI antigens.
- The crosslinking of surface Ig molecules with or without TH cell help (for B2 or B1 B cells, respectively) leads to the activation, proliferation, and differentiation of B cells.
  - Memory B cells are activated by smaller amounts of antigen than naïve B cells and can further differentiate to plasma cells. They are less dependent on T cell help, and express Ig molecules of a higher affinity for the antigen; these molecules are of an isotype other than IgM.
  - Plasma cells are end cells responsible for antibody production.

As they mature, B cells migrate from the subendosteum, adjacent to the inner bone surface, to the central axis of the marrow cavity. The CAMs (**C**ell **A**dhesion **M**olecules) on stromal cells form specific adhesion contacts with corresponding ligands on B cells to deliver the signals required for B cell development. Stromal cells also deliver other signals vital for this process. Thus, the interaction between SCF (**S**tem **C**ell **F**actor) on stromal cells and its receptor c-kit (CD117) on the lymphocyte is essential for the development of B (and T) cells. In addition, stromal cells also secrete growth factors — IL-7, secreted by stromal cells, is indispensable to B (and T) cell development (Figure 8.1). The later B cell developmental stages are much less dependent on such stromal contacts. The transcription factor Pax-5 has been identified as a key regulator of B cell development. It is essential for commitment to the B cell lineage and for the repression of alternative lineage differentiation. The major steps in the ontogeny of B cells are outlined below (Table 8.1, Flow chart 8.1).

- The **earliest committed B cell precursors are the pre-pro-B cells**. These cells have Ig loci in the germline[4] configuration. **They do not express components of BcR complex**, and express low levels of the proteins RAG-1 and RAG-2 (**R**ecombination **A**ctivating **G**enes-1 and -2). Cytokines induce the synthesis of RAG-1 and RAG-2 and the nuclear enzyme **T**erminal

[4] The Ig molecule consists of H and L chains. The V_H region is encoded by *V* (Variable), *D* (Diversity), and *J* (Joining) segments; the V_L region is encoded by *V* and *J* segments. In the germline, multiple loci exist for each of these segments, and a productive H or L chain is obtained through gene recombination joining one of the multiple *V* gene segments with any one of the multiple *D* (H chain only) and *J* segments (see chapter 6).

Table 8.1 B Cell Developmental Stages

Developmental stage	**Stem cell**	**Pre-pro-B cell**	**Early pro-B cell**	**Late pro-B cell**	**Large pre-B cell**	**Small pre-B cell**	**Immature B cell**	**Mature B cell**
	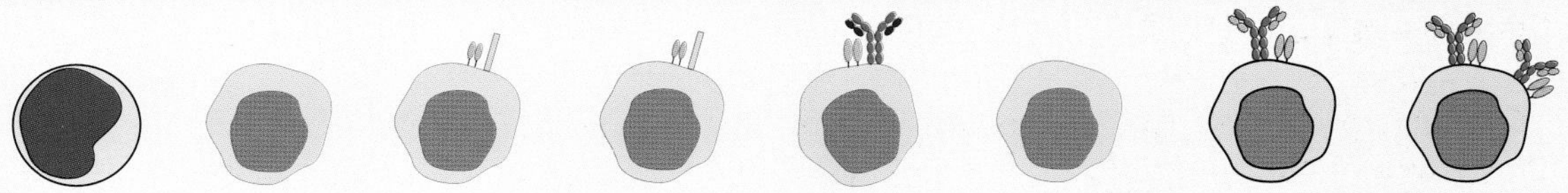							
H chain	Germline	Germline	*D–J* joining	*V-DJ* joining	*VDJ* re-arranged	*VDJ* re-arranged	*VDJ* re-arranged	*VDJ* re-arranged
L chain	Germline	Germline	Germline	Germline	Germline	*V–J* joining	*VJ* re-arranged	*VJ* re-arranged
RAG	–	±	+	+	+	+	+	–
TdT	–	±	+	+	–	+	+	–
Surrogate L chain (λL)	–	–	+	+	+	–	–	–
Surface Ig	–	–	–	–	μ chain in pre-BcR	Surface & cytoplasmic μ chain	mIgM	mIgM, mIgD
Ig-α, Ig-β	–	–	+	+	+	+	+	+
Other surface markers	CD34	CD34	CD34 B220 Class II*	B220 Class II* CD19 CD40	B220 Class II* CD19 CD40	B220 Class II* CD19 CD40	B220 Class II* CD19 CD40	B220 Class II* CD19 CD21 CD23 CD40

* MHC class II molecules

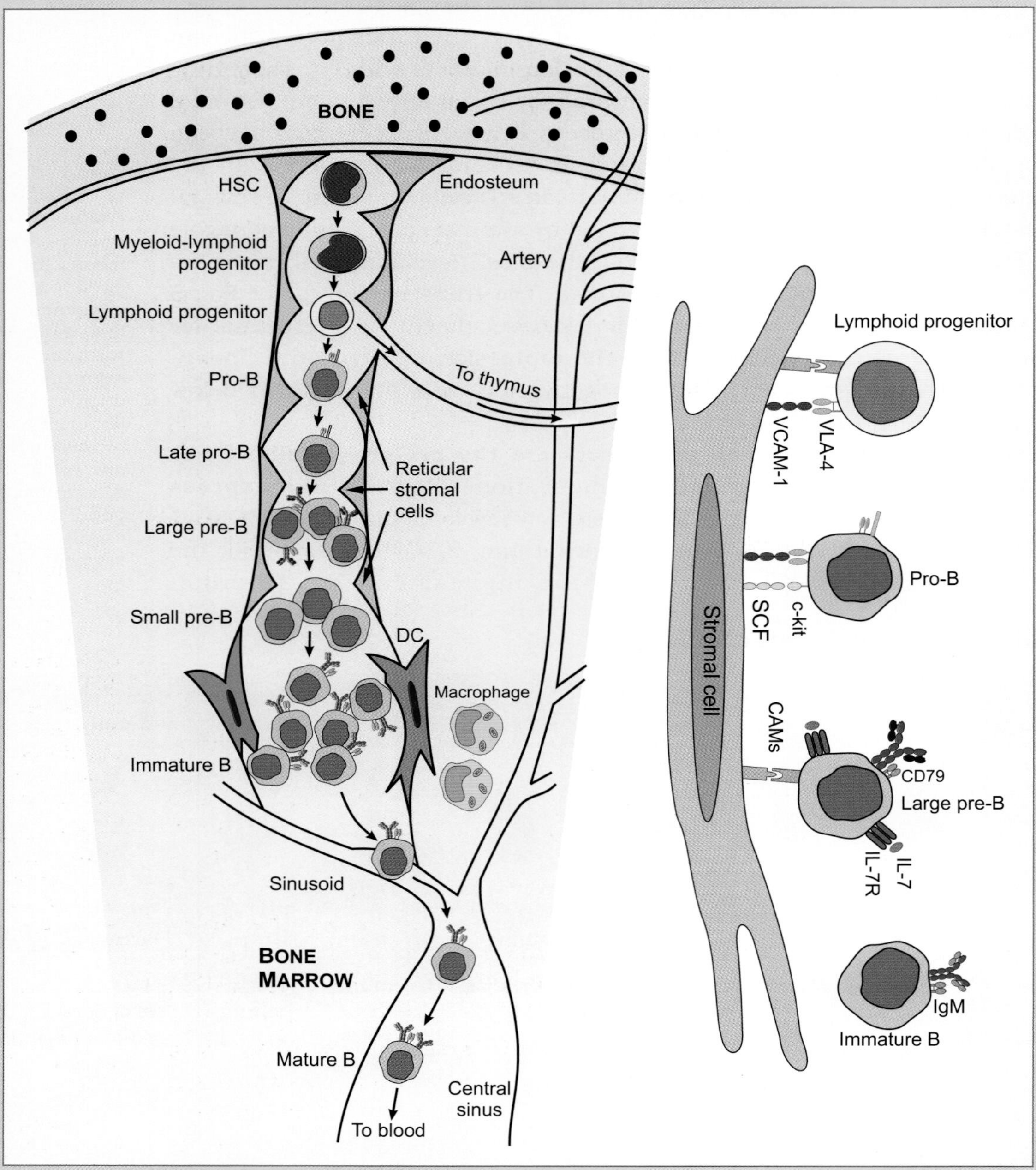

Figure 8.1 Stromal cells in the bone marrow are essential for B lymphocyte development. Haematopoietic stem cells (HSC) destined to become mature B cells migrate from the subendosteum, which is adjacent to the inner bone surface to the central axis of the marrow cavity. *En route,* they undergo various stages of development that are distinguished by expression of different cell-surface markers (left panel). Lymphoid progenitor cells destined to become T cells leave the bone marrow to travel to the thymus. Signals received from nonlymphoid stromal cells found in the bone marrow microenvironment are essential for the early stages of B cell development (right panel). The engagement of adhesion molecules on stromal cells by the corresponding ligands on B cells steers B cell development. The binding of the VCAM-1 on stromal cells to the VLA-4 on B cells, followed by signals delivered by interaction of SCF and CD117, is essential for pro-B cell development. IL-7, secreted by stromal cells, is the key cytokine needed for further development along this pathway. Later stages of B cell development are less dependent on stromal cell signals. B cells also undergo negative selection in the bone marrow, and self-reactive B cells undergoing apoptosis are phagocytosed by the macrophages present (not depicted here). Once development is complete, mature B cells exit to the periphery *via* the central sinus.

deoxynucleotidyl **T**ransferase (TdT) in pre-pro-B cells, resulting in their becoming early pro-B cells[5].

- **Early pro-B cells express a precursor BcR** composed of Ig-α, Ig-β and the chaperone calnexin. Ig gene recombination is initiated with *D–J* segment rearrangement on the H chain chromosome. Cells also begin expressing CD45 (B220) and MHC class II molecules.
- **In the late pro-B stage, V_H gene recombination is completed**. The assembling of the whole H chain completes the late pro-B cell stage. This recombination requires Pax-5 and IL-7.
- **The large pre-B cell has a functional H chain but lacks a functional L chain**. The *TdT* gene is switched off at this stage. Consequently, N-nucleotide additions to gene segment joints in L chains are not as common as in H chain sequences. *RAG-1* and *RAG-2* genes, needed for the rearrangement of L chain genes, remain operational. Cells start expressing pre-BcR, consisting of Ig-α, Ig-β, μ heavy chain, and two surrogate light chain molecules called λ5 and VpreB on their surfaces. The surrogate light chain λ5 resembles the constant region of the λ chain but is encoded by a different gene. λ5 associates noncovalently with VpreB, which resembles an Ig V domain. The pre-BcR is a key checkpoint regulator in B cell development. Lymphocytes that fail to assemble a pre-BcR fail to develop further and are deleted. Pre-BcR seems to
  - trigger B cell differentiation,
  - initiate clonal expansion, and
  - enforce H chain allelic exclusion[6].

  An individual B cell expresses only one species of functional H and L chains despite having two H chain alleles and four L chain alleles. This phenomenon is known as *allelic exclusion* (see chapter 6). Somatic recombination leads to allelic exclusion for both H and L chains in individual B cells, as each B cell productively recombines only one H chain and one L chain gene. In a heterozygote, each allele (allotype) is represented on about half the B cells and half the serum Ig molecules. Experiments have clearly shown that mIgμ expression in pre-B cells inhibits further H chain recombination, enforcing H chain allelic exclusion. Src kinases are believed to play a role in activating pre-B cell development and allelic exclusion.
- **Clonal expansion of large pre-B cells results in small pre-B cells**. These cells are arrested in the G1 phase[7]; Ig κ germline transcripts are expressed, and pre-B cells undergo *VJL* gene recombination. Successful L chain gene rearrangement leads to BcR assembly and replacement of the surrogate L chains in the pre-BcR by Ig κ or Ig λ chain. The cell is then referred to as an immature B cell.

  Allelic exclusion is operative in the L chains as in the H chains; however, the exact mechanism of this exclusion is not clear. L chains also show isotypic exclusion; an individual cell or molecule has only κ or λ chains. κ and λ are not represented equally on B cells or serum Igs. In humans, 65% Ig molecules have the κ light chain, whereas only 35% have the λ chain. In mice, 95% serum Ig is of the κ isotype; in cats it is 95% of the λ isotype. The ratio of κ to λ reflects the relative numbers of V region segments in each isotype and the relative efficiency of their recombination into functional L chain genes.
- **Immature cells are the first B lineage cells to express surface BcRs**. They display surface IgM but little or no IgD molecules. B cells remain in this stage for about 3-4 days. Immature B cells are particularly susceptible to BcR-induced

[5] Products of *RAG-1* and *RAG-2* are required in the recombination of *V*, *D*, and *J* gene segments for productive BcR and TcR gene rearrangements. TdT increases V gene diversity even further by adding N-nucleotides at the rearranging joints.

[6] Allelic exclusion is the expression of genes from the maternal or paternal chromosome, but not both, because of chromosomal inactivation. In B lymphocytes, allelic exclusion ensures that all antibodies expressed are derived from the same allele.

[7] Eukaryotic cellular activities are in the cyclic phases G1 to M. The G0 phase is the resting phase. In the G1 phase, the enzymes for DNA replication are synthesised with other proteins needed for cellular growth. In the S phase, DNA is synthesised with other related proteins. The G2 phase is recognised by the synthesis of proteins and the cellular material of the mitotic apparatus. Actual mitosis occurs in the M phase.

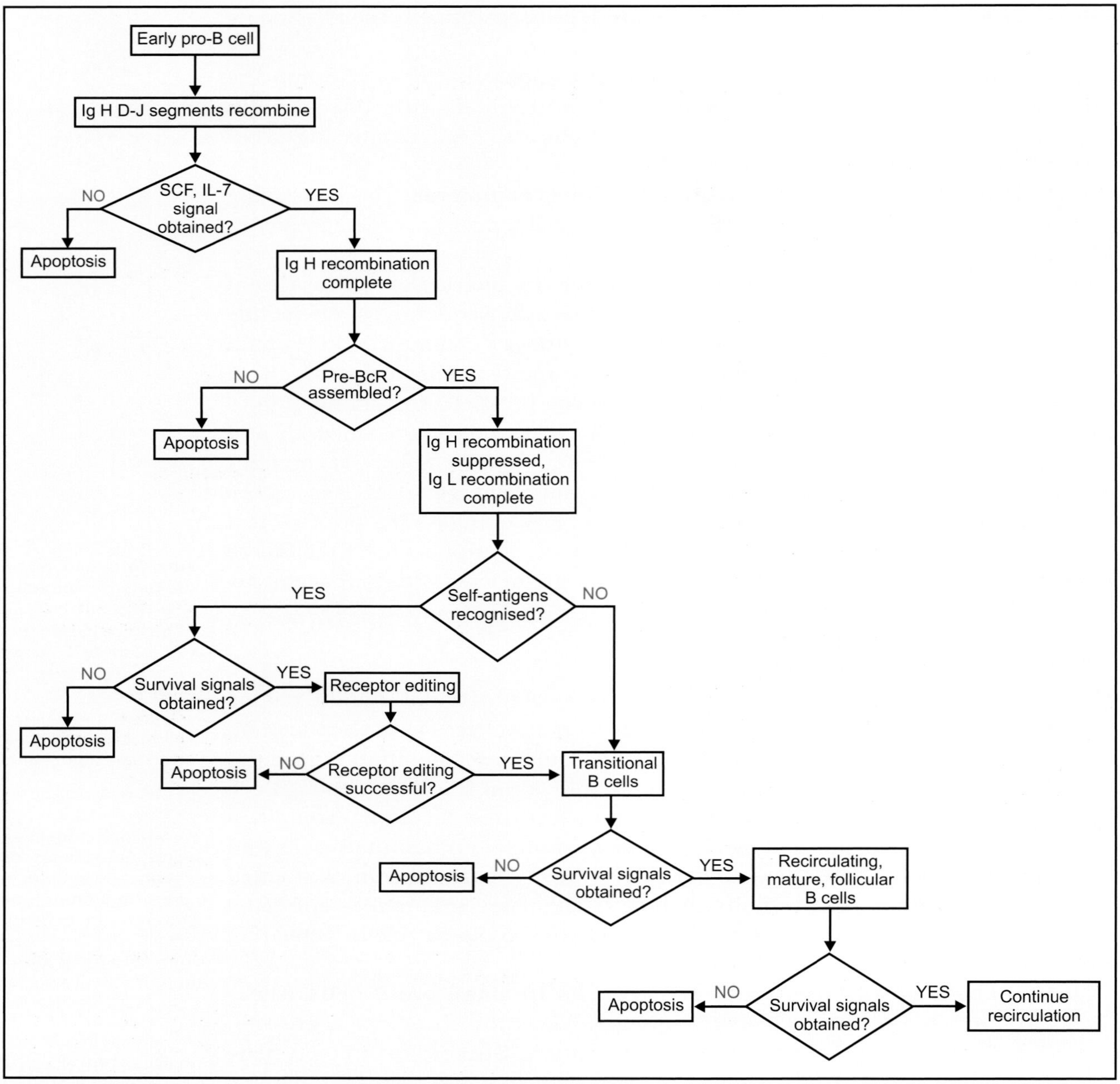

Flow chart 8.1 B cell development

apoptosis. They undergo negative selection at this stage, and self-reactive B cells are deleted or anergised. These cells must compete for survival signals in this stage. Those B cells that receive these signals enter the pool of circulating lymphocytes and remain quiescent until an encounter with their cognate antigens.

- Crosslinking of BcRs by self-antigens leads to the elimination of B cells. B cells that express high-affinity Ig molecules for self-antigens are sent down the road to apoptosis.
- B cells expressing BcRs with an intermediate affinity for self-antigens are anergised. Anergic B cells are short-lived and have difficulty in reaching maturity. BcR signalling seems to be downregulated in anergic cells, and chronic exposure to antigens can lead to decreased BcR expression in these cells.

Unwelcome and Unwanted B: B cell Lymphomas

B cell precursors go through several developmental stages before becoming mature B lymphocytes. Upon antigenic stimulation, mature B cells proliferate and differentiate to antibody-secreting plasma cells or memory cells; they also undergo affinity maturation. B cells thus undergo extensive gene rearrangements throughout their life cycles and are under strict regulatory controls. Proapoptotic signals ensure the homeostasis of the exploding B cell population. Malignant transformations of B cells can occur at any stage of their development — from early B cell progenitors to mature B cells and further differentiation to plasma cells. A B cell lymphoma is a cell population arising from a single B lymphocyte that gets arrested at a particular developmental stage. Lymphomas are pathological outcomes of genetic accidents occurring during lymphocyte development. They have greatly helped further the understanding of B cell development and biology. Amino acid sequencing and the elucidation of Ig domains was only possible because of the discovery of Bence Jones proteins secreted by patients suffering from multiple myeloma (a type of plasmacytoma). Plasmacytoma patients have large quantities of a single type of Ig in their serum (myeloma proteins) secreted by a clone of an immortalised plasma cell. They also excrete excess Ig L chains produced in their urine (called *paraproteins*). Because myeloma proteins and paraproteins from a single patient are of monoclonal origin, they are identical in their amino acid sequences. Analyses of these proteins therefore allowed the determination of the amino acid sequences of a complete Ig molecule, and the comparison of paraproteins from different patients allowed the study of variability in these sequences. Modern monoclonal antibody technology is based on the same principle of immortalising plasma cells.

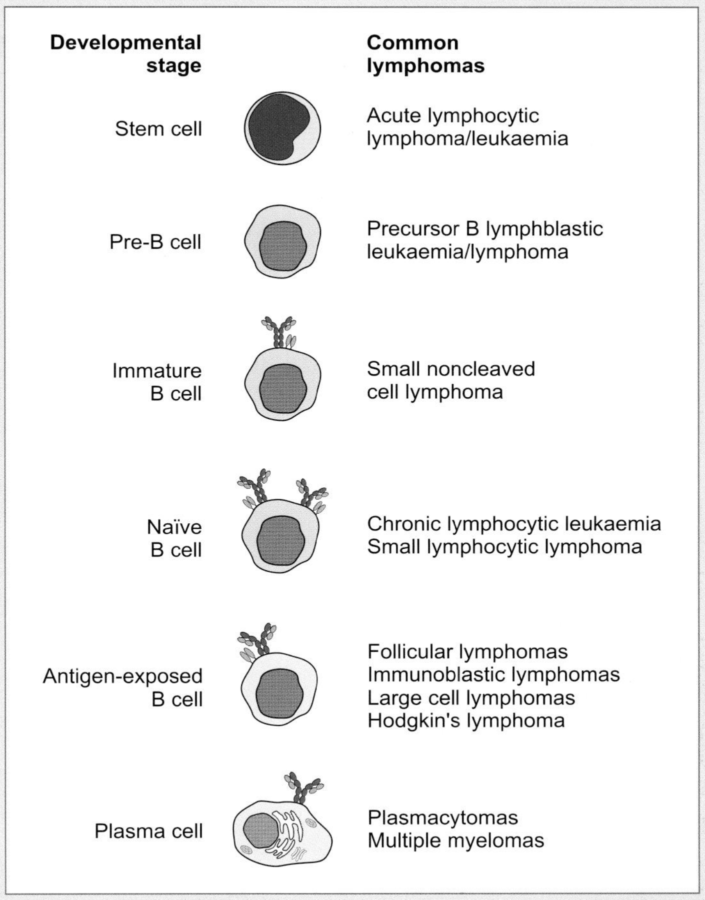

Figure 8.S1 B cell lymphomas can arise at various stages of B cell development.

Lymphomas are divided into two groups. Hodgkin's lymphomas are composed of unique malignant cells thought to be of B cell origin. Non-Hodgkin's B cell lymphomas arise during lineage differentiation, which is signalled by productive rearrangements of Ig genes. Mature B cell lymphomas often arise because of occasional errors in the rearrangement process that result in reciprocal chromosomal translocations that juxtapose Ig genes with novel genes. Many translocations place a proto-oncogene, such as *Bcl-2* or *myc*, next to the Ig locus. The juxtaposed gene is placed under the control of the Ig enhancer and becomes activated, dysregulating normal differentiation and initiating proliferation instead.

- Exposure to soluble antigens that bind to, but do not crosslink, BcRs also induces anergy in immature B cells.

Immature B cells retain their capacity for gene rearrangement even after the assembly of functional BcRs. Hence, although many self-specific B cells undergo clonal deletion or are anergised, some can escape elimination by further rearranging their BcRs. These secondary rearrangements, termed *receptor editing*, can occur at both H and L genes. The exact mechanism of receptor editing is still being investigated. A survival signal appears to be required for this second chance, and it is likely that receptor editing is influenced by interactions between small pre-B/immature B cells and bone marrow stromal cells. Whether due to successful receptor editing, or the elimination of self-reactive clones, surviving B cells lose the ability to bind self-antigens and start expressing mIgD molecules on their cell surfaces. These B cells lack the adhesion molecules necessary for extravasation and emigrate to the red pulp of the spleen to complete the final phase of development.

- **Newly formed B cells in the bone marrow or the red pulp of the spleen are known as *transitional cells*.** After lodging in the red pulp for about a day, these B cells pass through the T cell areas of the PALS to colonise the lymphoid follicles of the spleen and become $IgD^{hi}IgM^{lo}CD23^{+}$. They start expressing homing receptors that confer the ability to recirculate and become naïve follicular B cells. Survival signals received in the follicles enable them to enter peripheral B cell compartment; B cells that fail to lodge in the follicles die within a matter of days. The need for survival signals continues throughout the life of these B cells, so they must circulate through the follicles periodically. The signals promoting survival and differentiation are not well understood. Some survival signals are received through BcRs, and Btk (**B**ruton's **t**yrosine **k**inase)[8], and CD45 seem to be critical in this signalling. Knock out mice lacking either of these molecules involved in BcR signalling fail to develop mature B cells, although transitional B cells are formed. Another prosurvival factor is BAFF (**B** cell **A**ctivating **F**actor of the TNF **F**amily) secreted by macrophages and DCs. It is thought to promote the survival of transitional and mature B cells. Thus, only a fraction of transitional cells succeed in entering the long-lived, recirculating, follicular B cell pool.

There is method to this seeming madness of producing a large number of cells and killing of most of them. The immune system tries to balance opposing forces — the need to have a ready pool of naïve B cells at any time and the need to avoid the unnecessary build-up of mature B cells. The naïve B cell pool needs to be continually replenished, and the system continues to produce these cells in copious numbers. If all these cells mature, the body will be unable to support the humungous population. To maintain homeostasis, a large fraction of newly formed B cells have to be killed. Competition for survival signals between newly formed B cells and older B cells probably maintains this B cell homeostasis and prevents build-up. B cells that survive and enter the peripheral pool may still be sent down the road to death or anergy, this time to ensure the absence of self-reactivity. Because not all self-antigens are present in the bone marrow, such mechanisms ensure the silencing of autoreactive clones.

- Mature B cells that are exposed to multivalent self-antigens in the periphery are deleted or rendered anergic.
- Exposure of a mature B cell to high concentrations of soluble self-antigen can anergise the cell.
- Antigen recognition in the absence of T cells blocks B cell activation.

[8] A deficiency of this enzyme involved in BcR downstream signalling causes agammaglobulinaemia (complete absence of Ig).

A BAFFling, RANK-Ling TRAIL: The TNF Family

Proteins of the TNF and TNFR (**TNF R**eceptor) superfamily are important in the control of such disparate processes as death, proliferation, and the differentiation of immune cells; the modulation of the innate and adaptive responses; and the organogenesis of lymphoid organs. Research indicates that these proteins are involved in bone remodelling and lactation. Noteworthy members include TNF-α, TNF-β, CD95 & CD95L, CD134 & CD134L, CD40 & CD154, BAFF, TRAIL, RANK & RANK-L.

TNF receptors lack tyrosine kinase domains and do not interact with cytosolic proteins capable of phosphorylation. They are trimeric proteins defined by the homology of their extracellular domains, which has repeats of cysteine-rich domains. Cysteines allow the formation of an extended rod-like structure (called a jelly role) that is responsible for ligand binding. Receptor trimerisation enables the binding of cytosolic proteins called TRAFs (**T**NF**R A**daptor **F**actors) to the cytoplasmic domains of TNFRs and initiates different signalling cascades. Members of the family that induce apoptosis are known as the *death-receptors*. They have a **D**eath **D**omain (DD) in their cytoplasmic tail that is essential for transduction of the apoptotic signal. Upon activation, the DD serves as a docking site for DD-containing adaptor proteins. A **D**eath-**I**nducing **S**ignalling **C**omplex (DISC) is formed within seconds of receptor engagement, leading to the recruitment of caspase-8 or caspase-10. The initiation of the caspase cascade results in cell death. Trimerisation of other TRAFs that lack DDs results in cellular proliferation and may confer resistance to apoptosis.

- ❒ CD95 (Fas), involved in transducing the death signal, is expressed primarily on haematopoietic cells but can also be found on epithelial cells. CD95 expression can be boosted by cytokines, such as IFN-γ and TNF-α. In addition, lymphocyte activation upregulates expression of this molecule and downregulates its inhibitor, FLIP. CD95-mediated apoptosis is triggered by its natural ligand CD95L (FasL), which has a more restricted expression than CD95. Interestingly, cleaved, soluble human CD95L can also induce apoptosis, unlike murine CD95L. CD95 activation recruits the adaptor protein FADD (**F**as-**A**ssociated **D**eath **D**omain) and induces apoptosis. This is one of the major pathways of inducing death in target cells by cytotoxic lymphocytes (see chapter 11).
- ❒ Cell exposure to TNF-α has varied results.
  - ➢ The receptor TNFR-1 is expressed by all human tissues and is the major signalling receptor for TNF-α. Depending upon the adaptor proteins recruited, TNFR-1 can cause the induction of proliferative/inflammatory responses or apoptosis. Like Fas, TNFR-1 has a DD in its cytoplasmic tail. If it binds the adaptor molecule TRADD (**T**NF**R**-1-**A**ssociated **D**eath **D**omain), TNF-R1 causes apoptosis. Conversely, the recruitment of RIP (**R**eceptor **I**nteracting **P**roteins) or TRAFs is antiapoptotic. RIP activates the transcription factor NFκB; TRAF activates MAP kinases.
  - ➢ TNFR-2 is expressed mainly on cells of the immune system. It binds both TNF-α and TNF-β and lacks DD. Instead, it allows TRAF binding, which can lead to the recruitment of **I**nhibitors of **A**poptosis **P**roteins (IAPs) that may confer resistance to apoptosis.
- ❒ Activated T cells express TRAIL (**T**NF-**R**elated **A**poptosis **I**nducing **L**igand), yet another apoptotic member of this family. TRAIL is involved in cytotoxic lymphocyte-mediated apoptosis. It is also important in T cell-induced DC apoptosis. Thus, DCs presenting antigen to T cells are effectively eliminated, avoiding the excessive activation of the immune system. TRAIL is also implicated in neutrophil apoptosis.
- ❒ CD40 is widely distributed on the cells of the immune system. It is constitutively expressed on B cells, but its expression is lost at the plasma cell phase. CD154, the ligand for CD40, is expressed primarily on T cells. It is also found on cells such as NK cells, masts cells, basophils, monocytes, and DCs. CD40 ligation promotes B lymphocyte proliferation, survival, and affinity maturation. It also upregulates Fas expression, and unless the cell receives overriding signals, it can lead to apoptosis of the B lymphocyte.

- BAFF, expressed by DCs, macrophages, monocytes, etc., is a fundamental factor required for B cell survival. It enhances immune responses and controls B cell maturation in the spleen. BAFF receptors are believed to induce NFκB transcription and promote antiapoptotic signals in B cells.
- RANK-L (**R**eceptor **A**ctivator of **NF**κ**B** **L**igand) and its receptor, RANK (**R**eceptor **A**ctivator of **NF**κ**B**), are the latest additions to the TNFR stable
  - RANK–RANK-L (expressed by DCs and T cells, respectively) interactions regulate DC–T cell communications and DC survival.
  - Both RANK and RANK-L are crucial to lymph node organogenesis.
  - RANK-L and RANK are expressed in mammary gland epithelial cells and control the development of lactation.
  - These molecules are key regulators of bone remodelling and are essential for the development and activation of osteoclasts (cells involved in bone breakdown, i.e., bone resorption). Additionally, TNF-α and IL-1 induce osteoclast formation and cause bone resorption independently of RANK. Recent research indicates that a balance between RANK and RANK-L controls bone remodelling and bone loss. In animal models, systemic T cell activation resulted in RANK-L-dependent osteoclastogenesis, followed by bone loss, suggesting a role for T cells in the regulation of bone physiology.

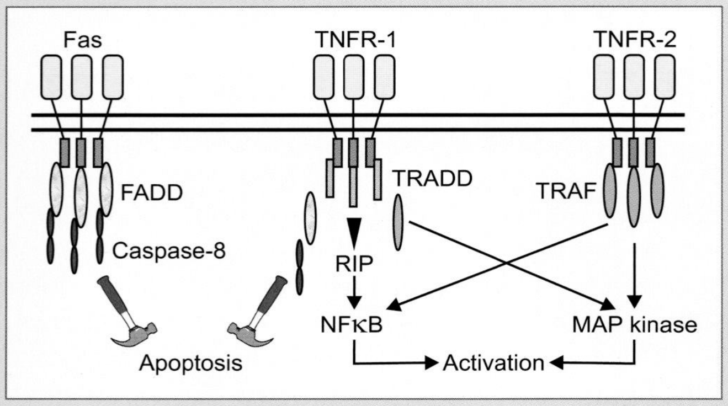

Figure 8.S2 TNF receptors are trimeric proteins lacking tyrosine kinase domains. Receptor trimerisation initiates different signalling pathways, depending upon the adaptor proteins recruited to receptors (adapted from Chan et al., *Immunity,* 2000, 13:419).

8.3 B Cell Heterogeneity

On the basis of phenotypic, topographic, and functional characteristics, three B cell subsets are recognised (Table 8.2). The exact lineage relationship between these subsets is not clear. Recent evidence points to B1 and B2 subsets arising from two separate lineages.

- **B1 B cells** have the $IgM^{hi}IgD^{-/lo}CD23^{-/lo}$ $CD21^{lo}CD5^{+}$ phenotype[9]. These are **self-renewing cells**; they give rise to mature naïve cells like themselves. By contrast, the pool of naïve conventional B cells is constantly replenished by the bone marrow. The majority of B1 cells develop from the foetal liver and are **found predominantly in the peritoneal and pleural cavity**, although they may also be found in the spleen and intestine. BcRs of B1 cells are produced preferentially from only some Ig gene segments and do not have additional N-nucleotides at the joints between segments. They are largely specific for common bacterial carbohydrate antigens. Thought to be responsible for natural IgM antibodies, B1 B cells predominantly respond to TI antigens and do not undergo affinity maturation. They can undergo limited isotype switching. They also seem incapable of differentiating to memory cells. BcRs of B1 B cells seem to recognise antigens on multiple pathogens with low affinity. Hence, these cells and their secreted antibodies are termed *polyreactive*. The role of B1 cells in immunity is discussed in chapter 2.
- **B2 B cells or follicular B cells are mature**, **recirculating CD5⁻ B cells** found in the periphery and located in the follicles of the lymph nodes and spleen

[9] hi: high levels of expression; int: intermediate levels of expression; –/lo: either absent or present in very low concentrations.

Table 8.2 B Cell Subsets

Characteristic	B1 B cells	B2 B cells	MZ B cells
Phenotype	IgM^{hi} IgD^{lo} $CD21^{lo}$ $CD23^{lo}$ $CD9^{lo}$ $CD5^{lo}$ $CD1^{+/-}$	IgM^{lo} IgD^{hi} $CD21^{int}$ $CD23^{hi}$ $CD9^{-}$ $CD5^{-}$ $CD1^{int}$	IgM^{hi} IgD^{lo} $CD21^{int}$ $CD23^{lo}$ $CD9^{lo}$ $CD5^{-}$ $CD1^{int}$
Recirculation in recirculate lymph	Do not	Recirculate to and from the blood and lymph	Essentially confined to the spleen; cells migrate to and from the PALS; plasma cells arising from these cells may exit the spleen
Renewal	Self-renewing	Replenished from the bone marrow	Replenished from the bone marrow
Response to antigen	TI: hi TD: +/–	TI: – TD: hi	TI: hi; especially to type 2 TD: lo
Binding of multiple ligands	Int	–	Int
Proliferation	LPS: hi Anti-IgM: +/– CD40: +/–	LPS: lo Anti-IgM: int CD40: lo	LPS: hi Anti-IgM: +/– CD40: –
Type of Igs produced	IgM, IgG3, IgA	IgG1	IgM, IgG3
Resistance to Fas-mediated apoptosis	Int	–	–
Apoptosis after anti-IgM treatment	–	Lo	Hi

Legend: Hi – high, Int – intermediate, Lo – low, +/– uncertain response, – negative or no response

(i.e., they are conventional B cells). They express the $IgM^{lo}IgD^{hi}CD23^{hi}CD21^{int}$ phenotype. B2 cells are capable of isotype switching and affinity maturation, and they are responsible for long-lasting humoral immunity. Unless otherwise specified, we use the term *B cell* in reference to these cells throughout this book.

- **MZ (Marginal Zone) B cells** are found in the splenic compartment located at the outer limit of the white pulp. The MZ is bordered by the MZ sinus on the inner side and by the red pulp on the outer side. The marginal sinus surrounds B cell follicles and the PALS of the white pulp. Several concentric layers of macrophages are found in the MZ. DCs and macrophages in this area filter particulate antigens from the blood and present them to T cells. **Long-lived, apparently naïve B cells that localise in close proximity to the concentric rings of macrophages are known as MZ B cells**[10]. The colonisation of the MZ by these B cells seems to occur after 1-2 years of age in human infants. These B cells are noncirculating B cells that have a partially activated phenotype ($IgM^{hi}IgD^{lo}CD23^{-/lo}CD21^{hi}$). Both B1 and MZ B cells express CD9, a molecule shown to be important in cell adhesion and migration, signal transduction, and cancer metastases in other cell types. Its role in B cells is yet to be

[10] Long-lived memory B cells responding to TD antigens also reside in this compartment for prolonged periods. As the central arteriole that carries blood to the spleen empties into the marginal sinus, newly formed B cells and different types of T cells can be found transiting through this zone.

established. BcRs of the overwhelming majority of these cells have germline V_H segments. Like B1 B cells, MZ B cells are thought to have originated from the foetal liver, although some transitional B2 cells may also become MZ B cells. They can present antigen efficiently, and they differentiate rapidly to plasma cells when activated. They seem to primarily respond to TI antigens but may also respond to TD antigens. MZ and B1 B cells respond most rapidly to antigens, giving rise to the earliest antibody response to immune challenge.

8.4 B Cell Activation

Mature B cells emerging from the bone marrow are in the G0 phase. Contact with antigen or polyclonal activators triggers their entry into the G1 phase. They then move through the S and G2 phases before undergoing mitosis and repeating the cycle. Activated cells thus undergo clonal expansion. Activation can be achieved by antigen-nonspecific and specific means. Two nonspecific B cell activators commonly used in research are as follows:

- **Lectins** are plant proteins that bind to specific cell-surface glycoproteins and deliver an activation signal. Lectins are *polyclonal activators,* as they activate B cells irrespective of their antigenic specificity.
- **Anti-IgM**[11] antibodies crosslink surface IgM to deliver a signal similar to antigenic stimulation. Cells expressing mIgM proliferate in response to anti-IgM antibodies. Anti-IgD antibodies may similarly be used to stimulate antigen-nonspecific B cell proliferation.

The mechanism of antigen-specific B cell activation is, to some degree, dictated by the nature of the antigen. **TI antigens can activate B cells independently of T cell involvement**. The strength of the signal generated by the engagement of multiple BcRs by repetitive epitopes of the TI antigen is enough to activate B cells. The cytokines secreted by T cells enhance activation by TI type 2 antigens. With TD antigens, however, activation is a two-signal process. The first signal is delivered by the crosslinking of BcRs by the antigen. Activated antigen-specific T_H cells deliver the second signal. Termed *cognate interaction*, this contact-mediated signal necessitates the close physical interaction of the two cells. Many receptor–ligand pairs contribute to cognate interaction.

- **Adhesion molecules**: Both B and T cells express a multitude of transmembrane adhesion molecules, such as CD54 (ICAM-1) and CD11a/CD18 (LFA-1), which mediate homotypic and heterotypic adhesion[12]. Apart from enhancing the physical association of the two cells, both LFA-1 and ICAM-1 transmit activation signals to B and T cells. Such signals increase antigen presentation by B cells, and they work together with CD40-mediated signalling.
- **MHC class II molecules**: The engagement of MHC class II molecules on B cells by TcRs stimulates early biochemical signalling events in B cells. This results in an enhanced B cell response to the activation signals emanating from T_H cells. MHC:peptide–TcR binding enhances the physical proximity of the two cells, assisting in the delivery of contact-mediated signals and soluble molecules. It also seems to stimulate the proliferation and differentiation of B cells. This interaction is reciprocal, as it also affects T_H cells. The TcR recognition of peptide:MHC complexes on B cells induces the transient expression of new molecules on these cells, allowing them to enter into a prolonged relationship with antigen-specific B cells.

[11] When murine IgM antibodies are injected in a rabbit, the rabbit's immune system responds by producing antibodies to the murine IgM. This rabbit serum can be used as a ready source of high-affinity anti-murine anti-IgM antibodies.

[12] Homotypic adhesion occurs when similar molecules are involved in adhesion (e.g., LFA-1 to LFA-1); heterotypic adhesion occurs when two different molecules participate in the adhesion process (e.g., LFA-1 and ICAM-1).

- **CD40–CD40L interaction**: The constitutively expressed CD40 is perhaps the most important molecule in cognate interaction. Engagement of this TNF family member by its ligand CD154 (CD40L) on activated TH lymphocytes promotes proliferation, cytokine production, antibody secretion, and isotype switching in B cells. It is also critical in upregulating molecules involved in antigen presentation by B cells. Interaction between CD40 and CD154 is essential in the development of germinal centres, B cell survival in the germinal centres, and affinity maturation. Interestingly, T cell expression of CD154 is upregulated upon T cell activation.
- **CD134L** (OX40L): This is another member of the TNF family that is important in cognate interaction. The engagement of CD134 (OX40) on TH cells with the ligand (CD134L) on B cells is thought to be important in isotype switching and secondary antibody responses.
- **CD72**: The engagement of the CD72 molecule on B cells by CD100 on T cells enhances B cell activation that is mediated by CD40. This interaction seems to be particularly important in developing high-affinity IgG antibody responses to TD antigens.
- **CD95/Fas**: Signalling through CD40 upregulates CD95 on B cells so that they become increasingly susceptible to the CD95L expressed on activated T cells; CD95 engagement leads to apoptosis unless the B cell receives overriding survival signals.

This feature of the humoral response, wherein both TH cells and B cells must recognise antigenic determinants on the same molecule for B cell activation, is termed *associative* or *linked recognition*. This requirement for cognate interaction may be advantageous in regulating the immune response. The elimination of self-reactive clones appears to be more efficient in T cells than in B cells. Thus, if a cognate antigen-specific T cell clone is not available, autoreactive B cell clones cannot be activated, decreasing the potential for autoimmune reaction. This also limits a bystander response because the B and T cells need to be activated and in physical contact for the second signal to be delivered. The requirement for antigen-specific T cell help also explains the carrier effect noted in chapter 10.

When B cells enter lymphoid tissues through HEV, they rapidly move through the T cell zone and enter the primary follicles in the B cell zone. By the second day after primary immunisation with a TD antigen, antigen-specific T cells are found within the PALS of the spleen and lymph nodes. By the fourth day, antigen-specific B cells migrate to the lymphoid follicles. The activated B cells start expressing adhesion molecules and chemokine receptors. The activated T and B cells migrate to the edge of the T cell–B cell zone, maximising their chances of interaction. The sequence of events in B cell activation and eventual differentiation is summarised below.

1. B cells internalise the antigen either by pinocytosis or (BcRs or other) receptor-mediated endocytosis.
2. The antigen is processed and loaded onto MHC class II molecules; MHC:peptide complexes appear on the surface of the B cell within 30-60 minutes of encountering the antigen.
3. B cells start expressing adhesion molecules and are trapped in the T cell zone of lymphoid organs.
4. Costimulatory molecules, such as CD80/86, are upregulated.

5. TcRs on TH cells recognise and bind the class II MHC:peptide complexes on B cells; CD28 on T cells binds CD80/86 on B cells. Close physical contact is established between the two cells.
6. A T cell–B cell conjugate is formed; the T cell golgi apparatus is reorganised.
7. The TH cell is activated; the expression of CD154, CD134, and CD100 is upregulated.
8. The ligation of various ligands on B cells (CD40, CD134L, and CD72, respectively) results in the activation of tyrosine kinases and ultimately results in the activation of the transcription factor NFκB; these signals also override the apoptotic signal generated by CD95 engagement.
9. There is a unidirectional release of cytokines from the TH cell to the B cell. The type of cytokine released is dictated by a number of factors, such as the type of TH cell, the nature of the antigen, the site of interaction, and may include IL-4, IL-5, IL-6, TGF-β, or IFN-γ.
10. Signals transmitted through the cytokine receptors support B cell activation and result in B cell proliferation. Thus, B cells establish a primary focus of clonal expansion.
11. T cells are also stimulated to proliferate so that both B and T cells proliferate at the border between the B and T cell zones.
12. After several days, the rate of B cell proliferation decreases, and the primary focus of proliferation begins to involute.
13. Some of the proliferating B cells differentiate to antibody-forming plasma cells and migrate to the red pulp of the spleen or the medullary cords of the lymph node. They are responsible for secreting Ig molecules with a relatively low affinity for the antigen that are found in the initial stages of the immune response.
14. Other B cells migrate into primary lymphoid follicles, where they continue to proliferate and give rise to germinal centres.
15. B cells undergo somatic hypermutation in the germinal centres, and affinity-based selection results in the survival of cells with an increased affinity for the antigen (explained in chapter 6).
16. The cells also undergo CSR, which results in B cells capable of secreting antibodies with the same antigenic specificity but different isotypes (see section 6.5.2).
17. Some germinal centre B cells undergo differentiation to plasmablasts that continue to divide rapidly but are already committed to becoming nondividing, Ig-secreting plasma cells.
18. Other germinal centre B cells differentiate to memory cells.

8.5 B Cell Differentiation

B cells encountering antigen and receiving appropriate T cell help undergo blast transformation. Cells enlarge, their nucleolus swells, polysomes, rough ER develop, and the rate of nucleic acid and protein synthesis increases rapidly. In the lymphoid organs, this is seen by the emergence of blast cells and formation of germinal centres. **Germinal centres are the sites of B cell differentiation**. They are also the sites for secondary genetic alterations in BcRs (see chapter 6). In the germinal centres, proliferating B cells multiply exponentially; within 72 hours their numbers increase 10,000-fold. The resting B cells not taking part in the immune response are pushed to the periphery and make up the mantle zone around proliferating B cells. The proliferating lymphoblasts begin to fill the FDC network of the lymphoid follicle and give rise to an organised structure termed the *germinal centre*. The main constituents of the germinal centre are activated B lymphocytes, FDCs, macrophages, and $CD4^+$ T lymphocytes.

Germinal centres are made of three distinct zones (Figure 8.2).

- **The dark zone**: In forming the germinal centre, B cell blasts cluster in the part of the follicle nearest the T cell zone. This area is known as the dark zone. The proliferating B cells, called *centroblasts*, undergo a number of changes in the dark zone. BcR expression is downregulated; the centroblasts appear to be mIg^-. They proliferate vigorously and undergo a diversification of their antibody repertoire through the somatic hypermutation of their Ig V region genes. The centroblasts give rise to nondividing cells, called *centrocytes*, which express surface Ig molecules and migrate to the light zone.
- **The light zone**: This is the site of isotype switching and the positive and negative selection of B cells — positive selection of cells expressing high-affinity receptors for the antigen that elicits the immune response and negative selection of cells that have lost their capacity to recognise that antigen. The B cells that receive the necessary survival signals eventually differentiate to plasmablasts (that give rise to plasma cells) and memory cells and leave the germinal centre.
  - FDCs are specialised cells present in the basal light zone. These cells can be distinguished by their long protrusions and relatively high expression of FcγRIIB and CR1 and CR2 (CD35 and CD21, respectively). These receptors

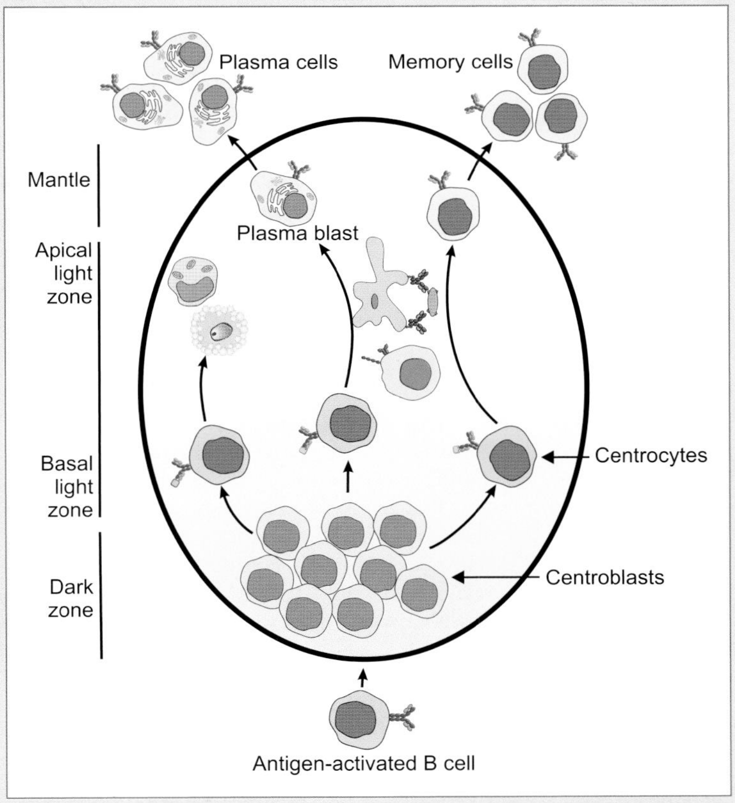

Figure 8.2 Germinal centres are discrete lymphoid compartments where antigen-activated B cells divide, switch isotypes, and differentiate. Germinal centres can be divided into three zones — the dark zone, the light zone, and the mantle. Antigen-activated B cells enter the germinal centre and undergo proliferation in the dark zone. The rapidly proliferating cells do not express mIg molecules and are called centroblasts. The centroblasts undergo hypermutation in their Ig V region genes and give rise to nondividing centrocytes that express mIg molecules and migrate to the light zone. The centrocytes have an apoptosis-sensitive phenotype. Only cells receiving survival signals delivered by FDCs and T cells are rescued from apoptosis. FDCs trap antigen–Ig complexes on their cell surface and are instrumental in selecting B cells with an increased affinity for antigen. Macrophages present in the light zone phagocytose dead or dying cells. Centrocytes receiving survival signals migrate to the apical light zone. The migrating centrocytes internalise, process, and present antigen to $CD4^+$ T cells. This cognate interaction results in the stimulation of B cells and promotes their clonal expansion, isotype switching, and differentiation to memory cells and plasmablasts. The differentiated cells eventually enter the follicular mantle and leave the germinal centre.

trap antigen in the form of immune complexes on the FDC cell surface; antigen–antibody complexes can be detected on FDCs months after immunisation.

- Adhesion molecules on B cells (e.g., **V**ery **L**ate **A**ntigen-4 (VLA-4) and LFA-1) and FDCs (VCAM-1 and ICAM-1) are crucial for intimate contact between the two cells. In humans, the interaction between the CD23 expressed on B cells and the CD21 on FDCs is also important in this context.
- The coligation of BcRs with the FcγRIIB expressed by B cells delivers a negative signal to B cells (Figure 6.3); however, the high density of the FcγRIIB expressed by FDCs results in the binding of all the available FcRs of Ig molecules in the immune complexes, ensuring the absence of such coligation.
- The FDCs convert immune complexes to iccosomes (immune complex-coated bodies). These are antigen-coated liposome-like particles of about a 0.25-0.38 μm diameter that are derived from FDC membranes. Iccosomes consist of antigens, C3b or its fragments, and Igs attached to FDC membranes. Iccosomes are released from FDC dendrites and rapidly endocytosed by germinal centre B cells. The presence of C3b (or its fragments) is thought to aid this rapid endocytosis. The antigen in the iccosomes is efficiently processed, loaded onto MHC class II molecules, and presented to follicular T_H cells. The ensuing cognate interaction is indispensable to the survival and differentiation of germinal centre B cells (Figure 8.3).
- Originally, FDCs were thought to be instrumental in the selection of B cells on the basis of their affinity for the antigen; competition for the limited antigen was postulated to result in only those B cells that expressed high-affinity BcRs binding FDC-trapped antigen. Recent experiments suggest that T cells may also be important in the selection process, and competition for the limited number of T cells that are present in the germinal centres may be essential in driving selection processes.
- Human germinal centre B cells show a typical apoptosis-sensitive phenotype. They express low levels of the antiapoptotic Bcl-2 and high levels of proapoptotic Fas and Bax, and they require specific signals to survive (see the sidetrack *A BAFFling, RANK-Ling TRAIL*). These cells are thought to contain preformed DISC (**D**eath-**I**nducing **S**ignalling **C**omplex) that causes the rapid activation of the enzymes involved in apoptosis[13].

[13] DISC formation and its role in apoptosis are explained in chapter 11.

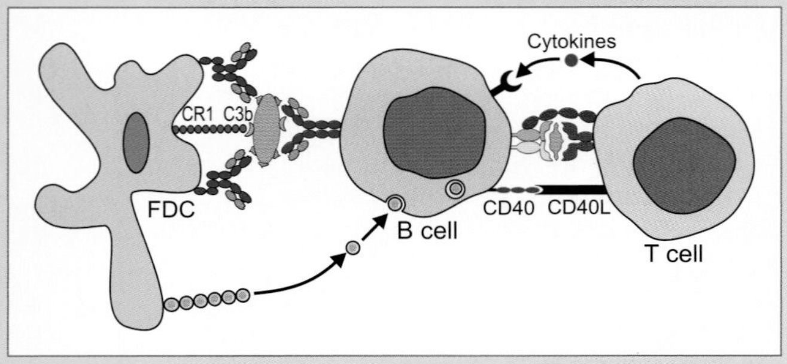

Figure 8.3 FDCs are vital to the affinity-based selection and survival of germinal centre B cells. FDCs are specialised cells found in germinal centres, which trap antigen in the form of immune complexes. Competition for the limited antigen is thought to allow only those B cells expressing high-affinity receptors for the antigen to bind FDC-trapped antigen and, in the process, obtain necessary survival signals. FDCs also convert immune complexes to iccosomes (shown here in yellow) that are antigen-coated liposome-like particles derived from FDC membranes. Iccosomes consist of antigen, C3b or its fragments, and Ig molecules attached to FDC membranes. Iccosomes released by FDCs are rapidly endocytosed by germinal centre B cells. The antigen in the iccosomes is efficiently processed, loaded onto MHC class II molecules, and presented to follicular T_H cells by B cells. This cognate interaction is indispensable to the survival and differentiation of germinal centre B cells.

- Whether they are delivered by FDCs or T cells, antiapoptotic signals rescue B cells from apoptosis and enable their migration to the apical light zone. The exact nature of the antiapoptotic signal is not clear, but it seems to prevent the rapid activation of two enzymes involved in apoptosis — caspase-8 and caspase-3.
- Centrocytes with low-affinity receptors are less likely to bind the antigen, and hence, they fail to get survival signals from FDCs. They are pushed on the road to apoptosis, or they may re-enter the dark zone. Thus, a large proportion of centrocytes that arrive in the light zone die there.
- Macrophages in the light zone phagocytose dead or dying cells. They are known as tangible body macrophages because they contain dense nuclear fragments of cells undergoing apoptosis.
- The antigen is internalised when BcRs are engaged. As B cells migrate to the apical light zone, they process and load the antigen onto MHC class II molecules.
- In the apical light zone, B cells encounter and present antigen to $CD4^+$ T cells. These T cells, called follicular helper T cells, are at the light zone's outer edge. They express the chemokine receptor CXCR5 and the costimulatory molecule ICOS.
- Antigen presentation by B cells and the engagement of ICOS induces T cells to express CD154. Follicular helper T cells have an intracellular store of CD154, which is expressed on the cell surface upon stimulation. The engagement of CD40 on B cells by CD154 delivers a second survival signal to B cells, resulting in cytokine secretion by T cells.
- The B cells that express MHC:peptide complexes that are not recognised by T cells (e.g., self-antigens) fail to engage in CD40–CD154 interaction and undergo apoptosis.
- Centrocytes expressing high-affinity BcRs for the antigen undergo clonal expansion and isotype switching; they also differentiate to memory cells and plasmablasts.
- Plasmablasts leave the germinal centre and migrate to the site of Ig production; those formed in the GALT germinal centres migrate to the lamina propria of the gut, and those formed in lymph nodes or the spleen migrate to other sites, such as the bone marrow, the medullary cords of the lymph nodes, or the red pulp of spleen.
- Memory cells may leave the germinal centre and re-enter circulation, or they may migrate to other sites (the MZ of spleen, the subcapsular sinus of lymph node, the intestinal epithelium under Peyer's patches, etc.). Alternatively, they may re-enter the dark zone for further mutations. Figure 8.2 summarises the events in the germinal centre.

❒ **The mantle**: The outer follicular mantle is the place where migrating B cells briefly reside before leaving the germinal centre.

The germinal centre reaction peaks in about 10-12 days after antigenic challenge and gives rise to

❒ small, relatively nondescript, Ig-nonsecreting memory cells, and

❒ large, rapidly dividing plasmablasts that give rise to Ig-secreting plasma cells (Figure 8.4).

Factors that determine the fate of the proliferating B cell (differentiation to memory cell or plasma cell) are not well understood. Some studies suggest that the

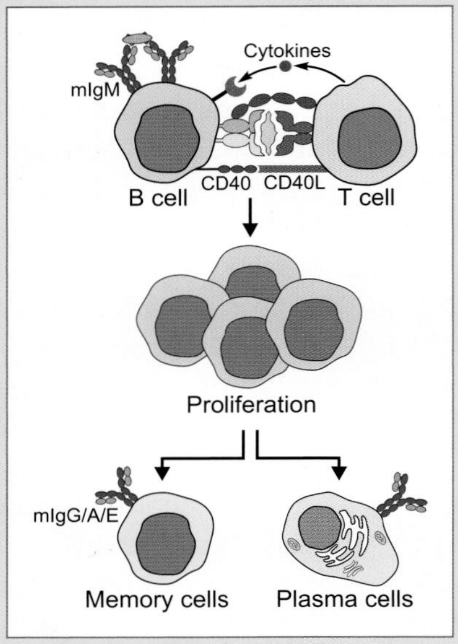

Figure 8.4 When antigen-activated B cells receive appropriate T cell help, they proliferate and differentiate to plasma cells and memory cells. Antigen-driven coligation of BcRs results in the upregulation of adhesion molecules and costimulatory molecules. The internalised antigen is processed and presented to TH cells. The recognition and binding of the MHC:peptide complexes by TcRs on TH cells leads to close physical contact between the two cells. The engagement of ligands, such as CD40, on B cells by their receptors on T cells delivers survival signals to B cells. The cytokines released by TH cells support B cell proliferation and differentiation. Such B cell differentiation results in the formation of plasmablasts that give rise to plasma cells and memory cells. The proliferating B cells undergo CSR and somatic hypermutation in their Ig V region genes. The plasma cells have a prominent, distended rough ER and are cellular factories of Ig production.

expression of high-affinity antibodies may favour plasma rather than memory cell development — clones having a high affinity for the antigen are actively recruited into the plasma cell pool throughout the duration of the germinal centre reaction. Experiments also indicate that signals received *via* CD40, OX40L (CD134L) and complement receptors CD21/CD35 favour entry into the plasma cell lineage. IL-6 is required for differentiation to plasmablasts and antibody secretion; IL-4 has been found to favour memory cell formation. It is unclear how the proteins drive this differentiation. Blimp-1 (**B l**ymphocyte **i**nduced **m**aturation **p**rotein-1), the master regulator for terminal B cell differentiation, is a key protein that drives activated B cells to become antibody-secreting plasma cells. It is present in all plasma cells, whether formed from naïve cells during a primary response or from memory cells in a secondary response. The continued expression of Pax-5, the transcription factor essential for commitment to B cell lineage, is essential for differentiation to memory cells. Blimp-1 is thought to directly downregulate Pax-5 expression and promote plasma cell formation.

Both memory cells and plasmablasts migrate from the germinal centre. Memory cells are found predominantly in the MZ of the spleen, the subcapsular sinus of lymph nodes, and under the intestinal epithelium in Peyer's patches and crypt epithelium of the tonsils; a few are also found in blood. The plasmablasts leave the germinal centre and develop into terminally differentiated plasma cells that secrete high-affinity antibodies. Some of these cells home to the bone marrow, where they receive survival signals from stromal cells through CAMs. These bone marrow plasma cells continue to secrete antibodies for many months. Other plasma cells may be found in the medullary cords of the lymph nodes, the red pulp of the spleen, and the mucosal lamina propria. The plasma cells found at mucosal sites are predominantly IgA-secreting plasma cells.

8.5.1 Memory B Cells

Although it is well-established that memory B cells are formed in the germinal centre, the exact developmental pathway that leads to memory B cell formation is not entirely clear. A degree of uncertainty also surrounds the phenotype of memory cells. It is a small, relatively undifferentiated cell. The vast majority of memory B cells express Igs other than IgM/IgD on their cell surfaces, although some may express either one or both of these isotypes. Memory B cells also express low levels of a heat-stable antigen and the adhesion molecule L-selectin (CD62L), whereas most naïve cells express high levels of both these molecules. The best hallmark of a memory B cell is a somatically mutated Ig gene that codes for high-affinity Ig molecules. Memory B cells are long-lived cells that survive even in the absence of antigens. They do not secrete antibodies. Upon leaving the germinal centre, memory B cells recirculate or home to draining areas of the lymph nodes and spleen. Following antigen encounter, memory cells are biased to become plasma cells.

Memory cells are more easily triggered by the immunogen than virgin B lymphocytes. This is because they express a different isoform of the CD45 molecule which is more efficient in signal transduction. They proliferate in response to lower amounts of antigen and can differentiate to plasma cells even in the absence of T cell help. Memory cells (whether B or T) are the cells that respond to subsequent antigenic challenges. The presence of memory cells inhibits the activation of naïve cells by the same antigen. This makes sense given that memory cells are better equipped to deal with the antigenic challenge. It would be waste of energy (and time) for naïve cells to respond to the secondary challenge. Many pathogens exploit this inhibition by undergoing antigenic variation (see the sidetrack *Trying to Fit In*, chapter 10).

Memory cell formation leads to the prompt and heightened antibody responses observed on secondary exposure to the immunogen. This is a simple yet elegant device to create a large pool of cells capable of responding to an immunogen that the body has encountered before (Figure 8.4). Upon first encounter, only a small fraction of the B cell repertoire will be capable of binding (either weakly or strongly) to epitopes on the immunogen. These B cells must receive appropriate T cell help. This can happen only after a series of events. First, antigen must be taken up and loaded onto MHC molecules by APCs and transported to the local (draining) lymph node. Second, APCs have to be scanned by thousands of naïve T lymphocytes before a specific reaction between an APC and a T cell can occur. Third, after contact with the peptide:MHC complex on the APC, the antigen-specific T cells must get activated. Fourth, activated T cells must then proliferate and differentiate to effector cells before they become capable of helping B cells. It is only after B cells differentiate, with T cell help, that Ig production begins. This process takes considerable time (about a week). After the first encounter, however, the animal will have a much larger repertoire of memory T and B cells capable of reacting to an antigen[14]. Also, thanks to affinity maturation, the antigen receptor of memory B cells will bind the antigen more strongly than the receptor of the naïve B cell. Hence, fewer antigen molecules will be sufficient to trigger the immune response, and the response will be faster.

8.5.2 Plasma Cells

Plasmablasts originating in the follicles of the Peyer's patches and mesenteric lymph nodes migrate by way of the lymph to the blood and eventually to the lamina propria of the gut or other epithelial surfaces. Those originating in the peripheral lymph nodes or splenic follicles migrate to the bone marrow. There, they differentiate to Ig-secreting plasma cells. **Plasma cells are terminally differentiated cells**

[14] During an immune response, lymphocytes undergo roughly 15-20 cell divisions. The number of cells of a given specificity are estimated to increase 10^5-10^6-fold at the height of the immune response.

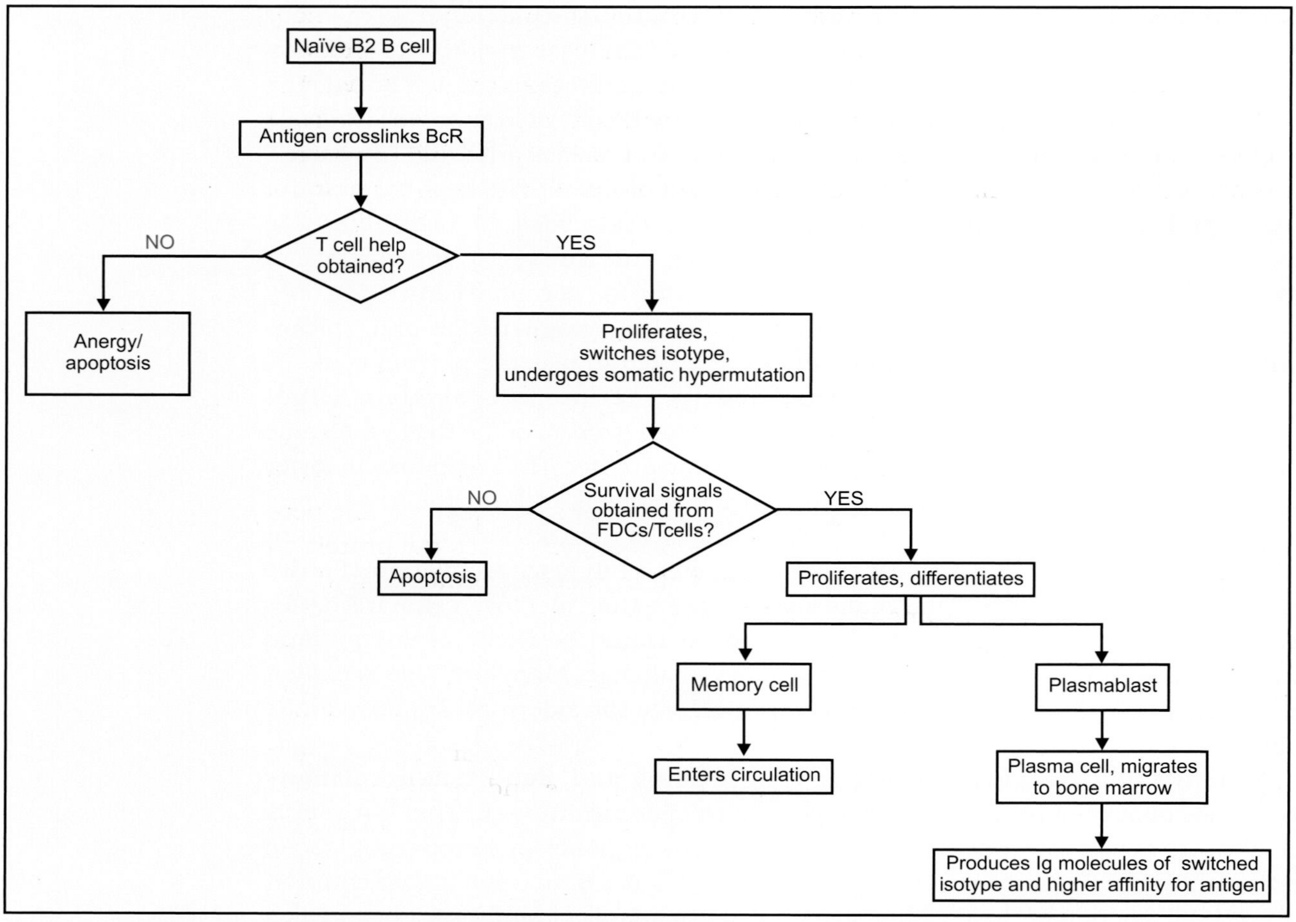

Flow chart 8.2 B cell differentiation

that are cellular factories for the manufacture and export of Ig molecules. They can secrete several thousand Ig molecules per second. Differentiation to plasma cells leads to the downregulation of numerous B cell-specific surface proteins. These include MHC class II molecules, CD45, CD19, and CD21. Chemokine receptors that allow homing to the lymph nodes and spleen are also downregulated. However, plasma cells continue to express CXCR4, which allows their movement out of the follicles. Histologically, the plasma cell is distinct from a small lymphocyte. The nucleus is eccentric, containing coarse radially arranged chromatin. The cytoplasm has a conspicuous and abundant rough and smooth ER, as well as a prominent golgi apparatus. The ER is usually packed into thin lamellae and is sometimes even distended with Ig molecules.

Two types of plasma cells can be distinguished based on their longevity and somatically mutated mIg molecules.

- Short-lived plasma cells seem to represent an early response to antigen exposure that is TH cell-dependent but germinal centre independent (e.g., those plasma cells formed at the primary foci of B cell proliferation). Consequently, these cells do not have a somatically mutated receptor.
- Long-lived plasma cells are products of the germinal centre reaction and show evidence of affinity maturation and isotype switching. These cells are thought to home preferentially to the bone marrow, where they secrete the high-affinity

antibodies typically observed in the second week of a primary response. They maintain significant levels of antigen-specific serum Igs, often for the animal's lifespan (Flow chart 8.2). It is possible (though not clearly established) that this pool of plasma cells is replenished by the occasional differentiation of memory cells. Although germinal centres last for only 3-4 weeks after initial antigen exposure, a small number of B cells continue to proliferate in them for months and may, in subsequent months or years, become precursors for plasma cells.

8.6 Ig Synthesis

Ig molecules are unique, in that they are simultaneously synthesised in both the membrane and the secreted form by plasma cells. Although both secreted and membrane proteins are synthesised at the ER, the sequence of events differs for the two. A hydrophobic sequence of amino acids at the N-terminus, termed the *leader sequence*, is present in both forms of the protein. This sequence causes the mRNA, ribosome, and translated leader sequence of the protein to bind to the pore structures in the membrane of the ER. The leader sequence causes the protein to thread through the pore; it is eventually cleaved off in secreted proteins, and the protein is released into the lumen of the ER. Membrane-bound proteins have a stretch of hydrophobic amino acids that stop the transfer through the ER membrane and cause the protein to remain trapped in the membrane. Thus, the presence of the hydrophobic stretch determines if the protein will be a membrane protein or a secreted protein. The translated proteins (whether membrane or secreted) are transported to the cell surface by way of the golgi apparatus and may be modified by carbohydrate addition. When vesicles are pinched off from the ER for transport through the golgi, membrane-bound proteins stay in the golgi membrane, while secreted proteins stay in the golgi lumen. The golgi vesicles ultimately fuse with the plasma membrane. The proteins in the vesicle are released to the exterior, whereas membrane-bound proteins become a part of the plasma membrane. Thus, the fate of the protein (i.e., membrane-bound or secreted) is determined by its amino acid sequence and is in turn sealed at the rough ER at the time of its synthesis. Needless to say, how the plasma cell manages to produce both the secreted and membrane-bound forms of the Ig molecule was a fascinating puzzle. This feat is achieved by differential RNA splicing. Two separate regions at the 3' end of the C_H region code for the two different forms of the Ig molecule. One codes a stretch of hydrophilic amino acids called the S (**S**ecreted) region; the two other exons code for a stretch of hydrophobic amino acids called the M (**M**embrane) segment. The primary transcript encodes both segments. Alternative splicing of the primary transcript in the nucleus yields transcripts that have either the M segment or have been cut after the S segment. The M segment containing H chain gets trapped in the membrane, and the Ig molecule becomes an integral membrane protein. The S segment-containing transcript encodes an Ig H chain that passes through the ER membrane into the ER lumen and becomes part of the secreted Ig molecules. The major steps in Ig synthesis are listed below (Figure 8.5).

- Ig L chains and H chains are synthesised separately; L and H chain genes are transcribed in the B cell nucleus to produce RNA transcripts. The introns are then spliced out to yield mRNA, which is translated by the ribosomes at the membrane of rough ER.
- Both chains are formed as larger precursors with an N-terminal leader sequence of about 20 amino acids.

- The leader sequence is translated by the ribosome in the cytoplasm; the translated leader sequence binds to a **S**ignal **R**ecognition **P**rotein (SRP). SRP binding blocks further translation.
- The complex, consisting of the mRNA, the translated leader sequence, the SRP, and the ribosome, is translocated to the ER membrane.
- The SRP binds to a protein known as the *docking protein* at a vacant site on the ER; if SRP is the lock that stops further translation, the docking protein is a key that permits further translation.
- The leader is cleaved during the process of chain transfer in the ER lumen.
- The new chain being synthesised at the ER traverses the membrane of the ER; as the chain elongates, it complexes with **Bi**nding **P**rotein (BiP) at the CH1 domain. BiP is a chaperone molecule that catalyses the folding of the growing H chain.

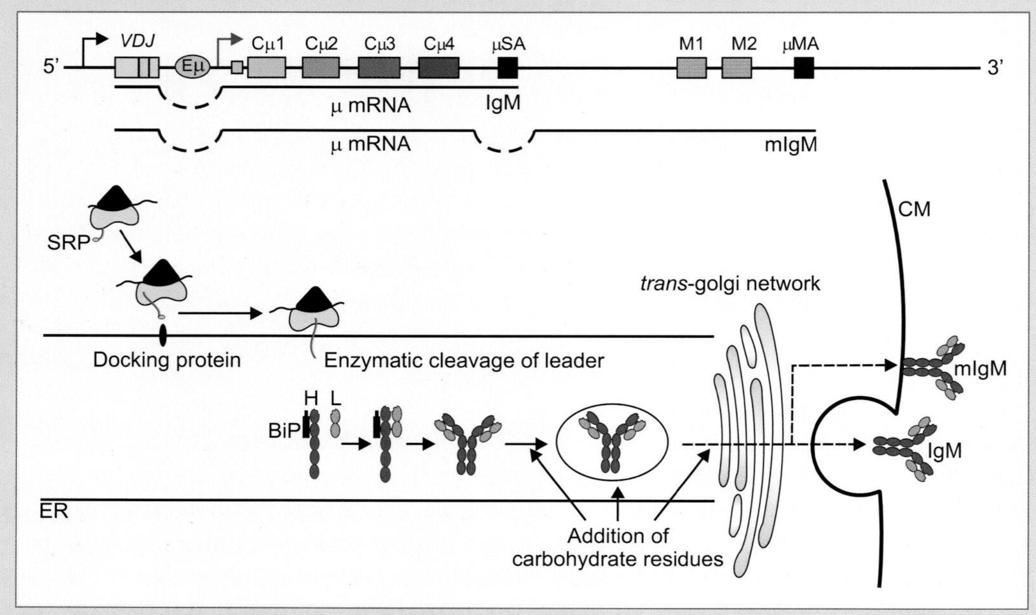

Figure 8.5 The differential splicing of the primary transcript allows plasma cells to simultaneously synthesise membrane and secreted forms of Ig molecules. The organisation of the human μ chain locus on chromosome 14 is shown in the top panel but is not to scale. The assembled V region (consisting of *V, D,* and *J* subexons) lies downstream of the promoter (depicted by a black arrow). Regions encoding the four constant domains (Cμ1 to Cμ4) lie 3' to the intronic enhancer (Eμ). Downstream to Cμ4 is the region encoding the carboxy-terminus of secretory IgM. This region has a polyadenylation site (μSA) that allows chain termination. Two separate regions encode the transmembrane region of mIgM (M1 and M2). They too have their own polyadenylation site (μMA). The primary transcript encodes both segments. Alternative splicing of the primary transcript in the nucleus yields transcripts that end at μSA or have Cμ4 joined to M1/M2 and μMA. Transcripts of membrane and secreted forms of other isotypes are produced similarly. H and L chains, whether of secretory or membrane Ig, are synthesised as larger precursors with N-terminal leader sequences. The translation of the leader sequence occurs in the cytoplasm (bottom panel; only synthesis of secretory Ig molecules depicted), but further translation is blocked by the binding of the leader sequence to SRP. The docking of SRP to its binding protein on the ER membrane permits further translation. The leader is cleaved during the translocation of the growing chain in the ER lumen. The elongating H chain binds to BiP and eventually combines with other H and L chains to form a complete Ig unit. Post-translational modifications, such as the addition of carbohydrate residues, occur in the ER as well as during transport through the *trans*-golgi network. The newly synthesised secretory Ig molecules are released to the external environment by reverse pinocytosis. The mIg molecules are expressed on the cell surface by the fusion of transporting vesicle membranes with cell membranes.

- The newly synthesised chain combines with other H and L chains to form the complete Ig unit. BiP is progressively displaced in this process.
- In the lumen of the ER, enzymes add carbohydrates to the newly formed Ig molecules.
- Ig molecules are transported to the cell surface through the *trans*-golgi network; further post-translational enzymatic modification occurs during this transport.
- Igs are released in the external environment by reverse pinocytosis; the transporting vesicle's membrane fuses with cell membranes to release the molecules.
- As explained, mIg molecules are inserted into the membrane of the ER by their hydrophobic sequences during synthesis. They are expressed on cell surface after the fusion of the vesicle membrane with the cell membrane.

Left Defenceless: Primary Immunodeficiencies

Primary immunodeficiencies are defined as genetic or developmental defects in the immune system. They are found in both the innate and adaptive arms of the immune system. Phagocyte or complement deficiencies lead to impaired innate immune responses and will not be considered here. Adaptive immune responses can be impaired because of either a solely B or solely T cell immunodeficiency or a *combined immunodeficiency*, where both compartments are defective. Combined immunodeficiencies are the most serious of all immunodeficiencies, with poor prognoses unless early interventions are undertaken.

B cell immunodeficiencies encompass a spectrum of diseases, ranging from a complete absence of mature recirculating B cells, plasma cells, and Ig molecules to the selective absence of some Ig isotypes. Human B cell immunodeficiencies are characterised by recurrent respiratory and gastrointestinal bacterial infections from around 3-4 months of age. However, patients often display a normal immunity to most viral and fungal infections. Noteworthy deficiencies include the following:

- **BcR deficiencies** are caused by mutations in the genes that encode Ig H or L chains or their associated signalling molecules and lead to agammaglobulinaemia or hypogammaglobulinaemia. Bruton's (X-linked) agammaglobulinaemia is caused by a defect in the gene coding for Btk. Btk plays a pivotal, nonredundant role in BcR signal transduction. A defect in this enzyme therefore results in severe malfunctioning of the B cell compartment. Btk maps to the X chromosome, and hence, the disease is almost exclusively found in males and results in a complete absence of Igs in serum.
- **X-linked hyper IgM syndrome** is caused by a defect in CD40L, which also maps to the X chromosome. As a result of this CD40L defect, B cells do not receive cognate help from T cells and are incapable of responding to TD antigens. Patients are, however, capable of responding to TI antigens. The lack of T cell help results in the absence of germinal centre formation and failure of the B cells to undergo isotype switching and affinity maturation. The end result is an absence of IgG, IgA, or IgE antibodies in serum, coupled with very high serum concentrations of IgM antibodies (IgM concentrations of 10 mg/mL are not unusual).
- **Common variable immunodeficiency** is characterised by a profound decrease in Ig molecules of all isotypes and recurrent infections usually accompanied by enlarged lymph nodes and spleen. The underlying defect that gives rise to this disease is unknown, and the condition generally manifests itself late in life.

T cell immunodeficiencies are more severe than B cell immunodeficiencies, as T cells affect both cell-mediated and humoral immunity. Defects in humoral immunity are characterised by recurrent infections with encapsulated bacteria, whereas cell-mediated immunity defects result in increased susceptibility to intracellular pathogens (whether bacterial, viral, protozoal, or fungal). Opportunistic organisms seem to pose a special threat in T cell immunodeficiencies. Important deficiencies are listed below.

- **DiGeorge syndrome** is caused by thymic aplasia — a quantitative decrease in functional thymic mass that is associated with the embryonic deletion of a region on chromosome 22. Interestingly,

T cell maturation is normal in patients of DiGeorge syndrome. The syndrome is characterised by distinctive facial abnormalities, cardiac malformation, and hypoparathyroidism. Not surprisingly, the syndrome results in a profound depression in T cell numbers and an absence of T cell responses. Although B cells are present in normal numbers, patients fail to respond to TD antigens. Prognosis is very poor, with long-term survival difficult even if patients are treated for immunological deficits.

- **T cell receptor deficiencies** can lead to **S**evere **C**ombined **I**mmuno**d**eficieny (SCID). SCID results in a combined B and T cell deficiency. Common features of SCID include recurrent opportunistic infections, diarrhoea, paucity of lymphoid tissue, and failure to thrive. The patient shows hypogammaglobulinaemia as well as decreased or absent T and B cell responses. Lymphocytes from these patients fail to respond to mitogens. Mutations in the tyrosine kinase ZAP-70 can result in SCID characterised by decreased serum IgG antibodies, impaired cell responses, and decreased or absent T cell function — particularly $CD8^+$ T cell function. Similarly, a mutation in *RAG-1* or *RAG-2* genes can result in the absence of TcR and BcR rearrangements, precluding the development of functional B and T cells.
- **X-linked SCID** is the most common form of SCID. It is caused by a deficiency in the functional common γ chain, a common subunit of the receptors for IL-2, IL-4, IL-7, IL-9, and IL-15. These cytokines are crucial for normal T cell proliferation and differentiation. A defect in the common γ chain receptor results in a profound perturbation in the T cell compartment. Patients show very low numbers of T and NK cells, with low to normal numbers of B cells.
- **Bare Lymphocyte Syndrome** (BLS) derives its name from a lack of MHC class I or II expression on haematopoietic cells. Two types of BLS are recognised. Mutations in one of several distinct genes involved in MHC class I synthesis and loading pathways can cause BLS type 1, and patients have a relatively better prognosis than for BLS type 2. Type 2 BLS patients have a mutation in one of the genes critical for the transcription of class II molecules. BLS type 2 patients have very poor prognoses and often die of progressive organ failure.

9 T Lymphocytes

Our Fathers were all soldiers,
Shall we be soldiers too
Fighting and falling like soldiers do

Nothing is clear in this tactical unclear war
I can't be bothered to find out
What we are fighting for
No one can win this war of the senses
I see no reason to drop my defences
So stand fast my emotions,
Rally round my shaking heart

Our Fathers were all soldiers,
Shall we be soldiers too
Fighting and falling like soldiers do

— Billy Bragg, *Like Soldiers Do*

9.1 Introduction

Lymphocytes that mature in the thymus after having originated in the foetal liver and adult bone marrow are known as T lymphocytes. The majority of T cells are small lymphocytes, possessing a large nucleus with very few intracytoplasmic organelles. They have a central role in immune responses and are functionally and phenotypically heterogeneous. Those T cells that express αβ TcRs are considered part of the adaptive immune response, and only these will be considered here. γδ TcR-expressing T cells are considered part of the innate immune response and have been discussed in chapter 2. αβ T cells can be divided into two major categories ($CD4^+$ or $CD8^+$), depending upon the coreceptor expressed[1]. Functionally, they fall into four broad groups.

- CTLs are involved in the lysis of altered self-cells (e.g., virus-infected cells, cells harbouring intracellular parasites, and tumour cells).
- T_H cells augment host defence by
  - cooperating with B cells in humoral responses,
  - collaborating with CTLs in cell-mediated responses, and
  - secreting macrophage-activating cytokines.
- T_{reg} cells control immune responses and help in the prevention of autoimmune diseases.
- NKT cells and lipid-specific T cells recognise lipid antigens and are important in defence against intracellular parasites and tumours.

Mature T cells show the presence of several distinctive markers on their cell surfaces.

- **TcR**: The most important distinguishing feature of a T cell is its antigen receptor complex.
  - As mentioned, T cells involved in adaptive immune responses express αβ TcRs; those expressing γδ TcRs are considered part of the innate immune system.
  - CD3 and CD45 molecules are part of the TcR complex and are found on all T cells.
- **CD4 and CD8 molecules**: CD4 and CD8 coreceptors are expressed by distinctive T cell subsets.
  - T_H and T_{reg} subsets express the coreceptor CD4, which binds to MHC class II molecules.
  - CTLs express the CD8 coreceptor, which binds to MHC class I molecules.
  - Some T cells express αβ TcRs and recognise lipid antigens in the context of CD1 molecules; they have either a $CD4^+$, $CD8^+$, or $CD4^-CD8^-$ phenotype.
  - NKT cells express αβ TcRs in addition to NK cell markers and recognise antigen in the context of CD1; NKT cells may express either CD4 or CD8 molecules, although some are both CD4 and CD8 negative.
- **CD7**: This is probably the first molecule to appear during T cell ontogeny. It is a pan-T cell marker expressed by prethymic, intrathymic, and post-thymic T cells. It is also expressed by most NK cells.
- **ANAE (α-Naphthyl Acid Esterase)**: Both human and murine T lymphocytes contain several lysosomal acid hydrolases. ANAE is localised in few regions of the cytoplasm that appear as dots upon cytochemical staining. This dotted cytochemical staining pattern is distinctive of T cells. Most B cells stain negative for ANAE; monocytes show a diffused staining pattern.

Abbreviations

AICD:	Activation-Induced Cell Death
AIRE:	Autoimmune Regulator
ANAE:	α-Naphthyl Acid Esterase
BCA-1:	B Cell Attracting chemokine-1
BLC:	B Lymphocyte Chemoattractant
CMI:	Cell-Mediated Immunity
CTLA-4:	Cytotoxic T Lymphocyte Antigen-4
DN T: cell	Double negative T cell
DP T: cell	Double Positive T cell
DTH:	Delayed-Type Hypersensitivity
EAE:	Experimental Autoimmune Encephalitis
ELC:	Epstein-Barr virus induced molecule 1-ligand Chemokine
ICOS:	Inducible T cell Costimulator
IL-1Ra:	IL-1 Receptor antagonist
MCF:	Macrophage Chemotactic Factor
MIF:	Macrophage Migration Inhibitory Factor
PD-1:	Programmed Death gene-1
SDF-1:	Stromal Cell Derived Factor-1
SLC:	Secondary Lymphoid tissue Chemokine
SP T: cell	Single Positive T cell
STAT:	Signal Transducer of Activation and Transcription

[1] γδ T cells do not generally express either CD4 or CD8.

- **CD2**: Human T cells express CD2 (LFA-2), an adhesion molecule that can fortuitously[2] also bind to ovine (sheep) RBCs. CD2 also augments TcR-mediated signalling.
- **CD28**: This is a costimulatory molecule found on naïve and activated T cells (see the sidetrack *Stimulating Company*).
- **Activation markers**: Activated T cells express a distinctive set of molecules.
  - Activation induces the expression of the α chain of the IL-2 receptor (CD25) and of CD69. Nevertheless, it must be noted that CD25 is constitutively expressed by a subset of T_{reg} cells.
  - CD154 (CD40L), a costimulatory molecule, is expressed only on activated $CD4^+$ T cells and NKT cells.
  - CTLA-4 (**C**ytotoxic **T** **L**ymphocyte **A**ntigen-**4**; CD152), expressed only on activated T cells, is a negative receptor that arrests the proliferation of activated T cells. It is detectable within 24 hours of T cell activation. T_{reg} cells are an exception to this in that CTLA-4 is constitutively expressed by these cells.
  - Resting human T cells do not express MHC class II molecules; their expression is induced upon activation. Murine T cells, whether resting or activated, do not express MHC class II molecules.

TECK: Thymus Expressed Chemokine
TIM: T cell Immunoglobulin- and Mucin-domain-containing molecule
VCP: Vaccinia Complement Protein

9.2 T Cell Development

T cells are derived from pleuripotent haematopoietic stem cells present in the foetal liver or adult bone marrow. These haematopoietic stem cells are confronted with successive cell-fate specification events, and the results (listed below) determine the type of lymphocytes they become and the functions that they perform (Flow chart 9.1).

- The cells take the path of common lymphoid progenitor cells. These cells have lost the potential to become erythroid or myeloid cells but retain the capability to give rise to lymphoid cells (B, T, NK) and, possibly, DCs.
- The cells become committed to the T cell lineage.
- The cells develop into αβ or γδ T cells.
- The αβ T cells silence either the CD4 or the CD8 genes to become $CD8^+$ or $CD4^+$ T cells.
- The cells commit to either a memory cell or effector cell fate. This last phase occurs only after antigenic challenge.

T cell development is thus very complicated. Factors and cells involved in this development are still being identified. Even less is known about the development of NKT cells. A simplified version of T cell development is described below. Students new to the subject are strongly encouraged to read these sections only after becoming familiar with thymic architecture, TcR structure and TcR gene rearrangement, and T cell functions. T cell development can be divided into two broad stages.

- In phase I (also called the early phase) lymphoid progenitor cells give rise to
  - γδ T cells or
  - αβ T cells expressing CD4 and CD8 (known as **DP** or **D**ouble **P**ositive cells).
- In phase II (the late phase), DP cells give rise to mature SP (**S**ingle **P**ositive; expressing either CD4 or CD8) αβ T cells that leave the thymus and migrate to the periphery.

[2] We say fortuitously because this property can be used to physically separate, relatively easily, human T cells from B cells, or other non-T cells.

9.2.1 αβ T Cell Development Phase I

In the initial stages of thymocyte differentiation, **committed lymphoid progenitor cells migrate to the thymus and start expressing αβ TcRs and both CD4 and CD8**. The major steps in this development are outlined below (Figure 9.1).

- $CD34^+$ committed lymphoid progenitor cells arise in the bone marrow and migrate to the thymus through blood. These cells can develop into T cells, B cells, NK cells, or DCs, depending upon the signals they receive. For the progenitors, the first step to becoming T cells is losing the potential to develop into NK cells, B cells, or DCs; the progenitors give rise to DN (**D**ouble **N**egative, i.e., $CD4^-$ and $CD8^-$) precursors. They do not express RAG-1 and RAG-2.
- DN cells express surface molecules that are characteristic of early-phase T cell development. A receptor for SCF, c-kit, is the first membrane molecule expressed. DN cells also express the receptor for IL-7. Both IL-7 and SCF are essential for early T progenitor maintenance. IL-7 appears to be indispensable for development from DN cells to SP cells. Signalling through IL-7 induces

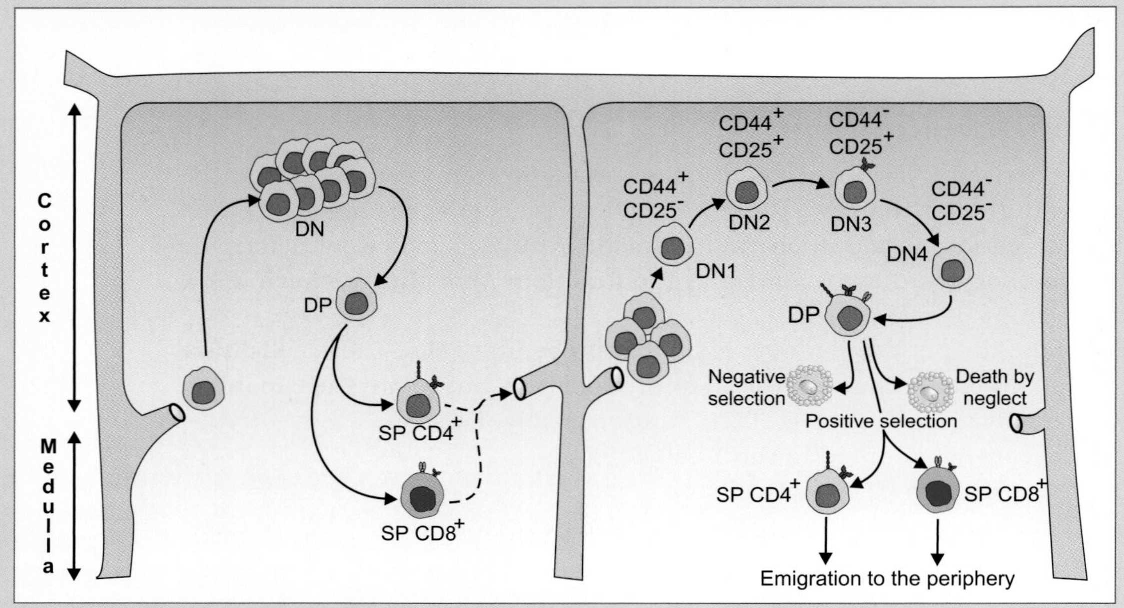

Figure 9.1 T lymphocytes undergo a complicated process of development in the thymus. $CD34^+$ lymphoid progenitors committed to the T cell lineage enter the thymus at the corticomedullary junction. These DN cells are $CD4^-CD8^-$. The left panel depicts the migratory path of thymocytes; the right panel depicts different stages of development. After entering the thymus, thymocytes migrate inwards to the cortex and start dividing on their way to the subcapsular cortex. Thymocytes undergo four phases of differentiation, distinguished by their CD44 and CD25 expression. DN1 cells express the adhesion molecule CD44. They enter the DN2 phase when they start expressing CD25. Cells in the DN3 phase lose CD44 expression but express pre-TcR, consisting of a pre-TcRα chain (yellow subunit) and a rearranged TcRβ chain (blue subunit). TcR expression is downregulated in the DN4 ($CD44^-CD25^-$) stage when the α chain undergoes rearrangement. The cells express rearranged TcRs and CD4 and CD8 molecules as they enter the DP phase. In the inner cortex, DP cells undergo further maturation and positive selection. Their development into mature SP cells is absolutely dependent on the ability of TcRs to interact with MHC class I and class II molecules on stromal cells. Those cells that fail to obtain survival signals undergo death by neglect. The surviving cells eventually lose expression of either CD4 or CD8 to become SP cells. Cells that bind self-ligands with too high an affinity or that bind MHC molecules with too low an affinity undergo negative selection. Cells recognising self-MHC molecules with moderate affinity are allowed to survive, and they enter the final stage of development before they exit the thymus and migrate to the periphery.

antiapoptotic Bcl-2 expression and rescues thymocytes from apoptosis. This signalling also seems to have a role in γδ T cell development.

- When thymocytes start expressing CD44 (a cell adhesion molecule), they enter the first of four stages (DN1 to DN4) of DN cell differentiation. These stages are distinguished by CD44 and CD25 expression.
- Cells in the DN1 phase of maturation are $CD44^+CD25^-$; these cells start expressing CD25 and enter the DN2 stage.
- $CD44^+CD25^+$ DN2 cells also express CD127 — those that are $CD127^{lo}$ enter the αβ lineage, whereas $CD127^{hi}$ cells are predisposed to γδ lineage.
- Cells of the αβ TcR pathway downregulate their CD44 and c-kit expression as they enter the DN3 stage. These cells have the $CD44^-CD25^+$ phenotype and start expressing a surrogate α chain — pre-TcRα. This is analogous to surrogate L chain expression by B cells. The pre-TcRα chain is encoded by a nonrearranging locus. The cells also express RAG-1 and RAG-2, the enzymes needed for TcR gene rearrangement.
- The TcRβ chain is the first to be rearranged. Pre-TcRα pairs with a rearranged TcRβ chain to give the pre-TcR that is expressed on the cell surface. This pre-TcR is also associated with the CD3 complex.
- To develop further, cells in the late DN3/early DN4 stage ($CD44^-CD25^+$pre-$TcR\alpha\beta^+$) must receive signals through the CD3 complex. Pre-TcR's interaction with an unknown ligand activates Src family kinases (Lck, Syk) and ZAP-70. As in B cells, signalling through pre-TcR suppresses further gene rearrangement of the β chain, ensuring its allelic exclusion.
- Eventually, CD25 expression is downregulated, and the cells enter the DN4 stage ($CD44^-CD25^-$).
- Cells undergo a short burst of proliferation (6-8 cell divisions) before the TcRα chain undergoes gene rearrangement. TcR expression is low at this stage as pre-TcRα disappears from the cell surface, and the rearranged TcRα chain is yet to be expressed. Thymocytes also begin to express coreceptors at this stage — first CD8, then CD4.
- The DN4 stage ends with the cells becoming DP and having a rearranged TcR associated with the CD3 complex.
- DP cells expressing αβ TcRs constitute 80-90% of the lymphoid compartment of the young thymus. Their development into mature cells completely depends upon their TcRs' ability to interact with MHC class I and class II molecules on stromal cells — especially thymic cortical epithelial cells (nurse cells).

***En route* maturation, thymocytes are subjected to four selective processes.** Only a small fraction (5%) survive these selective forces, and the survivors mature and migrate to the periphery (Figure 9.1)

- **Death by neglect**: Unlike in the other selective processes, death by neglect is caused by a failure to obtain survival signals. In order to survive and progress further from the DP stage, T cells must obtain survival signals through their TcRs. However, most DP cells fail to bind the MHC molecules expressed on thymic stromal cells. Such cells that do not encounter their restricting MHC molecule on the thymic epithelium undergo death by neglect within 3-4 days.
- **Negative selection**: Negative selection weeds out cells that bind self-ligands with too high an affinity. A transcription factor, AIRE (**Auto**i**mmune** **Re**gulator), expressed in the epithelial cells promotes the expression of tissue-specific proteins that are not normally expressed in the thymus. AIRE enables the

Not Immune to Death: Apoptosis in the Immune System

The immune system must balance the ability to respond to a multitude of antigens with the danger of autoimmunity. It must also maintain homeostasis, that is, ensure that the total number of lymphocytes remains constant and revert to a baseline level after antigen-induced clonal expansion. The immune system achieves this by fine-tuning the equilibrium between expansion and death. Generally, the immune system produces more cells than needed and eliminates them by apoptosis. Apoptosis is thus the immune system's central regulatory feature. A group of enzymes termed *caspases* are vital to the induction of death. These enzymes are normally present as proenzymes that have to be cleaved to yield the active form[3]. Two downstream pathways induce cell death, as explained in section 11.2.

B cell death: Three cell-surface molecules are key elements in the regulation of B cell survival and death — BcRs, CD95, and CD40. As a general rule, BcR activation induces apoptosis by the mitochondrial pathway. Stage of maturation, quality and quantity of signals provided, cytokines, and other microenvironmental factors are critical in dictating whether or not the death signal is over-ridden. The combined effects of BcR and CD40 engagement (by CD154 on T cells or macrophages) can rescue B cells in the germinal centre, as signalling through these molecules upregulates the antiapoptotic protein Bcl-XL. Other signals that promote B cell survival and differentiation are yet to be determined. Factors that can cause plasma cell death and the antiapoptotic signals that can rescue them also remain unknown.

T cell death: Only 3-5% of immature thymocytes leave the thymus as mature T cells. The rest undergo apoptosis through death by neglect or during positive and negative selection. The understanding of the molecular basis of this apoptosis is, however, fragmentary. Mature peripheral T cells undergo apoptosis. T cells that are insufficiently stimulated by growth signals undergo death by neglect. Death by neglect also occurs at the peak or down phase of the immune response, allowing a reduction in the number of effector cells and termination of the immune response. T cell activation induces CD95L expression. Like B cells, T cells can also be rescued from apoptotic death by costimulatory signals through CD28, and in the initial clonal expansion phase of the response, T cells are relatively resistant to CD95-induced cell death. Memory cells are also resistant to apoptosis. Effector T cells, on the other hand, are susceptible to apoptosis *via* the DISC pathway (see chapter 11), and Bcl-XL expression rescues them.

Apoptosis and disease: Apoptosis is fundamental to immune system regulation, and its derailment can lead to severe diseases.

- CD95 dysfunction causes the autoimmune lymphoproliferative syndrome in humans. Children with this disease show massive nonmalignant lymphadenopathy and severe autoimmunity. In many cases, the disease has been traced to mutations in the CD95 death domain.
- Apoptosis failure also leads to tumour survival. The translocation of *Bcl-2* into the Ig H-chain locus causes the deregulation of *Bcl-2* expression and of follicular lymphomas.
- HIV uses this apoptotic signal to advantage. Regulatory viral gene products (like Tat-1 of HIV-1) penetrate uninfected T cells and render them hypersensitive to TcR-induced CD95-mediated apoptosis. Binding of HIV gp120 to CD4 and crosslinking of the bound gp120 by anti-gp120 antibodies produced in response to infection further sensitises CD4$^+$ T cells. The massive death of CD4$^+$ T cells leaves the patient defenceless against opportunistic pathogens and neoplasms and results in patient death.
- In mice, the *lpr* (lymphoproliferation) and *gld* (generalised lymphadenopathy) mutations cause a disease similar to SLE. In lpr mice, a point mutation in the death domain of CD95 abolishes apoptotic-signal transmission; gld mice carry a point mutation in the carboxy-terminal of CD95L.

epithelial cells of the cortex and medulla to display a sampling of peptides derived from tissue-specific proteins that are normally expressed in various body tissues. Cortex and medullary epithelial cells also display peptides that are derived from proteins brought by the blood to the thymus and those derived from thymic cells. A high-affinity binding of TcRs to these MHC:peptide complexes generates an apoptotic signal in developing thymocytes.

[3] This is a common feature of all potentially harmful proteins (e.g., complement components, caspases, etc.) — it allows regulatory control on their activities.

- **Positive selection**: If the binding of TcRs to self-ligands is very weak, the TcR-generated signal cannot sustain survival, and the cell dies. Positive selection thus allows cells recognising self-MHC molecules to survive. Thymic epithelial cells play a major role in this positive selection[4].
- **Lineage-specific development**: In the late phase of development (see below), a DP cell loses one of its coreceptors and matures to become a $CD4^+$ or $CD8^+$ T cell. T cell function is determined by this development — $CD4^+$ T cells are destined for a helper/regulator role; $CD8^+$ T cells are destined to be cytolytic cells. If the coreceptor that is expressed mismatches the MHC molecule engaged (e.g., if the TcR binds best to a MHC class I molecule, but the cell expresses a CD4 coreceptor or *vice versa*), the TcR engagement generates a signal that is not sufficient to sustain survival, and the cell undergoes apoptosis. Optimal

[4] This is sometimes termed the *tickling rule*. Cells that are tickled too much or too little die; only moderately tickled cells survive to tell the tale. ☺

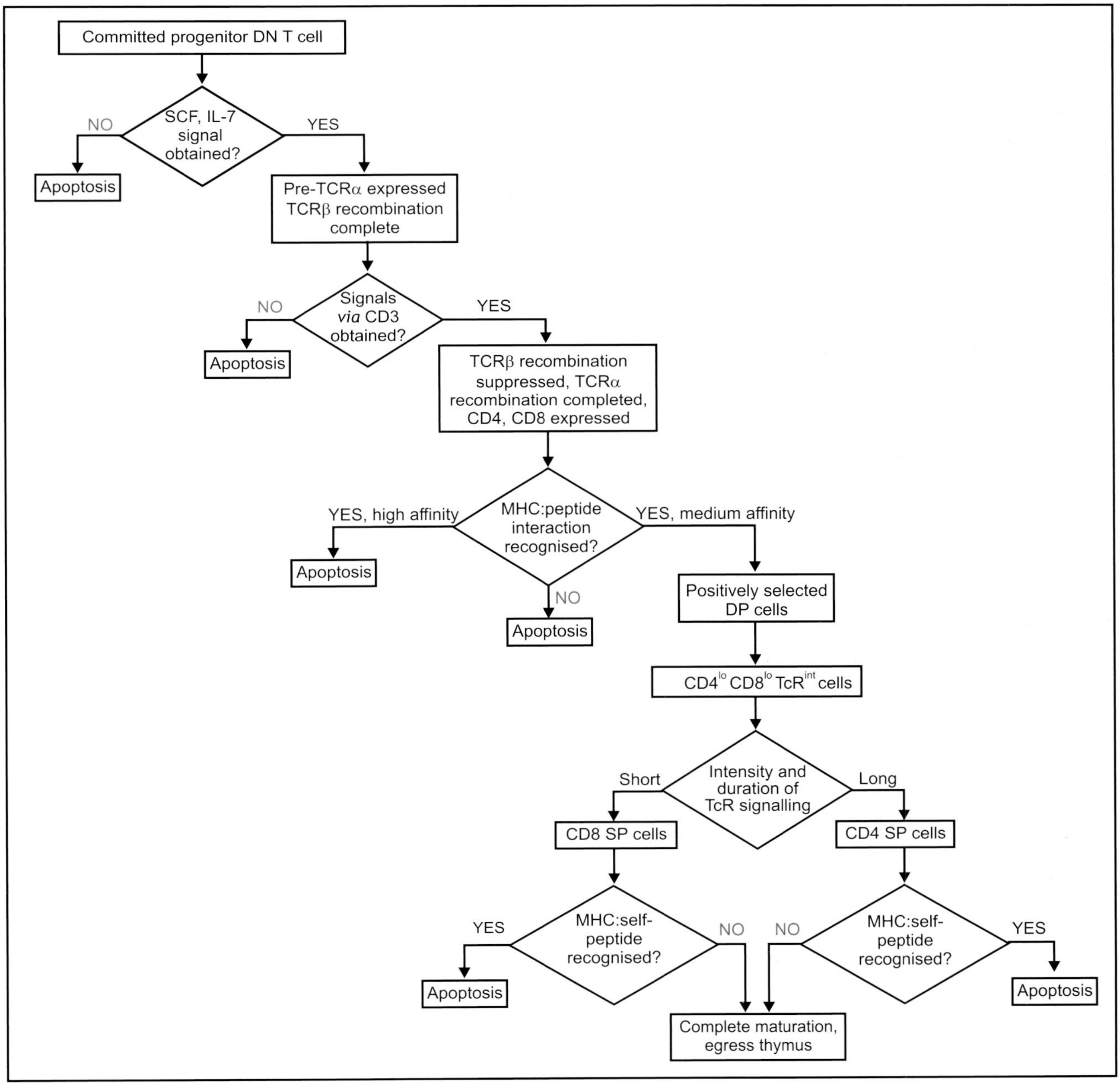

Flow chart 9.1 αβ T cell development

T cell function depends upon matching the MHC-class specificity of the coreceptor and the TcR.

Overproducing cells and then sending them to apoptosis may appear insanely wasteful. To appreciate the logic behind this process, we need to understand the constraints under which T cells operate. They must bind antigens in the context of self-MHC molecules, that is, they must respond to self-MHC molecules. Yet, high-affinity recognition risks autoimmune responses. T cells recognising self-peptides in the context of self-MHC molecules cannot be allowed to survive either. Although the unselected repertoire of TcRs is biased towards MHC recognition, the gene rearrangement and protein-pairing processes that generate TcRs are random. Thus, there is a slim likelihood that the produced TcRs will recognise MHC:peptide complexes in the individual. A large fraction of T cells will fail to bind MHC molecules and will undergo death by neglect. Conversely, some T cells that do recognise MHC:self-peptide complexes will bind self-ligands too well. Negative selection will remove these cells. Only cells that express TcRs that recognise self-ligands but generate signals with an intensity that is between these two extremes (no binding or high-affinity binding) can initiate the multistep positive selection process that results in lineage-specific differentiation for a cell to mature into $CD4^+$ or $CD8^+$ T cells. Nevertheless, selection procedures are imperfect, and autoreactive clones are found in the periphery. The absence of self-reactivity is therefore ensured through peripheral regulatory mechanisms.

9.2.2 αβ T Cell Development Phase II

The second phase in αβ TcR^+ thymocyte differentiation **involves a change from the DP to the SP state by silencing of the transcription of one coreceptor locus**. It is also accompanied by genetic events that determine the effector potential of the mature T cell. Progression from the DP to SP state (i.e., either $CD4^+CD8^-$ or $CD4^-CD8^+$) seems to comprise a series of intermediate states.

- ❒ DP cells initially downregulate both CD4 and CD8 to become $CD4^{lo}CD8^{lo}TcR^{int}$; they are uncommitted cells.
- ❒ Commitment to the CD4 lineage leads to the upregulation of CD4 expression, and cells become $CD4^+CD8^{lo}$. Eventually, these cells lose CD8 expression and emerge as $CD4^+TcR^{hi}$ SP cells.
- ❒ Cells committed to become $CD8^+$ T cells also show the $CD4^+CD8^{lo}$ phenotype before downregulating CD4 expression to become $CD4^{lo}CD8^{hi}TcR^{hi}$. Eventually, CD4 expression is completely silenced, and the cells become $CD8^+TcR^{hi}$ SP cells.

The development of SP cells from DP cells has been a matter of intense research, much debate, and some confusion. The *Strength of Signal Model* has general consensus. This model proposes that commitment to CD4 or CD8 lineage depends upon the intensity and duration of signalling from TcRs. Short duration signals lead cells to the CD8 pathway; more prolonged signalling leads them to the CD4 pathway. Information on how signalling strength dictates lineage choice is scarce, partly because knock out mice approach is unfeasible because many of these molecules have roles in the earlier stages of thymocyte differentiation. MAP kinase 1, NOTCH[5], and its ligands are known to be involved in this differentiation, but their exact roles are unclear.

[5] *NOTCH* genes encode highly conserved cell-surface receptors that regulate the development of a wide spectrum of cell types. NOTCH signalling functions in multiple cell-fate decisions during lymphocyte development, including commitment to T cell lineage and differentiation to the DN and SP stages.

9.2.3 γδ T Cell Development

The exact steps involved in γδ T cell maturation are still being defined. Following definitive commitment to the T cell lineage, thymocytes start expressing CD25,

RAG-1, and RAG-2, and enter the DN2 development phase. These cells start rearranging their β, γ, and δ chains. Those that successfully rearrange β chains enter the αβ lineage; those that rearrange γδ chains enter that lineage. The exact signals or events responsible for promoting these choices are unclear.

Despite similarities in the overall structure of their genes and polypeptide chains, γδ T cells are different from αβ T cells in their development, repertoire, and extrathymic tissue distribution. γδ T cells are found in adult peripheral blood and lymphoid tissues, as well as in epithelial tissues such as the skin. Distinct homing receptors probably determine the homing of γδ subset to peripheral organs. These cells are often $CD4^-CD8^-$ and can recognise both proteinic and nonproteinic antigens independent of the MHC (Table 9.1).

9.2.4 NKT Cell Development

NKT cells express αβ TcRs as well as markers characteristic of NK cells such as CD56, CD57 (in humans; murine NKT cells express NK1.1), and CD122 (IL-2Rβ). In mice, mature NKT cells constitute a small percentage of the T cells found in the thymus, spleen, and bone marrow, but a significant proportion (20-30%) of those are found in the liver. In humans, only 4% of hepatic T cells are NKT cells. NKT cells express a restricted T cell repertoire. The best characterised subset of human NKT cells expresses the NK cell marker CD161. The TcRs of these cells have an α chain consisting of a particular VJ segment (Vα24JαQ) with no junctional additions or deletions (hence, it is referred to as the *invariant α chain*). It is associated with a variant β chain (predominantly Vβ11).

Many aspects of NKT cell development are still being investigated. Majority of NKT cells develop in the thymus and branch out at the DP stage. In mice, conventional DP thymocytes seem to get diverted to the NKT lineage as a result of CD1d engagement by randomly generated semi-invariant TcRs. The molecular requirements and main events in the development and selection of NKT cells remain poorly understood.

Table 9.1 Distinguishing Features of B, αβ T, and γδ T Cells

Characteristic	B cells	αβ T cells	γδ T cells
Antigen receptor	Ig molecules	αβ TcRs	γδ TcRs
Antigen recognised	Proteinic and nonproteinic	MHC:peptide complexes	Proteinic and nonproteinic
MHC restriction	—	+	—
Coreceptors	CD19/21/CD81 complex	CD4 or CD8	Not known
Frequency in blood	5-10%	65-75%	5-10% 25-60% in gut
Effector mechanism	Ig production	CTLs: cytotoxic granules and death-receptor pathways TH1, TH2, TH17, or Treg cells; cytokine release	CTLs: cytotoxic granules and death-receptor pathways Cytokine release
Function	Humoral immunity	Immune protection, pathogen eradication, immune regulation	Immunosurveillance, immune regulation

9.2.5 Migration in the Thymus

On about the tenth or twelfth day of embryogenesis, lymphocytes or thymocytes from the foetal liver and yolk sac migrate to the rudimentary thymus. They colonise the thymus and ultimately give rise to T cells. In adults, T cell precursors, like B cell precursors, are formed in the bone marrow. These precursors enter the thymus at the corticomedullary junction and are recruited from the blood stream by chemoattractants that are produced by thymic epithelial cells. β_2-m has an important role in this recruitment. To enter the thymus, precursor T cells attach to the endothelium of thymic blood vessels. CD44, expressed by thymocytes, is thought to play a role in this entry. Other adhesion molecules such as LFA-1, ICAM-1, CD2, LFA-3, and CD90 (thy-1) are important in subsequent interactions with thymic epithelial cells. Thymic stromal cells (subcapsular epithelial cells, cortical epithelial cells, medullary epithelial cells, fibroblasts, macrophages, and thymic DCs) differentially influence the thymocytes on their developmental pathway. These stromal cells constitute the microenvironment where developing T cells receive signals to proliferate and mature or undergo apoptosis through cell–cell interactions and through contact with locally secreted hormones and cytokines. In addition, extracellular matrix proteins such as fibronectin, laminin, and collagen that are expressed by stromal cells are important in the adhesion and migration of thymocytes.

Instant Chemistry: Chemokines

Chemokines (chemoattractant cytokines) were originally discovered because of their chemotactic properties — they were found to regulate leukocyte transport by mediating leukocyte adhesion to endothelial walls, initiation of transendothelial migration, and eventually tissue invasion. It is now clear that chemokines are pleuripotent molecules with diverse functions. Aside from influencing the progress of the immune response, they also regulate processes such as angiogenesis, haematopoiesis, and the organogenesis of primary and secondary lymphoid organs. Chemokines are a family of structurally related single polypeptides (~70-100 amino acids in length). These heparin-binding proteins contain conserved cysteines at the NH_2-terminus. The number and spacing of the first two cysteines is used to characterise chemokines into subfamilies (C, CC, CXC, and CX_3C), where X stands for single amino acids.

Chemokines act through chemokine receptors. Coupled to small G proteins, these structurally homologous receptors have seven transmembrane domains. The receptors are named on the basis of the chemokine subfamily followed by a number (CCR1-9, CXCR5, etc.). Chemokines were referred to by multiple names, and as single chemokines bind multiple receptors, the confusion was compounded. Now, using standardised nomenclature, chemokines are named on the basis of the subfamily they belong to followed by L (for ligand) and a number.

Chemokines upregulate the expression of adhesion molecules on leukocytes, inducing their arrest and firm adhesion to the endothelium. Chief among secreted chemokines are IL-8, which acts on neutrophils; Monocyte Chemoattractant Protein-1 (CCL2), which induces the adhesion of monocytes; and eotaxin (CCL11), which induces eosinophil adhesion. Fractalkine (CX_3CL) and CXCL16, which both mediate the adhesion of NK cells and $CD8^+$ T cells, are important transmembrane chemokines.

Chemokines and their receptors are indispensable in T and B cell maturation. Both these cell types are produced in the bone marrow and must travel to other organs (e.g., the thymus and spleen) during their development and maturation. Chemokines are also important in B cell maturation and migration within the bone marrow (see Figure 8.1). The regulated expression of various chemokine receptors achieves this migration. The expression of distinct chemokines by different thymic areas also accomplishes the intrathymic migration of T lymphocytes. Chemokines also contribute to the sorting of positively and negatively selected thymocytes. Moreover, chemokines have a role in the movement and recirculation of mature lymphocytes. They also influence the differentiation of T cells to TH1 or TH2 subsets. Prominent chemokines include the following:

- SDF-1(**S**tromal cell **D**erived **F**actor-**1**; CXCL12), prevalent at the thymus's corticomedullary junction, is an important chemotactic factor for CD34$^+$ progenitor cells.
- TECK (**T**hymus **E**xpressed **C**hemo**k**ine; CCL25) stimulates the migration of DN cells from the outer cortex to inner cortical regions during their differentiation to DP thymocytes. TECK may act as a chemoattractant for newly selected SP cells as they pass from the cortex to the medulla after selection.
- ELC (**E**pstein-Barr virus Induced molecule 1-**L**igand **C**hemokine; CCL19) and SLC (**S**econdary **L**ymphoid tissue **C**hemokine; CCL21), expressed by medullary epithelial cells, have a role in the intrathymic migration of SP cells. They also have a role in the recirculation of lymphocytes in the secondary lymphoid organs (spleen, lymph nodes, and Peyer's patches). Naïve B and T cells express CCR7, the ligand for ELC and SLC, and home to the secondary lymphoid tissues expressing these chemokines. SLC, expressed by HEV cells, is responsible for T cell trafficking through lymph nodes.
- BLC (**B** **L**ymphocyte **C**hemoattractant), expressed by stromal cells in B cell areas, helps B cells localise to the lymphoid follicles of all secondary lymphoid organs.
- Chemokines regulate the movement of B and T cells from their respective areas to the B cell–T cell border.
  - BcR stimulation promotes the expression of CCR7, allowing the B cell to migrate towards T cell zones (rich in ELC and SLC) of the follicle.
  - Activated T cells express CCR5, the receptor for BLC. This furthers their propensity to move to the B cell area border and allows the rare antigen-specific B cell to present antigen to the rare antigen-specific T cell.
- CCL2 is found to be a positive regulator of TH2 pathways; it inhibits IL-12 production by DCs and promotes differentiation to TH2 type. Conversely, RANTES (CCL5), Macrophage Inflammatory Protein-3α (CCL3), and Macrophage Inflammatory Protein-1β (CCL4) promote IL-12 secretion by APCs and directly promote differentiation to TH1 type.

After entering the thymus, thymocytes move inward to the cortex. Here, they start dividing while migrating towards the subcapsular cortex, which is located just under the cortex's outer border (Figure 9.1). Thymus-expressed chemokines play a key role in both the initial colonisation of the thymus and cell migration within the thymus. In the inner cortex, DP cells undergo further maturation and positive selection. Thymic nurse cells found in the cortex are thought to be important players in this T cell thymic education (see section 5.2.1). Nurse cells express neuropeptides, such as vasopressin and oxytocin, and may also be important in neurohormonal regulation within the thymus. Thymic macrophages have a major role in clearing dead or dying cells and protecting the thymic microenvironment from the potentially harmful effects of their cellular contents. The surviving cells enter the medulla for the final stage of development, where they also undergo negative selection. Most mature SP cells are therefore found in the medulla. Naïve T cells leaving the thymus express homing receptors such as CCR7 (see the sidetrack *Instant Chemistry*) and CD62L, which allow their homing and entry into secondary lymphoid tissues.

9.3 T Cell Activation

Found predominantly in the T cell areas (paracortex) of the spleen, lymph nodes, and Peyer's patches, naïve T cells (whether CD4$^+$ or CD8$^+$) circulate continuously from and to these areas through the lymph and blood. They are emptied into the spleen's marginal sinuses with blood. They then move into the red pulp and finally to the PALS. On reaching the PALS or the paracortex, they remain there because they express CCR7, which binds to the chemokines CCL19 and CCL21 produced in T cell areas. If they do not encounter their specific antigen, these cells leave the lymphoid tissue after about 24 hours and return to the blood, again journeying to a different lymphoid tissue (Figure 9.2). Naïve T cells are characterised by the expression of low levels of most cell adhesion molecules. However, they express high levels of the homing receptor L-selectin (CD62L). This receptor enables the cells to bind and roll across HEV, allowing their migration to and from the lymph node and permitting them to remain as a part of the circulating lymphocyte pool. As they migrate through the lymph node's cortical region, T cells bind transiently to any APCs they encounter because of the adhesion molecules expressed by APCs. This gives the T cell enough time to sample the MHC molecules expressed by the APC for a specific peptide. T cells monitor APC surfaces for peptide:MHC complexes through transient binding and continuous circulation. Only 1 in 10^4-10^6 T cells are likely to be specific for a particular antigen. The continuous circulation therefore increases the probability that an antigen-specific T cell will come in contact with its specific peptide:MHC complex on an APC surface. During their normal pattern of blood to lymph recirculation, naïve T cells are metabolically quiescent and have a prolonged life span. This longevity requires at least two signals — contact with self-peptide:MHC complexes on DCs and IL-7. Recognition of these ligands probably delivers low-level signals that keep T cells metabolically active enough to avoid passive death.

The migratory properties of T cells prevent their entry into the initial site of infection. **Naïve T cells are programmed to recognise antigens only in T cell zones of lymphoid tissues**, and the transport of antigens to these areas from the site of infection initiates the immune response. Cells involved in this transport and initiation of the immune response are given below.

- **DCs are the principal cells involved in antigen transport**. DCs capture antigens from the site of antigen deposition by macropinocytosis and phagocytosis and transport them to the local draining lymph node. Resting or immature DCs are incapable of stimulating T cells and need to be activated for antigen presentation. DCs express a number of PRRs on their surfaces. LPS or lipoproteins on the surfaces of most infectious microorganisms provide the triggers needed for DC activation. Pro-inflammatory stimuli such as IFNs or cognate interaction with T cells by CD40–CD154 can also activate DCs. DCs seem to be able to present antigen from a variety of pathogens, whether viral, fungal, bacterial, or protozoan. DCs have an intrinsically high costimulatory capacity and may be particularly important in stimulating T cell responses to viruses that fail to induce costimulatory molecules in other types of APCs.

  DC activation

  - enhances the movement of DCs from nonlymphoid tissues to T cell areas;
  - causes DC maturation and increases the surface expression of the MHC class II molecules required for maximal T cell stimulation;
  - upregulates the expression of costimulatory molecules — the interaction of costimulatory molecules on T cells with their ligands on APCs (e.g., CD28–CD80/86, CD134–CD134L) can improve T cell activation by enhancing TcR signalling and may provide additional signals that increase T cell responses; and

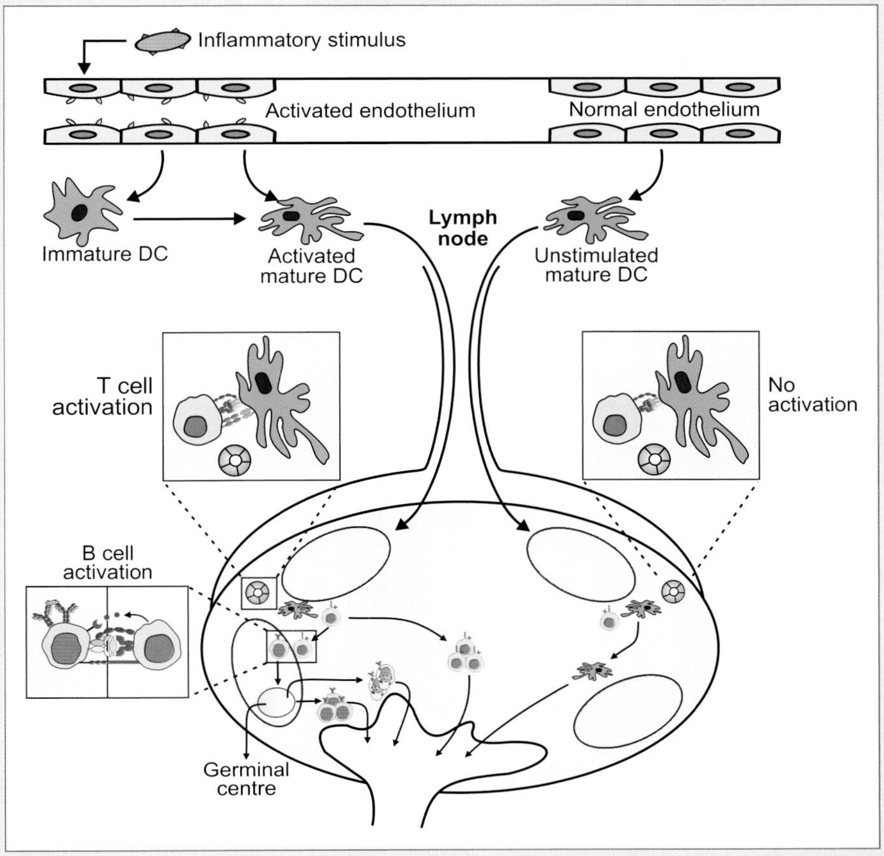

Figure 9.2 Unstimulated APCs do not express costimulatory molecules and do not trigger immune responses; exposure to pro-inflammatory stimuli induces costimulatory molecule expression and increases MHC class II expression, resulting in a productive immune response. Naïve T cells do not enter the site of infection. APCs transport antigen to the lymphoid tissues (here represented by the lymph node) and initiate an immune response. Naïve T cells express high L-selectin levels that enable them to bind and roll across HEVs and allow their migration to and from the lymph nodes. T cells bind transiently to any APCs they encounter as they migrate through the cortical regions of the lymph node and probe them for MHC:peptide complexes. As unstimulated APCs express low levels of costimulatory molecules, the recognition of self-peptide:MHC complexes on these cells does not activate naïve T cells and prevents autoimmune responses (right). The recognition of PAMPs or other pro-inflammatory stimuli activates APCs and results in a productive immune response (left panel). Activation enhances the APCs migration to the lymphoid tissues. They enter the lymph node *via* the HEVs. Activated APCs express high levels of costimulatory molecules and are efficient in antigen presentation. Antigen presentation triggers the activation, proliferation, and differentiation of naïve T cells to effector cells. The effector T cells migrate to the border zone between the T cell: B cell areas, where they encounter B cells. Antigen recognition and signals obtained from activated CD4$^+$ effector T cells result in the activation, proliferation, and differentiation of naïve B cells in germinal centres. Plasmablasts resulting from B cell differentiation give rise to plasma cells that secrete Igs. The end result of these series of events is a cell-mediated and humoral immune response.

- results in the synthesis and secretion of pro-inflammatory cytokines and chemokines by DCs and guides their movement to T cell areas; cytokines such as IL-6 or TNF-α produced by DCs also provide secondary signals that complement CD28–CD80/86 costimulatory signals and further augment T cell activation.

- **Macrophages** are involved in the removal of dead or dying cells. Resting macrophages express few MHC class II molecules and have no costimulatory capacity. This prevents them from activating self-reactive T cells and triggering autoimmune responses in spite of expressing self-peptide loaded MHC molecules. Macrophages express a number of receptors for microbial constituents (see chapter 2) and the ligation of these receptors activates them. PRR engagement upregulates MHC class II expression and also induces the expression of costimulatory molecules such as CD80/86, allowing activated macrophages to present peptides derived from the ingested pathogen to naïve T cells.

Stimulating Company: Costimulatory Molecules and T cells

TcR engagement by a peptide:MHC (or lipid:CD1) complex is the primary event required for T cell activation. As few as 1-50 MHC class I molecules can trigger $CD8^+$ T cell-mediated lysis. Experiments show that MHC molecules need not engage all TcRs simultaneously and may serially engage several TcRs, progressively amplifying intracellular signals emanating from the engagement. The ligation of costimulatory molecules, in addition to TcR engagement, is required for a productive immune response. Members of the CD80/86–CD28 superfamily are crucial to this costimulation.

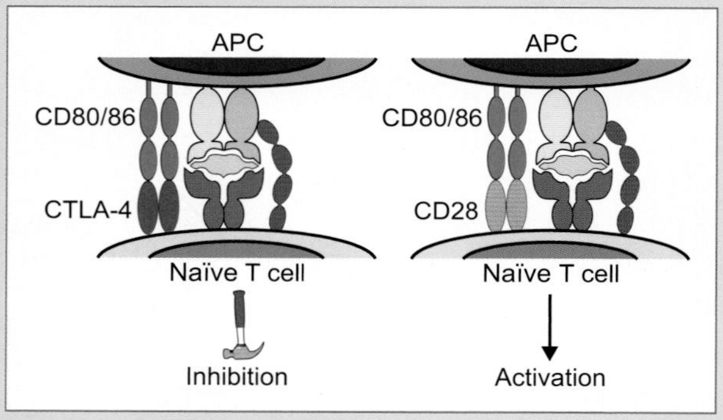

Figure 9.S1 The eventual outcome of antigen recognition by T cells is dictated by the engagement of costimulatory molecules on T cells by their ligands on APCs.

- **CD28**, a 44 KD glycoprotein that binds CD80 and CD86, is the most effective costimulatory molecule expressed by naïve and primed T cells. TcR engagement in the absence of CD28 ligation results in apoptosis or anergy. Anergic T cells do not produce IL-2 and cannot proliferate on subsequent stimulations. Besides increasing cytokine production, CD28 engagement also promotes T cell survival by inducing the upregulation of the antiapoptotic Bcl-x_L. CD28 ligation has been suggested to promote T_H2 pathways. Signals delivered through CD28 are also important for isotype switching in B cells. The cytoplasmic domain of CD28 is associated with Src kinases. Upon stimulation, the phosphorylation of CD28 recruits PI3-kinase to the cascade and ultimately leads to NFκB activation (see chapter 6), IL-2 production, and T cell proliferation.
- **ICOS** (**I**nducible T cell **Cos**timulator) is another stimulatory molecule expressed on activated T cells. Its ligand, ICOSL, is expressed on B cells, DCs, and macrophages. Signals through ICOS are thought to be important in cytokine production by newly activated and effector T cells. The ICOS pathway seems to enhance the production of both T_H1 and T_H2 cytokines, though some recent reports suggest that the engagement of ICOS preferentially induces IL-10 production.
- **CTLA-4** (**C**ytotoxic **T** **L**ymphocyte **A**ntigen-**4**; CD152) is an inhibitory receptor expressed on activated T cells. It shows ≈30% homology to CD28 and binds the same ligands as CD28 (i.e., CD80 and CD86), but with a higher affinity. CTLA-4 is involved in the downregulation of T cell responses, and CTLA-4 knock out mice develop fatal lymphoproliferative disease. CTLA-4 inhibits T cell activation by reducing IL-2 production and IL-2R expression and by arresting T cells in the G1 phase of the cell cycle. $CD4^+CD25^+$ T_{reg} cells constitutively express CTLA-4, suggesting a possible suppressive role for the molecule.
- **PD-1** (**P**rogrammed **D**eath gene-**1**) is an inhibitory molecule expressed by activated (but not resting) $CD4^+$ and $CD8^+$ T cells, B cells, and myeloid cells. PD-1 has an ITIM in its cytoplasmic tail. Its ligands are expressed not only by haematopoietic cells but also by cells in nonlymphoid tissues. The expression of ligands in nonlymphoid tissue has led to the suggestion that PD-1 may be involved in the regulation of self-reactive responses and/or regulation of inflammatory responses in the periphery.

- **B cells, unlike macrophages and DCs, are capable of presenting soluble antigens** internalised *via* BcRs. B cells constitutively express MHC class II molecules that allow them to display a sampling of the antigens internalised by the cell. However, they do not constitutively express costimulatory molecules. B cell presentation of self-antigens therefore fails to activate naïve T cells. It induces them to become anergic instead. Contact with microbial constituents, especially LPS, can induce the expression of costimulatory molecules, allowing B cells to become potent APCs.

The transient binding of adhesion molecules such as ICAM-1, ICAM-2, LFA-3, or DC-SIGN on APCs by LFA-1, CD2, or ICAM-3 on T cells is of a low affinity, allowing the cells to separate after brief contact. TcR–MHC:peptide binding results in the formation of a tight immunological synapse at the point of T cell:APC interaction within minutes.

1. The first direct consequence of such synapse formation is T cell immobilisation. The signals induced by TcR engagement result in a conformational change in the LFA-1 expressed by T cells, increasing its affinity for ICAM-1 and ICAM-2. This stabilises attachment between the cells and allows them to enter a prolonged association that can persist for several days. Antigen-specific trapping allows the continuous recruitment of newly arriving naïve antigen-specific T cells in the immune response without interfering with the circulation of other antigen-nonspecific T cells.
2. Synapse formation is associated with the rapid clustering of TcR molecules that have bound peptide:MHC complexes and the local accumulation of intracellular signalling molecules such as Lck and LAT in lipid rafts.
3. The formation of lipid rafts initiates downstream signalling events that cause T cells to proliferate and eventually differentiate into effector cells.
4. TcR/CD3 signalling is aided by CD4/CD8 coreceptors and the engagement of large numbers of costimulatory and adhesion molecules on T cells. Some of these costimulatory molecules (e.g., LFA-1–ICAM-1 interaction), stabilise synapse formation and/or recruit signalling molecules. Others (e.g., CD28) are important for inducing IL-2 synthesis by T cells; yet others, such as CD154, maintain or induce APC activation. Some interactions, such as CD154–CD40 engagement, also stimulate B cells during B/T collaboration. T cell activation requires two signals. Peptide:MHC complex recognition provides the first signal; the second is provided by the costimulatory signals generated when ligands on T cells (e.g., CD28 or OX40) interact with complementary molecules on APCs (CD80/86 and OX40L, respectively). Antigen presentation in absence of costimulatory signals anergises T cells. This two-signal model (whether for B or T cell activation) is a simple device that disallows the inappropriate stimulation of the immune system; it allows the immune system to distinguish between the harmful and the harmless so as to concentrate on the former. The costimulation requirement ensures that a T lymphocyte 'sees' not just any antigen, but only the antigen presented by appropriately stimulated professional APCs. Therefore, the need for the same cell to deliver antigen and costimulatory signals ensure that self-reacting T cells are not activated by tissue cells. This is crucial for survival because all self-reactive clones (especially for proteins expressed later in development) may not be deleted or anergised in the thymus.
5. APCs release stimulatory cytokines such as IL-1 or TNF-α that drive extensive T cell proliferation and promote their efficient differentiation into effector cells.

Thus, **professional antigen presentation triggers the entry of naïve T cells into the G1 phase**. The context in which the T cell recognises the antigen, the concentration of this antigen, and the duration of antigen exposure can all affect the speed and nature of the immune response. The intracellular pathways of activation involve a complex series of enzymatic steps that result in biosynthetic events. The ligation of the costimulatory molecule CD28 by CD80/86 switches on IL-2 production. Resting T cells express the β and γ chains of the IL-2 receptor, and this γβ heterodimer binds IL-2 with moderate affinity. Activation induces the upregulation of CD69 and CD25 (α chain of IL-2R). The expression of the α chain results in the formation of the high-affinity αβγ IL-2R heterodimer, making T cells responsive to very low concentrations of IL-2. Thus, activated T cells produce IL-2 as well as its receptor. This cytokine is needed for the rapid proliferation and differentiation of naïve T cells into effector cells and/or memory cells. IL-2 thus acts in an autocrine fashion, inducing clonal expansion and differentiation. In the lymph node, naïve T cells proliferate in response to antigen presentation at a very high rate for several days. Experimental evidence shows that $CD8^+$ T cells divide as often as every eight hours; $CD4^+$ T cells divide at a slower rate. Three days after the subcutaneous injection of a soluble antigen such as ovalbumin, there is a 10- to 20-fold increase in the number of antigen-specific naïve T cells in the lymphoid tissue. If an adjuvant (e.g., LPS or CpG DNA) is injected with the antigen, the increase can be upto 100-fold. The CD45 isoform changes upon antigenic stimulation so that the T cell becomes more sensitive to stimulation by low concentrations of peptide:MHC complexes. Late in the proliferative phase (4-5 days), T cells differentiate to effector cells that are capable of acting as helper, inflammatory, or cytotoxic cells.

9.4 T Cell Differentiation

In the early phase of infection, rapid pathogen replication leads to the continuous entry of large numbers of activated APCs into T cell zones. These APCs drive the responding T cells to proliferate rapidly, synthesise a wide range of cytokines, and differentiate either into CTLs ($CD8^+$ T cells) or TH cells ($CD4^+$ T cells). The destruction of the pathogen eventually leads to a lowered inflow of APCs in T cell areas. Cytokine secretion and differentiation into effector cells is diminished. This stage of the immune response is important for memory T cell formation. Thus, upon antigenic stimulation, T cells pass through three stages (Figure 9.3):

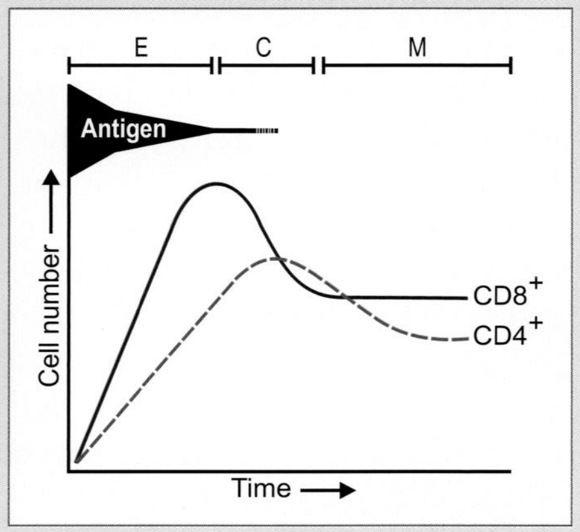

Figure 9.3 T cells pass through three stages — expansion (E), contraction (C), and memory (M) upon antigenic stimulation. In the early phase of the immune response, antigen-activated APCs drive responding T cells to proliferate rapidly. The proliferating cells differentiate to effector T cells. As the antigen gets cleared from the system, activated T cells enter a death phase because of AICD. The surviving antigen-specific T cells enter the memory phase. The number of memory cells stabilises in this phase, and cells can survive for years. Especially with viral challenges, the magnitude of the $CD4^+$ T cell response is normally smaller than that of the $CD8^+$ T cell response (adapted from Kaech et al., *Nature Reviews in Immunology*, 2002, 2: 251).

- The **expansion phase**, initiated in lymphoid tissue, in which the antigen-activated naïve T cells proliferate and differentiate to effector cells;
- A **death phase**, which occurs in the weeks after pathogen clearance — more than 90% effector T cells die in this contraction period; and
- **A memory phase**, wherein surviving antigen-specific T cells differentiate to memory cells — the number of memory cells stabilises in this phase and is maintained for a prolonged period.

9.4.1 Formation of Effector T Cells

Effector T cells are formed upon the antigen-induced proliferation of naïve T cells. These are large G1 phase cells incapable of further response without some stimulation. The initial interaction of the effector cell and its target is similar to that of naïve T cells; it is dependent upon adhesion molecules (LFA-1 and CD2 on

αβ T Cells

- αβ T lymphocytes mature in the thymus.
- Mature T cells
  - express either αβ or γδ TcRs and associated molecules,
  - generally express either the CD4 or CD8 coreceptor, and
  - express CD2, CD7, CD28, and MHC class I molecules.
- Activated T cells express CD25 and CD69 along with costimulatory molecules.
- T cells develop from $CD34^+$ common lymphoid progenitor cells.
  - In early phase development, DN cells give rise to DP cells expressing CD4 and CD8 coreceptors.
  - In the late phase of development, the transcription of one coreceptor gene is silenced to give SP T cells committed to the CD4 or CD8 lineages.
  - T cells undergo positive and negative selection during thymic development and emerge as self-MHC restricted and nonself-reactive mature cells.
- T cells are multifunctional cells.
  - T_H cells are $CD4^+$ T cells that cooperate with B cells in humoral responses and with $CD8^+$ T cells in CMI; they also help in macrophage activation.
  - T_{reg} cells are $CD4^+$ T cells that help in regulating immune responses and maintaining tolerance to self-antigens.
  - CTLs are $CD8^+$ T cells that kill infected, transformed, or mutated self-cells.
  - NKT cells and lipid-specific T cells recognise lipid antigens and are important in defence against intracellular parasites and tumours.
- Naïve T cells recognise antigens only in T cell zones of lymphoid tissues; APCs transport antigen to these areas and present them to T cells.
- The activation of naïve T cells results in their proliferation and differentiation to effector and memory cells.
  - Antigen-specific T cells recognise MHC:peptide complexes on APCs, and a tight synapse is formed at the point of contact.
  - Signalling through TcR-CD3 complexes, coreceptors (CD4/CD8), and costimulatory molecules induces IL-2 synthesis in T cells.
  - Cytokines released by APCs further aid in the process and drive T cell proliferation and differentiation to effector cells and memory cells.
- Effector T cells respond to antigenic stimulation with rapid cytokine production; they participate in cell-mediated and humoral immune responses.
- Compared to naïve T cells, memory T cells produce a broader array of cytokines and proliferate and generate effector cells more rapidly and at a lower antigen concentration.

T cells and ICAMs on APCs). This initial interaction allows more prolonged TcR–MHC:peptide engagement. Effector cells are not dependent upon costimulation. When appropriately stimulated by the antigen, these effector cells can rapidly synthesise high titres of cytokines and effector molecules, even without costimulatory signals. They also begin to synthesise DNA and progress through the cell cycle, either in response to the antigen or to cytokines such as IL-2 and IL-4. Depending upon their specialisation, these cells also participate in humoral responses ($CD4^+$ T cells) or **C**ell-**M**ediated **I**mmunity (CMI; $CD4^+$ and $CD8^+$ T cells). Effector T cells (especially $CD4^+$ T cells) are divided into type 1, type 2, type 17, or regulatory cells on the basis of their cytokine profile. We discuss them later in the section on T cell subsets. Effector T cells release cytokines and/or cytotoxins at the site of cellular contact between the target and effector cells, ensuring that effector molecules focus only on the antigen-bearing cell.

Once the immune challenge is eliminated, the generated effector cells rapidly disappear. Two mechanisms combine to bring about this disappearance.

- **Death**: A large number of effector cells are generated in the immune response; their elimination is important to avoid immune system overburdening (see the sidetrack *Giving Up and Giving In*). The mechanisms responsible for effector cell elimination are poorly understood. IFN-γ appears important to both $CD4^+$ and $CD8^+$ T cell elimination. $CD4^+$ T cells seem to be eliminated at a much slower rate than $CD8^+$ T cells, and their elimination may involve multiple cell-death-inducing mechanisms, including the Fas pathway. The absence of survival signals delivered by protective cytokines and/or molecules (e.g., CD40–CD40L interaction for $CD4^+$ T cells) may also induce death in both $CD4^+$ and $CD8^+$ T cells.

Giving Up and Giving In: AICD

The clonal activation of antigen-specific lymphocytes is central to an effective immune response. The removal of the expanded clones after the immune threat has been dealt with is equally important. Besides being a major drain on resources, the unchecked proliferation of these lymphocytes raises the risks of malignancy and autoimmunity. An immune response is therefore followed by the large-scale death of activated lymphocytes through a process termed ***A**ctivation-**I**nduced **C**ell **D**eath* (AICD). AICD is thought to be the default pathway of activated lymphocytes (whether B or T); it can be avoided only when the cells obtain the appropriate survival signals. Multiple pathways can lead to such default death.

- **Cytokine withdrawal**: After TcR ligation, T cells enter a phase of IL-2–dependent proliferation. As infection wanes, cytokines, such as IL-2, that support clonal expansion can become limiting. Insufficient IL-2 can lead to cell death unless cells are rescued by other cytokines such as IFNs or other members of the IL-2 family (i.e., IL-4, IL-7, and IL-15).
- **Fas/TNF-α pathway**: Resting T cells are relatively resistant to CD95 (Fas)-induced death. However, in the presence of IL-2, activated T cells become sensitive to CD95-induced death because of the downregulation of the CD95 inhibitor FLIP and increased expression of CD95L. Activated T cells also express TNF-α. The engagement of the death-receptors CD95L or TNF-α by their ligands (CD95 or TNFR, respectively) on adjacent cells results in cell apoptosis, or, essentially, suicide or homicide caused by adjacent cells.
- **Action of proapoptotic proteins**: The antiapoptotic Bcl-2 is downregulated in activated T cells. IL-2 family cytokines inhibit this downregulation to bestow a protective effect. Recent data suggests that most activated T cells die because of the activity of the Bcl-2 family protein, Bim, which may be aided by Bax and Bak in executing the apoptotic signal. This death signal can be over-ridden by Bcl-2.

- **Homing to nonlymphoid tissue**: Current dogma holds that memory T cells preferentially patrol the original site of infection, that is, they display tissue-specific homing. A combination of adhesion molecules, chemokines, and their receptors, expressed by endothelial cells and memory cells, are thought to result in this homing. T cell activation changes the surface expression of several molecules. Expression of the homing receptor L-selectin is lost in a subset of activated cells (termed *effector memory cells*; see below); instead, these cells express higher levels of adhesion molecules such as CD2 and LFA-1. Hence, such cells do not home to T cell areas of lymph nodes. Also, these cells express chemokine receptors such as CCR5 and CCR2 that allow them to home to the vascular endothelium at inflammatory sites. This ensures that effector T cells recirculate and scout for infections through peripheral tissues, where they may encounter sites of inflammation and where their armoury can be put to use.

9.4.2 Formation of Memory T Cells

Resting memory cells are found after the effector response has subsided. T cell memory, like B cell memory, is long-lived and can be evoked months or years after the initial antigenic encounter. However, factors that promote the maintenance of antigen-specific memory populations for prolonged periods of time remain largely unknown. Once formed, memory T cells, like memory B cells, inhibit the recruitment of naïve cells into the developing immune response. Thus, once memory cells are formed, all subsequent antigenic challenges only activate these cells. **Memory T cells are small quiescent cells that persist for months or years and respond more efficiently to antigenic challenge than do naïve cells**.

- Memory cells express the $CD44^{hi}$ marker that is usually associated with activated T cells and a different, low-MW isoform of CD45 (CD45RO).
- Upon antigenic stimulation, memory cells produce a broader set of effector cytokines, such as IFN-γ, IL-4, or IL-5, than do naïve cells; naïve T cells produce IL-2 and TNF-α in response to antigenic stimulation but produce almost no effector cytokines. The types of cytokines produced by memory cells depend upon the kind of cytokines present during initial stimulation.
- Memory cells can also proliferate and generate secondary effector cells more rapidly than their antigen-inexperienced counterparts; they show a shorter lag time for entering cell cycles, cytokine synthesis, and differentiation.
- Memory cells are also less dependent upon costimulation than naïve cells.
- Unlike BcR genes, TcR genes do not mutate after antigenic stimulus; hence, there is no affinity maturation in the conventional sense.

The source of memory T cells is unclear. Experiments have established that memory cells are generated from effector cells. The formation of memory T cells necessitates the survival of some effector T cells after pathogen clearance. Survival is promoted by administering adjuvants, but the mechanisms underlying escape from apoptosis are not clear. Thus, some normal effector cells may differentiate to memory cells. However, the factors affecting the transition of activated T cells to resting memory T cells have not been identified. There is some evidence to suggest that memory cells are derived not from normal effector T cells but from a separate subset of precursors; however, further experiments are needed to conclusively prove the ancestry of memory T cells. For years, there has been a controversy regarding the need for antigenic stimulation in memory T cell survival. The long-term survival of memory T cells has now been established to occur in the absence of antigen. IL-15 is important in this survival in the case of $CD8^+$ T cells. Survival of $CD4^+$ T cells, however, seems to be largely IL-15 independent and other cytokines may be important in their survival.

Memory T cells can be divided into two subsets on the basis of their location, proliferative responses, and activation markers. The exact relationship between these two subsets is unclear.

- **Effector memory cells**: These cells are in an overtly activated state and closely resemble effector cells. They have a rapid turnover and tend to express activation markers such as CD25 and CD69. Depending upon their polarisation status (i.e., T_{H1} or T_{H2}), they produce cytokines such as IFN-γ or IL-4. Also known as *activated memory cells*, these cells express chemokine receptors that allow them to enter nonlymphoid tissues (especially the liver, lungs, and gut). They are not found in the lymph nodes because they do not express the lymph node homing receptors CD62L and CCR7.
- **Resting or central memory cells** have a relatively slow turnover, lack activation markers, and closely resemble naïve T cells in terms of distribution. They produce mainly IL-2. These cells express CD62L and CCR7 and are found in lymph nodes. Though quiescent, these cells are activated more easily than naïve T cells.

9.5 T Cell Subsets

Conventional T cells express αβ TcRs and recognise antigens only in the context of antigen-presenting molecules. A majority of these are peptide-specific and recognise antigen in the context of MHC molecules. They fall into three functional categories — $CD4^+$ T_H cells, $CD4^+$ T_{reg} cells, and $CD8^+$ CTLs. A small fraction of αβ T cells are lipid specific and recognise antigen in the context of CD1. Those that express NK cell receptors and are called NKT cells. T cells expressing γδ TcRs recognise antigens independent of presenting molecules and are considered part of the innate immune system (Mind map 9.1).

9.5.1 T_H Cells

The naïve $CD4^+$ T cell is a multipotential precursor. Its antigen recognition specificity is fixed, but it has substantial plasticity, which allows it to develop along distinct effector or regulatory pathways. The microenvironment in which a T cell encounters antigenic stimulation guides this development, and signals from the innate immune system play a major role (Figure 9.4). **Three broad subsets of effector T_H cells are recognised — T_{H1}, T_{H2}, and T_{H17}** — on the basis of their cytokine-secreting patterns and functions (Table 9.2). The key word is *broad*. Not all T cells belong to one of these categories and may instead produce a mixed cytokine profile. The distinction between the subsets is essentially phenotypic and is based on the profile of the cytokines expressed — IFN-γ, TNF-α, TNF-β, and IL-2 are signature cytokines of T_{H1} cells; IL-4, IL-9, IL-10, and IL-13 are hallmark cytokines of T_{H2} cells although they also produce IL-5 and IL-6; and T_{H17} cells secrete IL-17, IL-21, and IL-22.

- Nonpolarised naïve T cells that have not yet developed into effector cells may also be referred to as T_{H0} cells.
- **T_{H1} cells are typically involved in CMI responses** against intracellular pathogens (e.g., *Listeria monocytogenes* or *Mycobacterium* spp.). If unchecked,

type 1 responses result in organ-specific autoimmunity and **D**elayed-**T**ype **H**ypersensitivity (DTH; described in chapters 15 and 16, respectively).

- **TH2 cells arm epithelial and mucosal sites against pathogens**. TH2 responses favour the elimination of parasites and helminths. If gone awry, type 2 responses culminate in immediate-type hypersensitivities (e.g., allergies and asthma, see chapter 16).
- TH17 cells are the latest addition to the TH cell stable. **TH17 cells and their effector cytokines mediate the host defensive mechanisms to various infections**, especially to extracellular bacterial infections. They are also involved in the pathogenesis of many autoimmune diseases.

In spite of extensive research, markers that can distinguish the three subsets have proven difficult to establish, and the patterns of secreted cytokines continue to be the basis for identifying them. A quantitative difference in the expression of some molecules has been reported for the TH1/TH2 subsets, although even that is not known for the TH17 subset. The expression of CCR5 and CXCR3 is associated with TH1 cells, whereas CCR3, CCR4, CXCR4, CCR8, and ICOS are thought to be expressed more by TH2 cells. Recent reports suggest that the differential expression of TIM molecules (**T** cell **I**mmunoglobulin- and **M**ucin-domain-containing molecule) can distinguish murine TH1 and TH2 subsets. Thus, murine TH1 cells express TIM-3, whereas TH2 cells express TIM-2. The human equivalent of TIM-2 has not yet been identified, and the role of TIM molecules in TH subset development and/or functioning is still being investigated.

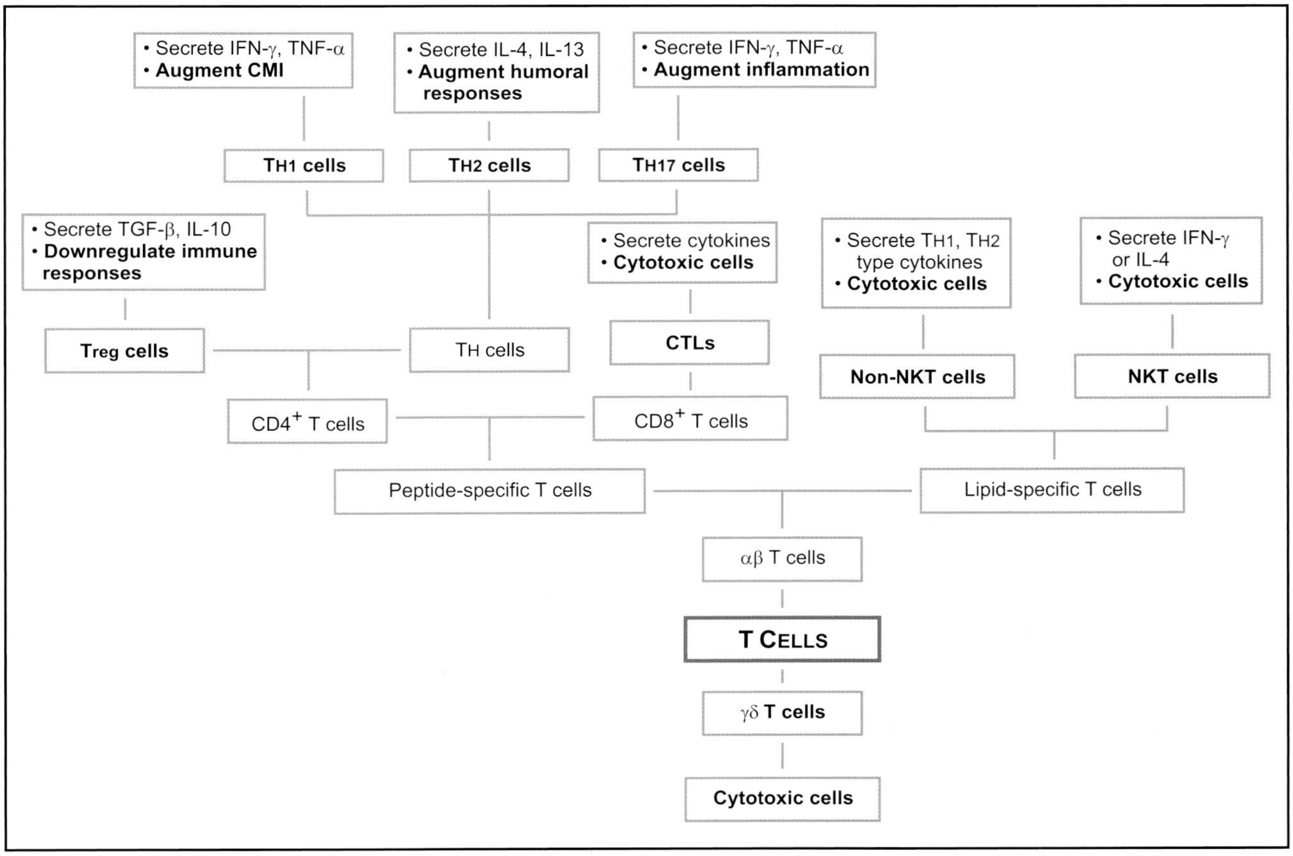

Mind Map 9.1 Types of T cells

Table 9.2 Features of TH Cell Subsets

Feature	TH1 cells	TH2 cells	TH17 cells
Signature cytokines	IFN-γ	IL-4, IL-5, IL-13	IL-17, IL-21, IL-22
Priming cytokines	IL-12	IL-4	TGF-β, IL-6
Autocrine cytokines	IFN-γ	IL-4	IL-21
Transcriptional regulators	T-bet, STAT-1, STAT-4	GATA-3, STAT-6	RORα, RORγt, STAT-3

In vitro, the type of TH cell generated is influenced by the cytokine cocktail in the microenvironment. **TH1 cells are preferentially obtained when CD4$^+$ T cells are cultured in the presence of IFN-γ or IL-12. IL-4 is required for the generation of the TH2 phenotype**. When both IL-12 and IL-4 are present in the same environment, the effect of IL-4 is dominant over that of IL-12. IL-4 partly induces this effect by reducing IFN-γ production. IL-13 is also a very powerful inducer of TH2 cells *in vitro*. **TGF-β and IL-6 are important for the development of murine TH17 cells; IL-23 and IL-6 (and not TGF-β) are needed for development of human TH17 cells.**

Development of TH Subsets

The recognition of peptide: MHC class II complexes by TcRs with the engagement of costimulatory molecules activates naïve CD4$^+$ T cells and results in their proliferation. Naïve T cells are uncommitted cells that can develop into TH1, TH2, or TH17 cells, depending upon the type of antigen being presented, the strength of the signal, the presenting cell, costimulatory molecules engaged, and the cytokine milieu (Figure 9.4). The innate immune system has a major role in this development, as the cells of the innate immune system dictate the antigen presentation, costimulation, and, to some degree, cytokine environment during the initial activation of naïve T cells. It seems probable that naïve T cells undergo a few rounds of proliferation before being committed to a particular phenotype by the cytokine milieu.

TH1 responses seem to be driven primarily by active signals derived from pathogen-activated innate immune sources, and there are indications that a strong TH1 response is maintained only in the presence of such signals. The major stages of TH1 commitment and development are given below (Figure 9.4)

- The invading pathogen activates cells of the innate immune system to secrete IFN-γ and IL-12.
  - IFN-γ, secreted by macrophages or NK cells, commits naïve cells to the TH1 phenotype.
  - IFN-γ acts through the STAT-1 (**S**ignal **T**ransducer of **A**ctivation and **T**ranscription-**1**) signalling pathway to increase expression of the transcription factor T-bet in proliferating cells.
- T-bet (T-box family transcription factor) is expressed in developing and committed TH1 cells. It is central to TH1 development.
  - The gene for IFN-γ expression is normally repressed in T cells; T-bet expression results in the remodelling of the gene encoding IFN-γ to an active status.

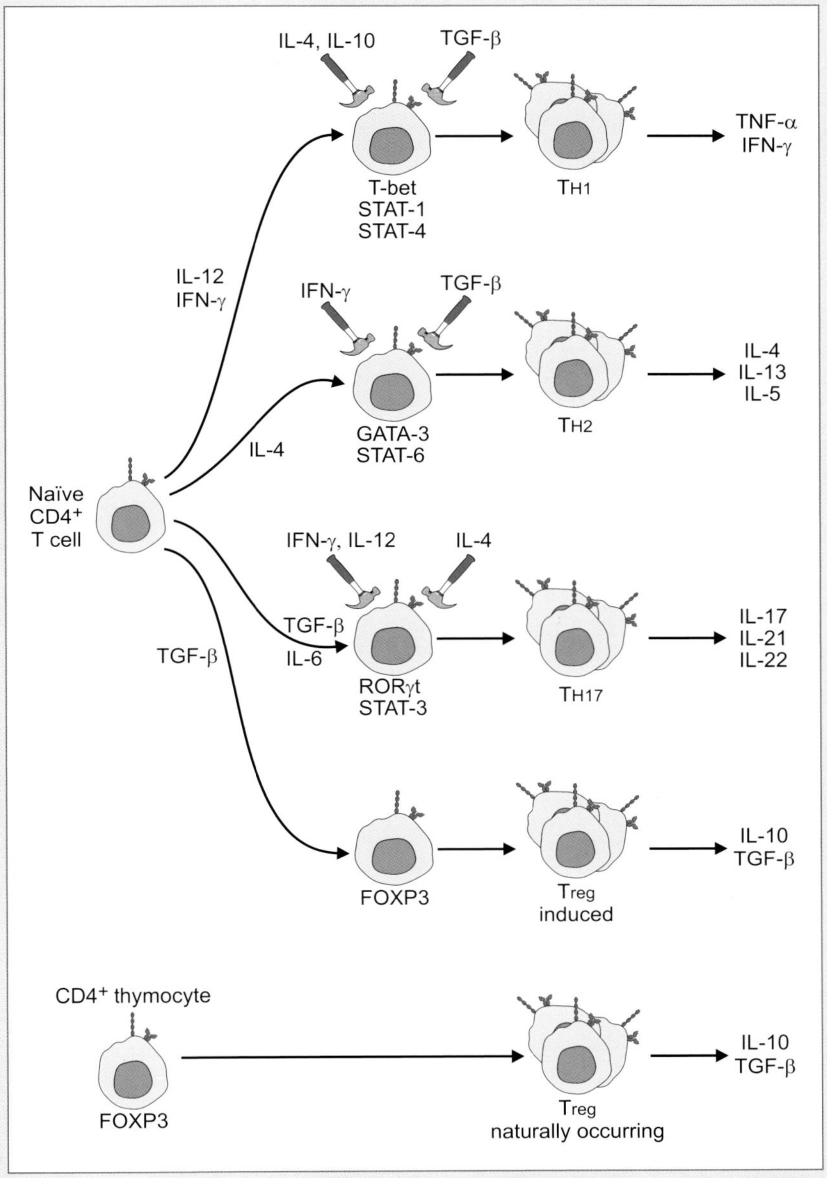

Figure 9.4 Antigen recognition, costimulation, and appropriate cytokine signalling trigger the formation of effector CD4+ T cells. Effector T cells are formed after repeated rounds of proliferation of antigen-activated naïve CD4+ T cells. Naïve CD4+ T cells are uncommitted, multipotent cells. Commitment to the TH1, TH2, TH17, or Treg lineages is dependent upon their cytokine milieu. IFN-γ and IL-12 promote the TH1 pathway by inducing T-bet transcription through the transcription factors STAT-1 and STAT-4. Cells eventually differentiate to produce IFN-γ and TNF-α. IL-4, IL-10, and TGF-β inhibit the commitment to this pathway. Commitment to the TH2 lineage occurs if cells are exposed to IL-4. IL-4 upregulates GATA-3, the master regulator of the TH2 pathway, through STAT-6. Effector TH2 cells produce IL-4, IL-13, and IL-5 upon stimulation. IFN-γ and TGF-β are antagonistic to the development of the TH2 pathway. The exposure of murine T cells to IL-6 and TGF-β commits them to the TH17 lineage. Cells produce IL-21 through STAT-3-mediated signalling. This cytokine acts in an autocrine fashion and promotes the transcription of RORγt, the master regulator of this pathway. TH17 cells produce the IL-17 family of cytokines (IL-17, IL-21, and IL-22). IFN-γ and IL-12 of the TH1 pathway and IL-4 of the TH2 pathway inhibit the development of Treg cells. Commitment to the Treg lineage may occur in the thymus (naturally occurring Treg cells) or in the periphery (induced Treg cells). FOXP3 is the master regulator of Treg cells. Whether naturally occurring or induced, Treg cells produce immunosuppressive cytokines such as TGF-β and IL-10. Cytokines produced by the different subsets cross-regulate each other's development and functioning.

- T-bet induces the expression of the IL-12 receptor and increases the responsiveness of activated T cells to IL-12.
- Subsequently, T-bet stabilises its own expression.

❒ IL-12 induces the transcription factor STAT-4. IL-12 engages IL-12R and phosphorylates STAT-4, resulting in the further augmentation of T-bet and IFN-γ transcription.

- Activated macrophages, DCs, and NK cells are major sources of IL-12. They can thus influence the polarisation of the T cell response.
- Human (but not murine) type I IFNs activate STAT-4 and augment T-bet and IFN-γ transcription.
- IL-12 acts directly on TH1 cells and their precursors; it also acts indirectly by inducing the production of IFN-γ by T and NK cells in cooperation with TNF-α and IL-1.
- IFN-γ, in turn, has a positive feedback effect on IL-12 production by monocytes and macrophages.

❒ In T cells that have already differentiated to the TH1 type, IFN-γ production can occur because of TcR engagement or through the combined stimuli of IL-12 and IL-18. These cytokines can induce IFN-γ synthesis even in the absence of TcR engagement. Such IFN-γ induction is strongly dependent on STAT-4.

❒ Recently, three other cytokines — IL-18, IL-23, and IL-27 — have been shown to be TH1-promoting factors.

- IL-12 induces the expression of IL-18R, making the cell responsive to IL-18. IL-18, produced primarily by mononuclear phagocytes, synergises with the IL-12 in committed TH1 cells to increase IFN-γ production. IL-18 and IL-12 can induce IFN-γ production from TH1 cells in the absence of TcR signalling.
- IL-23, an IL-12-related cytokine, seems to increase the antigen-presenting capacity of DCs. It also promotes IFN-γ production and proliferation in memory TH1 responses by the STAT-4 pathway.
- IL-27, produced by APCs, acts in conjunction with IL-12 in promoting IFN-γ production and is thought to be involved in early TH1 development.

Unlike commitment to the TH1 phenotype, commitment to and development of the TH2 phenotype is less dependent upon signals from the innate immune system. Whether any active innate signals drive this process is still uncertain. **TH2 development is thought to occur as a default pathway in the absence of inhibition by innate immune signals or in response to an extrinsic source of IL-4**. Activated DCs, macrophages, and NKT cells are the most likely source of external IL-4. Costimulatory signals provided by CD28 and ICOS are important in the development of TH2 pathways.

❒ IL-4 has been shown to be essential in the development of the TH2 subset; IL-4 induces the differentiation of naïve cells to the TH2 phenotype *via* STAT-6.

❒ Some alternative signals such as IL-6 or IL-13 may trigger the initial production of IL-4.

❒ IL-4 rapidly induces the expression of TH2 cell-specific factor GATA-3 through STAT-6 signalling.

- GATA-3 is the master regulator of TH2 differentiation, although GATA-3 also seems essential in normal thymocyte development and embryonic survival.
- Naïve T cells constitutively express GATA-3 at a low level, but its expression is dramatically increased in TH2 cells.

- GATA-3 induces remodelling of the IL-4 locus and promotes the expression of several TH2 cytokines (IL-4, IL-5, IL-9, and IL-13) through the coordinated expression of linked genes. This remodelling is heritable — daughter cells inherit the remodelled IL-4 locus and therefore remain of the TH2 type.
- CD28 costimulation augments the expression of GATA-3, whereas the engagement of LFA-1 inhibits its expression.
- GATA-3 has a transcriptional autoactivating property, which leads to a massive upregulation of *GATA-3* gene transcription by GATA-3 protein.

- Several members of the NFAT family seem to regulate the expression of TH2 cytokines after the triggering of differentiated TH2 cells through TcR engagement.
- IL-5 and IL-6, produced by TH2 cells, act synergistically with IL-4 in potentiating TH2 responses.

Relatively little is known about the developmental pathways of TH17 cells. As for TH2 cell development, costimulatory signals provided by CD28 and ICOS engagement seem essential for TH17 development. **In mice, TH17 development is induced by the combined action of IL-6 and TGF-β** and is further sustained and regulated by the IL-21 expressed by T cells in an autocrine manner. The crucial initiating cytokines for human TH17 cell development remain unclear. Although IL-23 was originally shown to be important for the proliferation of TH17 cells and for TH17 cell-mediated immune diseases, it does not seem to be necessary for the initiation of TH17 cell differentiation. **IL-6, IL-1, and IL-23 collectively mediate the differentiation of human TH17 cells.**

- The exact signalling events and their sequence for TH17 differentiation are not clear. In mice, naïve T cells activated in the presence of TGF-β and IL-6 begin differentiation towards the TH17 pathway by inducing IL-17. Treg cells are

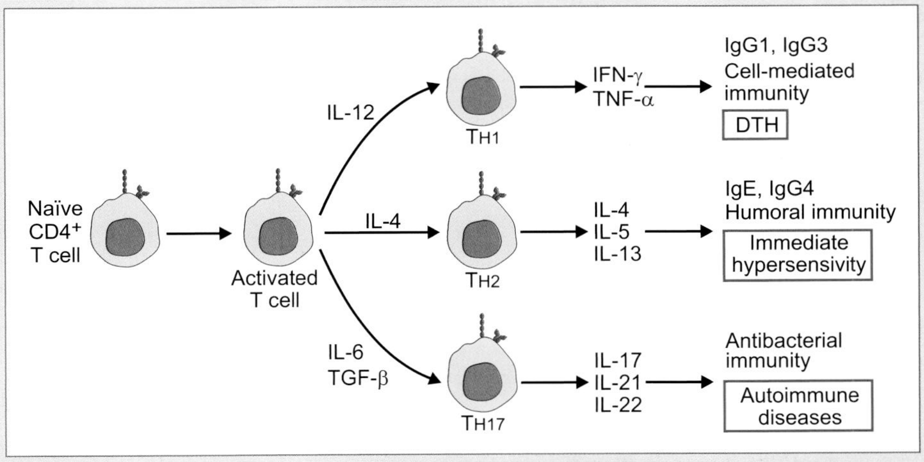

Figure 9.5 TH cells are multifunctional cells that are important in mediating a variety of immune functions. Three major subsets of $CD4^+$ T helper cells are recognised on the basis of their cytokine-secreting patterns and functions. IFN-γ, TNF-α, and TNF-β are signature cytokines of TH1 cells. These cells promote cell-mediated immune responses. They are important in the activation of macrophages and the generation of memory CTLs. In humans, they also promote switching to the IgG1 and IgG3 isotypes. If they go awry, TH1 type responses result in DTH responses. IL-4, IL-5, and IL-13 are considered signature cytokines of TH2 cells. They are associated with strong antibody responses and promote IgE and IgG4 isotypes. The hyperactivation of TH2 type responses leads to increased IgE antibody levels and hypersensitivity type I reactions. The IL-17 family of cytokines are the signature cytokines of TH17 cells. They seem to be important in defence against bacterial infections. Over-stimulation of these cells can result in inflammatory autoimmune disorders.

postulated to be the source of TGF-β, and this cytokine alone, rather than in combination with IL-6, results in differentiation to T_{reg} cells rather than T_{H17} cells.

- ❒ IL-6 upregulates the IL-21 and IL-23 receptors. After its induction by IL-6, IL-21 synergises with TGF-β to promote T_{H17} cell differentiation. IL-23 is thought to be involved in the final stages of these cells' differentiation and survival.
- ❒ STAT-3 is a crucial component of the IL-6-mediated regulation of T_{H17} cells. The IL-6-induced production of IL-21 depends on STAT-3 expression.
- ❒ RORγt, a member of the **R**etinoic-acid-receptor-related **o**rphan nuclear hormone **r**eceptor family, is the master regulator of T_{H17} differentiation. Mice deficient in RORγt are defective in T_{H17} differentiation. RORγt expression is induced by TGF-β and IL-6.

TH subsets act in an antagonistic fashion, suppressing cytokine induction in each other and cross-regulating each other's functions. The transcription factors T-bet and GATA-3, the master regulators of T_{H1} and T_{H2} subsets, operate intrinsically to suppress opposing cytokines. T_{H2}-inducing GATA-3 signalling extinguishes IL-12/STAT-4 signalling in T_{H1} cells. Conversely, T_{H1}-inducing T-bet extinguishes IL-4-STAT-6–mediated signalling in T_{H2} cells. Naïve T cells are capable of expressing both T-bet and GATA-3. With the loss of one of these master regulators,

TH Cells

- ❒ They express the αβ TcR and the CD4 coreceptor and therefore recognise antigen in the context of MHC class II molecules.
- ❒ Three broad subsets of TH cells are recognised on the basis of their cytokine profiles.
  - ➢ IFN-γ, TNF-α/β, and IL-2 are considered signature cytokines of T_{H1} cells.
  - ➢ IL-4, IL-10, and IL-13 are considered signature cytokines of T_{H2} cells.
  - ➢ T_{H17} cells produce cytokines of the IL-17 family.
- ❒ T cells get committed to the T_{H1} lineage because of the action of the IFN-γ and IL-12 produced by cells of the innate immune system.
  - ➢ Exposure to these cytokines results in the expression of the transcription factor T-bet.
  - ➢ T-bet remodels the IFN-γ gene to an active status.
  - ➢ IL-12 induces the transcription factor STAT-4, which further augments T-bet and IFN-γ transcription.
  - ➢ T_{H1} cells are pro-inflammatory cells that augment CMI. Uncontrolled T_{H1} responses may result in DTH and organ-specific autoimmunities.
- ❒ T_{H2} type cells develop in response to extrinsic IL-4 or as a result of a default pathway in the absence of inhibitory signals from cells of the innate immune system.
  - ➢ IL-4 induces the expression of GATA-3 *via* the transcription factor STAT-6.
  - ➢ GATA-3 induces remodelling of the IL-4 locus and promotes the expression of T_{H2} cytokines such as IL-4, IL-5, IL-9, and IL-13.
  - ➢ T_{H2} cells are associated with strong antibody responses; IL-4, IL-5, and IL-6 are important in B cell maturation and differentiation.
  - ➢ The hyperactivation of T_{H2} responses results in allergic responses and asthma.
- ❒ In mice, T_{H17} cells develop in response to TGF-β and IL-6 priming.
  - ➢ RORγt is the master regulator of T_{H17} differentiation.
  - ➢ IL-6 upregulates IL-21R and IL-23R.
  - ➢ IL-21 synergises with TGF-β to promote T_{H17} differentiation.
  - ➢ STAT-3 is a crucial component of IL-6 mediated T_{H17} signalling.
  - ➢ T_{H17} cells are important in defence against extracellular bacterial infections.
  - ➢ They are also important mediators of inflammatory and autoimmune diseases.

SMS, Cell-to-Cell: Cytokines

The term *cytokine* designates a heterogeneous group of nonenzymatic proteins produced by a wide variety of lymphoid and nonlymphoid cells. Cytokines are water-soluble, glycosylated, low MW proteins (<80 KD) generally secreted by cells to alter either their own functions (autocrine effect) or the functions of other types of cells (paracrine effect). Discovered because of their role in immunological phenomena, such as inflammation and hypersensitivity, cytokines have now been shown to be soluble mediators of inter- and intra-cellular communications. They regulate development, tissue repair, haematopoiesis, inflammation, and innate and adaptive immune responses. In many respects, cytokines behave like classical hormones. They act on a systemic level, affecting the neuroimmune network, growth, and development. However, classical hormones are produced by specialised glandular tissues and have a relatively small spectrum of target cells. Unlike hormones, diverse types of cells dispersed in the body produce cytokines, and they are effective on an equally wide spectrum of cells. Cytokines were originally isolated and named on the basis of their functions, with the result that the same cytokines have been given different names. To further add to the confusion, cytokines were also referred to as lymphokines, monokines, and so forth, to indicate their cellular sources. The terms *interleukin* and *chemokine* are hangovers of the same system. *Interleukins* was the name given to cytokines of leukocytic origin, whereas cytokines with chemotactic properties were called *chemokines*. Mercifully, a common nomenclature has been adopted. In appendix II, we give a brief overview of some important cytokines that we refer to in this book; it is by no means an exhaustive list. For convenience, we have also included some of the more popular aliases in this listing.

Cytokines regulate both the intensity and the duration of the immune response, and they mediate this effect through specific receptors on target cells. Cytokines have a very short half-life and are extremely potent — they are effective in picomolar (10^{-9} M) concentrations. Not surprisingly, the expression of most cytokines is strictly regulated. Although some cytokines are produced constitutively, the majority are produced only by activated cells and, even then, only in response to specific activation signals. Their expression is normally transient and can be regulated at all levels of gene expression (transcriptional and/or translational). Expression may also be differentially regulated, depending upon cell type and developmental stage. Cytokines have multiple overlapping effects on cells — they are pleuripotent. They are also redundant, that is, more than one cytokine can generate the same effect. Together, cytokines form an intermingling network wherein one cytokine can influence the production of and response to many other cytokines. Some cytokines are synergistic/additive in their action (e.g., IL-4 & IL-6 and IL-5 & IL-6). The presence of one induces receptors for the other or may even induce the production of the other. Others, such as IL-4/IFN-γ and IL-12/IL-4, are antagonistic. The type, duration, and extent of the cellular activities induced by a particular cytokine are influenced by several factors. These include the concentration of that cytokine, the microenvironment and developmental stage of the target cell, and other cytokines present in the microenvironment of that cell.

Cytokines bind to specific receptors expressed by target cells. Many of these receptors share a number of characteristics and subunits. They also share common signal transducing components, which explains, at least partially, their redundancy. Five broad classes of receptor families are recognised.

- The type 1 receptor family (no relation of the TH type 1 subset) has a conserved extracellular domain and shares sequence motifs. It includes IL-2, IL-3, IL-4, IL-5, IL-6, IL-7, IL-9, IL-11, IL-12, and GM-CSF.
- IFN-α, IFN-β, IFN-γ, and IL-10 are prominent type 2 cytokine receptors. For both type 1 and type 2 cytokine receptors, the cytokine-binding subunit (called the α subunit) is often different from the signal-transducing unit (referred to as the β or γ subunit). For example, the receptors for IL-2, IL-4, IL-7, IL-9, and IL-15 share the same γ subunit. From a cell's point of view, this system is economical as the same signal-transducing subunit can associate with different α (cytokine-recognising) subunits, allowing the same machinery to be used for different stimuli.
- Ig superfamily receptors have at least one Ig domain. The receptors for IL-1, M-CSF, and c-kit belong to this category.

- The TNF family of receptors consists of receptors for TNF-α, TNF-β, CD40, and Fas.
- Chemokine receptors exclusively bind chemokines and have a transmembrane domain that transverses the membrane seven times (see the sidetrack *Instant Chemistry*).

The receptors of most cytokines secreted by TH cells belong to type 1 or 2. Signal transduction through these receptors occurs by the same generalised pathway (often called the Jak/STAT pathway).

- The docking of the cytokine to its receptor causes the cytokine-induced di/trimerisation of that receptor.
- The intracellular domains of receptor subunits are often associated with different tyrosine kinases. Janus kinases are perhaps the most important of these. Named after the Greek God Janus*, kinases of this family have two sites — one that associates with the receptor and the other that acts as a tyrosine kinase when activated.
- Upon docking, **Ja**nus **k**inases (Jaks) phosphorylate receptor subunits.
- Members of a family of transcription factors termed *STATs* bind to phosphorylated tyrosine residues — STAT-1 is involved in IFN-γ signal transduction and STAT-4 in IL-12 signal transduction. STAT-6 transduces the IL-4 signal; STAT-3 is important in IL-6 signalling.
- Docked STATs undergo phosphorylation at a tyrosine residue.
- Phosphorylated STATs dissociate from the receptor docking sites, dimerise, and phosphorylate further at a serine residue.
- STAT dimers translocate to the nucleus, where they initiate the transcription of specific genes.

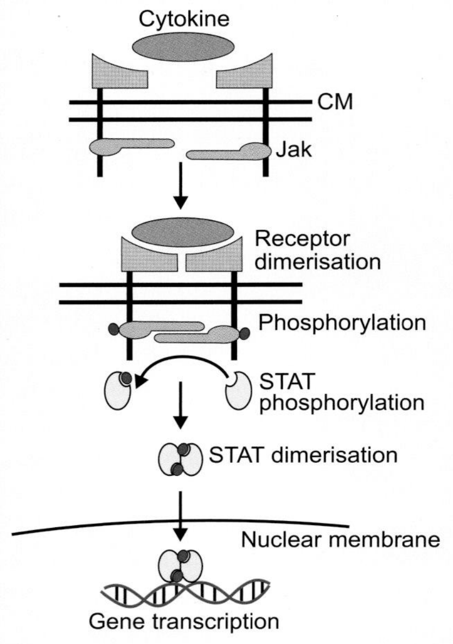

Figure 9.S2 The Jak/STAT pathway is a general pathway for cytokine signal transduction.

Cytokine inhibitors are found in the blood and extracellular fluids of healthy individuals. These proteins interfere with cytokine activity by binding either the cytokine itself or its receptor. They are being investigated for their therapeutic potential. The best-characterised inhibitor is the IL-1R antagonist that binds to the IL-1 receptor without activating the signalling cascade. It is thought to play a role in regulating the intensity of the inflammatory response. Many antagonists arise from the enzymatic cleavage of cytokine receptors. The released extracellular domains act as soluble receptors. The soluble receptor binds to the cytokine's receptor-binding site, inhibiting its binding to membrane-anchored receptors. Examples include IL-2, IL-4, IL-6, IFN-γ, TNF-α, and TNF-β receptors.

* These were originally named Jak (for **J**ust **a**nother **k**inase) because this family of tyrosine kinases was isolated in a random screen and the functions of these kinases remained unknown. Later, it was revealed that they are critical messengers in cytokine responses, and they were recast as Janus Kinases in reference to the double-headed Greek god of gates and doorways.

however, the capacity to sustain downstream cytokine gene expression is lost. The lineages become established in response to the remaining regulator, resulting in terminal differentiation into polarised TH subsets. Additionally, IL-10, produced by TH2 cells, suppresses IL-12 production of the APCs, inhibiting TH1 development. IL-10 can also suppress the production of effector molecules by macrophages and downregulate their MHC class II expression. The immunosuppressive TGF-β produced by Treg cells can suppress the expression of both T-bet and GATA-3, halting the development and commitment of naïve T cells. TH1 and TH2 cells also appear to negatively regulate the differentiation of TH17 cells. The addition of either IL-12, or IFN-γ, or IL-4 to cultures inhibits the differentiation of murine and human TH17 cells. IL-25 produced by TH2 cells and IL-2, required by most T cells for their proliferation, also inhibits TH17 development. Treg cells and their master regulator, FOXP3, also downmodulate TH17 development.

Functioning of TH Cells

An important TH cell function is recognising peptide: MHC complexes on antigen-specific B cells and helping antibody production. The anatomical separation of naïve B and T cells in the secondary lymphoid organs necessitates their movement to fulfill this function. Antigen-specific naïve $CD4^+$ T cells rapidly express the chemokine receptor CXCR5 in response to adjuvant-activated DCs. The expression of CXCR5 is associated with the downregulation of the lymphoid homing receptors CCR7 and CD62L. CXCR5 allows naïve TH cells to home to B cell areas that express the CXCR5 ligand BCA-1 (**B C**ell **A**ttracting chemokine-**1**), enabling T cells to provide the signals required for B cell stimulation.

As explained, **TH1 cells are pro-inflammatory cells that augment CMI**. Several TH1 cytokines activate cytotoxic and inflammatory functions and induce DTH responses. Studies provide strong evidence that $CD4^+$ T cells provide essential signals for the optimal generation of $CD8^+$ T cell memory; however, the precise nature of these signals remains elusive. Some TH1 cells have also been shown to be cytotoxic. **The activation of macrophages is the most important effector function of these cells**. Two signals are involved in this activation. IFN-γ, produced by TH1 cells and in some instances $CD8^+$ T cells, provides the first signal. The engagement of CD154 on the surface of TH1 cells by CD40, expressed on macrophages, provides the second signal. Membrane-bound forms of TNF-α or TNF-β may also deliver the second signal; antibodies to either can replace CD40 ligation in macrophage activation *in vitro*. LPS helps in activation by making the macrophages more responsive to IFN-γ. TH1-derived IL-3 and GM-CSF aid CMI by recruiting fresh macrophages to the infection site. TNF-α and TNF-β, also produced by these cells, increase the expression of CAMs on endothelial cells, promoting the adhesion of phagocytic cells to these endothelial cells. Other cytokines such as MIF (macrophage **M**igration **I**nhibitory **F**actor) and MCF (**M**acrophage **C**hemotactic **F**actor) promote macrophage chemotaxis and accumulation at the site of inflammation. Activated macrophages are potent antimicrobial cells capable of secreting a variety of effector molecules (Table 2.2). In addition, TH1 cells can induce apoptosis in target cells by the Fas–FasL pathway. They may thus have a role in maintaining T cell homeostasis. TH1 cells also promote opsonising antibody production by B cells (IgG2a and IgG2b in mice; IgG1 and IgG3 in humans), but higher TH1 numbers may suppress B cell responses.

TH2 cells are associated with strong antibody and allergic responses. TH2 cytokines encourage antibody production, particularly that of noncomplement-fixing IgG2a, IgA, and IgE isotypes in mice and IgG4, IgA, and IgE isotypes in humans. Type 2 cytokines such as IL-4, IL-6, and IL-10 are important in B cell maturation and differentiation. IL-5 and IL-6, produced by these cells, also enhance eosinophil proliferation and function. Both IL-4 and IL-13 enhance IgE-antibody production. IL-9, another TH2 cytokine, is a mast cell growth factor, and it synergises with IL-5 in promoting eosinophil maturation. Recent experiments demonstrate that macrophages can be activated by IL-4. Termed *alternatively activated macrophages* (to differentiate them from the classically IFN-γ activated macrophages), these cells seem to promote the resolution of inflammation and a reduction in T cell proliferation. The **hyperactivation of TH2-type responses leads to increased IgE antibody levels and hypersensitivity type I reactions**.

TH1/TH2 balance is important in the manifestation and progression of some diseases.

- Balb/c mice mount a predominantly T_{H2} type response to the intracellular parasite *Leishmania major*. The ensuing failure to activate macrophages results in susceptibility to the protozoan parasite. C57/Bl6 mice, on the other hand, give a predominantly T_{H1} response. CMI is activated and protects them from infection.
- Infection by *Mycobacterium leprae* in humans leads to tuberculoid leprosy if a T_{H1} type response is elicited. The infection results in tuberculoid granulomas, in which macrophages control the growth of the pathogen. Conversely, a T_{H2} type response results in a widely disseminated bacterial response. The predominantly humoral response is unable to control the spread of the pathogen, and the infection damages the nervous system.

Our understanding of T_{H17} cell functions is still rudimentary. These cells are the best-characterised source of IL-17 cytokine family members (IL-17A through F; IL-25 is also known as IL-17E). **T_{H17} cells and their effector cytokines have both pathological and protective roles during inflammation**. These cells mediate host defense to various infections, especially extracellular bacterial infections. They are also important mediators in inflammatory and autoimmune diseases such as **E**xperimental **A**utoimmune **E**ncephalitis (EAE; the murine equivalent of multiple sclerosis), SLE, and colitis (Figure 9.5).

9.5.2 T_{reg} Cells

T cells that could suppress immune responses were described in the early 1970s. However, the failure to clone factors that suppressed immune responses in an antigen-specific manner led to the demise of the entire suppressor T cell field. The discovery that immunosuppressive cells occurred naturally *in vivo* and that the adoptive transfer of cells suppressed immune responses in the new host rekindled interest. These cells are termed *regulatory* rather than *suppressor* T cells. Several *in vivo* and *in vitro* treatments generate T_{reg} cells, but many questions about them remain unanswered.

The literature currently describes two types of T_{reg} cells.

- **Naturally occurring T_{reg} cells develop in the thymus and are $CD4^+CD25^+$ T cells**. They display a diverse TcR repertoire that is specific for self-antigens. Such constitutive T_{reg} cells express FOXP3 (forkhead box P3), the key regulator for their development. FOXP3 knock out mice and individuals that lack FOXP3 develop a profound autoimmune-like lymphoproliferative disease, illustrating the importance of T_{reg} cells in the maintenance of peripheral tolerance.
- **T_{reg} cells can also be induced during inflammatory processes** in peripheral tissues. Two types of induced T_{reg} cells have been described.
  - **T_{R1} cells** downregulate immune responses because of their ability to produce high levels of IL-10 and TGF-β. The coculture of naïve T cells in the presence of T_{R1} cells and antigen-pulsed APCs leads to the suppression of naïve T cell proliferation.
  - **T_{R3} cells** are generated by the introduction of low doses of oral antigen. These cells are thought to act primarily by secreting TGF-β, although they may secrete small amounts of IL-10.

The mechanisms by which T_{reg} cells exert their effect is unclear. From a purely functional viewpoint, the suppressive mechanisms used by T_{reg} cells can be grouped as follows:

- **Suppression by inhibitory cytokines**: This is by far the most studied mechanism of suppression that is induced by T_{reg} cells. Both TGF-β and IL-10

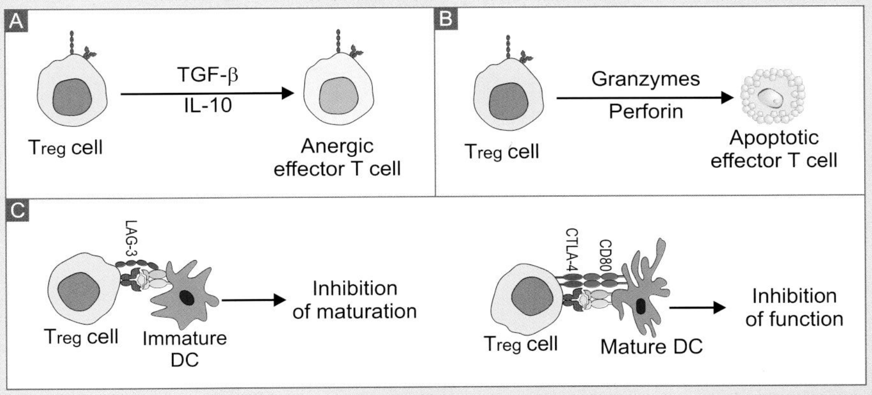

Figure 9.6 Treg cells exert their action in numerous ways. Antigen-activated Treg cells can secrete TGF-β or IL-10, cytokines that downmodulate immune responses in various ways (A). Treg cells can also induce cytolysis in B cells and effector T cells (B). Recent data suggests that Treg cells can prevent the maturation of DCs *via* LAG-3, a homologue of CD4 (C, left panel). The engagement of the CTLA-4 expressed on Treg cells by costimulatory molecules on DCs interferes with DC functioning (C, right panel).

have been implicated in Treg-mediated immunosuppression. IL-10 decreases the antigen-presenting capacity of APCs. It inhibits the secretion of cytokines and chemokines that influence T cell differentiation, proliferation, and migration. It also inhibits IL-2, IL-5, and TNF-α secretion by T cells. It can induce anergy in antigen-specific T cells. TGF-β is known to suppress IFN-γ production, decrease MHC class II expression, and limit NO and ROI production by macrophages. It can also inhibit T cell proliferation, cytokine production, and cytotoxicity. Both naturally occurring and induced Treg cells produce inhibitory cytokines in animal models of asthma and allergies. IL-10/TGF-β production by Treg cells has also been shown to be important in controlling infections such as tuberculosis and malaria and autoimmune disorders such as inflammatory bowel disease. Suppressive cytokine production by these cells may not always be beneficial to the host. Tumour microenvironment promotes the generation of FOXP3$^+$ Treg cells that mediate immunosuppression by IL-10 and TGF-β.

- **Suppression by cytolysis**: Treg cells induce cytolysis by the granzyme pathway in B cells. They can also induce apoptosis in effector T cells through the TRAIL pathway.
- **Suppression by DC maturation or function**: Treg cells are thought to induce DCs to express immunosuppressive molecules (e.g., LAG-3, a CD4 homologue). Several studies suggest that Treg cells may also downmodulate the capacity of DCs to activate effector T cells by downmodulating CD80/86 expression. Treg cells are also thought to prevent DC maturation (Figure 9.6).

9.5.3 CTLs

Cytolytic lymphocytes such as NK cells and CD8-expressing T cells provide a potent defence against viral infections and intracellular pathogens that are inaccessible to antibodies. CTLs are the body's major antigen-specific defence against intracellular parasites. **They are MHC class I restricted**; they recognise the antigen only in the context of MHC class I molecules. As virtually all nucleated cells of the body express MHC class I molecules, this restriction of CTLs ensures that any cell infected by a virus or parasite or that has undergone transformation (e.g., tumour cell) is a potential target for CTLs.

Lessons in Subversion: Virokines

Originally used to describe virally encoded cytokine homologues, the term *virokine* is now often applied to virally encoded proteins that are secreted from infected host cells. Most virokines are powerful immunomodulatory agents created by the capture and modification of the genes responsible for regulating host immune response. They are often homologous to, but more potent than, host proteins. This makes them attractive candidates for immunotherapy. Large DNA viruses such as poxviruses and herpes viruses are a virtual cornucopia of virokines. Virokines modulate different aspects of the host immune system.

- Virokines aid in immunoevasion by inactivating pro-inflammatory cytokines.
- They redirect the immune response away from the virus.
- They induce cell proliferation and migration, helping the virus spread.
- Virokines control homeostasis to enhance viral replication.

Virokines fall into five broad groups, depending upon their mode of action.

- **Complement regulatory proteins**: A complement regulatory protein produced by the vaccinia virus was one of the first virokines discovered. Known as the vaccinia complement protein, this virokine is perhaps the best characterised. It is functionally similar to regulators of the complement cascade (Factor H, C4bp, MCP, DAF, and CR1). The vaccinia complement protein has been shown to block antibody-mediated virus neutralisation. Through its blocking of complement activation, it reduces cellular influx and inflammation at the site of infection. VCP also binds heparin-like molecules on human endothelial cells, interfering with chemotactic responses and, possibly, with interaction with cytotoxic cells. Other complement regulatory virokines are listed in Table 3.2.
- **Cytokine homologues**: Several viruses encode cytokine homologues, although the advantage of producing these proteins is not always clear.
  - The human herpes virus 8, which causes Kaposi's sarcoma, encodes a homologue of human IL-6. It induces cell proliferation and is more efficient in signal transduction than IL-6.
  - The human herpes virus 8 also induces the secretion of a homologue of the chemokine MIP-2.
  - IL-10 homologues (vIL-10) are encoded by several other herpes viruses such as the Epstein-Barr virus and cytomegalovirus. vIL-10 mimics the immunosuppressive and anti-inflammatory activity of host IL-10 but lacks its immunostimulatory properties.
  - The *Molluscum contagiosum* virus encodes a homologue of the CCβ chemokine that blocks leukocyte trafficking to the site of infection.
- **Cytokine-binding proteins**: These often act as decoy receptors, mopping up the cytokine and blocking its effects.
  - The vaccinia, cowpox, and *M. contagiosum* viruses encode an IL-18-binding protein. IL-18 is a pro-inflammatory cytokine important in antiviral defence. It induces IFN-γ production, NK cell activation, and TH1 responses. vIL-18-binding protein binds IL-18, blocking its antiviral effect.
  - The parapoxvirus Orf encodes a protein that binds GM-CSF and IL-2 — cytokines integral to the development of adaptive immune responses.
  - The vaccinia virus encodes a vIFN-α-binding protein that can bind IFN-α from several species.
  - Several poxvirus family members encode the homologues of soluble cytokine receptors that often bind the cytokine with a greater affinity than their host counterparts. Homologues of TNFRs are encoded by the vaccinia, cowpox, and myxoma viruses. vTNFR binds to the soluble TNF secreted at the site of infection, blocking its antiviral and pro-inflammatory activity.
  - The variola, vaccinia, cowpox, and camelpox viruses encode a homologue of the IFN-γ receptor (vIFN-γR). vIFN-γR binds and blocks the action of IFN-γ , one of the most important host antiviral defences.
  - The vaccinia and cowpox viruses also produce vIL-1R. IL-1 is expressed early in viral infections. It induces a pro-inflammatory response, which includes the induction of the acute-phase reaction, pyrexia, and the growth and proliferation of T, B, and NK cells. vIL-1R binds IL-1 with a high affinity and blocks these antiviral pathways.
  - The cytomegalovirus encodes three chemokine receptor homologues that bind soluble chemokines (including RANTES and MIP-1) that are important in neutrophil chemotaxis.

- **Serpins**: Members of the poxvirus family also encode serine protease inhibitors or serpins. Serpins weaken the inflammatory response.
- **Growth factors**: Some viruses secrete growth factors for host cells, thus increasing the potential number of cells for them to infect. This is one of the reasons for the hyperplasia that is observed in such viral infections. For example, the vaccinia virus encodes a homologue of the epidermal growth factor, which causes substantial proliferation of epidermal layers near the site of infection and which has no known function in immune evasion.

Naïve $CD8^+$ CTL precursors lack cytotoxicity. They also have a greater requirement for costimulation than do naïve $CD4^+$ T cells. DCs are the most efficient activators of naïve CTLs as they have high intrinsic costimulatory activity. For APCs that do not have the required costimulatory potential, the recognition of peptides derived from the same antigen by naïve $CD8^+$ and effector $CD4^+$ T cells on the same APC's surface can compensate for inadequate costimulation. The $CD4^+$ T cell is thought to act by inducing increased expression of costimulatory molecules on the APC. TcRs recognise peptide:MHC class I molecules, and the required costimulatory signals induce the expression of receptors for cytokines such as IL-2 and IL-6, driving cytokine synthesis. The engagement of these receptors, in turn, leads to the synthesis of granules containing cytotoxic effector molecules. The activation process, which takes 1-3 days, also drives lymphocyte proliferation. CTLs kill infected cells through a mechanism similar to that of NK cells; we discuss it in detail in chapter 11. CTLs use two major pathways to bring about target cell death — the release of cytotoxins such as perforin and granzymes and the induction of death by the Fas pathway (Figure 9.7). In addition, most CTLs also release IFN-γ, TNF-α, and TNF-β. These cytokines augment host defence in multiple ways. As mentioned, IFN-γ directly inhibits viral replication. It also induces or upregulates MHC class I and II expression. Increased MHC class I expression increases the possibility of infected cells being killed by CTLs, whereas increased class II expression helps co-opt TH cells in the response. IFN-γ also activates macrophages, augmenting their phagocytic and antigen-presenting capabilities. TNF-α and -β, on the other hand, act synergistically with IFN-γ in macrophage activation. They also induce DC maturation.

Traditionally, $CD8^+$ cells were viewed only as IFN-γ and TNF-α producing cells. However, a number of studies clearly establish the existence of Tc1 and Tc2 subsets (cytotoxic T cell types 1 and 2) on the lines of TH1 and TH2 types. Tc1 cells produce IFN-γ and IL-2, whereas Tc2 cells produce IL-4, IL-5, and IL-10. Like the TH1 and TH2 subsets, these subsets probably represent the extremes of a whole spectrum of subsets. Once committed, the cells seem unable to revert to a different phenotype. Both subsets are involved in CMI and do not provide help to B cells except indirectly, through the production of cytokines. Factors that determine the polarisation of CTLs, their functions and roles in disease, and their interplay are still under investigation.

9.5.4 Lipid-specific T Cells

Lipid-specific responses, with other antigenic responses, are important for host defence against infections, tumour immunosurveillance, and autoimmunity. **Lipid-specific T cells recognise lipids in association with CD1 molecules**. Most do not express CD4 or CD8 molecules. Microbial lipids, lipoproteins/lipopeptides, glycolipids, phosphoglycerolipids, and so forth, whether of microbial or self-origin,

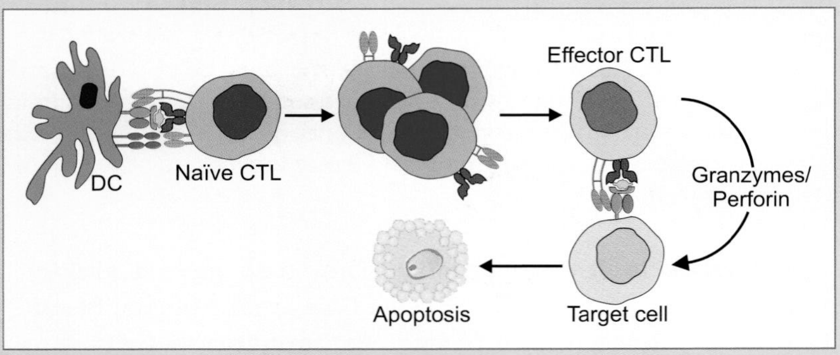

Figure 9.7 CTLs are important in antigen-specific host defence against intracellular parasites. The recognition of peptide:MHC class I complexes and appropriate costimulation together result in the activation of naïve CD8$^+$ T cells. Effector CTLs are formed after repeated rounds of the proliferation of activated naïve cells. Effector CTLs kill infected target cells by two pathways. They cause the lysis of target cells by releasing cytotoxins such as perforin and granzymes directly into the tight junction formed between the CTL and the target cell. Alternatively, they can induce apoptosis in the target cell through the Fas pathway.

can be recognised by these cells. The internalisation, digestion, and loading of lipids on CD1 molecules and their presentation by APCs is essentially similar to that observed for proteinic antigens. Although these cells were initially thought to occur at a very low frequency, recent studies have established this to be untrue. The frequency increases even further in multiple sclerosis and mycobacterial infections. Like their peptide-specific counterparts, these cells too can be divided into TH1 and TH2 types on the basis of their cytokine secretion patterns. Human lipid-specific T cells can be broadly divided into two groups:

- Those that have properties similar to peptide-specific T cells and express CD1a-c and CD1e and
- Those that express NK receptors, that is, NKT cells, and recognise antigen in context of CD1d. As mice express only CD1d, murine studies further the understanding of only NKT cells.

NKT cells have been shown to have cytolytic activity. TcR stimulation results in the rapid (within 90 minutes) induction of cytokine synthesis — predominantly IL-4 and IFN-γ — in these cells. The NKT-derived cytokines in turn recruit and activate several other cell types, such as NK cells, conventional T cells, macrophages, B cells, and DCs. The cytokines produced by NKT cells have been suggested to help polarise the subsequent adaptive immune response to TH1 or TH2 type. **NKT cells have been shown to be involved in host defence against infections caused by a number of intracellular parasites**, such as *L. major*, *Plasmodium falciparum*, *Trypanosoma cruzi,* and *M. tuberculosis*. This protective effect has been attributed to their IFN-γ production. **They are also believed to be important in antitumour immunity** by virtue of their cytotoxicity and their production of pro-inflammatory cytokines, such as IFN-γ and IL-12. Studies in mice also suggest a role for NKT cells in the control and prevention of autoimmune diseases (e.g., diabetes, EAE, and colitis). These studies suggested that NKT cells help downmodulate the inflammatory response by the secretion of IL-4 and IL-10. The potent cytokine-producing abilities and cytotoxicity of NKT cells has resulted in their being investigated for potential immunobased therapies.

Humoral Immunity 10

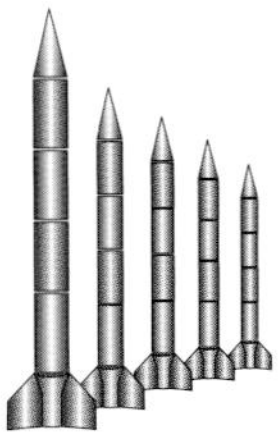

'Hey baby, thought you were the one who tried to run away.
Ohh, baby, wasn't I the one who made you want to stay?
Please don't bet that you'll ever escape me
Once I get my sights on you
Got a license to kill (to kill)
And you know I'm going straight for your heart.
(Got a license to kill)'

— Gladys Knight, *Licence to Kill*

Abbreviations

ADCC:	Antibody-Dependent Cell-mediated Cytotoxicity
CDR:	Complementarity-Determining Region
C: region	Constant region
Fab:	Antigen-binding Fragment
Fc:	Crystallisable Fragment
FcR:	Fc Receptor
FcRn:	FcR neonatal
IVIg:	Intravenous Ig
J chain:	Junction chain
mAb:	Monoclonal Antibody
pIgR:	Polymeric Ig Receptor
SC:	Secretory Component
VLR:	Variable Lymphocyte Receptor
V: region	Variable regions

10.1 Introduction

Humoral immunity is conferred by *body humours* (no, not the comedy variety; the Latin *humor* is any bodily fluid or semifluid), specifically, by Igs secreted by terminally differentiated B cells (i.e., plasma cells). This Ig response is unique in many respects.

- Humans can produce Igs against a virtually limitless array of antigens. These approximately billion Igs of differing antigen specificities are encoded by a genome of about 10^5 distinct genes. The genes coding for Igs are formed by random rearrangement of gene segments during lymphocytic development (chapter 6).
- Each B cell clone can yield progeny that produce Igs of a single antigenic specificity but of different isotypes. Naïve B cells express IgD and IgM molecules on their cell surfaces. Upon stimulation, the progeny of these B cells can switch the expressed Ig isotype (IgG, IgA, or IgE) and secrete antibodies of another class.
- A second or subsequent challenge with the same antigen generally results in a humoral response that is more rapid and specific than the primary response.
- B cells secrete Igs while expressing them as integral parts of their cell membranes. This is remarkable because membrane proteins differ from secreted proteins in their primary amino acid sequences. A stretch of hydrophobic amino acids in the transmembrane region anchors the protein in the cell membrane; however, this stretch must be absent for the protein to be secreted.

As discussed in earlier chapters, certain structural characteristics and genetic mechanisms produce the features unique to the humoral response. In this chapter, we focus on humoral responses: the types of responses, the characteristics of these responses, and the mechanisms by which antibodies confer immunity.

10.2 Primary Humoral Response

The preceding chapters (chapters 6 and 8) have already discussed the sequence of events leading to antibody secretion by B cells during the primary humoral response. These events are briefly summarised below.

- Upon primary challenge, the antigen (microbial, viral, etc.) is captured by DCs or macrophages and transported to the draining lymph node.
- APCs load digested fragments of the antigen onto MHC class II molecules and present them to naïve $CD4^+$ T cells.
- Recognition of the peptide:MHC complex (or lipid:CD1 complex) with the engagement of costimulatory molecules activates naïve T cells. The cytokines released by APCs further aid this activation. Activated T cells proliferate and finally differentiate to the helper phenotype; they also secrete chemokines.
- The ligation of BcRs by antigen stimulates B cells. They internalise antigen through pinocytosis (if it is soluble) or BcRs, digest it, and display parts of the antigen in context of MHC class II molecules.
- B cells migrate to the edge of the B cell zone in response to T cell chemokines and secrete chemokines such as BCL-1, resulting in the migration of T cells to the edge of B–T cell areas. The two cells engage in cognate interaction.
- Because of signals delivered by T_H cells, B cells undergo blast transformation and migrate from the primary foci of proliferating cells to secondary follicles, where they differentiate to plasmablasts and memory cells. Some proliferating B cells undergo affinity maturation and isotype switching in the germinal centre.

- Plasmablasts migrate to the bone marrow or medullary cords of the lymph nodes and give rise to plasma cells. These effector cells are capable of secreting Igs at the astounding rate of about 2000 antibodies per second!
- If the challenging antigen is of the TI type, T cells do not participate in the immune response; in such instances, the immune response primarily involves B1 and MZ B cells.
- Secreted Igs bind the antigen, resulting in a transient phase where antigens exist mostly as soluble antigen–antibody complexes. Phagocytes scavenge the complexes (this is also referred to as the immune-elimination phase).
- Free antibodies appear in the peripheral blood at the end of the immune-elimination phase and remain there until their eventual catabolisation.

The kinetics of the primary immune response and the concentrations of antibodies reached are dependent upon the nature of the antigen, the presence of adjuvants, the route of administration, and the species, strain, and age of the animal. Serum analysis for antigen-specific antibodies following antigenic stimulation reveals four phases (Figure 10.1).

- The initial **lag phase** lasts for 4-7 days, during which antigen-specific antibodies are not detected in serum. This time is required for the selection, clonal expansion, and differentiation of naïve B cells to plasma cells, as well as the immune elimination of the antigen.
- During the **log phase**, the concentration of antigen-specific antibodies increases exponentially with time, and the isotype of antibodies switches from IgM to IgG. Antibody production peaks within 7-10 days of antigen exposure. In this phase, the rate of antibody synthesis is far greater than that of its catabolisation, resulting in a logarithmic increase in serum antibody concentration.

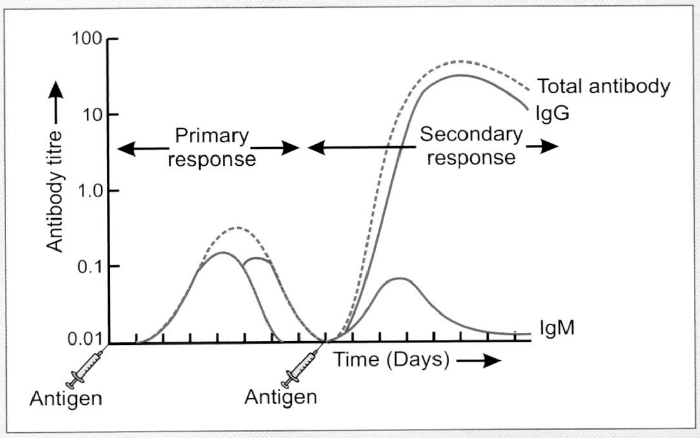

Figure 10.1 The secondary humoral response to TD antigens differs from the primary response in kinetics, concentration, affinity, isotype, and duration. The time required for the appearance of free antibodies in serum is much longer in primary humoral responses. This phase, called the lag phase, represents the time required for the selection, clonal expansion, and differentiation of naïve B and T cells and includes time required for immune elimination. IgM (represented by the blue line) is the predominant antibody produced in the early stages of the primary response, but it changes to IgG (red line) in the later stages. Because of clonal expansion, the frequency of the B and T cells recognising the antigen increases at the end of the primary response. The reprogramming of their antigen receptors also renders these cell types responsive to lower concentrations of antigen. Hence, secondary responses have faster kinetics. The total amount of antibodies secreted is much greater (green broken line) in secondary humoral responses; the response lasts longer and consists of predominantly IgG antibodies.

- In the **steady state**, antibody concentration reaches a plateau, as the amount of antibodies secreted is balanced by the amount catabolised.
- The **stage of decline** is the stage in which the rate of antibody catabolisation outstrips the rate of synthesis.

In Gandhi's Footsteps: Passive Immunity

Depending upon the mode of acquisition, adaptive immunity can be divided into active and passive immunity.

- **Active immunity** is acquired by an individual in response to antigenic challenge. If it results from a clinical or subclinical infection, the immunity is natural active immunity. Artificial acquired immunity, on the other hand, is a result of the deliberate inoculation of microbes or their products, as through vaccination.
- **Passive immunity** is acquired passively, without active involvement of the recipient's immune system (e.g., immunity transferred passively to the child *via* the placenta or breast milk). Artificial passive immunity is obtained by injecting the serum of an actively immunised animal into a recipient; this immunity is humoral as only antibodies are transferred from the donor. **Adoptive immunity** is the transfer of primed (antigen-stimulated) lymphocytes from an actively immunised donor to an unimmunised individual. Because lymphocytes are passively transferred, the recipient may show cell-mediated and humoral responses, depending upon the types of cells transferred.

  Passive immunity has a short duration (8-10 days), as the donor antibodies are eventually catabolised and destroyed. Passive immunisation is necessary under certain circumstances.
- **I**ntra**v**enous **Ig** (IVIg) is the treatment of choice for patients with antibody deficiencies. IVIg is administered at a dose of 200-400 mg/Kg of body weight every three weeks; a higher dose of IVIg (2 gm/Kg/month) is used as an immunomodulatory agent in a number of immune and inflammatory disorders (e.g., idiopathic thrombocytopaenic purpura).
- Passive immunity can be used as an emergency prophylactic measure to confer temporary protection. Passive immunisation is common in infections in which the major cause of host damage is a pathogen-secreted toxin. For example, in diphtheria, the toxin produced by the infection causing *Corynebacterium diptheriae* can prove fatal. Patients are therefore treated with diphtherial antitoxin, which neutralises the toxin and allows survival until the patient's immune system starts producing its own antitoxin. Similarly, the anti-tetanus serum is administered to patients with deep wounds conducive to the proliferation of the toxin-producing *Clostridium tetani.*
- Passive immunisation can also neutralise toxins from snakebites or scorpion bites.

IVIg is obtained from the pooled plasma of thousands of donors, and antitoxins are raised by repeated immunisations of animals. Collected sera are treated to ensure the absence of antibody aggregates, which can trigger large-scale complement activation and result in type III hypersensitivity reactions, kidney damage, and severe anaphylactic shock. The sera are also treated with detergents and solvents to minimise the risk of transfer of infectious agents. Repeated administrations of heterologous antibodies (i.e., from people with different genetic backgrounds or from other species) are likely to give rise to anti-isotypic or anti-idiotypic antibodies in the recipient. Anti-isotypic antibodies can lead to a hypersensitivity reaction called serum sickness (see chapter 16 for details). Anti-idiotypic antibodies, on the other hand, can neutralise antibodies in the antiserum and decrease its usefulness.

Serum antibodies are polyclonal and heterogeneous because they are a product of multiple clones of B cells. By contrast, Igs produced by a single clone of B cells are monospecific. Kohler and Milstein introduced **m**onoclonal **A**nti**b**ody (mAb) technology, in which plasma cells were fused with myeloma to yield hybrid cells (termed *hybridomas*[1]). Hybridomas have the Ig-producing capacity of plasma cells and the unlimited growth potential of myeloma cells. Antibodies produced by these hybridomas are monospecific and homogeneous. Originally, hybridomas were injected into laboratory animals, where they established tumours and produced Igs of a given specificity and isotype. Refinement of the technology has allowed for

[1] See appendix IV

the elimination of animals and large-scale production *in vitro*. The advent of this technology has allowed the development of engineered *humanised antibodies* and has opened a new era in passive immunotherapy. A number of mAb preparations have now been approved for therapeutic use. mAbs are being explored for the treatment of tumours, autoimmune diseases, graft rejections, and infectious diseases. The potential of this treatment can be gauged from the fact that more than a quarter of all biotech drugs in development are mAbs.

Technical difficulties (such as the lack of suitable myeloma cells and low rates of mAb secretion) resulted in the use of murine mAbs in initial clinical trials. However, murine antibodies have a low affinity for human FcRs (**Fc R**eceptors), reducing their effectiveness. Moreover, repeated administrations of murine antibodies give rise to human anti-mouse antibodies. The development of **chimeric antibodies** consisting of the Fab (antigen-binding) fragment of mouse antibodies, and the Fc region of human antibodies helped overcome some of these problems (Figure 10.S1). The creation of humanised antibodies by grafting CDRs from murine mAb onto human IgG molecules was a further refinement. The creation of bispecific antibodies was another approach. Bispecific antibodies can bind two different epitopes. Often, one epitope is a tumour-associated antigen, whereas the other is associated with an immune effector cell. The antibodies directly link the effector cell to tumour cells, enhancing immune responses. Epitopes of choice include FcγRI, expressed by granulocytes and macrophages, CD16, expressed by NK cells, and the pan-T cell molecule CD3. Fusion proteins consisting of radionuclides, toxins, or chemotherapeutic agents conjugated to the Fc region of human mAbs are currently being explored for therapeutics. Such immunoconjugates provide the advantage of targeting specificity. Additionally, because the toxic payload is delivered directly to the target cell, both dosage and side effects are reduced. Radiolabelled mAbs can also be used as diagnostic tools for detecting or locating tumour antigens, permitting early diagnosis of metastatic tumours. For example, ^{131}I-labelled mAbs have been used in the detection of breast cancer metastases. Some mAbs in clinical use are listed below:

- Muromonab, an anti-CD3 murine mAb preparation, has improved graft survival in some studies. Because it knocks out all CD3$^+$ T cells, this treatment increases the risk of infection. It is therefore used only in steroid-resistant patients.
- Anti-CD25 mAbs that bind IL-2Rs have the advantage of being effective against activated T cells. Basiliximab and daclizumab are chimeric humanised mAbs now used for prolonging graft survival in transplant recipients.
- Anti-TNF mAbs show potential for use in anti-inflammatory therapy. A human–mouse chimeric anti-TNF mAb, infliximab, has shown great promise in the treatment of rheumatoid arthritis and Crohn's disease. Similar anti-inflammatory activity has also been reported for the TNFR–IgG1 fusion protein, Etanercept.
- Rituximab, a chimeric anti-CD20 mAb, was the first mAb to be approved for the treatment of low-grade and follicular non-Hodgkin's lymphomas. This IgG1 mAb can activate complement and aid in ADCC (**A**ntibody-**D**ependent **C**ell-mediated **C**ytotoxicity). CD20 is expressed on pre-B and mature B cells. It also possesses anti-proliferative and apoptosis-inducing activity against B cells expressing CD20.
- Trastuzumab is a chimeric mAb directed against the HER2/neu receptor. Also known as ERBB2, this receptor is expressed on many breast and ovarian carcinomas; dimerisation of the receptor results in proliferation of the cancerous cells. Trastuzumab binds to the extracellular domain of the receptor and turns off the proliferation signal.
- MDX-210 is a bispecific mAb directed against ERBB2 and FcγRI. Used to treat ovarian cancers that overexpress ERBB2, this mAb is currently undergoing phase III clinical trials.
- Humanised anti-CD33 antibody conjugated to the toxin calicheamicin has been approved for the treatment of CD33$^+$ acute myeloid leukaemia.

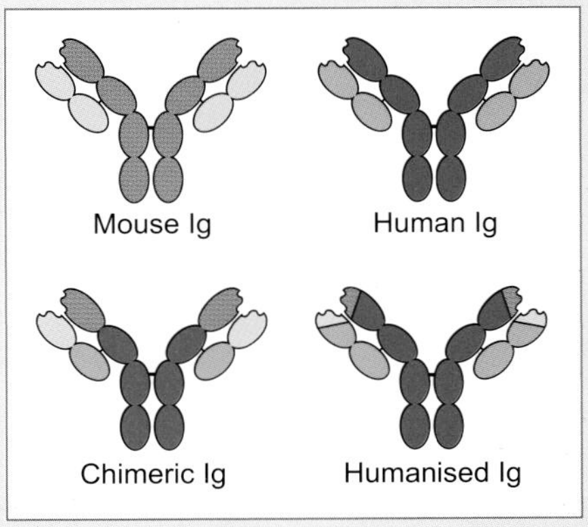

Figure 10.S1 Chimeric antibodies and humanised antibodies have helped reduce the complications arising from passive immunisation with murine antibodies.

10.3 Secondary Humoral Immune Response

The second encounter with an immunogen and encounters thereafter are *secondary humoral immune responses* or *anamnestic immune responses* (from *anamnesis*, meaning recall in Greek). The secondary responses elicited by TI antigens do not differ significantly from primary responses. Secondary responses to TD antigens, however, are more rapid and aggressive than primary responses because of a variety of factors.

- Clonal expansion during the primary response results in a much larger pool of antigen-specific B and T cells. Experimental animals have exhibited up to 1000-fold increases in the frequency of antigen-specific lymphocytes.
- Memory cells (B or T) formed at the end of the primary immune response are reprogrammed in their signal transduction capabilities and hence respond faster to immunogenic challenge.
- Because of affinity maturation, the BcRs of memory B cells bind antigen with a much higher affinity than the BcRs of naïve B cells; hence, fewer immunogenic molecules are required for activation (see chapters 6 and 8).

Secondary humoral responses differ from primary responses in several respects.

- **Secondary responses occur with faster kinetics**. The lag phase is shorter (about 1-3 days), the response peaks much earlier (3-5 days), and the steady state is maintained for a much longer period in secondary responses. The decline is also slower; therefore, secondary responses persist for prolonged periods.
- **They result in antibody titres of much higher magnitudes**. The large pool of antigen-specific B cells formed at the end of the primary response gives rise to an increased number of plasma cells in the secondary response, with a concomitant increase in antibody titres (concentrations).
- **Secondary responses are dominated by isotypes other than IgM** as a result of CSR (chapter 6).
- **Antibodies produced in the secondary response have a much greater affinity for the antigen**. An immune challenge, whether primary or secondary, results in a spectrum of antibodies with varying affinities for the antigen, reflecting the recruitment of multiple B cell clones in the immune response. Primary responses result in predominantly low-affinity antibodies. Only a fraction of secondary response antibodies are of a low affinity; the majority of secondary response antibodies display a high affinity for the antigen. This phenomenon, *affinity maturation*, is attributed to two processes.
  - During the proliferation of B cells in germinal centres, the DNA coding for the V region of Ig (i.e., the antigen-binding domain) undergoes point mutations. Those B cells expressing BcRs with higher affinity for the antigen are selectively allowed to undergo clonal expansion, whereas those expressing low-affinity BcRs are not allowed to survive. Antibodies produced later in the immune response therefore exhibit a higher affinity for challenging antigen (section 6.5.2).
  - During primary immune responses, antigen concentration far exceeds the cells capable of binding it. Hence, irrespective of affinity, all cells responding to the antigen are stimulated. On subsequent challenges, however, far more cells are capable of reacting with the antigen. The ensuing competition for the limited antigenic determinants leads to the selective stimulation of cells expressing high-affinity BcRs.

Secondary responses to hapten–carrier conjugates are dependent upon hapten-primed memory B cells and carrier-primed memory TH cells. If the

Trying to Fit in: Cross-reactivity

An antibody is said to be cross-reactive when it binds antigen other than the one used to elicit the immune response. This cross-reactivity might be caused by chemical relatedness or the presence of identical epitopes on unrelated antigens (i.e., because of shared epitopes). If chemical similarity is the basis for cross-reactivity, the cross-reacting antibody binds the homologous antigen better than the heterologous one. For shared epitopes, however, the antibody's affinity for both antigens is similar (Figure 10.S2). Although cross-reactivity is used in the diagnosis of infections, its disadvantages far outweigh that advantage.

Cross-reactivity has been implicated in many autoimmune diseases (see chapter 15). It is also the underlying cause of the phenomenon of *original antigenic sin*[2]. Also called epitope imprinting, this is the phenomenon in which exposure to an antigen influences subsequent responses to related antigens. It was originally described for the influenza virus, for which a second infection by an antigenically related strain resulted in antibodies against the earlier strain rather than a response to novel protective determinants on the second strain. Thus, the immune response to the first viral infection seems to have a dominating influence on subsequent immune responses to antigenically related viruses; the second virus often induces a response against the original strain. Epitope imprinting is attributed to the activation of memory B cells formed at the end of the first infection by cross-reacting antigens on the second viral strain. The antibodies produced are not efficient in neutralising the second viral strain, and the virus escapes. Antigenic variation is thus an efficient mechanism of immune evasion, and it is observed in a variety of infections, such as those caused by enteroviruses, reoviruses, paramyxoviruses, togaviruses, chlamydia, and *Plasmodium* spp. The phenomenon is exploited in epidemiological studies of these infections. Although initially observed for Ig responses, antigenic imprinting occurs for $CD8^+$ T cells as well. It proves especially deleterious in HIV infections. HIV is extremely prone to antigenic variation, and multiple variants of HIV are found to coexist in the same individual. Because of epitope imprinting, antibodies that can neutralise the initial virus or CTLs that can kill cells infected with the initial virus increase in numbers with time. However, these antibodies/CTLs are not effective against concurrent HIV isolates. Thus, the continued evolution of the virus, combined with original antigenic sin, allows HIV to escape adaptive immune responses (see chapter 13). Antigenic variation is also a major obstacle in designing vaccines. Immune responses generated by an infection will be against the strain in the vaccine, rather than against the variant causing the infection. Thus, the vaccine might *help* the infection, rather than checking it. One way to circumvent the problem is to vaccinate against newly emerging viral strains every year, as is done for the influenza virus. Inoculation with a cocktail of closely related viral peptides to elicit a more broadly reactive T cell response is an alternative approach now under investigation.

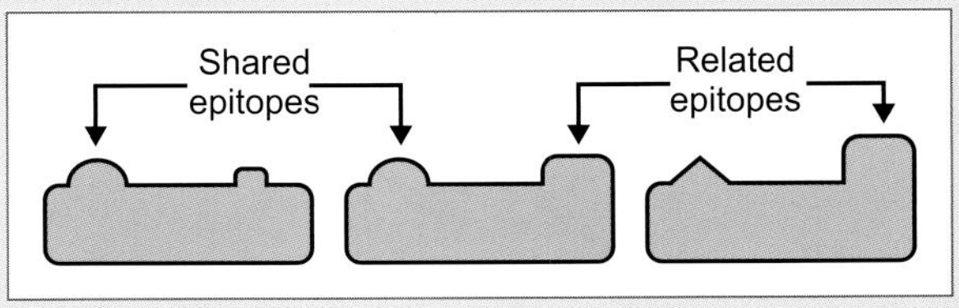

Figure 10.S2 Antibodies may cross-react with heterologous antigens because of the presence of shared epitopes or chemically related epitopes.

secondary challenge is from the same hapten used in the primary challenge but conjugated to a different carrier (e.g., a primary challenge with dinitrophenol conjugated to ovalbumin (DNP–OVA) and a secondary challenge with DNP–bovine serum albumin, (DNP–BSA)), the antihapten response is of the primary, not the secondary, type. This phenomenon is termed the *carrier effect*. As explained in chapter 4, hapten–carrier conjugates behave like TD antigens, resulting in an antibody response to the hapten and a T cell response to the carrier. Associative recognition necessitates that B and T cells recognise antigenic determinants on the same molecule. Therefore, DNP–BSA is unable to stimulate a secondary humoral response to DNP. However, if the animal is primed with BSA alone before the

[2] The name is a biblical reference, to the sin attached to humans because of Adam and Eve's original tasting of the forbidden fruit.

administration of DNP–BSA (i.e., an administration of DNP–OVA, BSA, and then DNP–BSA), BSA-primed memory TH cells will have been formed, and it will result in a secondary-type humoral response to DNP.

10.4 Immunoglobulin Structure

Igs are characterised by their specificity and sensitivity. Specificity is the ability of antibodies to discriminate between homologous epitopes (i.e., the epitopes against which antibodies were raised) and heterologous (i.e., related) epitopes. As we explained in chapter 4, antibodies can differentiate between three-dimensional conformations of closely related structures. This does not imply that antibodies cannot react with related ligands or cannot display cross-reactivity. However, the reaction between the antibody and the related ligand is generally weaker than that between the antibody and its homologous antigen[3]. Thus, specificity depends upon the relative affinity of the antibody for two closely related structures. The greater this difference, the more specific the antibody. Sensitivity refers to the ability to detect a particular substance (i.e., the antigen) in the presence of large amounts of other extraneous substances. An antibody's sensitivity is thus a consequence of its specificity.

10.4.1 Elucidation of Immuoglobulin Structure

When Tiselius and Kabat electrophoresed[4] serum at pH 8.5, they found that albumin, having the greatest negative charge, moves rapidly towards the positively charged cathode and is followed by three globulin fractions (called α, β, and γ globulins respectively). IgG, being positively charged, lies in the fraction closest to the anode (called the γ fraction; Igs are therefore called γ globulins). Although a major fraction of IgG lies in the γ fraction, significant amounts can also be found over the entire spectrum, extending from the α to γ fractions.

Elegant experiments by Porter, Nisonoff, and Edelman helped to establish Ig structure (and earned Porter and Edelman a Nobel Prize). Their approach, using digestion by proteases and reducing agents, is now commonly used to establish the structure of proteins.

- Porter subjected the low-MW fraction of globulins containing a γ globulin of 150 KD MW (known to possess antibody activity) to brief papain digestion. Prolonged exposure to this protease will completely digest practically any protein. During brief exposures, however, papain digests only the most susceptible IgG bonds, yielding two identical 45 KD **F**ragments that retain their **a**ntigen-**b**inding property (**Fab**) and one **F**ragment of 50 KD that **c**rystallises during low-temperature storage (**Fc**) but cannot bind antigen. Unlike whole Igs, Fab fragments cannot precipitate or aggregate antigen molecules, suggesting that the two Fab fragments need to be joined to participate in visible reactions (Figure 10.2).
- Nisonoff treated IgG molecules to brief digestion with pepsin. The resultant single 100 KD fragment could bind antigen to yield a visible reaction. Treating this fragment with very mild reducing agents yielded two identical fragments that, like Fab, reacted with the antigen but did not produce a visible reaction. The large fragment was therefore called $F(ab')_2$.
- Porter and Edelman also treated IgG molecules with the reducing agent β-mercaptoethanol and separated the fragments by chromatography. Disulphide bonds are irreversibly cleaved by β-mercaptoethanol. The experiment revealed

[3] Heteroclitic antibodies (Greek *heteros* — other, *klinein* — inclined) are an exception, in that heteroclitic antibodies bind heterologous antigens better than homologous antigens.

[4] Electrophoresis allows the separation of molecules on the basis of their electric charge and is explained briefly in appendix IV.

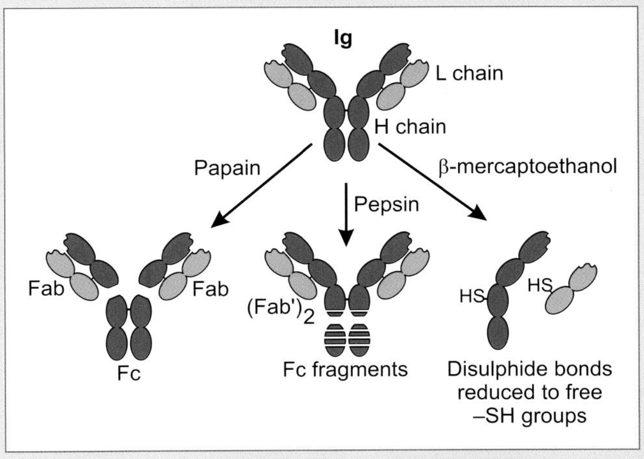

Figure 10.2 The four-chain structure of Igs was proposed on the basis of the experiments of Porter, Nisonoff, and Edelman. They treated Igs with two proteases (papain and pepsin) and a reducing agent (β-mercaptoethanol). Papain digestion yielded two identical fragments (Fab) capable of antigen binding but incapable of visible reaction. It also yielded another fragment that crystallised upon low-temperature storage (Fc). By contrast, pepsin digestion yielded a single fragment capable of both antigen binding and precipitation and multiple low MW fragments. On treatment with mild reducing agents, the large fragment generated two identical fragments that had a lower MW than the Fab fragments and were capable of antigen binding but incapable of producing a visible reaction. Hence, the fragments were called Fab′ and their dimer $(Fab')_2$. Treatment with β-mercaptoethanol resulted in the formation of two identical fragments consisting of H and L chains. The proposed four-chain structure was later confirmed by X-ray crystallography.

that an IgG molecule of MW 150 KD consisted of two H chains of MW 50 KD each and two L chains of MW 25 KD each.

- Porter established the relation between Fab and Fc fragments and H and L chains by raising antisera against Fab and Fc fragments. Antisera raised against Fab reacted with both H and L chains, but that raised against Fc reacted only with the H chain.
- On the basis of these experiments, Porter suggested a four-chain Y-shaped structure for IgG molecules that was later confirmed by X-ray crystallography.

10.4.2 IgG Structure and Function

The prototypical Ig, IgG, has a sedimentation coefficient of 7S. In chapter 6, we mentioned that IgG consists of a pair each of two polypeptide chains — L and H. The predicted MW of this four-chain structure is between 150-200 KD. However, the addition of carbohydrates at various sites on the H chains increases the weight of the molecule. Each L chain is bound to the H chain by a disulphide bond and by noncovalent interactions (e.g., salt linkages, hydrogen bonds, and hydrophobic interactions; see Table 4.1). Similar disulphide bridges and noncovalent interactions link the two H–L subunits to give a four-chain structure. The Ig molecule is thus a dimer $(H–L)_2$ of a dimer (H–L). The whole molecule may be visualised as a Y-shaped structure (Figure 10.3). Each arm of the Y consists of a dimer comprised of the entire L chain and half an H chain. The tail of the Y is formed by a dimer of the remaining half of the H chain. The arms of the Y contain the antigen-binding site; the tail is involved in effector functions. The Ig molecule is flexible and changes its tertiary structure upon antigen binding. When not bound to the antigen, the arms of the Y lie relaxed, covering the CH2 domain. When in contact with the antigen, the arms swing out. The disulphide bonds between the two H chains facilitate this movement, and hence, this region is termed *the hinge region*.

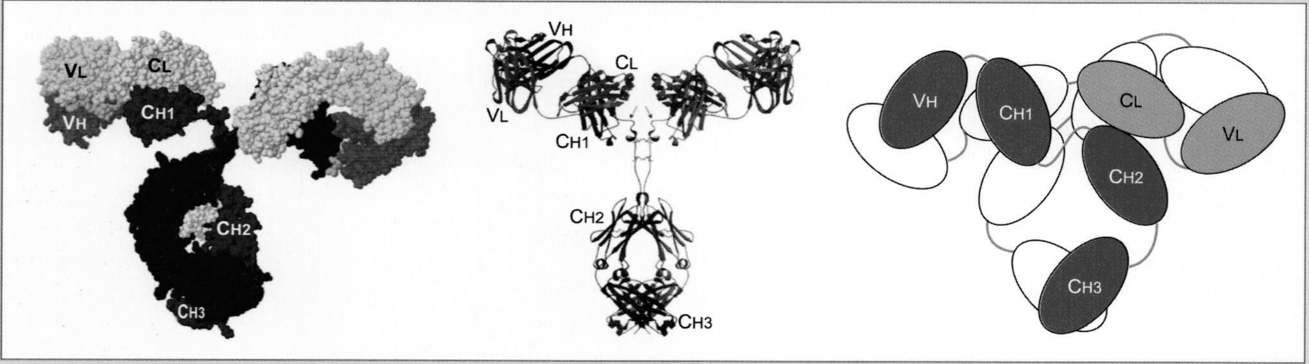

Figure 10.3 The four chains of IgG molecule are linked by inter- and intra- chain disulphide bonds to yield a Y-shaped molecule. The panel on the left is a space-filling model of an IgG molecule based on X-ray crystallography. The two antigen-binding sites (Fab) lie at the tips of the arms of the Y; both H and L chains contribute to the antigen binding. The main stem of the Y is formed by the H chains (Fc). The two arms of the Y are attached to the stem by a flexible hinge region. Note that when not bound to the antigen, the arms of the Y lie relaxed, covering the CH2 domains. The central panel is a ribbon diagram of the molecule, whereas the right panel is a schematic depiction of the molecule showing the different Ig folds (courtesy the Department of Immunology, Erasmus University, Rotterdam, The Netherlands).

The determination of the amino acid sequence of IgG was delayed because of the unavailability of sufficient amounts of homogenous IgG. The discovery of IgG in the blood and urine of multiple myeloma patients finally made IgG sequence determination possible. As we explained in chapter 6, the amino acid sequencing of L and H chains revealed the presence of a NH_2-terminal **V**ariable (V) region that differed considerably in amino acid sequences between different IgG samples and a -COOH terminal **C**onstant (C) region with much less variability. Therefore, the carboxy-terminal 110-amino acid region of the L chain is denoted CL (**C**onstant **l**ight) and the amino-terminal half VL (**V**ariable **l**ight) region. Similarly, the 110-amino acid domain at the NH_2-terminal of H chain is called the VH region and the rest (~ 330 amino acids), the CH region. The CH region consists of three separate domains — CH1, CH2, and CH3.

Examinations of the amino acid sequences of a large number of Ig molecules have established that variability is not evenly distributed within V domains[5]. Some short segments show a high degree of variability and are termed hypervariable regions. These hypervariable segments are located near amino acid positions 30, 50, and 95 in both the H and L chains, and together, they constitute about 15-20% of the human and murine V domain. These regions are directly involved in the formation of the antigen-binding site and are referred to as CDRs (**C**omplementarity-**D**etermining **R**egions). The intervening peptide segments are called FRs — **F**ramework **R**egions — and are responsible for maintaining the architecture of the molecule. Each Ig chain (L or H) contains three CDRs (CDR1, CDR2, and CDR3) and four FRs (FR1–FR4). The FRs form the basic β-pleated sheet structure of the Ig-fold of the V region (chapter 6), whereas the H and L CDRs are located on loops that connect the β-sheets of the VH and VL domains. The hypervariable regions are thus exposed in three separate but closely disposed loops because of the folding of the V domain, much like three fingers of a hand, and the FRs form the scaffold holding these loops together. The six hypervariable loops (three each of the VH and VL domains) contribute to the conformation of the antigen-binding site (Figure 6.1).

[5] In hindsight, this distribution of variability seems perfectly logical. As Igs can react with a virtually unlimited number of epitopes of different shapes and a close fit is a pre-requisite for epitopes and paratope interaction, the antigen-binding sites of different Igs had to have different shapes. In other words, to allow for the different configurations needed to interact with different antigens, the amino acid sequence of the Fab region of the molecule had to be highly variable. However, as the basic architecture of the molecule needs to be maintained, the V region had to have portions that were relatively less variable.

Variation is the Spice of Life: Tylopoda IgG and Lamprey VLRs

Antibodies are generally accepted to be tetrameric molecules consisting of two H and two L chains. The Ig molecules of camels are exceptional, in that 75% serum antibody (IgG2 and IgG3) molecules in camels are devoid of L chains. Llamas and dromedaries also have these L chain lacking antibodies, but they form a smaller percentage (< 45%) of the total serum Igs. These two-chain antibodies, *heavy-chain antibodies*, are nonetheless *bona fide* antibodies that can bind complement, are capable of agglutination, and have a comprehensive binding repertoire. They bind homologous antigens with a high degree of specificity and high affinity. The antigen-binding site of these antibodies consists of only a single domain, V_{HH}. The V domain of naturally occurring heavy-chain antibodies is very similar to conventional Igs and consists of three CDRs (**C**omplementarity-**D**etermining **R**egions). However, these antibodies have a mutation in key residues in the region corresponding to the side of V_H that interacts with V_L. The normal hydrophobic amino acids are substituted by hydrophilic amino acids. Consequently, V_{HH} domains and heavy-chain antibodies are highly soluble. V_{HH} domains also have additional cysteine residues, which allow the formation of an interloop disulphide bond that stabilises this domain. The V_{HH} region otherwise adopts a typical Ig-fold and superimposes perfectly on the conventional V_H structure. Heavy-chain antibodies differ from conventional Igs in one more respect — they lack a C_{H1} domain (Figure 10.S3).

Immunisation studies indicate that the type of Ig response (whether normal or heavy chain) is dictated by the type of antigen. Heavy-chain antibodies seem to be preferentially formed against active sites of enzymes or small haptens. For example, in llamas, the administration of hapten–carrier conjugates gives rise to conventional anti-carrier IgG antibodies, but anti-hapten heavy-chain IgG antibodies. This preferential binding to haptens or active sites of enzymes reflects the propensity of heavy-chain IgG antibodies to bind to grooves or cavities of proteins that is attributed to an inherent structural property of the single domain V_{HH}. The occurrence of heavy-chain antibodies raises many questions regarding the function of the L chains, isotype switching, and the role of various domains of conventional antibodies.

It is well known that adaptive immune responses based on Ig-like rearranging genes have not been observed in the earliest living vertebrates — jawless fish such as lampreys and hagfish. However, extensive studies of lampreys over the last few years have revealed a set of variable, lymphocyte-associated antigen receptors, termed VLRs (**V**ariable **L**ymphocyte **R**eceptors), that mediate adaptive humoral immune responses to repetitive carbohydrate and protein antigens. These VLRs are not in the least related to TcRs or Igs, but have leucine-rich repeat regions that give them a typical horseshoe shaped structure similar to TLRs (chapter 2). Interestingly, similar to B cells, VLR-bearing lymphocytes are stimulated by antigenic challenge to differentiate into plasmocytes that secrete multimeric forms of VLR. Although they are leucine rich, VLRs are structurally different from TLRs; lampreys have both VLRs and TLRs. VLRs are in fact more related to human platelet receptors. The diversity of VLRs arises from a combinatorial shuffling of a large number of cassettes, and utilizes an enzyme similar to the AID involved in CSR and somatic hypermutation!

We hope that this sidetrack drives home the point that there is no room for complacency in immunology — no tenet of immunology can be regarded as absolute and unquestionable. Surprising and fundamental discoveries are constantly being made, and immunologists may be justified in expecting a life of excitement and adventure (!).

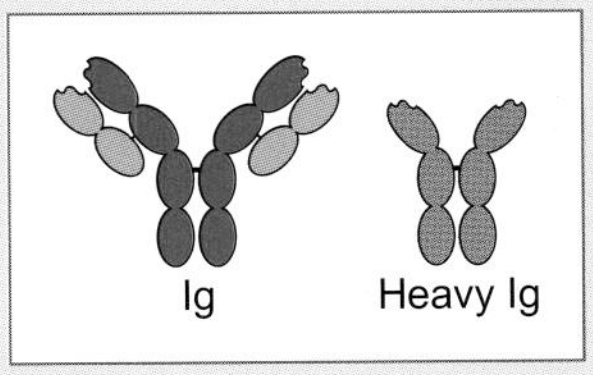

Figure 10.S3 Heavy-chain Igs, found in camels, are devoid of L chains, unlike normal Ig molecules, which consist of two H and L chains each (adapted from Muyldermans et al., *Trends in Biochemical Sciences*, 2001, 26:230).

It may seem surprising that these small hypervariable regions are responsible for generating the large variety of antigen-binding specificities. However, a little reflection will show why this is possible. Variations in the lengths and amino acid sequences of antibody molecules are responsible for the wide range of their specificities. There are three CDRs each in the V_H and V_L regions. Even if each hypervariable region were assumed to consist of a stretch of 12-15 amino acids, a large repertoire of possible combinations would become available, as a change in a single amino acid position can change the conformation of the antigen-combining site.

The recognition of individual domains in the IgG molecule led to the suggestion that these domains had evolved for particular functions.

- V_L and V_H domains together constitute the antigen-binding site (also called the antibody paratope).
- Both C_{H1} and C_L domains help to hold the V domains together by their disulphide linkages; they also extend the antigen-binding arms of the molecule, facilitating interaction with the antigen and increasing the maximum rotation of the arms.
- The C_{H1} domain binds the complement component C4b.
- The hinge region, an extended peptide sequence between the C_{H1} and C_{H2} domain of IgG (as well IgD and IgA), is a proline-rich region that does not have any homology with other domains. It is responsible for the flexibility of these molecules, and allows them to assume various angles with respect to each other upon antigen binding.
- The lower hinge region of the C_{H2} domain of IgG is the site of **Fc R**eceptor (FcR) binding.
- The C_{H2} region interacts with the C1q component of complement.
- A site for binding to FcRn (**FcR n**eonatal; section 10.7) lies at the C_{H2}–C_{H3} interface; this interaction is thought to determine the half-life of serum IgG antibodies.
- The C_{H3} domain is thought to be necessary for optimal complement activation.

Immunoglobulin Structure and Function

- The prototypical Ig molecule has a four-chain structure consisting of two H and two L chains.
  - Each L chain is bound to the H chain by a disulphide bond and noncovalent interactions.
  - The two H–L dimers are linked by disulphide linkages, yielding the Y-shaped Ig molecule.
  - Disulphide linkages in the hinge region bind the two H chains and allow the molecule to change its conformation upon antigen binding.
  - The H chain consists of one Ig domain of high sequence variability (V_H region) and three Ig domains of lesser variability (C_{H1}, C_{H2}, and C_{H3} regions).
  - The L chain has one Ig domain each of variable and constant sequences (V_L and C_L, respectively)
  - The V region contains three hypervariable CDRs that form the antigen-binding pocket and four FR regions of less variability; FR regions are responsible for maintaining the architecture of the molecule.
- Five classes of Igs, termed isotypes, are recognised according to the immunogenicity of H chains; two classes are recognised according to immunogenicity of the L chains.
- The allelic variants of isotypes that are found in some members of a species but not others are called allotypes.
- The immunogenicity of the antigen-binding site of the antibody gives rise to Ig idiotypes.
- Igs have a role in the neutralisation of toxins, enzymes, and viruses, as well as in opsonisation, complement activation, degranulation of mast cells, and ADCC.

10.5 Immunoglobulin Types

Igs, being large glycoproteins, are by themselves good antigens and can induce potent immune responses. Such anti-Ig antibodies are powerful tools used to dissect and understand B cell development and humoral responses. Antigenic determinants are found on the entire molecule, and differences in antigenicity reflect differences in the amino acid sequences of these molecules. Depending upon the location of their epitopes, Igs are classified as isotypes, idiotypes, and allotypes.

10.5.1 Isotypes

Isotypic determinants are present in the C region of H and L chains and they are used to define H or L chain subclasses. The C regions of both H and L chains can be further divided into isotypes on the basis of their amino acid sequences. The C_H regions reveal five basic sequence patterns, μ, δ, γ, ε, and α, yielding the isotypes IgM, IgD, IgG, IgE, and IgA, respectively (Figure 10.5). The C_L region consists of two isotypes, κ and λ. Either type of L chain can pair with any of the H chains. However, in a single molecule, both the L chains are identical — Igs have either two κ or two λ chains. A separate C region gene encodes each isotype. All individuals of a species carry the same C region genes, and all isotypes are expressed in normal individuals. This implies that isotypic determinants are species specific and that the administration of a particular isotype in an unrelated species will elicit an immune response. Antisera raised against isotypes of various species are routinely used as diagnostic and research tools. The various human H-chain isotypes are discussed below (Table 10.1, Figure 10.4).

- **IgM**: Accounting for about 5-10% of serum Igs (mean serum concentration of about 1.5 mg/mL), **IgM is the predominant antibody produced early in the primary immune response**. Monomeric IgM is an integral part of the BcRs of naïve B cells. Serum IgM is a 19S pentamer with a MW of about 970 KD. It consists of five monomers held together by disulphide bonds at the Fc region; the Fc regions are placed at the core of the pentamer, and the 10 Fab regions at the periphery. Each Fab can bind antigen, so IgM antibodies have a theoretical valency of 10. However, steric hindrance allows IgM to simultaneously bind only five or fewer molecules of large antigens. This multivalency makes it highly efficient in complement activation and opsonisation, but its large size hinders its diffusion in intercellular tissue fluids and largely confines it to the intravascular pool.

  The IgM pentamer dissociates to monomers after treatment with mild reducing agents. The H chain of monomeric IgM is longer and heavier than that of IgG. The presence of an extra domain of about 130 amino acids gives it a MW of about 65 KD. The IgM H chain thus has five domains — V_H and C_{H1}–C_{H4}. Amino acid sequencing shows that the additional domain is placed at the C_{H2} position and replaces the hinge region. The C_{H1}, C_{H2}, and C_{H3} domains of the γ chain thus correspond to the C_{H1}, C_{H3}, and C_{H4} domains of the μ chain. In addition, the μ chain contains a C-terminal, 18 amino acid-long secretory tailpiece. Pentameric IgM antibodies have a small polypeptide unit (15 KD, 129 amino acids) called the J (**J**unction) chain that is bound to the first and fifth μ chain. The J chain forms disulphide linkages with cysteine residues on the secretory tailpiece. The J chain is not mandatory for polymer formation but regulates the structure and function of formed polymers. Similar J chains are

Table 10.1 Properties of Human Ig Isotypes

Property	IgM	IgD	IgG	IgE	IgA
MW (KD)	900	150	150	190	150-600
Form	Pentamer (serum) Monomer (mIg)	Monomer	Monomer	Monomer	Monomer/ polymer
Carbohydrate content (%)	12	11	3	13	7
Subclasses	–	–	4 (IgG1–IgG4)	–	2 (IgA1, IgA2)
Adult serum concentration (mg/mL)	1.5	0.04	13	0.0003	3.5
Biological functions					
Binds FcRs on phagocytes	?	–	+++	–	+
Virus neutralisation	+	–	++	–	+++
ADCC	–	–	+++	–	–
Degranulation (mast cells, basophils, eosinophils)	–	–	–	+++	–
Complement activation pathway	Classical	–	IgG1 classical IgG2 classical IgG3 classical IgG4 alternative	Alternative	IgA1 classical IgA2 alternative
Transcytosis	+	–	–	–	+
Placental transfer	–	–	IgG1 + IgG2 +/– IgG3 +++ IgG4 –	–	–

Legend: +++ – high, ++ – medium, + – low, – negative, +/– – negligible, ? – unclear,

also found in dimeric IgA antibodies. The J chain is synthesised by IgM-secreting or IgA-secreting plasma cells in secretory tissues. By contrast, antibody-secreting cells in the bone marrow, lymph nodes, or spleen do not secrete the J chain. The gene for the J chain does not lie on the same chromosome as those for antibodies. The J chain enables IgM to bind to receptors on secretory cells that facilitate its transport across the epithelial lining to mucosal secretions (see IgA, below).

- **IgD**: A 7S monomer of 184 KD, IgD differs in structure from IgG in two respects — it has a longer hinge region consisting of about 64 amino acids, and it has an unusually high carbohydrate content that is responsible for its high MW. Along with **mIgM**, **mIgD is expressed on the cell surface of all naïve B cells and is the major component of BcR complex**. Because of mechanisms not yet understood, the L chain of mIgD is of the κ type but that of serum IgD is

λ type. As with other isotypes, the majority of mIgD molecules are anchored in the plasma membrane by a transmembrane domain (chapter 6). However, a small fraction of mIgD molecules are linked to the plasma membrane by a glycosylphosphatidylinostiol anchor. The function of this form of mIgD is not known.

IgD-secreting plasma cells are very rare in the bone marrow or digestive mucosa. There are many more IgD producing plasma cells in the lymphoid tissues of the upper respiratory tract, such as in the nasal mucosa, salivary glands, lachrymal glands, adenoids, and tonsils. As many as 20% of tonsil plasma cells produce IgD antibodies, which suggests that it plays a role in defending the upper respiratory tract.

- **IgG**: This is the major Ig in serum, constituting 70-75% of total Igs. Its mean adult serum concentration is about 13 mg/mL, and **it is the major serum antibody in secondary immune responses**. Found in intra- and extra-vascular pools, IgG antibodies are important in the defence of the internal milieu. IgG has a very low carbohydrate content (2-3%, compared to 7-14% found in other isotypes). Four subclasses of IgG are recognised in humans, depending upon their amino acid sequences (IgG1–IgG4). They differ from each other in the sizes of their hinge regions and the numbers of inter-H chain bonds. IgG1 is the most common, accounting for 70% of serum IgG molecules; IgG4 is the least common (3%). The four subclasses also differ in their functions.
    - Except for IgG2, IgG isotypes cross the placenta and protect the newborn in the first few months after birth.
    - IgG3, IgG1, and IgG2 antibodies can fix complement by the classical pathway (in that order of efficiency), but IgG4 cannot. Complement-fixing isotypes are responsible for lysing microbes that have entered the body.
    - IgG1 and IgG3 antibodies can opsonise microbes because they can bind to high-affinity FcRs (FcγRI) expressed on phagocytic cells. By contrast, IgG4 antibodies are not as efficient in opsonisation, as they bind the receptor with intermediate affinity. IgG2 antibodies have very low affinity for FcγRI, and hence, plays a negligible role in opsonisation.

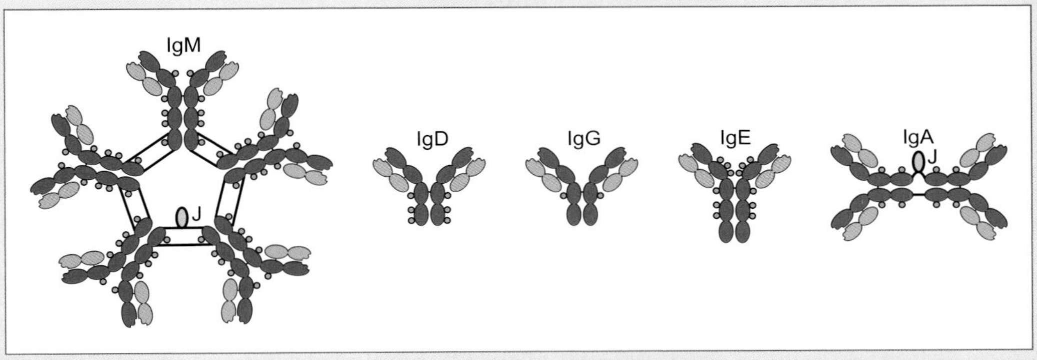

Figure 10.4 Five classes of Igs are recognised on the basis of their H chain amino acid sequences. The classes also differ in their fine structure, carbohydrate content, and functions. Serum IgM antibodies are pentamers that can dissociate to monomers upon treatment with mild reducing agents. The μ chain consists of five Ig domains (four C_H and one V_H). The first and fifth μ chains of pentameric IgM are bound by the J chain. IgD, IgG, and IgE antibodies are monomers having the prototypical four-chain structure. Like the μ chain, the ε chain also consists of five Ig domains. IgA antibodies are found both as monomers and dimers. Dimeric IgA antibodies are present predominantly in exocrine secretions. The two monomers of dimeric IgA are linked to each other by the J chain.

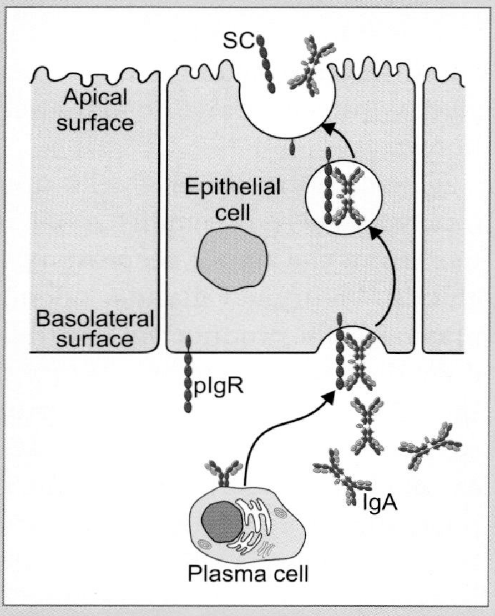

Figure 10.5 pIgR mediates the transcytosis of polymeric IgA or IgM. Secretory epithelial cells express pIgR, a receptor for the J chain of polymeric Igs, on their cell surface. pIgR that are expressed on the basolateral surfaces of these cells bind, internalise, and transport, IgA and IgM polymers to the apical surface of the cells. The antibodies eventually get released at the apical surface by the cleavage of pIgR. In addition, epithelial cells also secrete a free, soluble form of pIgR (termed SC) by proteolytic cleavage of the extracellular domain of the molecule.

- **IgE**: Although it is **present at extremely low concentrations in normal serum** (0.03-0.05 μg/mL), IgE antibodies have potent biological activity. **They are responsible for immediate hypersensitivity disorders**, such as asthma, hay fever, and urticaria. Basophils, mast cells, and eosinophils express FcεRI, a high-affinity receptor for IgE antibodies. Ligation of the receptor-bound IgE molecule by the antigen causes the degranulation of these cells. Pharmacological mediators released by these cells are responsible for the symptoms of allergic diseases (e.g., coughing, sneezing, and vomiting; see chapter 16). IgE has a sedimentation coefficient of 8S and a MW of 188 KD. It is heat-labile and can be destroyed by heating at 56°C for 30 minutes. Like the μ chain, the ε H chain has five domains, the additional domain being placed at the C_{H2} position, replacing the hinge region. IgE antibodies are thought to confer immunity against intestinal parasites.
- **IgA**: This molecule accounts for 15-20% of serum Igs and has a mean serum concentration of about 2.5 mg/mL. The bulk of the body's IgA-producing plasma cells are concentrated along mucosal and exocrine sites, especially along the intestinal tract, and most of them produce antibodies against environmental antigens. **IgA is the main Ig in exocrine secretions**, such as saliva, tears, colostrum, milk, and genitourinary and tracheobronchial secretions. It is not the major isotype in serum, even though it is the major antibody isotype synthesised in the body.

  In human serum, more than 80% IgA antibodies occur as monomers; the rest are dimers or even higher polymers. In human external secretions, however, polymers predominate. In other mammals, IgA is mainly dimeric (whether in serum or secretions). **IgA does not bind complement and hence protects mucosal surfaces without triggering an inflammatory response**. Two subclasses of IgA antibodies are recognised in humans — IgA1 and IgA2. About 80% serum IgA antibodies are of the IgA1 type, whereas both subclasses are equally represented in exocrine secretions. The secretory IgA is 11S dimer with a MW of about 385 KD. The Cα chain, like the Cμ chain, has a secretory tailpiece and a J chain. The IgA dimer is formed by disulphide linkages between the

J chain and the cysteine residues on the secretory tailpiece. Cysteine residues in the α CH3 regions of the two monomers are also directly linked by disulphide bonds, resulting in tail-to-tail dimers with Fab regions that are free to react with the antigen (Figure 10.5). A glycoprotein called the **S**ecretory **C**omponent (SC) was originally found to be associated with IgA molecules from exocrine secretions. Later, SC was established to occur in the free, unassociated form as well. Secretory epithelial cells express the molecule on their cell surfaces and secrete it into the surrounding milieu. SC is a transmembrane epithelial receptor for the J chain containing dimeric IgA and pentameric IgM antibodies, and it is now called by its functional name, **p**olymeric **Ig R**eceptor (**pIgR**). These antibodies bind to pIgR on the surface of epithelial cells through the J chain. The pIgR–antibody complexes get internalised by endocytosis at the basolateral surface of the cell and are eventually released to the apical surface[6]. Thus, receptor-mediated endocytosis by pIgR–J chain interaction enables antibodies to be transcytosed across epithelial cells and released in exocrine secretions (Figure 10.5).

[6] Epithelial cells have an apical surface that faces the outside (that is, the lumen) and a basolateral surface that makes contact with adjacent cells and the underlying connective tissue.

Milk and Tears: Protective Secretions

Lachrymal glands, responsible for the secretion of tears, are crucial to the immunological protection of ocular surfaces. These glands are the predominant source of sIgA antibodies in tears. They contain T cells, B cells, DCs, and macrophages, along with an extraordinarily high density of IgA^+IgD^+ plasma cells. Once produced, sIgA antibodies are secreted into the tear film. The defensive role of sIgA antibodies seems to be especially important during prolonged eye closure and at night, when sIgA represents almost 80% of the total tear protein. Contact lens use has been associated with decreased sIgA, IL-8, and PMN levels in the tear film, and it may predispose the wearer to eye infections.

The milk producing capability of mammals bestows a major survival advantage. Besides providing balanced nutrition, breast milk is important in the defence, growth, and development of the newborn. Important components of breast milk include the following:

- **IFNs**: Colostrum, the scant, sometimes yellowish milk produced during the first few days after birth, is particularly rich in IFNs and provides strong antiviral defence.
- **Leukocytes**: In the first 10 days after birth, there are more leukocytes per millilitre of breast milk than of blood.
  - Macrophages and neutrophils are the most common leukocytes in human milk and confer immunity through their phagocytosing capabilities.
  - Macrophages also secrete lysozyme, an enzyme that can disrupt the cell walls of Gram-positive bacteria.
  - Breast milk has both B and T cells. Milk lymphocytes seem to behave differently from blood lymphocytes — they proliferate in response to bacteria such as *E. coli* that cause life-threatening illnesses in babies, but they are less responsive to other nonthreatening organisms. Milk lymphocytes also secrete chemokines and cytokines such as IFN-γ that help strengthen the infant's immune response.
  - HIV-positive mothers have $CD8^+$ T cells in their breast milk. This adoptive transfer of cells to infants helps reduce the viral load and may help in controlling virus transmission from mother to infant.
- **Igs**: Although IgM, IgD, IgG, and IgA antibodies are all found in milk, sIgA is by far the most abundant. sIgA is synthesised and stored in breast tissue. sIgA antibodies are produced by the mother in response to pathogens in her (and hence the child's) immediate environment, and these are especially useful in

protecting the child. As the mother's immune system is tolerised to normal gut flora, the antibodies are not against this flora and do not interfere with colonisation of the infant's gut. sIgA is a noncomplement fixing antibody, and therefore, its binding to antigen does not trigger inflammation or damage the delicate mucosal membranes of the infant gut.

- **Oligosaccharides and mucins**: Both of these intercept bacteria and form harmless complexes that are excreted. They prevent microorganisms from attaching to mucosal surfaces.
- **Lactoferrin and vitamin B_{12}-binding factor**: These factors bind iron and vitamin B_{12}, respectively, making them unavailable to microorganisms.
- **Bifidus factor**: This is one of the oldest known components of human milk. It promotes the growth of *Lactobacillus bifidus* in the infant gut, crowding out potential pathogenic varieties of microorganisms.
- **Erythropoietin**: This erythropoiesis-regulating hormone is secreted by mammary epithelial cells. Milk erythropoietin may play a role in infant erythropoiesis, neurodevelopment, gut maturation, apoptosis, and immunity.
- **Hormones and growth factors**: These breast milk components help the infant in multiple ways. The gut membrane of newborn babies is termed *leaky* — it allows easy access to potentially harmful agents. Breast milk contains hormones (e.g., cortisol), and growth factors (such as epidermal growth factor, nerve growth factor, insulin-like growth factor, and somatomedin C), that help in the maturation of this mucosal lining, so that it becomes relatively impermeable to pathogens or harmful agents. Some hormones are thought to stimulate the production of lactoferrin, lysozyme, and sIgA in the mucosal lining of the baby's urinary tract, inducing local immunity.

10.5.2 Allotypes

All individuals of a species inherit the same set of C region genes. However, multiple alleles for these genes code for subtle amino acid differences, resulting in antigenic determinants termed *allotypic determinants*, which occur in some, but not all, individuals of the species. The sum of all the allotypic determinants on an Ig molecule is its allotype. **Allotypes are therefore allelic variants of isotypes that are found in some members of a species but not others** (from the Greek *allos*, meaning others). Allelic variation is caused by mutations in the corresponding structural genes that occur mainly in C regions of the molecule. Each allotypic determinant represents a difference of 1-4 amino acids encoded by different alleles. It is possible to find multiple allotypes in a heterozygote but not on the same Ig molecule. Most Ig allotypic variations do not seem to affect the antigenic specificity or the effector functions of the Ig molecule. Allotypes have been identified in both L and H chains and are designated by their class and subclass followed by the allele number, for example, G1m(1) is a IgG1 allotype 1. In humans, twenty five γ chain allotypes, two $\alpha 2$ chain allotypes, and three κ chain allotypes have been identified. Injecting antibodies from one member of a species into another member of the same species can yield antibodies to allotypic determinants, provided the two individuals differ in these determinants. Allotypic antibodies can arise during blood transfusions or pregnancy (when the mother produces antibodies to paternal allotypic determinants on foetal Igs).

10.5.3 Idiotypes

An individual can produce antibodies to virtually limitless antigenic determinants. These antibodies differ in their antigen-binding sites because of subtle differences in the amino acid sequences of the V region. **These unique amino acid sequences in the molecule's antigen-binding region are themselves**

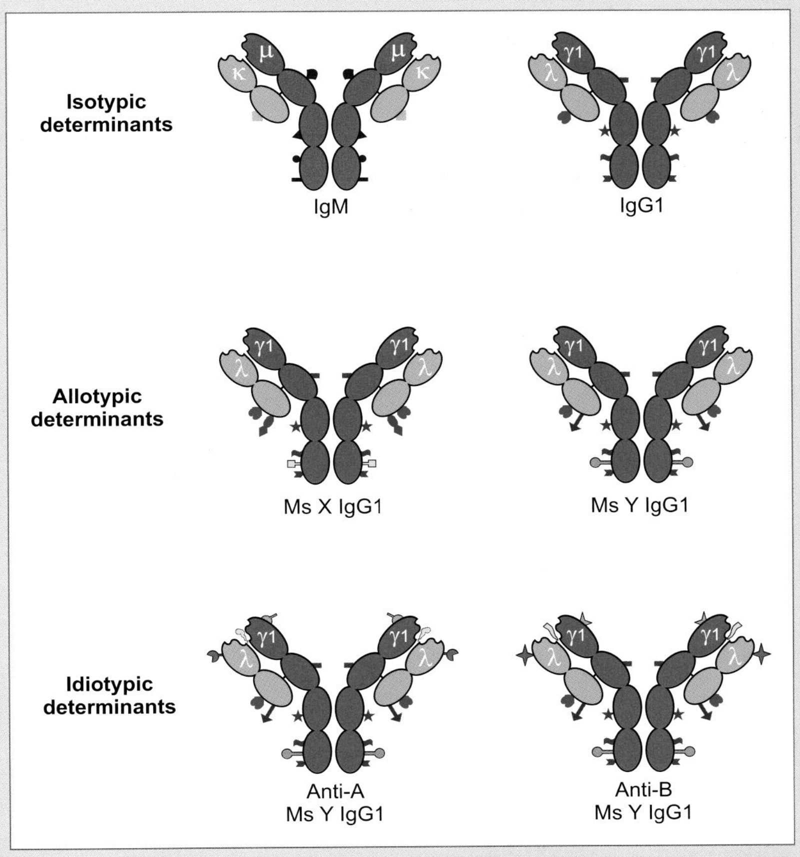

Figure 10.6 Different classes of Igs can be recognised based on the presence of antigenic determinants on different regions of the Ig molecules. Isotypic determinants are present on the C regions of the H and L chains of the same individual. Allotypic determinants are encoded by different alleles of the C region genes; different individuals (represented by Ms X and Ms Y) are therefore likely to have different allotypic determinants. Idiotypic determinants are present on the rearranged V_H and V_L genes. Anti-A Igs will therefore have idiotopes different from anti-B Igs.

unique antigenic determinants and are termed *idiotypic determinants* (Figure 10.6). Both L and H chains contribute to the conformation of the antigen-binding site and therefore to the idiotype. The antigen-binding site and V region sequences outside this site can also contribute to idiotypic determinants. A single determinant is called the idiotope, and the sum of all the idiotopes on a single V region is an idiotype. Because a clone of B cells secretes Igs with identical V regions, these Igs have the same idiotype. As it is difficult to determine if a given antiserum detects one or more idiotopes, such antisera are referred to as anti-idiotypic antisera. Anti-idiotypic reagents can be used to follow the inheritance of genetic markers on Ig V regions, map V region genes, and measure the regulation of V gene expression.

10.6 The Role of Immunoglobulins in Immunity

Igs bind antigen through their Fab regions. This binding is in some instances sufficient to confer a degree of resistance to the host. For example, the binding of Igs to viral epitopes interferes with the virus' attachment to its target cell receptors and helps in antiviral resistance. An antibody may also disrupt the viral outer coat structure, preventing its interaction with cell surface receptors or interfering with

Linking Back: The Network Hypothesis

Jerne propounded the network hypothesis to explain self-tolerance in 1974, proposing that the immune system is a network of idiotypes and anti-idiotypes. According to this hypothesis, when an antibody is produced in response to an antigen, the V region, consisting of many idiotopes, can itself act as an antigen and stimulate the production of an antibody, the anti-idiotype. The idiotopes on the anti-idiotype lead to the production of anti-anti-idiotype and so on, until a network of idiotypes and anti-idiotypes is established. Some of these anti-idiotypes are an internal image of the idiotype, that is, an image of the antigen. The original hypothesis postulated the network to be multibranched, with each idiotype-producing cell controlled by several anti-idiotypes. The whole network was thus dependent on all its parts for control. It was proposed that the equilibrium is disturbed when an external antigen is added to the network, and appropriate cells are stimulated to produce antibodies to the antigen, the idiotype, and so forth, to regain balance. Thus, antibody production continues until equilibrium is established again.

This concept is plausible and very attractive. However, experimental proof — the real test of any hypothesis — is still elusive. Any experiments designed to prove the existence of networks have to be conducted under conditions so far removed from normality that there are theoretical objections to extrapolating the evidence to *in vivo* physiological conditions. Anti-idiotypes seem to be produced in normal immune responses and to coexist in serum with their specific idiotype, probably in the form of immune complexes, but the presence of an extensive network of idiotypes–anti-idiotypes has not yet been proven. The fact that anti-idiotypes can modulate (transiently suppress or enhance) immune responses supports the network theory. This modulation can be observed even when the anti-idiotype is injected in amounts likely to be present in normal serum. Moreover, T_{reg} cells that are idiotype- and not antigen-specific have been found in normal immune responses. There is also some evidence to show that idiotypic T_H cells are formed and needed in the development of normal immune responses.

The hypothesis continues to arouse the passions of immunologists. The antigen appears to be of prime importance in stimulating and regulating an immune response, but idiotypic regulation may direct the response toward a particular spectrum of idiotypes. In the resting stage, before the immune system comes in contact with the antigen, idiotype–anti-idiotype regulation is probably important in determining the initial state in which the immune system encounters the antigen.

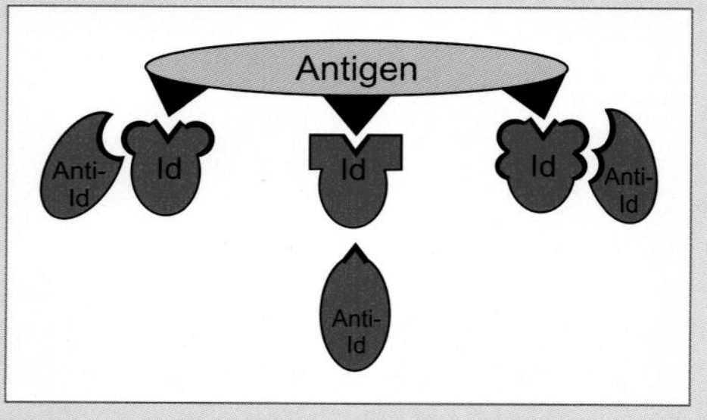

Figure 10.S4 The network hypothesis proposes that antigenic challenge gives rise to a network of antibodies consisting of idiotypes (Id) and anti-idiotypes (Anti-Id), and the whole network is dependent on all its parts for control.

the fusion of the virus membrane and the cell membrane. In either case, the antibody interaction neutralises the infecting potential of the virus, and this process is termed *virus neutralisation*. Antibodies can similarly neutralise the action of toxins or enzymes. However, the mere binding of Igs to a pathogen is insufficient in killing it or removing it from circulation. **To be effective instruments of defence, antibodies must not only bind the target antigen but also invoke effector functions that will remove the antigen and/or kill the pathogen**. The Fc region of the antibody invokes these functions by interacting with participating cells or proteins.

The tissues and cells of the body express specific receptors for FcRs of different antibody isotypes and are described in section 10.7. Ig–FcR interaction triggers the following effector functions, depending upon the type of FcR and the antibody's isotype.

- **Opsonisation**: Coating of the antigen by IgG antibodies promotes its phagocytosis. Macrophages and neutrophils express FcRs that can bind the Fc region of most IgG subclasses. Particulate antigens or pathogens coated with IgG antibodies bind to these FcRs. Although the strength of the interaction between individual FcR and IgG may be weak, the simultaneous binding of multiple Igs to the same target produces a signal of considerable strength. These interactions immobilise the pathogen to the phagocyte's surface. The clustering of FcRs leads to the activation of the cell, phagocytosis of the immobilised pathogen–antibody complex, and the pathogen's eventual destruction. An antigen internalised *via* FcR is loaded more efficiently on MHC molecules than that internalised by other receptors, and such an antigen is more efficient in triggering the immune response.
- **Complement activation**: The interaction of IgM and most IgG isotypes with homologous antigen activates complement. Activated complement components are multifunctional molecules with powerful immuno-modulatory and effector functions (see chapter 3). For example, the complement component C3b and its products promote opsonisation, virus neutralisation, B cell activation, and so on. C3a and C5a are chemotactic agents that attract neutrophils and macrophages to the site of antigen–antibody interaction and promote inflammatory responses. The C5b6789 complex is responsible for the lysis of target cells.
- **ADCC**: The ligation of FcRs on K cells by pathogen- (or antigen-) bound antibodies activates them. Activated K cells release cytotoxins that bring about lysis of the target cell by ADCC (chapter 11).
- **Degranulation**: The ligation of FcεRI on the surface of mast cells, basophils, and eosinophils by parasite- (or antigen-) bound IgE antibodies releases pharmacological mediators that can kill the parasite. These mediators are also responsible for the unpleasant effects of immediate hypersensitivity.
- **Transcytosis**: IgA (and IgM) antibodies are transported across the epithelial cell barrier as a result of pIgR-mediated endocytosis; IgA antibodies present in exocrine secretions defend mucosal surfaces against pathogenic invaders.

10.7 Immunoglobulin Receptors

Physiologically, antibody molecules are adaptors that link the target (antigen) to effectors and mediate target elimination. Therefore, by definition, receptors for Igs must be expressed by a variety of effector cells. These receptors comprise a family of molecules that bind the Fc region of the Ig molecule, with each member of the family recognising Igs of one or a few closely related types. FcRs are expressed by many cells of the immune system — macrophages, neutrophils, basophils, mast cells, eosinophils, lymphocytes (K, T, and B cells), DCs, and FDCs.

Most FcRs are members of the Ig superfamily. **FcRs are membrane-associated glycoproteins that mediate a vast array of functions**, including positive and negative regulation of immune cell responses, triggering internalisation of opsonised cells, and transcytosis. FcRs have been identified for all Ig istoypes except IgD (Figure 10.7). Different types of FcRs are found for a single Ig isotype, and these differ in functions as well. Generally, type I receptors bind Ig molecules with high affinity, and type II or III receptors bind Igs with low affinity. Thus, of the three types of FcRs that bind IgG, FcγRIs are high-affinity receptors that bind monomeric IgG, whereas FcγRII and FcγIII are low-affinity receptors that can only bind antigen–

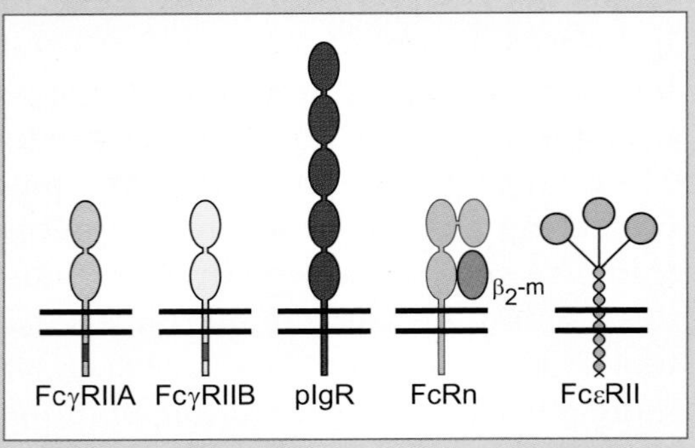

Figure 10.7 FcRs are membrane-associated glycoproteins that differ greatly in their structure and in the functions they mediate. Most FcRs have an Ig-fold and belong to the Ig superfamily. Some FcRs can activate or inhibit cellular functions when the Igs bound to them are ligated by the antigen. FcRs having an ITAM (or ITAM-like motif) in their cytoplasmic tails deliver an activatory signal; those having an ITIM (or ITIM-like motif) deliver negative signals. Others, like pIgR, trigger the internalisation and transport of bound Igs from basolateral to apical surfaces. FcRn is unique, in that it is structurally similar to MHC class I molecules and is associated with β2-m. It is thought to be involved in the transfer of maternal IgG antibodies across the neonatal intestinal epithelium and in IgG homeostasis in adults. Unlike other FcRs, the low-affinity receptor for IgE, FcεRII, does not have an Ig-fold. It is a C-type lectin adhesion molecule expressed in different isoforms on different cells.

IgG complexes. Similarly, FcεRI, present on eosinophils and mast cells, is a high-affinity IgE receptor; FcεRII, present on a variety of cells, binds IgE molecules with low affinity.

Whether FcRs deliver an activatory or an inhibitory signal depends upon the motif in their cytoplasmic domains. In mice, FcγRI, FcγRIIA, FcγRIIC, and FcγRIIIA deliver activating signals. Most activatory FcRs are multicomponent units. The component chains of these receptors are named α, β, and γ. The α chain is involved in specific recognition and is unique to each receptor. The other chains are involved in signal transduction or intracellular transport. The γ chain, termed the FcR common γ chain, is related to the TcR CD3ζ chain and contains an ITAM. This common γ chain is also necessary for the assembly and surface expression of FcγRI, FcγRIII, FcαRI, and FcεRI. Clustering of the activating receptors results in phosphorylation of the tyrosine residues in ITAMs by Src family kinases. Further events result in the phosphorylation of Syk and activation of PLCγ and PI3-kinase. This activates MAP kinases, mobilises Ca^{2+}, and reorganises the cytoskeleton of the activatory FcR expressing cells (these signalling pathways are explained in chapter 6). Signalling through ITAM containing FcRs leads to an oxidative burst, cytokine release, phagocytosis by macrophages, ADCC by K cells, and degranulation in mast cells.

FcγRIIB (incidentally, the only FcγR expressed by B cells) is a single-chain molecule that has an ITIM in its cytoplasmic domain. It is also expressed on myeloid and monocytic cells, and it acts as a negative regulator of immune complex-triggered activation. The coligation of BcR with FcγRIIB by antigen–IgG complexes blocks the downstream biological responses of B cells (including activation, antigen presentation, proliferation, and antibody production). The initial event in this inhibitory signalling is the phosphorylation of tyrosine in the ITIM by the Src family kinase Lyn. This signalling interferes with translocation of the Btk kinase that is essential in BcR signalling. It also suppresses antiapoptotic signals that are necessary for the proliferation and survival of activated B cells. In addition, the crosslinking of FcγRIIB by IgG molecules is thought to deliver an apoptotic signal to germinal centre B cells. Thus, when the concentration of serum IgG exceeds a certain threshold, FcγRIIB seems to switch off Ig synthesis and downregulate the humoral response.

FcεRI is of special interest because of its importance in type I hypersensitivity reactions (see chapter 16). It binds IgE antibodies with a very high affinity and has a limited distribution. The classical tetrameric form of this molecule is constitutively expressed on effectors of anaphylaxis (i.e., mast cells, basophils, and activated eosinophils); a trimeric form of the molecule is found on APCs, such as monocytes, DCs, and Langerhans cells. It is one of the earliest markers expressed on mast cells. The α chain is responsible for IgE binding, the β chain increases stability and signalling capacity, and the shared γ chain dimer has ITAMs that are responsible for downstream signalling. The binding of polyvalent antigens by receptor-bound IgE molecules causes receptor aggregation and triggers cellular activation, degranulation, and the release of pharmacological mediators by mast cells and basophils. The activation of eosinophils by FcεRI provides defence against parasitic infections. On APCs, FcεRI delivers IgE-bound antigen to MHC class II presentation pathways. FcεRII (CD23) is the low-affinity receptor for IgE. CD23a is expressed constitutively on B cells; the IL-4 induced CD23b isoform is found on T cells, macrophages, monocytes, DCs, platelets, eosinophils, etc. Unlike other FcRs, CD23 is a C-type lectin adhesion molecule that can bind CD21 (CR2). Proteolysis of CD23 leads to the release of a soluble form of FcεRII that can be further degraded. All except the smallest 12 KD fragment retain IgE-binding capacity. In rats, CD23 has been shown to be involved in the transepithelial transport of IgE. A similar role is suspected in humans. CD23 has also been implicated in IgE synthesis and IgE-mediated immune and inflammatory functions.

FcRn is structurally similar to the MHC class I molecule, except that it has a closed peptide groove. In neonatal rodents, it is involved in the transfer of maternal IgG antibodies across the neonatal intestine. In humans, it may have a similar role in early post-delivery days, when colostrum has a high percentage of IgG. Adult human enterocytes express high levels of FcRn and have been shown to be involved in bidirectional transcytosis of IgG molecules across intestinal epithelial monolayers *in vitro*. FcRn is also thought to have a role in IgG homoeostasis through its regulation of the rate of IgG catabolisation by endothelial cells. These cells are postulated to take up IgG molecules by nonspecific pinocytosis, as the pH of blood is not suitable for FcRn–IgG interaction. The internalised IgG molecules enter the acidic endosomal/lysosomal pathway, where they can bind FcRn and get rescued from catabolisation. IgG that fails to bind to FcRn may be destroyed. Support for

Immunoglobulin Receptors

- FcRs bind Fc regions of Ig molecules.
- Many receptors act as adapters that link Ig molecules to target cells.
- Multiple types of receptors can bind Igs of a single isotype; those that bind Igs with high affinity are called type I, and those with low affinity are called type II/III FcRs.
- Coligation of FcRs delivers an activatory or inhibitory signal, depending upon the presence of ITAMs or ITIMs in the cytoplasmic tails of the FcRs.
- Important Ig receptors include:
  - FcεRI, involved in the degranulation of mast cells, eosinophils and basophils;
  - FcγRIIB, which acts a negative regulator of B cells;
  - FcRn, involved in the transfer of maternal IgG antibodies and possibly IgG catabolisation; and
  - pIgR, responsible for the transcytosis of IgA and IgM antibodies.

this theory comes from work in knock out mice that are deficient in β2-m (required for FcRn assembly). These mice show dramatically reduced levels of serum IgG despite an apparently normal B cell compartment, suggesting abnormally high degradation of IgG.

Like FcRn, **pIgR** is responsible for the transcytosis of polymeric IgA antibodies (and IgM, if present) across epithelial cells. Non-IgA-bound pIgR is also efficiently transcytosed, leading to the release of the free cleaved extracellular domain of pIgR, SC, in lumenal secretions. Recent evidence suggests that the carbohydrate-rich SC may act as a scavenger molecule in its own right, helping to mop up enteric pathogens in lumenal secretions. Thus, pIgR seems to have a dual function — it can behave as a FcR and as a PRR.

Linking Back: Theories of Antibody Formation

As the field of immunology grew, a theoretical framework was needed to explain antibody formation. In the early years, theories focused entirely on antibodies because the two types of immune responses (cell-mediated and humoral) were yet to be recognised. Two schools of thought emerged, each with strong adherents.

- **The instructive theorists** proposed that antibody formation can occur only after the antigen provides the information necessary for antibody synthesis.
- **The selective theorists** believed that information required for antibody synthesis pre-exists in the cell; the immunogen merely selects and stimulates the appropriate cell.

The instructive theories and the template hypothesis

The best known of the instructive theories was the template hypothesis. Formulated by Haurowitz, Mudd, and Alexander in the early 1930s, it was later modified by the renowned scientist Linus Pauling. The theory proposed that the antigen acts as a mould or template that can enter any Ig-producing cell and modify the pattern of laid-down amino acids to fit the template. This results in the synthesis of a molecule with a spatial configuration complementary to that of the antigen molecule. Thus, the template hypothesis suggested that the specificity of the Ig molecule is not determined by the primary amino acid sequence but by the process of moulding the nascent molecule around the antigenic determinant; the immunogen is a template at the level of protein synthesis. To account for the continued production of antibodies, either the antigen or a part of it was presumed to remain in the cell to direct the configuration of future antibody production. A modified form of the theory proposed that the antigen modifies genetic information in the DNA of the cell so that the cell and its progeny continue to produce Igs of a particular specificity.

These theories became popular because they seemed to fit with immunochemical discoveries of the specificity of the immune response, especially those in Landsteiner's experiments, outlined in chapter 4. Also, little was known about macromolecular synthesis at that time. Once it was established that the shape, as well as the activity of a protein, was determined by its amino acid sequence, which in turn was determined by its DNA sequence, the template theory was abandoned. The discovery that the three-dimensional structure of antibodies is reversible, that is, the molecule unfolds on denaturation but refolds under appropriate conditions, rang the death knell for the template hypothesis. The idea that a molecule can act as a template for protein synthesis may sound ridiculous today, but countless papers were published about and tempers were frayed over this theory in the early 1940s.

The selective hypothesis and the clonal selection theory

The clonal selection theory of Jerne and Talmage stated that an individual possesses an immensely diverse pool of cells, with each cell capable of responding to only one or a few related antigens. According to this theory, when an immunogen penetrates the body, it selects only those few cells that possess Ig molecules specific to the antigen — the antigen is only the trigger inducing cells to produce antibodies.

In 1959, Burnet proposed that cells of the antibody forming system arise from random mutations, resulting in the emergence of a small number of cells, or clones of differentiated cells, capable of producing one or a few specific antibodies. He suggested that the contact of such differentiated cells with self- or foreign immunogens during foetal life — before the cells reached maturity — suppresses rather than stimulates antibody formation, possibly because of clonal abortion. He later added that this clone censoring could occur in the thymus. In an immunocompetent individual, however, the contact was thought to trigger the proliferation and generation of a clone of effector cells capable of antibody secretion. Some of these lymphocytes were said to circulate in the body and give rise to an increased response upon secondary immunisation. The clonal selection theory explained most of the phenomena known at the time and was generally accepted. The reality, as we now know, although very close to the proposed theory, is much more complicated.

11 Cell-mediated Immunity

Goldeneye, I found his weakness
Goldeneye, he'll do what I please
Goldeneye, no time for sweetness
But a bitter kiss will bring him to his knees

— Tina Turner, *Golden Eye*

11.1 Introduction

The term **cell-mediated immunity (CMI) indicates an immune response in which cells are the effectors of immunity, and antibodies have only a minor role** (if at all; Table 11.1). It is the most primitive form of immunity; CMI evolved before humoral responses. Effectors of CMI help eliminate foreign cells and infected or altered self-cells. Both the innate and adaptive arms of immunity are involved in CMI and include

- CTLs, NK cells, and γδ T cells involved in the elimination of infected and neoplastic cells;
- macrophages and neutrophils that are both important in the destruction of bacteria, viruses, and other intracellular pathogens; and
- eosinophils, mast cells, and basophils, engaged in the destruction of helminths.
- **CTLs**: As explained in section 9.5.3, CTLs are MHC class I restricted $CD8^+$ T cells that represent the cytotoxic arm of adaptive immunity (Table 11.2). Like humoral responses, secondary CMI responses — mediated by CTLs and aided by $CD4^+$ T cells — are more rapid and aggressive than primary immune responses. These rapid CTL responses (with the Ig response) help control and/or fully eliminate intracellular pathogens. As with all secondary adaptive responses, the swift secondary response is partly a consequence of the clonal expansion of antigen-specific $CD4^+$ and $CD8^+$ T cells and partly due to the reprogramming of the gene-expression profile that differentiating T cells undergo during the primary encounter. As a result of clonal expansion, the number of CTLs capable of responding to the pathogen increases. Furthermore, unlike naïve CTLs, effector and memory $CD8^+$ T cells have elevated levels of mRNA for IFN-γ and cytotoxic molecules, such as perforin and granzyme B. Although the synthesis of these proteins occurs only upon antigen contact, these elevated mRNA levels endow memory $CD8^+$ T cells with the capacity to produce larger quantities of proteins at a more rapid pace than naïve T cells. Memory $CD8^+$ T cells (like their $CD4^+$ counterparts) also express a different pattern of surface proteins involved in cell adhesion and chemotaxis than do naïve $CD8^+$ T cells. These surface proteins allow them to extravasate into nonlymphoid tissues and mucosal sites. Lastly, the number of memory $CD8^+$ T cells is maintained for a long time because of homeostatic cell proliferation, which occurs at a slow yet steady pace. IL-2, IL-7, and especially IL-15 are thought to be involved in this slow proliferation.
- **NK and related cells**: NK cells are large, granular lymphocytes comprising 5-15% of peripheral blood lymphocytes. They are also found in peripheral tissues, such as the liver, peritoneal cavity, and even the placenta. Resting NK cells circulate in the blood. Upon cytokine activation, they are capable of extravasation and infiltration into tissues. They are a vital component of the innate arm of immunity that mediates the killing of tumour cells or virus-infected cells and are discussed in chapter 2. IL-2 has been shown to enhance the cytotoxic activity of NK cells. IL-2-activated NK cells, known as LAK cells, are used in cancer immunotherapy. The morphologically similar K cells are considered a subset of NK cells. They express FcγRIII (the low-affinity receptor for IgG; CD16) and, in some instances, FcμR. These cells are effectors of ADCC.
- **γδ T cells**: These cytotoxic T cells kill target cells by the perforin-granzyme pathway and are described in chapters 2, 4, and 9.

Abbreviations

BCG:	Bacillus of Calmette and Guerin
CAD:	Caspase-Associated DNAase
CMI:	Cell-Mediated Immunity
DTH:	Delayed-Type Hypersensitivity
DTP/: DTaP	Diphtheria, Tetanus, Pertussis
DISC:	Death-Inducing Signalling Complex
FADD:	Fas-Associated Death Domain
FLIP:	FLICE-Inhibiting Protein
IAPs:	Inhibitors of Apoptosis
IPV:	Inactivated Polio Virus
MMR:	Measles, Mumps, Rubella
OPV:	Oral Polio Vaccine
RNI:	Reactive Nitrogen Intermediates
ROI:	Reactive Oxygen Intermediates
TRAIL:	TNF-Related Apoptosis-Inducing Ligand

Table 11.1 Comparison of Cell-Mediated and Humoral Immunity

Characteristic	CMI	Humoral immunity
Effector cells	CTLs NK cells, NKT cells, $\gamma\delta$ T cells Phagocytic cells: macrophages, neutrophils Eosinophils, and mast cells	B cells
Mechanism of action	Antigen-nonspecific effector molecules released by effector cells in the immunological synapse kill target cells. Effector molecules released include • cytotoxic molecules, such as perforin and granzymes released by CTLs, NK cells, and $\gamma\delta$ T cells • pharmacologically active mediators released by eosinophils and mast cells • microbicidal products, such as ROI and RNI produced by macrophages and neutrophils Induction of apoptotic pathways by CTLs and CD4$^+$ T cells	Antibody molecules bind specifically to epitopes on antigens, and this results in the • neutralisation of viruses, enzymes, and toxins; • opsonisation of the antigens; • complement activation; • ADCC; and • release of pharmacological mediators from eosinophils, mast cells, and basophils.
Recognition system	PRRs, NK cell receptors (KIR, NKG), TcRs, and FcRs	BcRs (and TcRs)
Time taken for manifestation	NK cells act within minutes of activation CTLs usually take between 24-72 hours following secondary antigenic challenge to exert their effects ADCC can be manifested from within seconds to minutes of secondary antigenic challenge	Response may occur from within seconds to minutes of secondary antigenic challenge
Role of CD4$^+$ T cells	Help in activating phagocytic cells Augment CD8$^+$ T cell responses Some CD4$^+$ T cells can contribute to CMI directly *via* the death-receptor pathway	Cognate interaction with helper CD4$^+$ T cells is indispensable for humoral responses to TD antigens Cytokines produced by CD4$^+$ T cells augment humoral responses to TI type 2 antigens

❒ **Phagocytic cells**: Forming a link between specific and innate immunity, phagocytic cells are one of the most important watchdogs of the body and have been extensively discussed in chapter 2. They have a central role in innate, cell-mediated, and humoral immune responses (Table 11.3).

Macrophages and neutrophils are normally quiescent cells responsible for the removal of tissue debris and dead or dying cells. Phagocytes express a broad spectrum of receptors, such as those for complement components, integrins, scavenger receptors, and PRRs, as described in chapter 2. The engagement of these receptors with microbes or their products results in the mobilisation of the phagocytic membrane and reorganisation of the actin cytoskeleton. As a consequence, the microbe is internalised and eventually digested in specialised structures called phagolysosomes. Concurrent exposure to IFN-γ (produced by NK cells or T cells) and/or contact with pathogen-derived molecules activates phagocytic cells. Activated phagocytes have an increased number of lysosomes and secrete enzymes and cytokines that contribute to inflammation. Cytokine exposure also increases the production of microbicidal molecules such as superoxide anion, H_2O_2, RNI (**R**eactive **N**itrogen **I**ntermediates), and ROI (**R**eactive **O**xygen **I**ntermediates; Table 2.1). Activated cells are efficient effectors of CMI and are important in inflammatory and DTH (**D**elayed-**T**ype **H**ypersensitivity) responses. By virtue of their FcR expression they are also effectors of ADCC.

Eosinophils, basophils, and mast cells produce several pharmacologically active substances that are effective in killing large parasites that cannot be phagocytosed (chapter 2). In addition, mast cells have been shown to phagocytose microbes that are bound to their cell surfaces *via* complement receptors and FcγRs. All three cell types express FcεRs. The crosslinking of FcεR-bound IgE antibodies by a multivalent parasite leads to the degranulation of these cells. Thus, these cells can also participate in ADCC.

Table 11.2 NK Cells and CTLs in CMI

Characteristic	NK cells	CTLs
Target range	Effective against tumour cells and virus-infected cells	Effective against tumour cells, virus-infected cells, and cells harbouring intracellular pathogens
Target recognition	Specific antigen-recognition or MHC restriction is absent A balance of activatory and inhibitory signals determines fate of target cell Recognition of antigenic epitopes by FcγRIII-bound IgG antibodies activates K cells	TcR recognition of pathogen-derived peptide:MHC class I complexes
Nature of responses	Effectors of the innate arm of immunity Activation does not lead to proliferation of cells; memory responses are absent	Effectors of adaptive immunity Activation leads to proliferation and differentiation of cells; secondary responses are rapid and more aggressive than primary responses
Mechanism of killing	Exocytosis of cytotoxic granules containing perforin and granzymes Induction of apoptotic pathways Granules exist preformed during NK cell development; killing can occur within minutes of activation	Exocytosis of cytotoxic granules; cytotoxic granules also contain granulysin in addition to perforin and granzymes Induction of apoptotic pathways Naïve $CD8^+$ T cells are not cytolytic; they must undergo differentiation to effector CTLs
Cytokines	IFN-γ is the major cytokine produced	Two broad types of CTLs are recognised; type I produce IFN-γ , type 2 produce IL-4 and IL-10

Table 11.3 Macrophage Cytokines and Immune Responses

Cytokine	Effects
TNF-α	Multipotent cytokine Increases respiratory burst and RNI production Increases expression of adhesion molecules Increases expression of FcRs on macrophages Induces IFN-γ release from NK cells Causes acute-phase reaction and shock in conjunction with IL-1 and IL-6
IL-1	Activates NK cells, macrophages, and neutrophils Induces IL-2R expression and IL-2 synthesis in activated T cells Aids in B cell differentiation
IL-6	Release of acute-phase proteins by hepatocytes Aids in B cell differentiation
IL-8	Causes chemotaxis of neutrophils and T cells
IL-10	Promotes TH2 responses Suppresses macrophage function
IL-12	Promotes TH1 pathways Induces IFN-γ release by NK cells
IFN-α and -β	Have antiviral effects; important in CMI Increase MHC class I expression
CSFs (Colony Stimulating Factors)	Promote growth of granulocytes (G-CSF), monocytes (M-CSF), and granulocyte/monocytes precursors (GM-CSF) GM-CSF promotes differentiation of monocytes to DCs

11.2 Mechanisms of Cell-Mediated Cytotoxicity

Cytotoxic cells employ two death-inducing strategies to bring about the contact-dependent death of target cells — the Ca^{2+}-dependent exocytosis of cytotoxic granules and the engagement of death-receptors. The first pathway appears to be predominant in NK cells and $CD8^+$ T cells, whereas the second seems to be of special importance for TH1 effector cells. The cytotoxic granules of CTLs and NK cells are complex organelles that combine specialised storage and secretory functions with the degradative functions of typical lysosomes. FcR-expressing effector cells with a cytotoxic potential, such as macrophages, neutrophils, eosinophils, mast cells, basophils, and K cells, induce death by a different mechanism — ADCC. The effectors of CMI also produce a variety of cytokines, such as TNFs and IFN-γ, which when secreted in the vicinity of target cells have cytotoxic actions; these will not be discussed here.

11.2.1 Perforin and Granzyme Pathway

Perforin/granzyme-mediated apoptosis is the principal pathway used by cytolytic lymphocytes, such as NK cells, NKT cells, γδ T cells, cytolytic $CD4^+$ T cells, and CTLs to eliminate target cells. Vesicles containing these cytolytic molecules are formed during NK cell development and thus are present in mature NK cells. CTLs, by contrast, synthesise cytotoxic vesicles within a day of T cell activation, that is, only effector CTLs have cytotoxic vesicles. Cytotoxic vesicles containing perforin and granzymes reside in the cytoplasm of NK cells and effector

Cheating Death: Viral Evasion of Cytotoxic Lymphocyte-Mediated Apoptosis

The detection and elimination of pathogen-infected cells is perhaps the most important function of CMI. One obvious strategy the intracellular parasite can use to avoid death by CTLs is the downregulation of MHC class I molecules on the infected cell. However, this strategy makes the infected cell susceptible to NK cell attack. Hence, viruses use other strategies to evade/survive CMI. These strategies are now being investigated for their therapeutic potential.

- The cowpox virus and the baculovirus encode caspase inhibitors that block caspase activity and allow the survival of infected cells.
- A large number of viruses encode functional Bcl-2-like proteins that allow them to evade intrinsic death pathways. M1 1L, from the myxoma virus, and UL37, from the human cytomegalovirus, inhibit apoptosis by blocking the release of cytochrome *c* from mitochondria.
- The adenovirus protein L4-100K directly inhibits granzyme B.

CTLs, and these vesicles migrate to the site of contact upon appropriate stimuli. The engagement of activatory receptors and costimulatory molecules causes these vesicles to migrate to the contact site. The vesicles fuse with the cell membrane, and their contents are secreted into the tight intracellular junction (termed *immunological synapse*[1]) formed between the two cells. The cytotoxic granules of NK cells and CTLs contain several toxic constituents.

- **Perforin is a pore-forming, membrane-disrupting protein**, like the complement component C9. It was originally thought to be the chief architect of NK cell-/CTL-mediated cell death. Upon release from granules, perforin rapidly polymerises in the presence of Ca^{2+} to form a ring-like structure with a central pore which gets inserted into the target cell membrane. It is now recognised that perforin-induced cell damage alone is insufficient to bring about cell death, and perforin is suggested to be an entry portal for other cytotoxic molecules (such as granzymes) that cause cell death. Studies in knock out mice clearly show the importance of perforin in lymphocyte-mediated cytotoxicity; such mice show a deficiency in all aspects of granular killing. The exact role of perforin in cell death is unclear, and the delineation of its functions remains contentious. Perforin is rapidly inactivated in the presence of high lipid concentrations and Ca^{2+}, and the presence of either of these factors in the immunological synapses helps cells escape perforin-inflicted damage.
- **Granzymes are serine proteases found in their active, processed form in cytotoxic granules**. Five granzymes (A, B, H, K, and M) are found in humans. **Perforin and granzymes are thought to act cooperatively in causing target cell apoptosis** (Figure 11.1). Granzymes' entry into target cells is a pivotal step in cell death. The mechanism by which granzymes enter a cell is not clear. The older model postulating that granzymes enter target cells through perforin pores is increasingly questioned. Recent experiments suggest that at least some granzymes may enter a cell by way of specific receptors. Granzyme B is the most extensively studied granzyme. It cleaves target cell proteins at specific aspartate residues and is a potent activator of apoptosis. It can cleave procaspases-3 and -8 to release the active forms and initiate target cell death (see next section). It also activates Bid, a proapoptotic member of the Bcl-2 family.

[1] The immunological synapse is a distinct region formed at the contact zone between the cytotoxic lymphocyte and target cell because of the specific reorganising of cell-surface membrane proteins.

Together this results in the leakage of proapoptotic mediators (e.g., cytochrome *c*) into the cytosol. Granzyme B is also thought to be involved in DNA fragmentation, which is the hallmark of apoptosis. Granzyme A, by contrast, is a tryptic protease that does not activate caspases but kills cells by directly cleaving nuclear proteins and facilitating the formation and accumulation of single-stranded DNA breaks. The role of other human granzymes is still being investigated. Because granzymes and perforin are released in immunological synapses, cytotoxic cells need protective mechanisms to avoid being killed by their own toxins. CTLs and NK cells are thought to avoid such granzyme B-induced suicide by expressing protease inhibitor-9 in their cytoplasm and nucleus. A similar inhibitor for granzyme A has not been reported yet. Antithrombin A and α-2 macroglobulin have been shown to interact stably with granzyme A, neutralising it. Their role in cell protection is still being investigated. Cathepsin B which is present in the granules affords protection against perforin.

- **Other constituents** of cytotoxic granules include the following:
  - Several glycosaminoglycan complexes, including **serglycin**, are thought to act as a scaffold for the packaging of highly positively charged granzymes and may also act as a chaperone for secreted proteases. The recent discovery of macromolecular complexes of serglycin, perforin, and granzymes that bind to target cell surfaces has led to the suggestion that toxic complexes

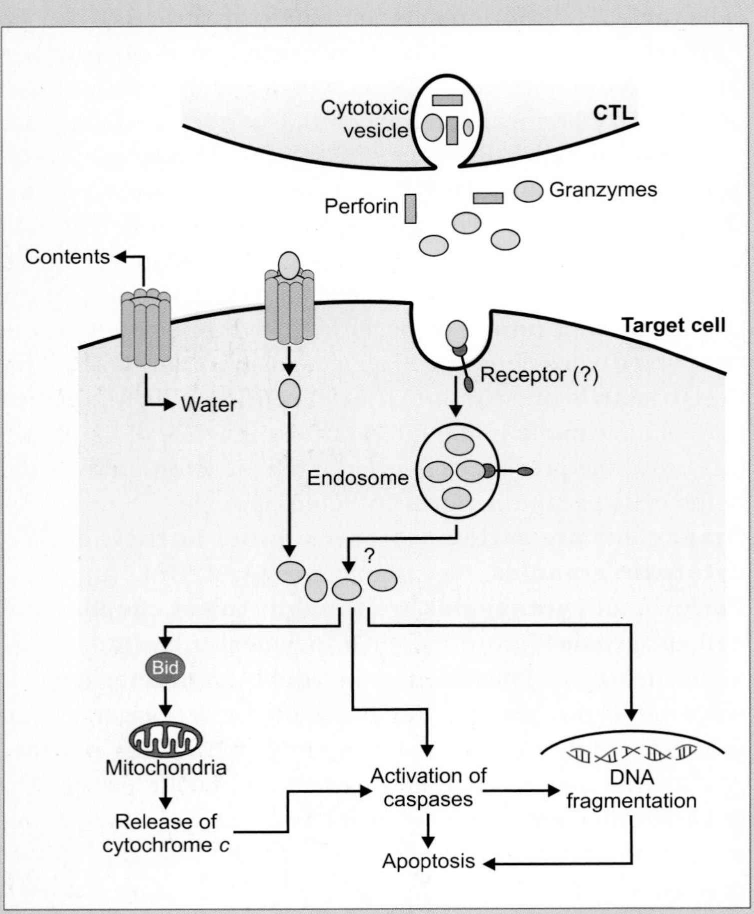

Figure 11.1 The perforin and granzymes that are released by CTLs and NK cells kill target cells by activating their apoptotic pathways. Cytotoxic vesicles fuse with the cell membranes of NK cells or CTLs and release their contents into the tight intracellular junction formed between the two cells. The released perforin rapidly polymerises to form a ring-like structure with a central pore. It gets inserted into the target cell membrane, causing an efflux of cytoplasmic contents and an influx of water. It is unclear how granzymes enter target cells. It has been suggested that they may enter through perforin pores or through receptor-mediated endocytosis. The nature of the receptor and the mode of release from endosomes are not known (denoted by question marks). Granzymes can activate the caspase pathway directly or activate the proapoptotic protein Bid. The activation of Bid results in the leakage of cytochrome *c* from mitochondria, which in turn activates caspases. Granzymes can also directly participate in fragmentation of nuclear proteins or DNA.

get internalised by target cells through receptor-mediated endocytosis. The internalised perforin and granzymes then cause the death of target cells.

- **FasL** is the ligand for CD95 or Fas. Its importance in apoptosis of the target cell is unclear.
- **Granulysin** is synthesised by human CTLs during their activation. It has potent antimicrobial activity against many extracellular pathogens, including bacteria, fungi, and parasites. A combination of perforin and granulysin has been shown to be effective against intracellular *M. tuberculosis* as well. Granulysin can cause target cell membrane damage and induce apoptosis through mitochondrial polarising and release of cytochrome *c*. It also activates caspase-3, a key enzyme involved in apoptosis.
- **Calreticulin** is a chaperone molecule that is a normal ER resident. It is thought to act as a regulatory molecule that dampens the effect of perforin.
- **Cathepsin B**, unlike other constituents of cytotoxic granules, is an enzyme involved in the protection of effector cells. It gets sequestered on the effector cell membrane after degranulation and can inactivate perforin molecules that diffuse back to the effector cell.

11.2.2 The Death-Receptor–Ligand Pathway

The induction of death in unwanted cells is essential for the normal development and functioning of a multicellular organism. As explained in chapter 8, **death pathways are based on protein–protein interaction domains that lead to the activation of a cascade of proteases**. These proteases, termed *caspases*[2], have a cysteine residue in the active site and specificity for cleavage at the aspartic residues. Proteins, such as FLIP (**FL**ICE-**I**nhibiting **P**rotein) or **I**nhibitors of **Ap**optosis (IAPs), can modulate the activation of caspases and protect the cell. Two pathways can induce cell death in vertebrates.

- **The intrinsic pathway of cell death** is triggered by the loss of trophic receptor stimulation (e.g., through withdrawal of growth factors), DNA damage (e.g., because of irradiation), inappropriate loss of contact with neighbouring cells,

[2] The name *caspase* is derived from *c* (for *cysteine*) *asp* (for *aspartate*) and *ase* (signifying enzymes).

CELL-MEDIATED IMMUNITY

- It is immunity mediated by effector cells; antibodies have only a minor role.
- CMI cannot be transferred passively by transferring sera.
- Effectors of CMI include CTLs, NK cells, K cells, macrophages, neutrophils, eosinophils, basophils, and mast cells.
  - CTLs are MHC class I restricted $CD8^+$ T cells that kill target cells by releasing cytotoxins, such as perforin, granzymes, serglycin, and granulysin. They can also cause cell death by inducing apoptosis in target cells.
  - NK cells are a part of the innate immune mechanism, yet they share the CTLs' killing mechanisms. K cells, which form a subset of NK cells expressing FcγRIII, can also bring about ADCC.
  - Macrophages and neutrophils kill target cells by secreting several microbicidal molecules, such as peroxides, superoxide anions, and RNI. They are also effectors of ADCC.
  - Eosinophils, basophils, and mast cells synthesise and store many pharmacologically active substances that are effective in killing large extracellular parasites. These three types of cells can also participate in ADCC.
- CMI is important in defence against virally transformed cells, helminths, intracellular parasites, and tumour cells.

or contact with glucocorticoids. This evolutionarily conserved pathway ensures the appropriate development of organs and the elimination of cells with genetic errors or with abnormal development. This intrinsic pathway is also known as the mitochondrial pathway of apoptosis. Bcl-2-related proteins have a regulatory role in the pathway; they control mitochondrial membrane permeability and have either antiapoptotic or proapoptotic activities (chapter 14). These proteins fall in three classes. The first class consists of antiapoptotic mitochondrial membrane proteins, such as Bcl-2 and Bcl-XL; the second and third groups are proapoptotic. Bax and Bak belong to the second group, whereas the third group consists of proteins such as Bid, Bim, Bad, and Bik. Triggering of the intrinsic pathway leads to the activation of proapoptotic members of the Bcl-2 family. The exact mechanism of the action of these proteins is not clear, but proapoptotic Bcl-2 family proteins cause mitochondria to swell, leak, and release cytochrome *c*. An apoptosome containing cytochrome *c*, procaspase-9, and other proteins is formed. The formation of the apoptosome activates procaspase-9 and yields caspase-9, which recruits other members of the caspase family, such as caspase-3 (Figure 11.2). Activated caspase-3 then cleaves an inhibitor molecule tightly associated with the nuclease CAD (**C**aspase-**A**ssociated **D**NAase). Free CAD initiates DNA fragmentation and cell death. Antiapoptotic Bcl-2 proteins, such as Bcl-2 and Bcl-XL, block mitochondrial swelling, and hence, the release, of cytochrome *c*.

Figure 11.2 Two pathways can induce cell death by apoptosis — the Fas-mediated, or extrinsic, pathway and the mitochondrial, or intrinsic, pathway. Various apoptotic stimuli, such as withdrawal of growth factors, DNA damage, and the loss of contact with neighbouring cells, activate proapoptotic cellular proteins, such as Bax and Bak, and trigger the intrinsic death pathway. Antiapoptotic proteins, such as Bcl-2 and Bcl-XL, can inhibit this activation. The activation of Bax and Bak causes the release of cytochrome *c* from mitochondria. The Fas-mediated pathway is triggered by Fas–FasL engagement. It causes the intracellular aggregation of FADD. FADD associates with procaspase-8 to form DISC. FADD and Fas interaction activates procaspase-8 and results in the formation of caspase-8. This activation is inhibited by the protein FLIP. Caspase-8 activates a cascade of downstream caspases that results in the activation of caspase-3. The activation of caspase-3 irreversibly commits the cell to apoptosis. Activated caspase-3 cleaves an inhibitor molecule tightly associated with the CAD nuclease. CAD initiates DNA fragmentation. Caspase-8 also activates the pro-apoptotic protein Bid. Bid causes mitochondria to swell, leak, and release cytochrome *c*, thus linking the extrinsic pathway to the intrinsic death pathway. An apoptosome containing cytochrome *c* and pro-caspase-9 is formed, and this results in the activation of caspase-9. Caspase-9 recruits other caspases, including caspase-3, into the cascade. IAPs can modulate the action of caspase-9 and -3. The activation of caspase-3 eventually leads to DNA fragmentation and cell death.

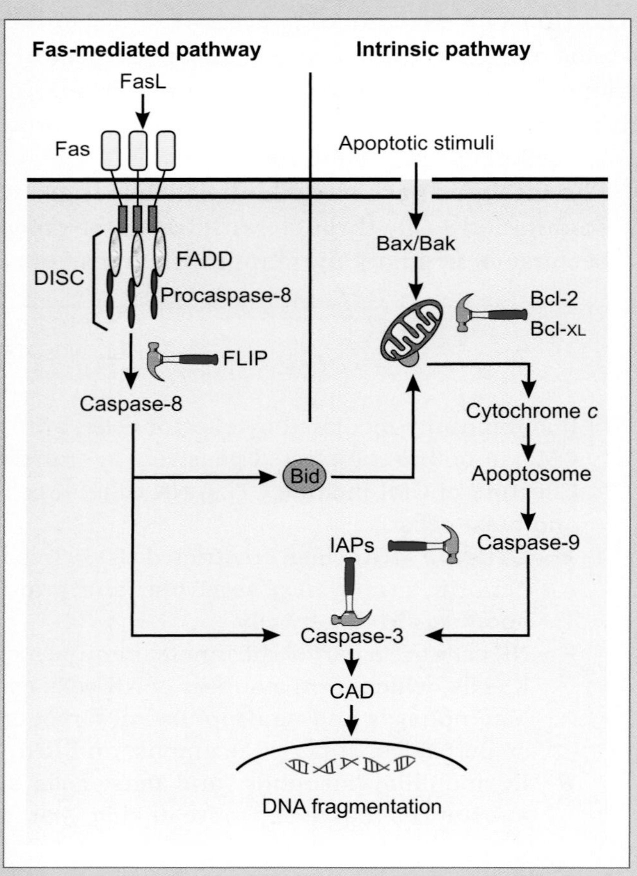

- **The Fas–FasL-mediated death pathway** is thought to be unique to vertebrates. This pathway is initiated by the engagement of the Fas receptor (a member of the TNFR family) expressed on target cells by its ligand (FasL). FasL is a membrane-associated ligand synthesised within several hours of cytotoxic lymphocyte activation. The interaction of the FasL expressed on CTL/NK surfaces with Fas receptors on target cells results in the aggregation of FasL's intracellular death domain and the recruitment of FADD (**F**as-**A**ssociated **D**eath **D**omain). FADD and procaspase-8 form a **D**eath-**I**nducing **S**ignalling **C**omplex — DISC. In the DISC, FADD and Fas interact to activate procaspase-8. Activated caspase-8 activates additional downstream caspases, eventually leading to the activation of procaspase-3. Once caspase-3 is activated, the extrinsic and intrinsic pathways merge. Although the Fas-mediated pathway is the most important extrinsic pathway of apoptosis induction, experimental evidence suggests that cytotoxic lymphocytes can also induce cell death *via* TRAIL (**T**NF-**R**elated **A**poptosis-**I**nducing **L**igand) and TNFR (see *A BAFFling, RANK-Ling TRAIL*). The relative roles of these different pathways in maintaining homeostasis, cancer prevention, autoimmune diseases, and transplantation immunology are still being investigated.

11.2.3 ADCC

Effector cells having cytotoxic potential and expressing FcRs can bring about ADCC. Although ADCC is most studied in NK cells, it may also be induced by several morphologically and developmentally distinctive cell types. These include macrophages, PMNs, eosinophils, and mast cells. ADCC is dependent upon the presence of appropriate antibodies but independent of the presence of complement and of phagocytosis. The exact mechanism of the process is poorly understood.

- **K cells expressing CD16** (FcγRIII) **are capable of ADCC**. IgG–antigen complexes bind to CD16 expressed on K cells. The ligation of multiple FcγRIIIs activates K cells and results in target cell destruction. ADCC is thought to be the dominant component of antibody-mediated antitumour activity.
- **Eosinophils and mast cells** (and possibly basophils) **use ADCC to kill parasites that are too large to be phagocytosed**. When IgE antibodies combine with epitopes on the cuticles of worms, their Fc portions protrude from these surfaces and bind to the FcεRI expressed by eosinophils or mast cells. The ligation of multiple FcεRI activates these cells. Activated cells flatten out along the cuticle

Linking Back: Theories of Immunity

Cellular theory of immunity: In 1884, Metchnikoff postulated that leukocytes play an important role in defending the body. His suggestion that phagocytic cells had a protective function and were the most important contributors to immunity (whether natural or acquired) resulted in a major controversy. Doubts about his theory increased after the discovery of two humoral components — complement and antibodies — involved in host defence. The death blow to the cellular theory of immunity came in 1897. In that year, Ehrlich gave convincing proof of humoral immunity. He demonstrated that a specific antitoxin could protect an animal against a 100-time lethal dose of diphtheria toxin, thus proving that immunity was of a humoral and not cellular origin. It was only much later — the mid 1950s — that the concept of cellular immunity once again gained credence, and it was accepted that both body humours and cells were accepted to be involved in immunity.

Ehrlich's side-chain theory: Paul Ehrlich's 1887 theory attempts to explain antibody formation by body cells. It postulates that body cells possess natural side-chains or receptors consisting of groups or subgroups having an affinity for foreign material. Each cell was assumed to produce a variety of such receptors (Figure 11.S1; top left panel). Chemically different subgroups were presumed to combine with different antigens, that is, each cell was thought to be able to combine with more than one antigen. Ehrlich proposed that the complementarity of the receptors with analogous structures on the antigen permitted specific interaction (Figure 11.S1; top panels). He further hypothesised that such a combining of side-chains with the antigen rendered the cell useless for normal functions. Instead, the cell was stimulated into producing and secreting more receptors (Figure 11.S1; bottom left panel), some of which were cast off into the blood stream (Figure 11.S1; bottom right panel). These circulating receptors protected the body by combining with any antigen that entered the blood. This hypothesis was discarded in the 1920s, when antibodies were produced against synthetic agents. At the time, the pre-existence of receptors for synthetic substances or to a large array of substances seemed inconceivable. Nevertheless, considering that it was formulated at the beginning of the nineteenth century when not much was known about the functioning of the immune system, the theory is remarkable in its boldness and vision.

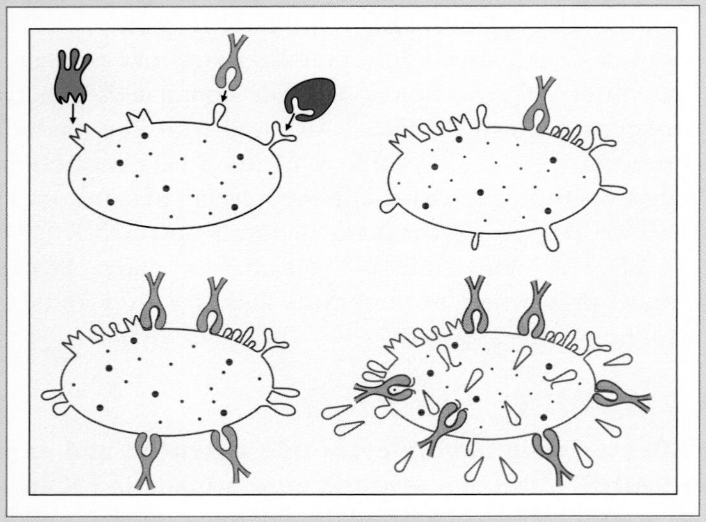

Figure 11.S1 Ehrlich's side-chain theory attempted to explain antibody formation by host cells.

to form a large attachment zone and spill their cytoplasmic granules into the intervening space through exocytosis. These granules collectively help in destroying helminths and their larval forms. Recent research indicates that eosinophils express FcαR. IgA antibodies may therefore have a role in eosinophilic ADCC.

- **Neutrophils** have been shown, *in vitro*, to lyse tumour cells without phagocytosis. They appear to kill tumour cells through release of cytotoxic granules, but their role in ADCC is as yet unclear.

Immune Tolerance 12

One love, one blood, one life, you got to do what you should.
One life with each other: sisters, brothers.
One life, but we're not the same.
We get to carry each other, carry each other.

— U2, *One*

Abbreviation

BBB: Blood–Brain Barrier

12.1 Introduction

The human body produces thousands of proteins. It also comes in contact with a wide range of potential pathogens and their products. Even then, the immune system can usually distinguish between pathogens and self-tissues, attacking the former while ignoring the latter. This ability to not respond to self-antigens is termed *self-tolerance*. **Immunological tolerance can be defined as an antigen-induced block in the development, growth, or differentiation of specific lymphocytes.** It is thus an active and specific state of unresponsiveness to a particular antigen that is induced by prior exposure to that antigen. The main features of tolerance are as follows:

- **It is a specific state of unresponsiveness** toward a particular antigen. It is different from the general inability to respond to antigens that stems from immunosuppression or immunodeficiency.
- Tolerance is specific, **it can be induced only in cells expressing specific antigen receptors** (i.e., in antigen-specific B and T cells).
- **It requires prior exposure to the antigen**.
- **It can be either natural or induced**. Natural (or self-) tolerance is attained during lymphocyte development and leads to the absence of an immune response to self-antigens. Induced (or acquired) tolerance to external antigens is often a result of manipulating the immune system.
- **It is induced more easily in immature lymphocytes** than in mature lymphocytes.

Tolerance to self-antigens is central to survival, and multiple mechanisms at the central (i.e., in the primary lymphoid organs — the thymus and bone marrow) and peripheral levels ensure this continued tolerance. Central tolerance mechanisms ensure that very few self-reactive clones escape to the periphery, whereas peripheral tolerance mechanisms ensure the continued silencing of escaped self-reactive clones. The major mechanisms for achieving tolerance include the following:

- **Clonal deletion** — the removal of self-reactive clones in the primary lymphoid organs
- **The induction of clonal anergy** — causing self-reactive clones to become incapable of responding to antigenic stimulus
- **Active suppression** — wherein any surviving self-reactive clones are kept under strict regulatory controls to prevent them from responding to self-antigens

12.2 Central Tolerance

This tolerance operates in the primary lymphoid organs and induces tolerance in T and B cells during their development. Clonal anergy, clonal deletion, and negative selection are important central mechanisms of T and B cell tolerance induction. During thymic maturation, T cells that express TcRs binding self-MHC molecules with a high affinity are induced to undergo apoptosis (chapter 9). Similarly, developing B cells that bind self-antigens with a high affinity apoptose in the bone marrow. Clonal anergy and/or maturational arrest silence B cell clones that bind self-antigens with a moderate affinity or those B cells that encounter soluble self-antigens in the bone marrow (chapter 8). Because B cells require T cell help to respond to most antigens, clonal deletion of self-reactive T cells further ensures the absence of B cell responses to self-antigens. Nevertheless, as we explain below, tolerance induction in the primary lymphoid organs operates under certain constraints, and some self-reactive clones inevitably escape to the periphery.

- The success of central deletion mechanisms depends upon the expression of all self-antigens in sufficient concentrations in the primary lymphoid organs — especially the thymus. If the concentration of the relevant antigen is low, the number of MHC:peptide complexes on thymic epithelial cells will be too low to be detected even by high-affinity clones. Such clones could escape negative selection. Moreover, some proteins are expressed only after the animal attains maturity (such as antigens involved in breast or sperm development). Yet others are expressed at a certain phase of the individual's life cycle (as during pregnancy and lactation). Even if the *AIRE* gene promotes the expression of tissue-specific proteins that are not normally expressed in the thymus, such proteins may not be expressed in the developing lymphoid organs of the embryo. Hence, lymphocytes recognising those proteins will be found in the periphery.
- TcR recognition is promiscuous by nature. Therefore, more stringent the negative selection, greater the risk of dangerously narrowing the T cell repertoire and of allowing potential pathogens to escape. It is more efficient for the system to let some thymocytes with a degree of self-reactivity escape central deletional mechanisms and then to regulate those escaped self-reactive cells in the periphery.

12.3 Peripheral Tolerance

Peripheral mechanisms that ensure the absence of self-reactivity are crucial for several reasons. First, mechanisms enforcing central tolerance are leaky. Second, because genes encoding BcRs undergo mutations during affinity maturation, there is a substantial chance that some clones generated after encountering a foreign antigen will cross-react with self-antigens. Such clones need to be deleted or strictly regulated. Third, the body is constantly exposed to innocuous environmental antigens — proteins from commensals colonising the airways and intestines, food antigens, and so forth — to which it must remain tolerant. Peripheral tolerance-inducing mechanisms ensure the silencing of clones that can respond to such antigens and the absence of destructive immune responses. Peripheral tolerance can be induced at T or B cell levels. As T cells are multifunctional cells involved in humoral and cell-mediated immune responses, tolerising T cells switches off both these responses. Therefore, T cells are more strictly regulated than B cells.

- **Peripheral T cell tolerance** operates at several levels to silence autoreactive T cell responses (Figure 12.1). Some mechanisms can be considered intrinsic to T cells — they act directly at the level of T cells — whereas others exert their effects through other cells and are considered extrinsic to T cells.
  - ***Ignorance*** is a term used to describe the state of self-reactive T cell clones that do not come in contact with the relevant antigen and are therefore never activated. In other words, ignorance of self-antigens occurs when T cells fail to encounter the relevant self-antigen[1]. This ignorance is achieved by various strategies. First, trafficking of naïve T cells is restricted to the blood or lymphatics so that these T cells do not enter any and every tissue; only memory cells are allowed to wander. This ensures that naïve T cells encounter only antigens that are transported to the lymphoid organs by professional APCs. Second, some antigens are sequestered in sites that are not easily accessible to the blood-borne or lymph-borne immune system (e.g., brain, testes, and eyes). In reproductive organs that express new proteins post-puberty, such sequestration helps maintain ignorance of self-

[1] Ignorance really *is* bliss!

Takes Guts to Tolerate: Tolerance and Mucosal Immunity

The gastrointestinal tract (specifically, the GALT) has a difficult balancing act to conduct. It must tolerate food antigens and normal gut flora while mounting potent immune response to pathogens. To avoid constant inflammatory responses to innocuous antigens, a variety of strategies are operative in the GALT.

- The epithelial layer is covered with mucous and retains most bacteria, preventing their entry into inner layers.
- A breach of epithelial barriers results in immune responses dominated by IgA rather than IgG antibodies. IgA antibodies are noninflammatory — they do not bind complement. GALT is known to be a major source of polymeric IgA antibodies (unlike systemic IgA, mucosal IgA is polymeric).
- APCs in the GALT express low levels of the costimulatory molecules CD80/86. MHC:peptide recognition in the absence of adequate costimulation tolerises T cells.
- T cells found in the GALT express CTLA-4, which is also a ligand for CD80/86. It competes with CD28 to further reduce available costimulatory signals.
- TGF-β is abundantly expressed in the GALT.
  - TGF-β induces the generation of T_{reg} cells that promote tolerance to food antigens.
  - TGF-β is vital to IgA secretion because it is essential for isotype switching to IgA.

The phenomenon of *oral tolerance* was described a century ago, when the oral administration of soluble protein was observed to induce systemic tolerance. The default state of tolerance of the GALT and the possibility of inducing oral tolerance make success difficult with orally administered vaccines, although oral administration is the route of choice to elicit mucosal immunity. However, animals can be actively immunised against orally administered antigens by combining soluble proteins with strong mucosal adjuvants, such as the cholera toxin or a heat-labile derivative of the *E. coli* enterotoxin. Such adjuvants can overcome oral tolerance and induce an efficient immune response, and they are now being explored in the administration of oral vaccines.

reactive T cell clones (see the sidetrack *Privileged Being, Privileged Seeing*). Last, the threshold of the antigen required to trigger a T cell response is such that although some self-antigens may enter circulation, they will fail to elicit an immune response because their concentration fails to reach the required threshold.

- **Anergy induction** causes the functional inactivation of T cells following self-antigen encounter. TcR engagement in the absence of CD28 costimulation has been shown to induce anergy in T cells *in vitro*. This requirement for the simultaneous delivery of antigen-specific and costimulatory signals in naïve T cell activation ensures that only professional APCs initiate T cell responses. By contrast, research suggests that the engagement of CTLA-4 and PD-1 induces anergy. CTLA-4, a CD28 homologue with an inhibitory function, binds the same molecules (CD80/86) as CD28, but induces anergy in CTLs *in vivo*. PD-1, another member of the costimulatory molecule family, is thought to control T cell unresponsiveness by inhibiting cytokine secretion or by causing cell-cycle arrest (see the sidetrack *Stimulating Company*, chapter 9).
- **Phenotypic skewing** is another important tolerance-induction mechanism. Regulating the chemokine receptors that self-reactive T cells make and therefore where they go can prevent autoimmune destruction in the periphery. Alteration in chemokine or cytokine expression patterns can

avert an autoreactive T cell response in spite of self-reactive cell activation. For example, T cells need to express CXCR5 to migrate to the B cell areas of the secondary lymphoid tissues (section 9.5.1). The induction of CXCR5 is dependent on CD28 ligation, and it may not be induced in the absence of an adjuvant. Thus, even if self-reactive T cells proliferate in response to a self-antigen, their migration to the B cell areas of the secondary lymphoid tissues is defective in the absence of CXCR5 expression, and an autoimmune response is averted.

- **Apoptosis** induction in self-reactive T cell clones by AICD is perhaps the most effective way of preventing autoimmunity. Self-antigens cannot be cleared easily, and they lead to repetitive T cell activation. Such repetitive activation that is characteristic of self-antigen–T cell encounters is thought to trigger AICD. Recent evidence suggests that IL-2 may have a crucial role in AICD. IL-2 is thought to promote apoptosis through the induction of the Fas ligand. It also downregulates the expression of the inhibitory protein FLIP. We have already discussed this mechanism of Fas-mediated cell death at some length in chapters 9 and 11.
- **Tolerance induction by APCs** is a simple but elegant method for ensuring the absence of self-reactivity. APCs have long been accepted to influence the nature of the immune response (tolerance *versus* initiation). DCs are key APCs in both these processes. The currently accepted hypothesis postulates that DCs present a sample of proteins that are in their immediate environment to T cells. Such antigen presentation by immature DCs tolerises T cells — probably through a lack of costimulation and/or because of the kind of chemokines expressed by immature DCs. However products of microbial origin such as LPS, peptidoglycan, and CpG DNA cause DC maturation. Similarly, the by-products of stressed or necrotic cells (possibly, heat shock proteins or mitochondrial by-products) also induce DC maturation. Mature DCs are potent APCs; they express an array of costimulatory and adhesion molecules and produce a number of cytokines. Antigen presentation by mature DCs therefore results in an immune response. Certain DCs are also thought to induce tolerance through their ability to generate T_{reg} cells. Experiments in which repetitive stimulation of allogeneic naïve cord blood $CD4^+$ T cells with immature DCs resulted in a nonproliferating population of T_{reg} cells support this theory. It is not yet clear whether a separate subset of DCs or whether DCs at distinct developmental stages are responsible for T_{reg} cell generation. The therapeutic potential of the possible influence of DCs on the type of immune response is being investigated.
- **T_{reg} cell generation** has been shown to be an important method of tolerance induction. T_{reg} cells have a critical role in the generation and maintenance of tolerance, but many questions regarding T_{reg} cell lineages, differentiation factors, antigen specificity, and mechanisms of action remain unanswered. Although many T_{reg} cell subsets have been characterised in different experimental systems, their relationship to each other is unclear. *In vitro* studies show that instead of proliferating, T_{reg} cells suppress the proliferation of other T cells in response to antigenic stimulation. The possible mechanisms of immune suppression include signalling through CTLA-4, the production of cytokines, such as TGF-β and IL-10, and interference

with DC functioning (chapter 9). Recent data suggests that signalling through T cell expressed CD2 in the absence of additional costimulatory signals may also be a mechanism of tolerance induction. A better understanding of Treg cell effector functions will help understand these relationships and allow for their exploitation in immunotherapy.

- **Peripheral B cell tolerance** is achieved partially through the absence of appropriate T cell help, which leaves B cells 'helpless' in responding to many antigens. Exposure of peripheral B cells to antigens in the absence of T cell help can result in anergy induction and exclusion from the primary lymphoid follicles. Mechanisms effective in inducing T cell tolerance are therefore indirectly operative at the B cell level as well. Moreover, exposure to multivalent antigens in the absence of T cell help may lead to clonal deletion or clonal anergy in mature B cells. Mature mIgM$^+$ mIgD$^+$ cells lose surface IgM expression upon exposure to high concentrations of soluble antigens (as happens with self-antigens). The critical parameter seems to be receptor occupancy; greater than 5% mIgM occupancy leads to anergy in experimental systems. Such cells exposed to high levels of soluble antigen are unable to enter the primary lymphoid follicles, and they become anergic and are eventually lost. Low levels of autoreactive IgG antibodies are observed in healthy animals. It is hypothesised that such antibodies could be formed by B cells that have escaped central tolerance mechanisms. The crosslinking of BcRs and B cell-expressed FcRs by autoreactive IgG–antigen complexes can also result in silencing of self-reactive B cell clones (Figure 6.3).

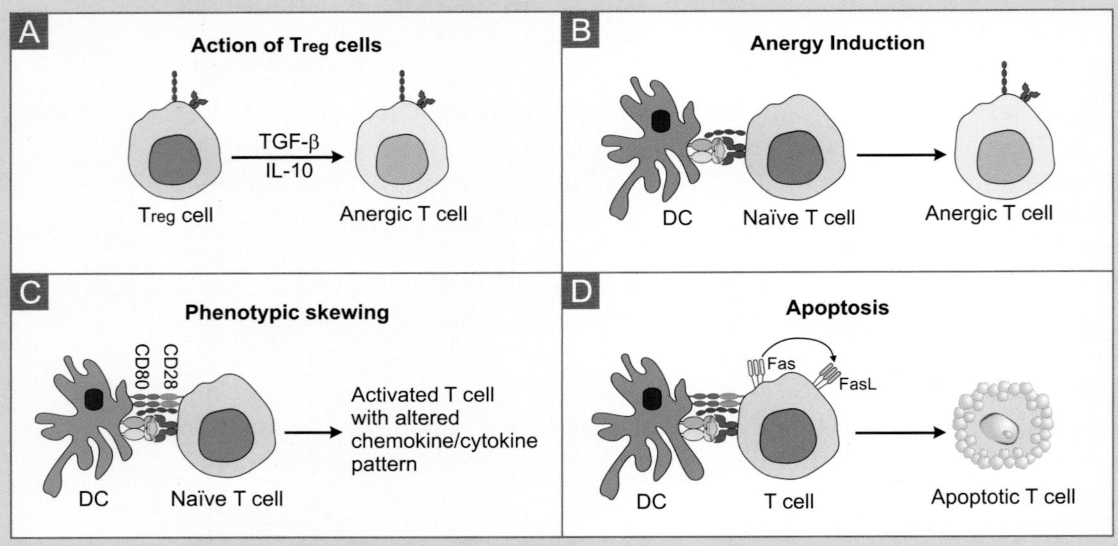

Figure 12.1 Peripheral T cell tolerance to self-antigens is induced by multiple means. A. Treg cells have a critical role in the generation and maintenance of tolerance. These cells produce cytokines, such as TGF-β and IL-10, in response to antigenic challenge, causing effector T cells to become anergic. B. The recognition of MHC:peptide complexes by TcRs in the absence of costimulation can also render peripheral T cells anergic or functionally ineffectual. C. The trafficking pattern of lymphocytes or the nature of the immune response (destructive *versus* nondestructive) can be controlled by altering the pattern of the cytokines/chemokines secreted. Hence, phenotypic skewing can avert destructive autoimmune responses, even if self-reactive T cells get activated. D. Inducing apoptosis in self-reactive T cells is the most efficient way of inducing tolerance. T cell activation causes the upregulation of FasL. It also induces Fas expression and downregulation of the inhibitory protein FLIP, causing cells to undergo AICD.

Mechanisms of Self-tolerance

- Self-tolerance inducing mechanisms are operative at the central and peripheral levels.
- Mechanisms important in central tolerance include clonal abortion, clonal deletion, and clonal anergy.
- Peripheral T cell tolerance inducing mechanisms include the following:
  - Clonal ignorance,
  - Induction of anergy in self-reactive clones,
  - Phenotypic skewing,
  - Induction of apoptosis in self-reactive clones,
  - Skewing of immune response by APCs,
  - Suppressive action of T_{reg} cells.
- Peripheral B cell tolerance is induced by
  - Controlling T cell responses and
  - Induction of anergy in B cells.

12.4 Tolerance Induction

Although tolerance to self-antigens is integral to survival, it sometimes breaks down, resulting in autoimmune diseases (chapter 15) and may necessitate tolerance (re-)induction. The induction of tolerance to foreign antigens, on the other hand, happens naturally in gut colonisation with commensal flora and is artificially brought about to allow transplant survival (chapter 17). Antigens that induce tolerance are termed *tolerogens*. Depending upon how the same molecule is presented to the immune system, it may behave as an immunogen or a tolerogen. Treatments that induce immune responses include aggregation of the molecule, conjugation to carrier proteins, and administration with adjuvants. Such treatments are contraindicated for tolerance induction. Here, it is important to note that tolerance is a type of immune response and hence is induced against particular epitopes. Thus, if an animal is tolerised to a particular epitope, it will show tolerance to all antigens that have that epitope as their major antigenic determinant or immunodominant epitope, and the tolerance will be maintained as long as the antigen persists in adequate concentrations. *Split tolerance* is unresponsiveness to some epitopes on an antigen but not others on the same antigen. As a rule, T cell tolerance induction is comparatively easier, requires smaller amounts of antigen, and lasts longer than B cell tolerance induction. Multiple factors affect the nature and duration of tolerance induction and are outlined below.

- **Nature of the antigen**: Generally, monomeric or soluble antigens are tolerogenic. Monomeric antigens cannot crosslink BcRs and therefore do not stimulate B cells. Also, monomeric or soluble antigens are taken up predominantly by pinocytosis and do not induce the expression of costimulatory molecules. Polymers of D-amino acids are also effective tolerogens — D-amino acids can bind BcRs but cannot be digested and presented to T cells. B cells, therefore, get tolerised in the absence of T cell help. Hapten density also seems to influence tolerogenicity. Higher densities of haptens on the carrier molecule have been shown to induce tolerance, whereas lower densities are immunogenic.
- **Dose of antigen**: Experiments in mice aimed at determining the relation between antigen dosage and tolerance induction show the presence of three distinct zones.

- High concentrations of antigen induce tolerance (known as high-zone tolerance). High-zone tolerance is considered important for maintaining tolerance to abundant, ubiquitous self-proteins, such as plasma proteins.
- Medial concentration ranges induce an immune response.
- Concentration ranges much below the immunising dose induce low-zone tolerance.

- **Route of administration**: The route of administration influences both the magnitude and type of immune response. An antigen given orally may induce tolerance, but it may prove immunogenic when given intradermally or subcutaneously. Systemic tolerance can be induced by the oral administration of haptens or small, soluble molecules. Tolerance can also be induced to soluble proteins delivered by aerosols through airway mucosa. Injecting the antigen in portal circulation has induced tolerance in animal models.
- **Age and immune status of the animal**: Immature lymphocytes are tolerised more easily than mature cells. Neonatal or foetal exposure to the antigen induces tolerance, and the introduction of an exogenous antigen into the foetus tolerises the animal to that antigen. It is much more difficult to induce tolerance in an adult animal. Previously immunised animals are refractory to tolerance induction because they have a much larger repertoire of memory B and T cells capable of recognising antigens. Moreover, memory cells are more resistant to tolerance induction than naïve cells. Nevertheless, tolerance induction is still possible in a mature immune system that is compromised by irradiation or drugs.
- **Nonspecific immune depression**: Generally any treatment that depresses immune responses nonspecifically facilitates tolerance induction. Such treatments include immunosuppressive drugs, X-irradiation, and the use of anti-lymphocytic sera. Anti-idiotypic cytotoxic antibodies can also be used to eliminate specific B cell clones. Drugs, such as cyclophosphamide, help in tolerance induction by delaying the regeneration of mIg receptors on B cell surfaces and by increasing response time.

Privileged Being, Privileged Seeing: Immunoprivileged Sites

The term *immune privilege* is derived from the work of Medawar, who won a Nobel Prize for his pioneering contribution in the early 1950s. He found that grafts placed in certain areas of the eyes (cornea or anterior chamber), brain, or reproductive organs were *privileged* — they survived for prolonged periods of time. Immune privilege is thus the property of some sites in the body where immune responses are limited or prevented. Immune responses in these sites are termed *deviant* because they are different from those in other sites. Privileged sites generally mount predominantly noninflammatory immune responses consisting of non-complement-fixing antibody isotypes; CTL responses are limited or absent. Such deviant responses probably reflect an evolutionary adaptation to protect vital structures from the damage that normally results from pro-inflammatory immune responses. Originally, the sequestration of antigens was thought to be the major contributor to immune privilege. It is now clear that an active rather than a passive process maintains immune privilege.

The Brain: An immune response in the brain — perhaps the most important and most vulnerable organ of the body (*brain dead* is, after all, the clinical definition of death) — is bad news. Multiple factors are responsible for the brain's immunoprivileged status. A specialised **B**lood–**B**rain **B**arrier (BBB) secludes brain parenchyma from circulating blood. BBB is the result of special characteristics of the walls of the capillaries in the brain that prevent potentially harmful substances from moving out of the blood stream and into the brain or cerebrospinal fluid. BBB protects the brain from surging fluctuations in blood constituents (e.g., hormones, amino acids, and K^+ ions). The endothelial cells of cerebral blood vessels constituting BBB have tight junctions. These junctions consist of rings of proteins that seal the epithelium and restrict the passage of molecules into and out of the brain. Most transport across the BBB is an active process involving specific carrier proteins; only lipophilic substances can diffuse across this barrier. BBB thus tightly regulates molecular and cellular traffic into the brain. The brain is further isolated from immune effectors by the absence of organised lymphatic drainage. Furthermore, brain cells are typically either MHC negative or MHC low; normal immune responses cannot occur in the absence of MHC molecules. The brain parenchyma also lacks DCs. Lack of these professional APCs significantly hampers elicitation of an immune response. Gangliosides and cytokines such as TGF-β further augment the immunosuppressive environment of the brain. The downside of these multiple barriers to immune responses is acutely felt in the treatment of brain infections or tumours, as it is difficult to deliver antibiotics, chemotherapeutic agents, or vaccines to the brain. The brain's immunoprivileged status does not imply that it is devoid of an immune response. Although immune reactivity is not constitutive in the brain, CNS parenchyma can be induced to support local immune responses.

The Eyes: Much of what is known about immune privilege comes from studies of the eyes. The preservation of vision necessitates the protection of the eyes from invading pathogens. However, immune responses can cause collateral tissue damage as a result of nonspecific inflammation. The eyes' delicate microanatomy makes them particularly vulnerable to distortion from relatively trivial amounts of intra-ocular inflammation, and the regulation of the immune response is critical. Regulatory molecules expressed by cells of the eye modulate both the induction and expression of immunity to self-antigens, resulting in the virtual elimination of immunogenic inflammation in the eyes. This regulation unfortunately also renders the eyes vulnerable to those pathogens whose eradication requires the participation of inflammatory molecules and cells.

A variety of mechanisms contribute to the immune-privileged status of the eye (specifically, the anterior chamber, vitreous cavity, subretinal space, and corneal stroma). They include a lack of lymphatic drainage, the presence of a physical barrier between the blood and the eyes (termed the blood–ocular barrier), low expression of MHC class II molecules, increased expression of CD59 that inhibits complement activation, local production of immunosuppressive neuropeptides and cytokines (e.g., TGF-β), and constitutive expression of FasL. Expression of FasL is known to induce death in Fas-expressing lymphoid cells. Studies have shown that FasL expression results in apoptosis of antigen-specific cells that enter the eye, preventing an immune response and promoting tolerance. Uptake of apoptotic cells by local DCs is also thought to favour the activation of T_{reg} cells, further strengthening a tolerant response. In fact, FasL expression seems to be common in many privileged sites, including the brain, testes, and thyroid, and it may be responsible for the immune deviation (shift from cell-mediated to humoral immune responses) observed in some of these sites.

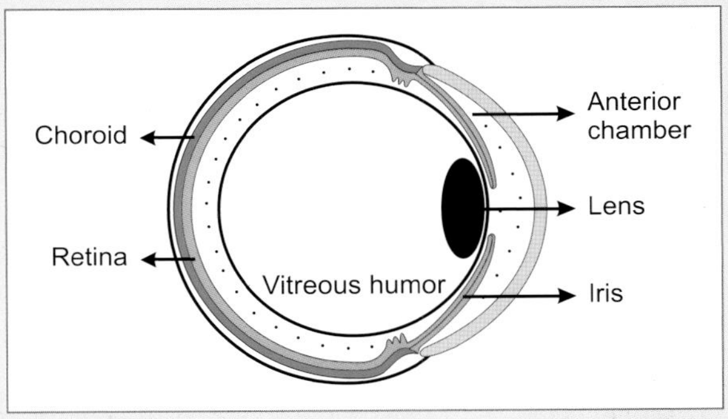

Figure 12.S1 The anterior chamber, vitreous cavity, subretinal space, and corneal stroma of the eye are immunoprivileged.

Immune Tolerance

- Immune tolerance is an antigen-induced block in the development, growth, or differentiation of specific lymphocytes; it results in an absence of, or an active suppression of, an immune response to specific antigens.
- It requires prior exposure to antigens.
- Tolerance induction is dependent upon
  - the antigen — dose, nature, physical state, and route of administration; and
  - the animal — its genotype, age, and immune status.
- The nonspecific depression of the immune system helps in tolerance induction.

– ☮ –

13 Infections and the Immune System

I ain't got a fever got a permanent disease
It'll take more than a doctor to prescribe a remedy
I got lots of money but it isn't what I need
Gonna take more than a shot to get this poison out of me
I got all the symptoms count 'em 1, 2, 3

— Jon Bon Jovi, *Bad Medicine*

Abbreviations

AIDS:	Acquired Immuno-deficiency Syndrome
DTH:	Delayed-Type Hypersensitivity
G6PD:	Glucose-6-Phosphate Dehydrogenase
HIV:	Human Immuno-deficiency Virus
IBD:	Inflammatory Bowel Disease
WHO:	World Health Organisation

13.1 Introduction

In the preceding chapters, we explain the principles and processes by which the human immune system identifies, targets, and destroys that which it deems dangerous. This danger may be in the form of microbial invaders, toxic chemicals, or transformed self-cells. In this chapter and the next, we describe how the immune system deals with these challenges. Using selected examples, we provide a broader perspective on how the immunological principles and concepts described earlier collectively operate. This chapter focuses on how the immune system copes with the myriad microorganisms it continually encounters. We will not describe the different infections afflicting humanity; we have chosen some of the biggest global infectious threats — HIV/AIDS, tuberculosis, and malaria — to illustrate various defence mechanisms and to exemplify the strategies used by infectious agents to subvert the immune system (see the sidetrack *Hiding in Plain Sight*).

An immune response to a pathogen is a defensive, protective mechanism. Yet, in some infections, the symptoms of the disease are caused not by the pathogen but by the immune response it arouses. An inappropriate or excessive immune response can cause as much, if not more, damage than the pathogen itself. As explained in chapter 2, infection results in the release of pro-inflammatory cytokines. The pathogen-stimulated excessive production of cytokines results in the symptoms of TSS (chapter 4), food poisoning, and septic shock. Septic shock results in a disastrous reduction in blood pressure, breathlessness, and leukocytosis. LPS in the cell walls of Gram-negative bacteria engages TLRs (especially TLR-4) on cells of the innate and adaptive immune system, causing a large-scale release of IL-1 and TNF-α. These cytokines are responsible for symptoms of septic shock. Similarly, the unarousable coma observed in cerebral malaria is thought to result from the overproduction of TNF-α and IFN-γ, rather than from parasite activities (see section 13.6). One of the protective responses induced during chronic infection by intracellular pathogens is the formation of granulomas. Granulomas can interfere with the normal functioning of the involved organ, and they cause more damage than infection. For example, adult parasites of *Schistosoma mansoni* migrate to mesenteric veins and lay hundreds of eggs per day. Some eggs are trapped in the microvasculature of the liver and induce a vigorous granulomatous response. The subsequent fibrosis and portal hypertension are the primary causes of morbidity in infected individuals and may prove fatal. Consequently, many of the symptoms of schistosomiasis are attributed to the egg-induced granulomatous inflammatory response and the associated fibrosis.

Hiding in Plain Sight: Subverting the Immune System

Any organism (whether micro- or macro-) trying to gain a foothold in our body is thwarted by the physiological barriers and cells of the innate immune system. If these initial defences fail and the pathogen establishes a focus of infection, the adaptive immune system takes over. The pathogens are not passive players in this battle for a warm and nutrition-rich ecological niche. Evolution has ensured that they have developed various mechanisms evading both the innate and the adaptive arms of the immune system. Noted examples include the following:

- **Surface expression of immune modulators**: The external surface of pathogens is the central interface — the battlefield — where host and pathogen meet. Pathogens express adhesins or receptor ligands to anchor to host surfaces; the recognition of the exposed surface by the immune system initiates microbial clearance. The camouflage of its surface such that it is not recognised by host surveillance systems and the use of its surface to alter or avoid host immune responses is crucial to the pathogen's survival. One of the best examples of this strategy is LPS of Gram-negative bacteria. The outer part of LPS is made of highly variable carbohydrates (O antigens). It allows different strains of the same species to (re)infect the same host. The bacterial capsule, on the other hand, helps in hiding the complex surface of proteins and carbohydrates from immune surveillance and TLR recognition while allowing pili and flagella to protrude and help in invasion. Many viruses display host-derived proteins on their surfaces — CD-receptors (e.g., gp120 of **H**uman **I**mmunodeficiency **V**irus – HIV), complement inhibitors, signalling ligands, adhesion molecules, and so forth. Host-derived proteins help them in evading host immune machinery. Some parasites (e.g., *Toxoplasma gondii* or the sporozoites of malarial parasite) cover themselves with host molecules for the same effect.
- **Secretion of immunomodulators**: Immunomodulators help dampen or alter the host immune response and promote survival of the pathogen. Viruses in particular induce host cells to secrete a spectrum of proteins with immunomodulatory properties. These include superantigens, immune cell ligands, receptor mimics, CD-homologues, complement inhibitors, binding proteins that sequester cytokines, and regulators of activation — particularly of NK cells, T cells, DCs, and macrophages (see the sidetracks *Cheating Death*, chapter 11 and *Lessons in Subversion*, chapter 9, for some examples).
- **Variation in immunodominant molecules (antigenic variation)**: Given that acquired immunity relies on memory of previous exposure to antigens, antigenic variation is especially suited to circumventing humoral and cellular responses. It is a classic mechanism used by viral, bacterial, and parasitic pathogens. Viruses utilise error-prone replicases and DNA-repair mechanisms that induce mutations and result in antigenic variation (termed *antigenic drift*). Antigenic hypervariation has been more effectively adopted by RNA viruses than by DNA viruses. For HIV, the antigenic drift rate is so rapid that it effectively outpaces the development of an effective immune response in the infected host. This antigenic drift thus confounds attempts to develop prophylactic vaccines. A variety of molecular mechanisms are used by bacteria and parasites for antigenic variation. The three major ones listed below are all used by *Plasmodium* spp., to the clear advantage of the parasite:
  1. Having multiple but different copies of a molecule, each of which is under an independent on/off switch,
  2. Having one expression locus plus many silent copies of the gene and constantly changing the gene that is expressed, or
  3. Having a highly variable, constantly changing region in a molecule (termed *antigenic polymorphism*).
- **Subversion of phagocytes**: A central component of the innate response is the deployment of phagocytes to physically remove infectious agents that may have breached initial physical barriers. Phagocytic cells not only internalise and kill microbes, but they also recruit additional immune cells to amplify the innate response. Cytoskeletal proteins, such as actin, are central to phagocytosis and hence are a key target for many bacteria. Other strategies of subverting phagocytic function include the inhibition of phagocytosis by producing a capsule, avoiding entry into the phagolysosomal compartment (e.g., *T. gondii*), escaping from the phagosome (e.g., *Shigella* spp.), blocking phagosome–lysosome fusion (e.g., *Mycobacterium tuberculosis*), or the use of mechanisms that allow survival in phagolysosomes (e.g., *Leishmania* spp.). The malarial parasite *Plasmodium* uses a novel mechanism for safe delivery from the liver to the blood. As explained in chapter 2, dead and dying cells flip their cell membrane, and the exposed inner leaflet delivers a positive signal for phagocytosis. *Plasmodium*-infected hepatocytes fail to flip this inner leaflet, even after they are dislodged from the liver, allowing them to deliver their cargo to the blood without being phagocytosed.
- **Infecting immunoprivileged sites or tissues**: Many viruses and parasites infect immunoprivileged sites (e.g., liver, lungs) or proliferate within host cells where they can be sequestered from immune attack (e.g., erythrocytes, FDCs, T cells).
- **Interfering with MHC class I and II pathways**: Many pathogens have coevolved with their human hosts to develop immunoevasion strategies that involve disruption of the intracellular MHC class I

and class II pathways. The subversion of phagocytosis automatically subverts cross-presentation by the class I pathway and class II pathways. Viruses must subvert CTL and NK cell surveillance for survival. Hence, they have developed elaborate strategies to interfere with MHC class I pathways, such as the inhibition of class I assembly and transport, and enhanced class I downregulation (see chapter 7). Obligate intracellular parasites more often disrupt the endocytic pathway and hence the MHC class II pathway. Examples include the internalisation and degradation of host class II molecules (*Leishmania* spp.), interference with endocytic membrane trafficking (*Helicobacter pylori*), interference with MHC class II to export (*Chlamydia trachomatis*), and downregulation of surface MHC class II and CD1 expression (mycobacteria).

- **Inhibition of cytokines/chemokines**: Cytokines and chemokines are involved in myriad immune processes (such as lymphocyte migration, activation, T and B cell differentiation, proliferation, and survival). Viruses in particular have evolved an array of strategies to subvert the action of cytokines and chemokines (see chapter 9). Many pathogens (e.g., mycobacteria, *Bordetella pertusis, Leishmania* spp.) can induce the production of anti-inflammatory cytokines, such as IL-10 and TGF-β, which dampen the immune response. Nematodes produce proteins that modulate cytokine responses, the most prominent trait being the upregulation of anti-inflammatory IL-10 by macrophages. Conversely, *Listeria monocytogenes* induces the expression of the chemokines IL-8 and macrophage chemoattractant protein-1. This attracts circulating phagocytes and increases diapedesis, facilitating the spread of bacteria-infected macrophages to other tissues.
- **Manipulation of apoptotic pathways**: Bacteria, viruses, and parasites can either induce or prevent apoptosis to augment infection. The induction of apoptosis in macrophages and neutrophils is of obvious benefit to pathogens — it gets rid of cells that can kill pathogens. Many intracellular pathogens also prevent or control the time of apoptosis during infection because inhibiting apoptosis allows them to establish a replicative niche in the host. *M. tuberculosis* delays early apoptosis in infected cells to its own advantage (see section 13.4.1). *Chlamydia* and *Neisseria* spp. prevent apoptosis by preventing cytochrome *c* release from mitochondria. Caspase-3 is central to apoptosis; many viruses (e.g., Epstein-Barr virus, herpes virus, Kaposi's sarcoma virus), bacteria (*Shigella, Legionella*), and parasites (*T. gondii, Leishmania* spp.) target this enzyme. In addition, intracellular pathogens also downregulate death-receptor signals on infected cells. This allows the cell to escape CTL or NK cell-induced apoptosis.
- **Interference with complement pathways**: Complement plays a major role in innate and adaptive immune responses and becomes an attractive target for microbial subversion. Chapter 3 lists a number of instances of the evasion of the complement cascade by various pathogens.

13.2 Immune Responses to Prions

A prion (a proteinaceous infectious particle) is a unique infectious agent thought to cause transmissible spongiform encephalopathies — invariably fatal neurodegenerative diseases[1]. According to the *Protein-only* or *Prion Hypothesis*, the key event in transmissible spongiform encephalopathy aetiology is the conversion of normal, cellular PrP (PrP^{C}) to an alternate isoform (PrP^{Sc}). This conversion event involves only conformational change, with no detectable difference in the primary structure of the protein. PrP^{C} is a host-encoded, glycophosphatidylinositol-anchored sialoglycoprotein that has an α-helical structure. By contrast, PrP^{Sc} is the abnormal, misfolded, and aggregated conformational state that is structurally dominated by a β-sheet. This isoform is partially resistant to digestion by proteinase K and is detergent insoluble. PrP^{Sc} is considered to be the main component of the transmissible agent that causes prion diseases. These diseases are characterised by typical lesions in the CNS with spongiform vacuolation, neuronal-cell loss, microglial activation, and astrocyte proliferation. Prion diseases include the Creutzfeldt-Jakob disease, familial fatal insomnia, and Kuru in humans and scrapie

[1] Prions (pronounced *pree-ahns*) are intriguing, in that they enter cells and apparently convert normal proteins found within the cells into prions just like themselves. Normal cell proteins have all the same 'parts' (i.e., the same amino acids) as the prions but fold differently, that is, they have a different tertiary structure. They are much like the 'Transformer' toys of the 1980s; a Transformer robot, for example, could also become a train engine, without any addition or subtraction of parts.

and bovine spongiform encephalopathy (mad cow disease) in animals. Prions can be generated sporadically, as a result of an as-yet-uncharacterised stochastic event causing the conversion of PrP^{C} to PrP^{Sc}. Alternatively, dominant mutations in the gene that encodes the protein may alter it such that it readily undergoes spontaneous conversion to PrP^{Sc}. Prion disease is also thought to be caused by infections with exogenous prions that induce conformational changes in host-encoded PrP^{C}.

Prion proteins are highly conserved in mammals. However, the normal function of the protein remains unclear. Expression levels are particularly high on neurons and lymphocytes, APCs, DCs, and monocytes. The expression is upregulated in T cell activation. Furthermore, antibody crosslinking of surface PrP^{C} modulates T cell activation and leads to rearrangements of lipid raft constituents and increased phosphorylation of signalling proteins. The marked lack of an immune response to the conformationally altered PrP^{Sc} suggests tolerance to the infectious form. In fact, the immune system is thought to contribute to pathogenesis by amplifying prion load in lymphoid compartments, facilitating efficient neuroinvasion. Studies in $PrP^{-/-}$ mice and conditional knockouts should help in understanding the function of the protein and its role in the immune system.

13.3 Immune Responses to Viral Infections

As obligate intracellular pathogens, viruses are the simplest and most intimate of the life forms that are programmed to live in, or on, humans. Consisting of a nucleic acid (RNA or DNA) core and a protein or lipoprotein coat, these obligate parasites depend upon the host's biosynthetic machinery for replication and protein synthesis.

The innate immune response to viral infections is multilayered. A number of naturally occurring cytoplasmic proteins (including the Fv protein mentioned in chapter 4) directly target and restrict viruses, especially retroviruses. Another factor crucial to limiting and eliminating viruses is complement. Activated complement opsonises viruses and facilitates their removal by phagocytes. It also causes the lysis of enveloped viruses. The envelope of the virus is largely derived from host cell membrane, and hence is susceptible to MAC-formed pores. A viral infection also results in the induction of multiple cytokines (IL-6, TNF-α, and IL-12). However, the hallmark of innate antiviral immune response is the production of type I interferons (IFN-α, IFN-β) and NK cell activation. Type I IFNs are produced by all nucleated cells in response to a viral infection. Viruses must travel through the cytosol on the way out of the cell. Nucleated cells possess receptors and signalling pathways to induce IFN-α, IFN-β gene expression in response to cytosolic viral presence. Most viruses also encounter the endocytic pathway on their way in and out of cells, either because they infect cells *via* endosomes or because they bud into those compartments after their replication cycle. TLRs (TLR-3, -7, -8, -9) and other receptors in the endosomes act as sensors that alert cells to viral presence. Signalling through the TLRs also results in the production of type I interferons. Type I IFNs bind to a common heterodimeric receptor and exert their effect by way of the Jak/STAT pathway and NFκB (chapter 9). The binding of IFN-α/IFN-β to their receptors initiates or upregulates the expression of over 300 genes. Many of these genes encode PRRs, heightening antiviral detection in the cell. Some of the expressed proteins, including proteins that catalyse cytoskeletal remodelling, that induce apoptosis, that regulate post-transcriptional events (splicing, mRNA editing, RNA degradation, and the multiple steps of protein translation), and those involved in post-translational modification have the potential for direct antiviral activity. Mice with mutations or deficiencies in key steps of the pathways triggered by these proteins show an increased susceptibility to viral infection.

Although NK cells were initially identified because of their ability to kill tumour cells, studies from several laboratories showed their importance in antiviral defence. There is clear evidence that NK cells are important in protection against infections caused by the cytomegalovirus, the influenza virus, the hepatitis virus, and the **H**uman **I**mmunodeficiency **V**irus (HIV). Viral infection results in NK cell proliferation and NK cell recruitment to infected tissues and organs. As mentioned in chapter 2, mature NK cells constitutively express transcripts for IFN-γ and contain preformed cytolytic mediators (granzymes and perforin) stored in intracellular granules. Despite being armed for attack, NK cells require activation by type I interferons or pro-inflammatory cytokines, such as IL-12 and IL-15, to become fully functional effector cells. Yet, the activation of NK cells by cytokines expressed as a result of viral infection does not fully explain their antiviral role. The cognate recognition of a virus-infected cell by NK cell receptors is considered to be important to host protection.

Humoral immunity, in particular neutralising antibodies, are central to protection against acutely cytopathic viruses (e.g., poliovirus, rabies virus, and smallpox virus). Antibodies that bind viral surface antigens prevent the binding of the virus to specific host cell membrane molecules. Such binding also activates complement. Virus opsonisation by antibodies (and complement) facilitates their uptake by phagocytic cells. They are thus vital in preventing viral spread during acute infection. The humoral immune system provides a first line of defence against initial viral infection because of the presence of natural antibodies secreted by B1 B cell-derived plasma cells, independent of T cell help. They limit dissemination of the pathogen, form immune complexes that activate the adaptive immune response, activate complement, and recruit antigen into germinal centres. Antibodies present at the site of viral entry are particularly important in preventing the establishment of infection. If that line of defence is crossed, T cell-dependent immune responses generate a humoral response, consisting of specific antibodies with increased affinity and the appropriate isotype needed to afford protection (e.g., IgA if the infection is at mucosal sites). Secreted protective memory antibodies provide an efficient line of defence against reinfection and are backed up by specific memory B and T cells.

Although antibodies are important in preventing the spread of the virus, they cannot eliminate viruses that have entered host cells. CMI has a crucial role in containing infection at this stage, and the cross-presentation of viral antigens is vital to switching on CMI (chapter 7). Virus-specific CD8$^+$ T cells and TH1 cells constitute the major mechanism for virus clearance once infection is established. Activated TH1 cells are a major source of IFN-γ and TNF-α — cytokines vital to antiviral defence. IFN-γ in particular is important in inducing cells into an antiviral state. IL-2, produced by TH1 cells, is important in converting activated CTLs to effector cells. Both these cytokines are important in activating NK cells — the major players in viral defence in the initial stages of infection, when virus-specific CTL response, which takes about 3-4 days to initiate and peaks in 7-10 days of primary infection, is yet to develop.

As viruses are completely dependent on the host cell molecular machinery for the replication and assembly of new viral particles, they are experts in manipulating the host system. HIV is one of the most successful examples of such manipulation.

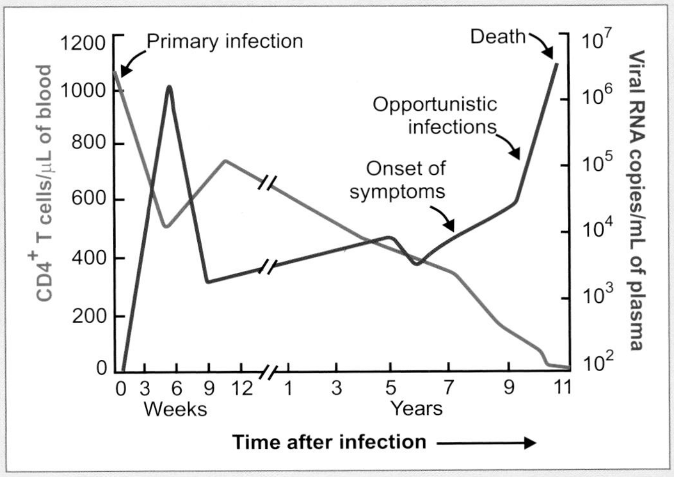

Figure 13.1 HIV infection is characterised by a rapid decline in CD4⁺ T cells, followed by a brief rebound before they resume their downward trend. Primary infection is followed by a burst of viraemia, when the virus disseminates widely throughout the body, with a concomitant and rapid decline in CD4⁺ T cells. As an immune response ensues, the numbers of CD4⁺ T cells recover but never reach pre-infection levels. Although culturable viruses are absent from the plasma at this point, viral mRNA can be easily detected in the plasma. The number of T cells declines slowly but steadily, with an attendant increase in plasma viral mRNA. The patient is largely asymptomatic in this phase and can remain so for years. With the onset of symptoms, the patient enters the clinical phase of infection. There is a rapid increase in the number of copies of viral mRNA and detectable viraemia. The patient becomes susceptible to opportunistic infections and continues to deteriorate (adapted from Pantaleo et al., *New England Journal of Medicine*, 1993, 328:327).

13.3.1 HIV/AIDS

AIDS (**A**cquired **I**mmunodeficiency **S**yndrome) is a fatal immunodeficiency that, unlike SCID, is acquired and not inherited. It leads to a progressive decrease in the efficiency of the victim's immune system and leaves the patient susceptible to opportunistic infections and neoplasms. HIV infection is called a *syndrome* because of the wide spectrum of signs and symptoms that characterise it. HIV is present in the secretions of infected persons — blood, semen, breast milk, saliva, urine, and tears; the concentration of HIV in saliva, urine, tears, and sweat is too low to be of clinical importance. Primary HIV infection is followed by a burst of viraemia; following this burst, the immune system keeps the virus in control until CD4⁺ T cells decline to a critical level (200 cells/μL; Figure 13.1). After that, the patient suffers from full-blown clinical AIDS. Although AIDS is primarily a sexually transmissible infection, contaminated blood or blood products, infected needles, or breast milk can transmit it. The current AIDS epidemic surfaced predominantly among intravenous drug abusers and homosexual males in the USA. Nevertheless, heterosexual contact (unprotected sex involving the exchange of body fluids) is now the major mode of transmission globally. The sex of the victims has also shifted. More women than men are now infected, with a concomitant rise in vertical transmission (from mother to child) occurring during childbirth or *via* the placenta or breast milk. Almost three decades since its recognition, HIV now infects about 33 million people worldwide. AIDS has debilitated a whole continent (Africa), altered medical practice, forced an examination of the concept of individual responsibility, and changed the way we live.

Life cycle: HIV is a human retrovirus. The mature virus is about 90-100 nm in diameter. It consists of a bar-shaped electron-dense core containing the viral genome and enzymes (reverse transcriptase, protease, ribonuclease, and integrase) enclosed in an outer lipid envelope (Figure 13.2).

HIV enters the body through the exchange of bodily fluids and infects mainly TH cells, macrophages, microglial cells, and DCs. The major steps in HIV infection of host cells and its replicative cycle are outlined below.

- HIV infection occurs predominantly through mucosal surfaces of the genital tract. DCs and Langerhans cells in mucosal tracts express the chemokine receptor CCR5 needed for the entry of the virus (see below). These cells act as both vectors and reservoirs of infection. Submucosal DCs express DC-SIGN, a C-type lectin that binds HIV gp120 with high affinity. However, this interaction does not trigger the conformational changes necessary for viral fusion with the DC membrane. Instead, the virus is internalised and subsequently displayed on the DC surface after DC migration and maturation. The virus's transit through acidic DC compartments is thought to enhance its ability to fuse with T cells, and it can easily infect any T cell to which the DC is presenting antigen. Thus, DCs expressing DC-SIGN act as 'Trojan horses', facilitating the spread of the infection from mucosal surfaces to lymphatic organs.
- Productive infection occurs when HIV enters the cell by the engagement of gp120 on the viral surface with CD4 on host cells (Figure 13.2). gp120 binds CD4 and undergoes conformational changes to express additional sites for chemokine receptors. CCR5 and CXCR4 seem to be important for cell entry *in vivo*.
  - CCR5 is expressed on macrophages, DCs, and T cells, and isolates that bind this receptor are called R5 (the earlier M trophic) isolates. R5 viruses mediate both the mucosal and intravenous transmission of HIV infection.
  - CXCR4 is expressed mainly on T cells, and isolates that preferentially bind it are called X4. Early in infection, only R5 isolates are found in the infected individual, but X4 dominate the later phase.

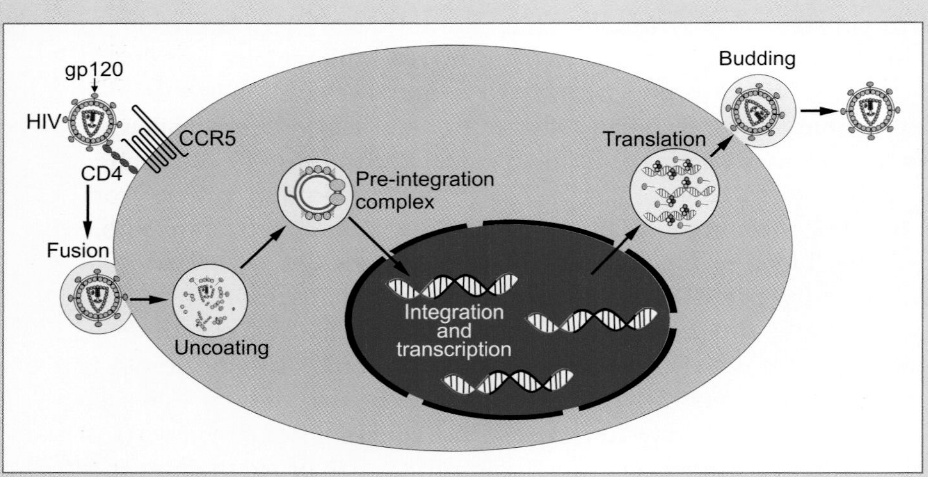

Figure 13.2 The replicative cycle of HIV begins when the virus gains entry into the cell by binding to CD4 and CCR5 receptors. The mature virus consists of a bar-shaped electron-dense core containing the viral genome, some enzymes, and regulatory and accessory proteins enclosed in an outer envelope of matrix proteins. The virus is encased in a host cell membrane-derived envelope. gp120 molecules, important for gaining entry into the cell, protrude from the surface of this envelope as projections. The sequential engagement of gp120, CD4, and the chemokine receptor CCR5 leads to the fusion of viral membrane to the target cell membrane and the release of the viral core into its interior. The virus rapidly uncoats following entry. Completion of reverse transcription yields a pre-integration complex comprising double stranded viral cDNA, viral enzymes, and some viral and host proteins. Upon reaching the nucleus, viral cDNA is transcribed, and the transcripts are rapidly transported to the cytoplasm, where they are translated to structural and enzymatic proteins. The final phase of the assembly takes place at the cell surface. The assembled virion buds through lipid rafts, yielding virions with cholesterol-rich envelopes.

- The sequential engagement of gp120, CD4, and chemokine coreceptors prompts a conformational change in the viral protein gp41 and promotes the fusion of the virion and the target-cell membrane, releasing the HIV viral core into the cell interior.
- Viral uncoating occurs once the virion enters the cell and generates the viral reverse transcriptase complex.
- Completion of reverse transcription yields the HIV pre-integration complex composed of double-stranded viral cDNA, enzymes (reverse transcriptase and integrase), Vpr, matrix protein, and some host proteins.
- Integrase plays a central role in the integration of viral cDNA into host DNA. The HIV provirus can integrate into many different chromosomal locations although it has a preference for active genes. Integration can lead to either latent or transcriptionally active forms of the virus. Most infected cells contain more than one provirus, and the possibility that at least one of them will be transcriptionally active is high. The formation of the latent provirus helps HIV escape from the vigorous immune response mounted in the initial phases of infection. It also allows the re-emergence of the virus when host defence gets weaker. This ability to establish transcriptionally latent forms helps HIV escape potent antiviral therapies.

13.3.2 HIV and the Immune System

HIV has several features that make it a formidable pathogen:

- Its persistence in the host, which is due to an ability to integrate irreversibly in the host genome and remain latent for extended periods of time; the host remains asymptomatic for years, ensuring spread of the virus
- Its ability not to kill host cells indiscriminately; HIV seems to strike the host in a target- and time-specific manner, allowing the host survival for a long time; death is due to opportunistic infections and neoplasm, not the virus
- Its propensity to hijack the immune system for its own proliferation
- Its capacity to evade and destroy the immune system

Dynamics of HIV infection: HIV's primary target is the immune system, which gets slowly destroyed as the disease progresses. HIV infects cells expressing CD4 and the chemokine receptors CCR5 or CXCR4. Although $CD4^+$ T cells are the main targets of the virus, various accessory cells are also infected and have a role in disease progression.

- DCs' ability to migrate to secondary lymphoid tissues and their close association with T cells make them important in the initial seeding of secondary lymphoid tissues with HIV.
- FDCs play a role in pathogenesis. FDCs trap antigen–antibody complexes for prolonged periods, with the antigen maintained in its native form (chapter 5). HIV trapped within virion–antibody complexes not only remains infectious but is more virulent than the free virion. FDCs thus transmit the trapped virus to $CD4^+$ T lymphocytes as they migrate through germinal centres. The continued presence of the virus in the lymph nodes results in the destruction of its architecture, although the exact mechanism of destruction is not known.
- Even though monocytes and macrophages harbour large quantities of the virus, they appear relatively resistant to HIV killing. Because these cells travel throughout the body, they disseminate the infection to various organs, such as the lungs and brain.

- Microglial cells are thought to be the major reservoir of the virus in the brain. Patients with clinical AIDS often show CNS and peripheral nervous system abnormalities. The brain is heavily infected; histological evidence points to extensive viral replication. It is not clear if AIDS-related dementia is a consequence of viral replication or of an immune response to this replication.

How HIV evades and survives effectors of the immune system has been a matter of intense scientific interest. The immune system has evolved a two-pronged system for detecting and destroying intracellular pathogens — one antigen-specific (by virtue of CTLs) and other antigen-nonspecific (*via* NK cells; chapter 11). The recognition of pathogen-derived peptides loaded onto MHC class I molecules results in the antigen-specific activation of $CD8^+$ T cells. The downregulation of MHC class I expression can allow a pathogen to escape CTL recognition. However, such downregulation leaves the infected cells susceptible to NK cell attack. Nevertheless, HIV-encoded proteins reshape the cellular environment of the virus and allow it to evade the immune system.

Blast from the Past: CCR5 and the Black Plague

HIV-1 entry into mucosal cells depends upon the surface expression of CD4 and CCR5; many viruses use such chemokine receptors to enter cells. Upto 10% of the individuals of Northern European descent have a mutation in the *CCR5* gene. These individuals have a 32 basepair deletion in the region encoding the second extracellular loop of CCR5. Termed Δ32, this deletion prevents cell-surface expression of CCR5. By contrast, less than 2% of central Asians carry this mutation, and it is completely absent among East Asians, Africans, and Native Americans. Individuals homozygous for Δ32 show an almost complete resistance to HIV infection, underscoring the importance of CCR5 in the person-to-person spread of HIV. Heterozygous Δ32 is found in 10-15% of Caucasians, and cell-surface expression of CCR5 is reduced in these individuals. The entry of the R5 strain of HIV-1 is reduced but not prevented, the viral load is decreased, and progression to AIDS is delayed in such individuals.

Although the origin of the mutation is obscure, it appears to have suddenly become relatively common among Northern Europeans about 700 years ago — coinciding with the biological catastrophe called the *Black Death*, which decimated about a quarter to a third of the European population between the years 1347 and 1350 AD. The mutation is therefore thought to have arisen during this infection and conferred a selective survival advantage. People with this mutation may have survived the Black Death, or bubonic plague, caused by *Yersinia pestis,* and passed the gene on to their progeny. This sidetrack would have ended here, if two scientists — Susan Cameron and Christopher Duncan — from the University of Liverpool who now challenge this theory had not added an interesting (yet controversial) twist. They do not question the survival advantage of Δ32CCR5 but rather the assumption that the Black Death was the bubonic plague. After sifting through historical evidence, they suggest that the infection was probably an Ebola-like virus that spread through person-to-person contact. If indeed the history books have it wrong, the theory raises the scary possibility of other severe outbreaks of viral infections. In support of their theory, Cameron and Duncan point to the following facts:

- Quarantine measures used to contain the infection would not have been successful if rat-borne fleas spread the infection. Rats have no respect for quarantines.
- The large-scale death of rats was not reported before the outbreak; in any case, black rats were not introduced to Europe until 50 years after the Black Death.
- The symptoms described included the black splotches typical of Ebola-like haemorrhagic fever, not of the bubonic plague.

- **HIV reverse transcriptase is error-prone**. The low fidelity of the enzyme allows the virus to mutate at extremely high rates; the enzyme is estimated to introduce a mutation once per 2000 nucleotides incorporated. Because of epitope imprinting, this high variability allows newly developing strains to escape effectors of adaptive immunity (see the sidetrack *Trying to Fit In*, chapter 10).
- **Tat interferes with MHC class II transcription** in infected monocytes and macrophages, interfering with their ability to present antigen to $CD4^+$ T cells and to generate an adaptive immune response.
- **HIV induces the death of uninfected immune effectors**. The crosslinking of CD4 by the HIV envelope in the presence of soluble Tat can induce FasL expression and the apoptosis of uninfected cells. Similarly, the interaction of the HIV envelope with CXCR4 on macrophages induces TNF expression in these cells and leads to the death of bystander $CD8^+$ T cells who start expressing TNFR.
- **HIV-encoded Nef makes virus-infected cells invisible to CTLs and NK cells**.
    - Nef ensures that MHC class I molecules do not reach cell surface. In the presence of Nef, MHC class I molecules (HLA-A and HLA-B) are diverted from the cell surface to the endosomes and eventually to the *trans*-golgi network, where they get trapped.
    - Nef does not decrease the expression of HLA-C and HLA-E, which bind inhibitory effectors on NK cells.
- **Nef has a major role in HIV-mediated destruction**. HIV commandeers the immune system for its spread and in the bargain destroys it.
    - **Nef triggers accelerated endocytosis and the subsequent lysosomal degradation of CD4** in infected cells. It ensures that released virions do not rebind CD4 on infected cells but remain free to bind CD4 and chemokine receptors on fresh cells.
    - **Nef induces the upregulation of FasL expression** on infected cell surfaces. FasL interacts with neighbouring cells that express Fas (such as virus-specific CTLs) and triggers their apoptosis.
    - **Nef ensures infected cell survival by blocking mitochondrial apoptotic pathways**. It thus prevents the premature death of infected cells and facilitates the completion of the viral replicative cycle.

Pandora's Box: Moral and Ethical Issues Raised by the HIV Epidemic

HIV/AIDS is scary not only because it starkly exposes and manipulates our vulnerabilities — both biological and sociocultural — but also forces us to face up to the realities of the society we live in. HIV/AIDS is biologically potent, but its potency is magnified by its exposing of our illusions of invincibility, our prejudices, and our societal denial.

The current HIV/AIDS pandemic surfaced amongst gay men and drug abusers in North America and became associated in the public mind with homosexuality and drug abuse — issues normally swept under the carpet. In Africa and Asia, the epidemic has spread through heterosexual contact — mainly through prostitutes or people, especially men, with multiple partners. For these reasons, HIV/AIDS came to be conveniently viewed as a disease of the depraved. Early in the pandemic's history (and arguably even now), most countries adopted a moralistic attitude; HIV/AIDS was a problem faced by *other* societies and cultures. The rapid spread of HIV/AIDS amongst heterosexual populations shatters our illusions. It makes us take a long, hard look at the grey areas in our morality. It throws up evidence of poverty, promiscuity, prostitution, oppressive double lives, and rampant drug abuse. Even more scarily, it shows us conclusive evidence of the magnitude and frequency of their prevalence. But it does not stop there.

Many of its victims unknowingly contract the disease, and this is not counting blood transfusion recipients. Amongst heterosexuals, women are more likely to catch the infection than men. What makes it worse is the fact that the vast majority of women in the developing world have very little say in sexual matters. Whether they are prostitutes or housewives, most women cannot insist that their partners use condoms. In countries such as India, women in monogamous relationships have been found to contract the infection from promiscuous partners. Even sadder, they often pass it on to their children during childbirth, and whole families are afflicted by the disease.

The only way the epidemic can be controlled is through education. Education can dispel the many myths surrounding the infection — bizarre myths that run the gamut from ideas that HIV does not cause AIDS to claims that having sex with a virgin cures the infection. The resistance to open discussion about sexual practices is a major barrier to stemming the spread of the epidemic. Sex has always been a touchy issue in many cultures. The resistance of the religious right (of whichever religious denomination or country) to sex education or condom use plays right into the hands of the virus. No method of sexual protection can guarantee an HIV/AIDS-free life. Even abstinence does not guarantee freedom from receiving infected blood during transfusions or from accidentally acquiring the disease, such as through a prick from a tainted needle. If we could better control our impulses, sexually-transmitted infections would not have survived since pre-historic times. At least condoms reduce the risks of catching the disease by offering 90% protection — which is much better than no protection at all.

HIV/AIDS has, in a matter of decades, become mainstream. The stigma attached to the infection, however, remains. A lack of awareness about the disease has resulted in individuals infected with HIV facing varying degrees of discrimination. The fear of contagion has sometimes even led to discrimination against people infected with HIV by medical personnel, and legal struggles to rectify this situation continue. At the family level, when an adult member is infected with HIV, this translates into a severe crunch on resources because the person's capacity to work is reduced while the cost of treatment multiplies. The high cost and long-term nature of treatment makes AIDS a death sentence for poorer sections of the society. Thus, another consequence of the infection is that families often abandon the infected — whether babies or adults.

Discrimination is only one of the ethical issues that HIV has raised. It has also thrown up the issue of personal choice *vis á vis* risk to society. The high viral mutational rate makes it necessary to use a combination of three or more drugs, and this may cost up to USD 10,000 or more annually. To make matters worse, multiple drug therapy involves taking a minimum of eight HIV fighting pills (and frequently many more) a day, on an often complex schedule, in addition to any other medicines the individual might need. Trying to adhere to this schedule while trying to find and hold down a job is a daunting task. Patients who fail to follow the instructions of their prescriptions risk encouraging the proliferation of drug-resistant viral strains, making subsequent treatment more difficult and increasing the risk of infecting others with resistant strains.

The current HIV/AIDS pandemic seems to successfully emphasise the weaknesses of our societies — in terms of moral issues, notions of sexuality and sexual repression, gender and class inequalities, egocentrism, ignorance, discrimination, and denial. HIV has an advantage over the human population not only in the biological domain but also in the psychological domain. Unless we face the realities exposed by the disease, we may never be able to win the biological battle. The costs of this disease are not limited to individuals that are infected or even their families. Ultimately, they will be borne by whole sections of the societies and economies that it has spread to. Although the current pandemic originated in the USA, the global burden of HIV/AIDS will be overwhelmingly borne by people in the developing world, and inequalities of gender, race, class, and wealth will dictate its future course.

13.4 Immune Responses to Bacterial Infections

The innate immune system is largely successful in keeping out potential pathogens. Bacteria that successfully breach or subvert the innate arm of the immune system induce an adaptive immune response, which relies on a different set of strategies to contain them. Antibodies are mainly involved in eliminating infections by extracellular bacterial pathogens; CMI is important in the elimination of intracellular pathogens. Antibodies help destroy extracellular pathogens in multiple ways.

- They opsonise bacteria and promote their removal by phagocytic cells.
- Antitoxin antibodies efficiently neutralise toxins secreted by the pathogen. Phagocytic cells are highly efficient in removing the toxin–antitoxin complexes that are formed.
- The binding of antibodies to the bacterial cell envelope activates complement. This results in the localised production of inflammatory mediators that help in the development of an amplified and more effective immune response.
  - The released C3b augments the opsonisation of bacteria and promotes their uptake by phagocytes.
  - Complement activation results in the lysis of pathogens (especially Gram-positive bacteria) in the immediate vicinity.
  - Anaphylatoxins (C3a, C5a) released during activation induce local mast cell degranulation and the release of bioactive mediators. Some of these are powerful chemoattractants for neutrophils and macrophages. The released vasodilators also cause the extravasation of neutrophils from blood and tissue spaces.

Although NK cells provide an early defence against intracellular pathogens, adaptive cell-mediated immune responses are enlisted to control intracellular pathogens. Many intracellular bacteria have evolved mechanisms that allow them to survive within infected cells. Such pathogens cause the chronic activation of antigen-specific CD4$^+$ T cells. This chronic activation results in TH1-cell driven **D**elayed-**T**ype **H**ypersensitivity (DTH; chapter 16) responses and results in granuloma formation. Although these TH1 cell-driven responses are effective in controlling the spread of the pathogen, they may also result in tissue damage.

Licensed to Kill: Vaccination

Vaccination has traditionally been defined as the deliberate introduction of a pathogen or its products into the body to confer resistance to a specific infection. The term comes from the work of Edward Jenner, who showed that inoculating humans with skin lesions caused by the vaccinia (cowpox) virus protected them from smallpox. The basic principle of vaccination is to actively immunise the subject and induce a *primed state*, so that exposure to the pathogen results in a rapid secondary immune response which leads to the accelerated elimination of the organism and protection from clinical disease. The first administration of the vaccine is termed the *primary* vaccination, and subsequent administrations aimed at inducing a lasting secondary response are known as *booster* doses. The induction of memory cells (B and T) and presence of neutralising antibodies in sera determines the success of the vaccine.

Traditional approaches to vaccine design are outlined below.

- **The introduction of live organisms into the host**: Given that the organism replicates in the host, live vaccines generate both humoral and cell-mediated responses, and immunity is long-lasting. It is essential that the organisms in the vaccine remain viable for vaccination to succeed. Live vaccines therefore need to be appropriately stored. Three types of live vaccines are commonly used.
  - Attenuated vaccines consist of organisms whose virulence has been reduced by various means. However, the introduction of live organisms presents the potential threat of a rare reversion to a virulent strain. There is also the risk that such occurrences might destroy the general populace's faith in vaccines.
  - Heterologous vaccines make use of a closely related organism that shares antigens with but is of lesser virulence than the original pathogen. These vaccines are safer than attenuated vaccines because there is no danger of reversion. The most famous example is Jenner's use of the cowpox virus to immunise against smallpox.
  - Live recombinant vaccines are produced through genetic engineering. Genes coding for the immunogenic proteins of a pathogen are introduced into a harmless organism. The recombinant organism replicates in the host and expresses the recombinant protein, eliciting a cell-mediated and/or humoral immune response.
- **Inactivated vaccines**: These vaccines are used when attenuation fails or when there are concerns about the safety of the live vaccine. Inactivation of the organism can be achieved by heat or through chemical means. Instead of using the whole organism, its subcellular fractions may also be employed in the vaccine. Another method is to express the immunogenic protein in an expression vector such as *E. coli* and purify the protein of interest for use in vaccines (e.g., the Hepatitis B vaccine). Although safer than live vaccines, inactivated vaccines often have lower immunogenicity. They result in a predominantly humoral response and may fail to invoke CMI.
- **Toxoids**: These are used to confer immunity to infections, such as diphtheria and tetanus, in which manifestations and mortality are due to the toxins produced by pathogens. Toxoids are produced by denaturing toxins such that toxicity is lost but immunogenicity, retained.

To be effective, a vaccine must fulfil some basic criteria.

- **Efficacy** is the primary criterion. The vaccine should ideally elicit an appropriate long-term protective response. For example, to protect against intracellular pathogens such as *M. tuberculosis,* the vaccine must elicit a cell-mediated response; respiratory pathogens, such as the influenza virus, require a mucosal antibody response; a robust serum antibody response confers protection against most common bacterial and viral infections.
- **Safety** is paramount. As vaccines are administered to large populations, even low-level toxicity is unacceptable. Therefore, neither the constituents nor the additives used for stabilising or preserving the vaccine should have undesirable or toxic side-effects.
- **Stability** is a particular concern for developing countries of the hot, equatorial belt. Vaccines ought to be stable (i.e. retain immunogenicity) over a range of temperatures.
- **Cost-effectiveness** is a consideration as vaccinations are perhaps the cheapest forms of large-scale health care. For vaccines to be successfully used by the developed and developing worlds, they need to be affordable and relatively cheap.

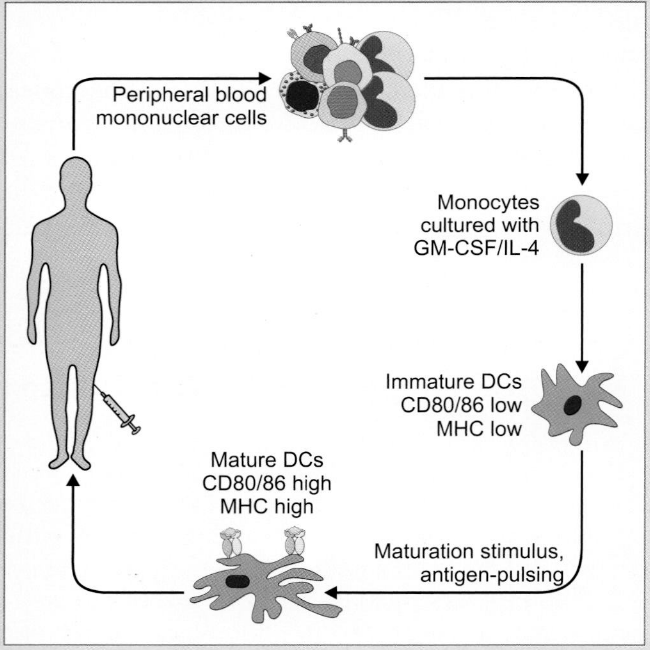

Figure 13.S1 Autologous vaccines can help overcome the defective DC functions observed in certain cancers and can aid a potent antitumour response.

- **Ease of administration** is especially desirable as infants and young children are usually at the receiving end of vaccines.

DNA and gene-based vaccines have opened up a new era in vaccine technology. These vaccines introduce the DNA encoding the antigen, not the antigen itself, into the patient. The earliest type of DNA vaccine consisted of attenuated viral vectors (such as the vaccinia virus or an adenovirus) engineered to contain DNA encoding antigenic epitopes from a pathogen. Currently, genes coding for antigenic determinant(s) of the pathogen are inserted into a plasmid, and the genetically engineered plasmid is injected into the host. Intramuscular injections or a device called the gene gun, which propels plasmids into cells near the body surface (e.g., skin or mucous membranes), are used to deliver these vaccines. The mode of vaccine introduction appears to influence type of response — intramuscular injection seems to enhance TH1 responses, the gene gun favours a TH2 type response. Once inside cells, the recombinant plasmid enters the nucleus, and genes encoding the antigen are transcribed and translated. The protein is eventually internalised by APCs, loaded on MHC molecules, and presented to T cells, eliciting an immune response. A combination approach, where the immune system is primed with the DNA vaccine and boosted with recombinant viruses, seems to hold the most promise — it generates a more potent immune response than when either is used alone. The use of cytokines in enhancing the efficacy of DNA-based vaccines is also under investigation.

Autologous vaccines are being investigated for antitumour therapy. DCs are the body's most potent APCs. Nevertheless, DC function is defective in many types of cancers. The generation of autologous DCs *ex vivo* is being attempted to circumvent this problem (chapter 14). Thus, monocytes (DC precursors) obtained from a patient are allowed to differentiate to DCs, matured through the use of cytokines or CD154, pulsed with tumour antigens, and then re-injected into the patient. Initial studies have shown that such vaccines induce a potent antitumour response. Expense is however an issue, as the vaccine has to be prepared individually for each patient.

Newer approaches to vaccine design: As Table 13.1 shows, many infections have been successfully tackled with vaccines that mimic the natural immune response. Unfortunately, with chronic infections, such as HIV/AIDS and mycobacterial infections or cancers, the natural immune response cannot afford protection. Therapeutic, rather than prophylactic, vaccinations are therefore being explored for these infections. New strategies are aimed at increasing immunogenicity, influencing type of response, inducing local mucosal immunity, or decreasing inhibitory immune mechanisms.

- Epitope enhancement involves the subtle alteration of T cell epitopes of a pathogen (usually a virus) to make them more immunogenic. Epitopes can be modified to achieve different goals.
  - Peptide sequences can be modified to increase their affinity for MHC molecules. Anchor residues that bind the peptide to the MHC molecule can also be altered to enhance epitope affinity. This approach has been used for HIV and cancer vaccines; however, it requires a detailed knowledge of viral or tumour T cell epitopes that can bind multiple MHC haplotypes.
  - A second approach consists of increasing the affinity of the peptide:MHC complex for the TcR, which allows the stimulation of CTLs expressing low-affinity TcRs. This approach is especially useful in the case of tumour antigens, where high-affinity T cells are often eliminated or rendered tolerant because of self-antigen recognition.
  - A third approach is to induce a broadly cross-reactive T cell response that can recognise multiple strains of the virus. This approach has been used in the development of HIV vaccines — the appropriate manipulation of T cell epitopes allows the development of broadly cross-reactive CTLs that recognise multiple strains of HIV. For administration, modified epitopes can be conjugated to a carrier protein or genes encoding the protein can be introduced in a viral vector.
- The incorporation of cytokines or costimulatory molecules into a vaccine can increase its efficacy.
  - IL-2, given as a fusion protein with the Fc segment of IgG, greatly enhanced the protective immunity of a vaccine for the simian immunodeficiency virus.
  - GM-CSF has been found to be the most broadly useful cytokine. It can increase the range of T cell responses and augment both humoral immunity and CMI.
  - Cytokines can skew the immune response and may be used to selectively induce CTL or humoral responses. IL-12 and IL-15 have been found to promote CTL memory and are being investigated as vaccine adjuvants.

- Costimulatory molecules enhance vaccine efficacy and may be used synergistically with cytokines. A triple combination of CD80, ICAM-1, and LFA-3 expressed in a recombinant poxvirus has recently been found to be synergistic for the induction of CTL responses and antitumour immunity.

- The natural transmission of many viruses (e.g., influenza and HIV) occurs at mucosal surfaces, and hence, vaccines aimed at preventing mucosal transmission are being developed. The route of administration is important in these vaccines. The normal subcutaneous route gives rise to a systemic response, leaving mucosal sites immunologically naïve and susceptible to infection. By contrast, mucosal immunisation (either intranasally or intrarectally) gives rise to CTLs in mucosal sites, such as the Peyer's patches, as well as in systemic sites such as the spleen. Peptides linked to heat shock proteins have been found to be effective mucosal vaccines, and CpG oligonucleotides have been found to be effective mucosal adjuvants.
- The discovery of Treg cells has stimulated research aimed at preventing their induction in diseases such as cancer. Most research concentrates on the regulatory pathway involving CTLA-4. CD80/CD86 has two ligands — CD28, which gives an activating signal to the T cell, and CTLA-4, which gives an 'off' signal (see the sidetrack *Stimulating Company*, chapter 9). CTLA-4 is important in anergy induction, and preventing its interaction with CD80/86 can reverse $CD8^+$ T cell tolerance. The blocking of CTLA-4 has been shown to enhance tumour immunity and is being considered for application in tumour therapy.

13.4.1 Tuberculosis

M. tuberculosis, which causes an illness that affects the respiratory system, is a global health threat. On the basis of skin tests, about 30% of the world's population (i.e., about one third of humanity) is estimated to be infected with this pathogen. The WHO (**W**orld **H**ealth **O**rganisation) estimates that the infection kills between 1.4 to 2.8 million people annually.

The most common form of the infection is initiated following inhalation of droplets containing a few tubercle bacilli. The normal course of events following the inhalation of microbes is outlined below.

- Once in the lung, bacilli are internalised through phagocytosis by the resident alveolar macrophages of the lung. Macrophages (including alveolar macrophages) are equipped with a full range of TLRs and other PRRs capable of recognising and inducing a preliminary inflammatory response against the invading microbes. The engagement of these receptors activates macrophages.
- The activated macrophage shows increased microbicidal activity through the expression of components of the ROI and RNI pathways and is transformed into a potent APC.
- The phagosome containing the internalised microbe fuses with a highly degradative endocytic vesicle — the lysosome — to give rise to the phagolysosome. For most microbes, the acidic, highly hydrolytic environment of the phagolysosome is sufficient to kill. In that sense, a macrophage is like a good household disinfectant — it kills 99.9% of household germs quietly and efficiently.
- Fragments of the internalised microbe are loaded onto MHC class II molecules and presented to T cells.
- Infection by intracellular pathogens also initiates apoptosis in macrophages. This default pathway can be viewed as altruistic suicide; it leads to apoptosis of the macrophage but proves helpful to the host. Because the membranes of cells that are dying by apoptosis remain intact, infected apoptotic macrophages are able to contain their intracellular bacteria. These bacteria are ultimately

Table 13.1 Vaccines in Common Use

Infections	Vaccine preparation	Comments
Anthrax	Extract of attenuated bacteria	Primarily for veterinarians and military personnel in the developed world; for animals and farmers in some parts of India and other parts of the globe
Cholera, plague	Crude fractions of organisms	
Diphtheria and tetanus	Purified toxoid	Often given with pertussis as a single dose (DTP, or, as now called, DTaP)
Haemophilus influenzae type b	Capsular polysaccharide conjugated to protein	Prevents ear infections in children
Hepatitis A	Inactivated virus	Available in a single shot with HBsAg
Hepatitis B	Purified recombinant surface antigen (HBsAg)	
Influenza	Haemagglutinins from type A and type B viruses	High mutation rates cause rapid antigenic variation; new vaccines containing antigens derived from strains in current circulation need to be produced every year
Measles, mumps, rubella	Attenuated virus	Often given as a mixture (MMR)
Menigococcal disease	Purified polysaccharides	
Pertussis	Killed bacteria (P) Purified components (aP)	
Poliomyelitis	Attenuated virus: **O**ral **P**olio **V**accine (OPV; Sabin) **I**nactivated **P**olio **V**irus (IPV; Salk)	
Rabies	Inactivated virus	A vaccine prepared from human diploid cell culture has replaced the earlier duck vaccine
Smallpox	Attenuated virus	Use discontinued after the world was declared smallpox-free; production has been restarted because of the threat of bioterrorism
Tuberculosis	Live attenuated cells (**B**acillus of **C**almette and **G**uerin or BCG)	
Typhoid, paratyphoid	Three types available: killed bacteria, oral live attenuated vaccines, and polysaccharides conjugated to proteins	
Yellow fever	Attenuated virus	

killed by effector molecules associated with apoptosis or scavenged by other activated phagocytes, limiting spread of infection.

- In addition, smaller fragments of apoptotic cells that are shed as blebs are efficiently taken up by DCs, which stimulates adaptive immunity through the priming of antigen-specific T cells.
- Thus, the major functions of the macrophage are all dependent on the degradative capacity of the phagosome, and they lead to the death and digestion of most microbes.

M. tuberculosis, however, has evolved tactics that allow it to evade macrophages, and it uses the cell for survival and dissemination. Its cell wall composition makes this microbe unusual. A large part of the *M. tuberculosis* genome is devoted to the production of molecules involved in lipid synthesis, and this may be related to the many functions lipids play in the survival of the organism. Glycolipids in the mycobacterial cell wall can interfere with phagosome–lysosome fusion and allow survival of the pathogen in macrophages. The capacity of *M. tuberculosis* to block transfer to lysosomes is operational almost exclusively in unactivated macrophages. In activated macrophages, bacteria are rapidly transferred to lysosomes, where they are destroyed. Not surprisingly, mycobacterial glycolipids also play an important role in the subversion of macrophage activation. Lipoarabinomannans, found abundantly in the mycobacterial cell walls, can modulate signalling pathways that induce macrophage activation, including IFN-γ-mediated gene expression, TLR activation, and phagolysosome formation. In addition, the pathogen has evolved means of using host molecules for its own survival. By mechanisms that are not entirely clear, *M. tuberculosis* recruits host proteins such as coronin-1, which inhibit the phagosome–lysosome fusion. Recent research also shows that virulent strains of the organism block early apoptosis of infected macrophages but cause them to undergo necrosis-like cell death (Figure 2.3). This cell death produces a permeable cell membrane that enables bacterial escape and spread. By blocking cell apoptosis early after infection, *M. tuberculosis* also prevents or delays antigen presentation and thus succeeds in delaying effective T cell responses. The infected macrophages that fail to kill the tubercle bacilli either remain in the lung or are disseminated in the body.

Despite the bacterium's immunoevasion tactics, most infected individuals (<10% of infected people) fail to progress to full-blown tuberculosis because in most healthy individuals, the immune system can keep the pathogen in check. Young children and elderly people have the highest risk of developing not only active tuberculosis but also the disseminated form of the infection. Protective immunity to tuberculosis is T cell mediated and requires the secretion of IFN-γ, TNF-α, IL-12, and the production of RNIs. The role of $T_{H}17$ cells in protective immunity to tuberculosis is not clear. Studies show that although IL-17 is not important during a primary immune response and may even hamper the induction of a memory response against the pathogen. $T_{H}17$ cells may have a role in secondary responses to *M. tuberculosis*. Tuberculosis infection follows a relatively well-defined sequence of events. Alveolar macrophages internalise microbes, and a major fraction of the internalised bacteria are killed, although a few survive. The priming of naïve T lymphocytes against mycobacterial antigens is thought to occur in the proximal draining lymph nodes. After their priming in lymph nodes, memory $CD4^+$ and $CD8^+$ T cells become central components of antituberculosis defence. *In vitro* activated $CD4^+$ and $CD8^+$ T cells have been shown to be cytotoxic for mycobacteria and the macrophages containing them. Activated T cells migrate back to the lungs through blood. Here, they recognise

M. tuberculosis-derived peptides presented on the surface of infected macrophages. CD4$^+$ T cells that are *M. tuberculosis* specific produce primarily TH1 cytokines (IFN-γ, IL-2 and TNF-α). The few tubercle bacilli that survive in the macrophages cause chronic antigenic stimulation and the accumulation of T cells around the infected macrophages. The chronic stimulation also results in the recruitment of mononuclear cells.

In an attempt to contain the infection, multiple T cells and macrophages organise themselves into a classic structure termed a *granuloma*, or *tubercle* — the hallmark of tuberculosis. The granuloma's primary function is containment. It prevents mycobacterial dissemination and is the product of a robust cellular immune response to bacterial components (chapter 16). Even though the granuloma may not succeed in killing the infecting organism, it prevents further replication. Granuloma formation also occurs within regional lymph nodes to control organisms that have been carried to this site by infected macrophages. Thus the tubercle bacilli are 'walled off' by the immune system inside a tissue nodule. Granuloma

Cellular Ghettos: Granulomas

A granuloma is a localised inflammatory reaction that represents a form of DTH. Granulomas are complex, organised immunological structures comprising differentiated, interdigitated macrophages (also known as *epithelioid cells*), T cells (CD4$^+$, $\gamma\delta$, and NKT), B cells, and NK cells. T cells are indispensible for the formation of the DTH granuloma. The central feature in granuloma formation is the persistence of antigen that cannot be easily cleared by phagocytic cells. In infection-induced granulomas, the lesions represent localised interfaces between infectious agents and the immune system. Pathogen persistence in macrophages causes their chronic activation. Such activated macrophages fuse to form characteristic epitheloid and multinuclear giant cells around the source of irritation, i.e., the pathogen. An extracellular matrix shields this cellular conglomerate from healthy tissue.

Macrophages are the prominent cell types in all granulomas. Macrophages are attracted to the persistent inflammatory stimuli and begin to nucleate the granulomatous lesion, and the basic structure of the granuloma is formed. T cells provide the TNF-α that is crucial for the initiation and maintenance of granulomas. T cells are also the source of the chemokines required for macrophage recruitment and are also involved in complex regulatory circuits that determine every aspect of these localised inflammatory lesions. CD4$^+$ T cells that accumulate at the site recruit other effector cells, including macrophages, eosinophils, $\gamma\delta$ T cells, and NKT cells. Such granuloma formation is protective for the host because the inflammatory reaction gets isolated and insulated while adjacent tissue remains healthy. The expansion and spread of infectious agents is also inhibited; compromised granuloma formation results in disseminated infection and host death. However, the view that infection-induced granulomas (such as seen with *M. tuberculosis* infection) serve as an indestructible, encircling barrier is thought to be simplistic. Granulomas are now believed to represent a mutually advantageous compromise between some infectious agents and their hosts. They ensure a longer life for the host, and the infectious agent is allowed to survive while being subjected to a restrictive and harsh environment. Previously, pathogens were thought to remain dormant in the granuloma until the individual was immunocompromised; it was then that the pathogen could establish a focus of infection. However, new evidence shows that *M. tuberculosis* does not exist in a dormant state in the granuloma; instead, it expresses an array of granuloma-specific genes that are actively induced by this specific microenvironment. DTH responses and granuloma formation were initially suggested to be a TH1 response. Both TH1 and TH2 cells are now known to be involved in granuloma formation. TH1 cells are involved in granulomas caused by bacterial, viral, and fungal pathogens (e.g., tuberculosis, leprosy, listeriosis, Q fever, blastomycosis, histoplasmosis, and infectious mononucleosis); TH2 cells are involved in granulomas caused by helminths, such as schistosomes and ascaris.

formation is typically seen in the *containment phase* of the infection, in which there are no overt signs of disease and the host does not transmit the infection. In the later stages, the granuloma develops a marked fibrous sheath. Granulomas fail to form in the absence of adaptive IFN-γ secreting $CD4^+$ T cell responses.

Although *M. tuberculosis* can cause primary disease, the most common manifestation of tuberculosis in adults is the reactivation of a pre-existing, chronic infection. Successful containment within granulomas results in latent disease, which occurs in up to 95% of cases after infection. In about 5% cases, containment within the granuloma is unsuccessful. Any condition that affects the functioning of $CD4^+$ T cells, for example, old age, malnutrition, and HIV co-infection, can precipitate active disease. The bacteria outcompete the immune response, and their intracellular replication causes widespread macrophage death. The granuloma decays into a structureless mass of cellular debris termed *caseation*, which consists of the remains of multiple dead macrophages at the centre of the granuloma. It ruptures to spill thousands of viable, infectious bacilli into the airways and results in the development of a productive cough. In this stage, the person becomes infective. The local spread of the organism, triggering formation of more granulomas, is typical of lung tuberculosis. Extensive granuloma breakdown causing widespread tissue destruction characterises adult tuberculosis, which is visible on chest radiographs as cavities. The importance of host factors in restricting the spread of infection from the granuloma is evidenced by a vastly increased rate of reactivation in HIV-infected people whose $CD4^+$ T cells, critical for protection against disease, are depleted.

13.5 Immune Responses to Fungal Infections

Most fungi are saprophytic, and relatively few fungal species are associated with serious human diseases when compared to protozoan parasites. Fungal diseases, or mycoses, can be superficial, cutaneous and subcutaneous, or deep and systemic. The skin, hair, and nails are affected in cutaneous and subcutaneous infections. Deep mycoses may involve lungs, bones, the viscera, or the CNS. Serological and skin reactivity assays indicate that fungal infections are common, but clinical disease is rare in healthy individuals. The clinical relevance of fungal diseases has increased enormously in the second-half of the twentieth century, mainly because of an increasing population of immunocompromised hosts, including individuals infected with HIV, transplant recipients, and cancer patients.

The immune system does not have as many targets to use in the recognition of eucaryotic pathogens as it does for viruses and bacteria. Innate immune barriers help resist most potential fungal infections. Antimicrobial factors secreted by normal commensal flora also contribute to the host's antifungal response. The destruction of this flora by prolonged antibiotic treatment can often lead to infection by opportunistic fungal pathogens, such as *Candida albicans*. Host defence against fungal infection relies on an integrated immune response; the mononuclear phagocyte system has a predominant role in initial pathogen recognition, and neutrophils have an adjunctive role in fungal killing. In fact, neutrophils are so important in this defence that neutropaenia (lowered neutrophil count) is known to increase susceptibility to fungal infections. Complement too has a major role in antifungal resistance. Components of fungal cell walls activate complement by the alternative and MBL pathways, resulting in the opsonisation and removal of these pathogens by phagocytic cells.

Immune Response To Infections

- Prions are unique infectious agents that cause transmissible spongiform encephalopathies. These conserved proteins do not elicit an immune response, probably because they are recognised as self.
- Viruses are the simplest and most intimate of the life forms that are programmed to live in, or on, humans.
  - The innate immune system has a major role in antiviral defence, especially IFNs, complement, and NK cells.
  - Activated complement opsonises viruses, facilitates their removal by phagocytes, and causes the lysis of enveloped viruses.
  - Antibodies are central to protection against acutely cytopathic viruses and preventing the spread of infection, but they cannot eliminate viruses from infected cells.
  - CMI has a crucial role in eliminating infected cells; virus-specific $CD8^+$ T cells and TH1 cells constitute the major mechanism for virus clearance.
- Antibodies are mainly involved in eliminating infections by extracellular bacterial pathogens; CMI is important in the elimination of intracellular pathogens.
- Innate immune barriers help resist most potential fungal infections.
  - Antimicrobial factors secreted by the normal flora are also an important contributor to the host's antifungal response.
  - Mononuclear phagocyte and neutrophils have a predominant role in fungal pathogen recognition and killing.
  - Components of fungal cell walls activate complement by the alternative and MBL pathways, resulting in the opsonisation and removal of these pathogens by phagocytic cells.
- Surviving a protozoan infection requires the generation of a controlled immune response that recognises the invading pathogen and limits a potentially harmful host response.
  - A humoral response is most effective in the blood stage of the parasites.
  - CMI is important in controlling or limiting intracellular stages.
- Helminths are the most common infectious agents in humans.
  - TH2 type responses seem to have evolved specifically to deal with helminths.
  - TH2 type responses are often effective in protecting against parasites at mucosal surfaces but are less effective against tissue-dwelling parasites.

Antibody-mediated immune responses have a comparatively smaller role in antifungal immunity, although antibodies to common fungal pathogens, such as *Cryptococcus neoformans* are found in many healthy individuals. In contrast to what is observed for bacterial pathogens, B cell deficient mice do not exhibit increased susceptibility to fungal pathogens. The effective tissue response to invasion by many fungal pathogens is granulomatous inflammation, a hallmark of CMI. $CD4^+$ T lymphocytes play an important protective role against fungal infections, as judged by the susceptibility of HIV/AIDS patients to *Pneumocystis jiroveci* (formally *P. carinii*), *C. albicans*, *Histoplasma capsulatum*, *C. neoformans*, and *Aspergillus fumigatus*. It is generally believed that TH1-biased responses, characterised by TNF-α, IL-12, and IFN-γ production, are protective to fungal infection. By contrast, TH2-biased responses, characterised by IL-4, IL-6, and IL-13 production, are maladaptive or deleterious. Certain fungi have the capacity to alter T cell responses upon infection, creating favourable conditions for their persistence. For example, *C. neoformans* produces melanin, and melanin-producing strains elicit more IL-4 than nonproducing strains. Thus, melanin production might enhance *in vivo* survival by limiting the induction of TH1-biased responses and generating a TH2-biased response. Although the role of $CD8^+$ T cells in fungal infections has not been defined as clearly, recent studies demonstrate a protective role for CTLs in infections of *H. capsulatum*, *C. neoformans*, and *P. jiroveci*.

13.6 Immune Responses to Protozoan Parasites

Infection with protozoan parasites causes vast morbidity and mortality, especially in developing countries. Most protozoan parasites cause chronic rather than acute infections, such as amoebiasis (caused by *Entamoeba histolytica*), African sleeping sickness and Chaga's disease (caused by trypanosomes), Kala azar and leishmaniasis (caused by *Leishmania* spp.), and toxoplasmosis (caused by *T. gondii*). Malaria is a notable exception, in that it can be acute or chronic. Unlike bacteria and viruses, protozoans often differentiate within the host into discrete, morphologically and molecularly distinct forms, reflecting their unique ability to adapt to the host environment and establish a biological niche within the constraints of the immune system.

Surviving a protozoan infection requires the generation of a controlled immune response that recognises the invading pathogen and limits a potentially harmful host response. The type of response that develops and its effectiveness depend on the parasite's location within the host. Most protozoan parasites have a free blood stage in their life cycle; an antibody response is most effective at this stage of infection. As most of them also have an intracellular stage, cell-mediated responses are important in controlling or limiting parasites in these stages of infection.

13.6.1 Malaria

Human malaria is caused mainly by four species of parasites belonging to the genus *Plasmodium* — *P. falciparum*, *P. vivax*, *P. malariae*, and *P. ovale*. Of these, *P. falciparum* is the most virulent. By adhering to the endothelia of diverse host organs, *P. falciparum* infected RBCs can be sequestered from the peripheral circulation. The obstruction of blood vessels in the brain, together with aberrant immunological responses, can cause cerebral malaria. The obstruction of blood vessels in other vital organs can also damage these organs and lead to patient death.

Malaria accounts for a major portion of the global disease burden, with about 40% of the world's population at risk of infection. Of these, greater than 500 million become severely ill, and the annual mortality rate is about one million. Although the burden of cases and deaths is highest in Africa, malaria also affects Asia, Latin America, the Middle East, and parts of Europe. In areas of high disease prevalence, malaria contributes significantly to anaemia in children and pregnant women. In such regions, it is also responsible for adverse birth outcomes, low birth weights, and high infant mortality. The clinical manifestations of malaria range from a mild, asymptomatic parasitaemia to severe and often fatal syndromes, such as cerebral malaria and multiorgan failure.

Life cycle: The malarial parasite has a complex life cycle that is completed in two different hosts — mosquitoes and mammals. It consists of an exogenous sexual stage in the *Anopheles* mosquito and an asexual reproductive stage in the human host (Figure 13.3). The cycle starts when a previously uninfected female *Anopheles* mosquito bites an infected individual; only female mosquitoes of this genus transmit the parasite. With the blood meal, the mosquito also ingests the parasite's gametocytes. Shortly after ingestion, in the mosquito midgut, male and female gametocytes develop into gametes that undergo fertilisation. The newly formed zygotes eventually give rise to oocysts. The parasite then undergoes multiple rounds of proliferation to give rise to thousands of haploid sporozoites. About 14-16 days after infecting the mosquito, the sporozoites are released into the mosquito haemocoele. They then migrate to and invade the mosquito salivary glands, where they mature into infective sporozoites. The infective sporozoites gain entry into a new host when the mosquito vector bites a new vertebrate host.

In the mammalian host, the initial phase of development occurs in the liver (exoerythrocytic phase) and lasts for a week. The sporozoites infect a hepatocyte, undergo multiple rounds of nuclear divisions and a series of complex transformations, and finally differentiate into thousands of first generation merozoites. Ultimately, the merozoite-filled hepatocyte gets dislodged and enters the bloodstream, where it releases its cargo. The parasite develops further in the blood during the erythrocytic phase, and merozoites invade RBCs initiating the erythrocytic cycle. The clinical symptoms of malaria — from periodic fever cycles

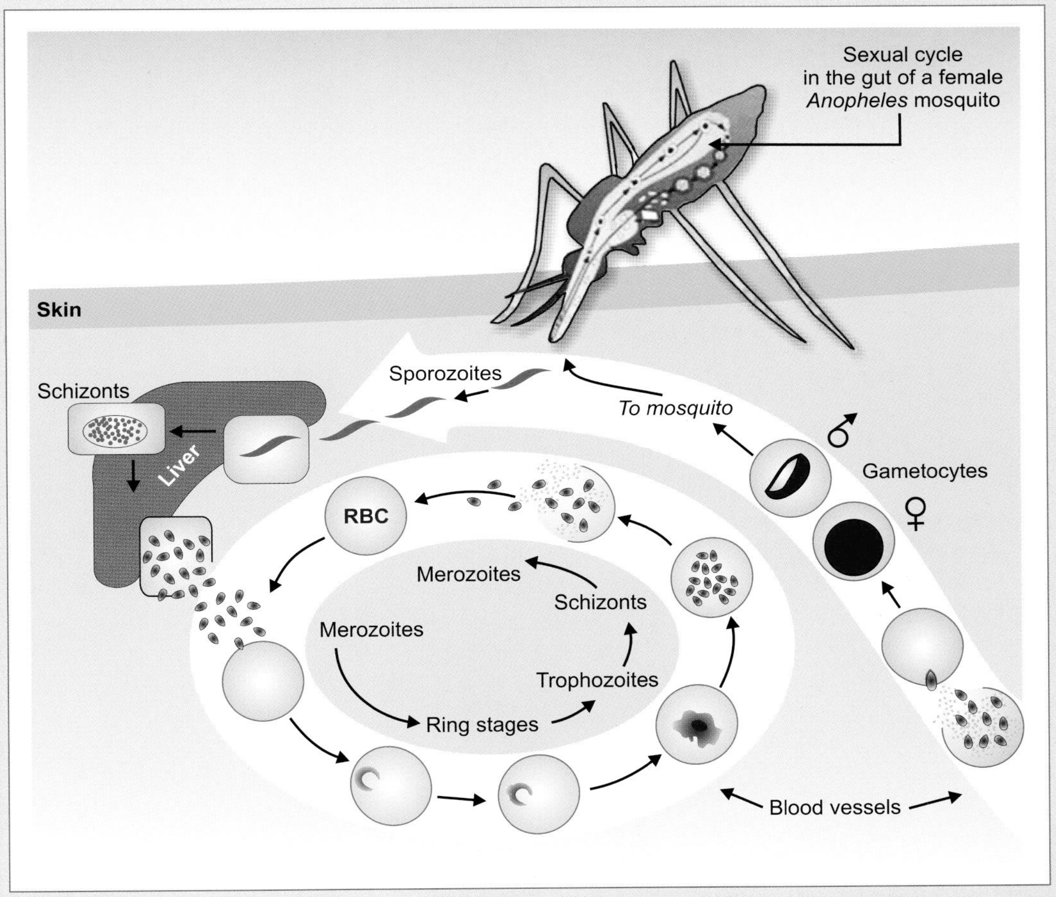

Figure 13.3 The life cycle of the malarial parasite is completed in two different host species. The cycle starts when a previously uninfected female *Anopheles* mosquito bites an infected human and ingests the parasite's gametocytes. The gametocytes develop into gametes that undergo fertilisation in the mosquito midgut, undergo transformation, and proliferate to give rise to sporozoites. The sporozoites gain entry into a new host when the mosquito vector bites a new human host. The sporozoites infect a hepatocyte, undergo multiple rounds of nuclear division and a series of complex transformations, and finally differentiate into thousands of merozoites. The merozoite-filled hepatocyte gets dislodged and enters the bloodstream. The merozoites invade RBCs, initiating the erythrocytic cycle during which they undergo a complex series of transformations to give yield to new merozoites. Some of the invading merozoites give rise to male and female gametocytes, which, when ingested by the mosquito, perpetuate the cycle (courtesy Prof. S. Sharma, Department of Biological Sciences, TIFR, Mumbai, India).

Mal-Adjusted: Malaria and Human Genetic Traits

Malaria is one of the major causes of child mortality in areas of high-transmission. Therefore, it is not surprising that it is considered to exert the strongest known selective pressure in the recent history of the human genome. Apart from its effects on genes directly involved in immune responses, malaria is the evolutionary driving force behind the spread of RBC-associated diseases, such as sickle cell anaemia, α1-thalassemia, and glucose-6-phosphatase deficiency.

An erythrocyte is essentially a bag filled with haemoglobin and circumscribed with a membrane that expresses a variety of proteins. Most of the malarial parasite's life cycle takes place in the erythrocyte. The parasite is dependent upon a number of membrane proteins for invasion and upon the haemoglobin environment for its metabolic activities. Alterations in surface proteins or haemoglobin can affect both invasion and the biochemical and cellular machineries of parasite development. Such alterations will be beneficial to the human host if they reduce parasite invasion or multiplication, and they are selected for even if it means suboptimal functioning of RBCs.

One of the best-documented examples of the effect of the malarial parasite on RBC-membrane biology is the loss of the Duffy blood group antigen in the West African population. The Duffy antigen is a chemokine receptor (see chapter 17) that is essential for RBC invasion by *P. vivax*. Absence of the antigen completely prevents invasion. Most of the population of sub-Saharan Africa is resistant to *P. vivax* infection, unlike other human populations. The absence of the Duffy group antigen accounts for the noteworthy absence of this parasite species in parts of Africa where other *Plasmodium* species are extremely common. Similarly, the loss of glycophorin C — responsible for the Gerbich blood group — is common in coastal areas of Papua New Guinea. This protein serves as one of the receptors for *P. falciparum* invasion, and its loss confers increased resistance to the infection. Another mutation in erythrocyte-membrane protein that seems to have been selected for resistance to malaria is an anion exchanger known as band 3 protein. A mutation in this gene results in a form of ovalocytosis[2] that is common in parts of Southeast Asia. It appears to be protective against both malarial infection and cerebral malaria, although the mechanism of resistance is unclear.

The haemoglobin molecule is a heterotetramer of $(\alpha\beta)2$ globin-associated nonprotein haeme groups. Malaria has exerted tremendous selective pressure on the structure and regulation of both α and β globins. The allele that results in sickle-cell anaemia is termed HbS (sickle haemoglobin or haemoglobin S) and is found across a large part of sub-Saharan African, as well as in parts of the Middle East. HbS tends to polymerise at low oxygen concentrations, causing the RBC to deform into a sickle-like shape. HbS homozygotes have sickle-cell disease, a debilitating and often fatal disorder. The heterozygous state is not generally associated with any clinical abnormality, and it confers ~10-fold increase in protection from severe forms of malaria, with a lesser degree of protection against milder forms of the disease. Haemoglobin C is another variant found in several parts of West Africa. Homozygotes have relatively mild haemolytic anaemia because of the abnormal rupture of RBCs, whereas heterozygotes do not appear to have anaemia. Both heterozygotes and homozygotes are protected against severe malaria. Thalassemias are a group of clinical disorders that result from the defective production of α- or β-globin chains. Generally, homozygous individuals have severe thalassaemia disease, and it can be fatal. Evidence from population studies seems to suggest that heterozygotes with defective α-globin are protected against malaria, but again, the exact mechanism of this mutation's protective action is unclear.

The sheer volume of haemoglobin in RBCs can restrict the space available for the parasite; haemoglobin breakdown becomes mandatory for the parasites growth. This breakdown releases toxic by-products — particularly iron, which can cause oxidative stress. The enzyme **G**lucose-6-**P**hosphate **D**ehydrogenase (G6PD) helps reduce this stress by producing the electron-donor nicotinamide adenine dinucleotide phosphate. A G6PD deficiency leads to haemolytic anaemia. Deficient G6PD activity has been shown to correlate with protection against severe malaria. Reduced parasite replication in G6PD-deficient erythrocytes is thought to be the mechanism of protection.

[2] Ovalocytosis is an inherited condition in which a person's RBCs, which are supposed to be round, have a slightly oval or elliptical shape instead (also known as hereditary elliptocytosis). This condition is associated with mild haemolytic anaemia.

and rigors to an enlargement of the spleen or other severe complications — are primarily a consequence of this asexual life cycle. The merozoite undergoes a complex series of transformations in RBCs to give rise to new merozoites. Eventually, RBCs rupture to release 8-32 merozoites that rapidly invade fresh erythrocytes, and the cycle is perpetuated. Some of the invading merozoites give rise to sexual forms (male and female gametocytes) instead of merozoites. These gametocytes are also released in blood when the erythrocytes rupture. The ingestion of viable gametocytes by feeding *Anopheles* starts a fresh round of infection.

13.6.2 Immune Responses to Malaria

Clinical studies of malaria in the field and laboratory studies that use rodent models have provided insights into host immune responses that protect against the disease. Nevertheless, factors that regulate antiparasite immunity (i.e., those processes that kill parasites and reduce parasitic biomass) are not completely understood. Even the definition of antimalarial immunity has different contexts. People living in areas of intense malaria transmission build up partial immunity to malaria after recurrent infections and no longer suffer from serious disease. These individuals commonly tolerate parasites in their blood that would render others symptomatic. This immunity is often described as *clinical immunity* and implies that the infection is controlled and tolerated rather than prevented or eliminated. Immunity that does not allow the parasite to persist in the body is called *sterile immunity* and is rarely observed as a result of natural infections.

Multiple factors contribute to impaired immune responses to the parasite. The parasite undergoes multiple maturational stages (sporozoites, merozoites, gametocytes), during which different antigenic molecules are expressed. The parasite is only very briefly exposed in the blood before it invades the liver — an immunoprivileged site. It then remains predominantly intracellular and invades RBCs — cells that do not express MHC class I and II molecules. The parasite is also capable of extreme antigenic variation, and almost all immunogenic epitopes on the parasite are highly polymorphic. The parasite profoundly affects many components of the adaptive immune system — it compromises DC function, results in the apoptosis of activated T and B cells, and interferes with macrophage function. Not surprisingly, protective memory responses to the parasite are short-lived and may be lost when an individual is not continually exposed to the parasite, such as when a previously immune person moves away from an area of intense malaria transmission.

Studies indicate that protection against malaria is dependent upon both cell-mediated and humoral mechanisms. Invasion of red cells is a key event in the establishing of a malarial infection, and red cells are therefore likely to be important targets for protective immune responses. However, in the erythrocytic stages of the infection, merozoites are predominantly intracellular. Although many parasite proteins are exposed to the host immune system only briefly, infection leads to the expression of parasite proteins on the RBC surface. Because RBCs do not express MHC molecules, infected RBCs cannot be a direct target of $CD8^+$ or $CD4^+$ T cell effector mechanisms. In the absence of a direct T cell-mediated immune response against infected cells, antibodies have a pivotal role in immunity against merozoites. Antibodies prevent invasion by opsonising merozoites and facilitating their phagocytosis by macrophages, inducing complement-mediated damage of merozoites, preventing processing of invasion proteins, or blocking their erythrocyte-binding sites. Agglutination studies on the sera of people in areas of high malaria transmission have shown the presence of antibodies to several variant antigens, which confer protective immunity to multiple strains.

As erythrocytes do not express MHC class II molecules and lack the apparatus essential for antigen processing and presentation, they cannot present antigen to T cells directly. Instead, they must depend on DCs to present parasite debris for the priming of naïve $CD4^+$ T cells. Because the parasite seems to compromise DC function, T cell activation is hampered. Nonetheless, $CD4^+$ T cells protect against blood-stage parasites by providing cognate help to B cells and by secreting macrophage-activating IFN-γ. Macrophage activation promotes the clearance of infected RBCs. Experiments in mice that are genetically incapable of making B cells have shown that $CD4^+$ T cells have a protective role in the blood-stage of the infection. The secretion of pro-inflammatory cytokines (such as IFN-γ, TNF-α) in the initial stages of infection seems to mediate a protective function. IFN-γ in particular is pivotal in this protection. It helps in the production of cytophilic antibodies important for controlling the infection. IFN-γ also induces IL-12 production by monocytes, macrophages, and DCs and is thus vital in amplifying the immune response. Nonetheless, IFN-γ-induced TNF-α and IL-1 are responsible for some of the clinical manifestations of disease, such as fever and nausea. These pro-inflammatory cytokines produced later in infection also contribute to pathological conditions, such as cerebral malaria. Recent experiments suggest that anti-inflammatory cytokines, such as IL-10, and TGF-β, downregulate the potentially pathogenic effects of the pro-inflammatory cytokines produced in early stages of infection. Thus, the pro- and anti-inflammatory cytokines produced by T cells play a decisive role in the final outcome of a malarial infection, and the initial T_{H1} response needs to be counterbalanced by a T_{reg} response in its later stages.

The unifying theme emerging from these studies is that the optimal immune response to blood-stage malaria infection is characterised by early, intense, pro-inflammatory cytokine-mediated effector mechanisms that help kill or clear parasite-infected cells. These seem to be equally rapidly suppressed by anti-inflammatory effectors once parasite replication has been brought under control. The antigenic overload caused by the expanding parasite population is thought to result in compromised DC function, the apoptosis of T and B cells, and the generation of T_{reg} cells in later stages of infection, and it may be the underlying cause for the impaired and short-lived immunity to a naturally acquired infection.

13.7 Immune Responses to Worms

Helminths are parasitic worms that have been human companions over many millennia. Eggs of intestinal helminths can be found in the mummified faeces of humans dating back thousands of years. Currently, helminths are the most common infectious agents in humans in developing countries, and they produce a global burden of disease exceeding that of malaria and tuberculosis. The WHO estimates that soil-transmitted worms, including hookworms, ascarids, and whipworms, pose a health risk to two billion people (about one-third of the world's population), and waterborne helminths, such as schistosomes, pose a health-risk to between 500 and 600 million people.

With the exception of *Strongyloides stercoralis*, helminths do not replicate within human hosts. Most parasitic helminths have at least one stage in their life cycle where they are free living. For example, the eggs of many nematodes are present in the faeces of infected hosts. These eggs develop through several larval stages outside the host, and ingestion of the larvae results in the infection of a new host. Helminths that enter orally live at the mucosal surface of either the gastrointestinal or respiratory tracts. They may sometimes cross these barriers temporarily, before returning to the mucosal surface, or permanently, to access other host tissues and sites. By contrast, *Schistosoma* spp. infect human hosts by boring through the skin and migrating through the lungs and liver to infect either intestinal mesenteric veins (*S. mansoni, S. japonicum*) or veins of the urinary bladder (*S. haematobium*). Most helminth infections, result in multiyear, chronic inflammatory disorders if left untreated. Some infections can result in extremely debilitating conditions. For example, onchocerciasis can cause blindness, and filariasis may culminate in elephantiasis. Chronic helminth infections can also lead to more insidious, persistent health conditions, such as anaemia, stunted growth, malnutrition, fatigue, and poor cognitive development.

The reduction in intensity of some human helminth infections with age is thought to be indicative of acquired host immunity. Immune responses to helminths are intriguing because a major branch of the mammalian immune system, the TH2 type response, seems to have evolved specifically to deal with these pathogens. Typically, helminths induce strongly polarised TH2 responses that are often effective in mediating protective immunity against parasites at mucosal surfaces but less so in protecting against tissue-dwelling parasites. In animal models, TH2 type immune responses can prevent the survival of infecting parasites during a homologous secondary infection, expel adult parasites from the gut, and allow host survival in a setting where the immune response cannot clear parasites. Such responses are characterised by eosinophilia, mastocytosis, and IgE production. One effect of such a response is the stimulation of the smooth muscles of the gastrointestinal tract, increased gut motility, and diarrhoea. Clearly, this helps

Worming into the History Books

Viral and bacterial infections are known to have changed the course of history. Thus, the Spanish conquest of Mexico in the 15th century was aided by the silent allies that the Spanish brought with them — diseases such as smallpox, mumps, measles, and others — which killed 18 million South American natives in the 16th century. The famous microbiologist Zinsser notes in his book *Rats, Lice, and History*: "Napoleon's retreat from Moscow was started by a louse!"

What is less well appreciated is that helminths too have had their fair share in moulding history. The best-documented parasitic disease is undoubtedly that which was caused by the nematode worm *Dracunculus medinensis* — also known as the *Guinea worm*. Descriptions of this worm are found as early as 1500 BC. Most parasitologists accept that the "fiery serpents" that struck down the Israelites in the region of the Red Sea after the exodus from Egypt somewhere about 1250 to 1200 BC were actually Guinea worms!

These same helminthiasis markedly altered the course of world history in the modern 20th century, especially in China during the Cold War — acute schistosomiasis sickened Mao's troops and forced them to abort their amphibious assault on Taiwan long enough for American ships to enter the fray. The schistosome consequently came to be known as 'the blood-fluke that saved Formosa' (Taiwan was historically known as Formosa).

expel parasites, but diarrhoea is also detrimental and is a pathological effect induced by the host immune response. Immune responses to chronic infections can also result in pathological fibrotic responses. Fibrosis is a manifestation of wound healing responses that are required on an ongoing basis in hosts chronically infected with pathogens, such as helminths, that cause large amounts of tissue damage.

It is unclear how the TH2 response is regulated during the course of long-term chronic helminth infections, during which antigen loads remain high and pathology is, at least in part, immune mediated. Treg cells have been hypothesised to play a role in these processes, and such regulation may underlie observed reverse correlations between helminth infection and asthma, allergy, and certain autoimmune diseases (see the sidetrack *Too Clean for Comfort*, chapter 16).

Worm-In: Helminths as Therapeutic Agents

Inflammatory **B**owel **D**iseases (IBDs), such as Crohn's disease and ulcerative colitis, are chronic immune diseases of the gastrointestinal tract. Although the aetiologies of these diseases remain unknown, IBD is currently hypothesised to result from an uncontrolled immune response to normal gut flora. The incidence of IBD, like that of other autoimmune diseases and allergies, is on the rise in the Western world, and it is often linked to increased sanitation and decreased exposure to parasitic infections (see the sidetrack *Too Clean for Comfort*, chapter 16). The current hypothesis holds that IBDs result from dysregulated gastrointestinal mucosal inflammation largely mediated by TH1 and possibly TH17 cells. Helminths, on the other hand, induce TH2 and Treg cells, both of which suppress TH1 effector cells.

The logical outcome of this idea was to test if parasitic infections ameliorate autoimmune diseases or allergies. Preliminary work in mice suggested that *S. mansoni* infection protects against trinitrobenzene sulphonic acid-induced colitis, and earnest work to explore helminth therapy (or worm therapy) began. For human studies, eggs of the pig helminth *Trichuris suis* have been used, as the worm does not usually infest and propagate in human intestinal tracts. Initial studies report that treatment results in the clinical amelioration of both Crohn's disease and ulcerative colitis. Nevertheless, the treatment of patients with living helminths, even pig helminths, has many drawbacks. Persistent infection and/or invasion of the parasite to other tissues in the human host, where they might cause pathology, is a major risk. The most promising approach is the identification and characterisation of helminth-derived immunomodulatory molecules that contribute to the anticolitis effect. We hope that in the future, we will be able to treat allergies and autoimmune disorders without resorting to worm-egg spiked drinks!

14 Cancer and the Immune System

I think I'll find another way
There's so much more to know
I guess I'll die another day
It's not my time to go

For every sin, I'll have to pay
I've come to work, I've come to play
I think I'll find another way
It's not my time to go

I'm gonna avoid the cliche
I'm gonna suspend my senses
I'm gonna delay my pleasure
I'm gonna close my body now

I guess, die another day
I guess I'll die another day
I guess, die another day
I guess I'll die another day

— Madonna, *Die Another Day*

Abbreviations

BRCA-1, -2:	Breast cancer-1, -2
IAP:	Inhibitor of Apoptosis
MAGE:	Melanoma Antigens
MIF:	Macrophage Migration Inhibitory Factor
TAA:	Tumour-Associated Antigen
TIL:	Tumour Infiltrating Lymphocyte
TRAIL:	TNF-related Apoptosis Inducing Ligand
TSA:	Tumour-Specific Antigen
VEGF:	Vascular Endothelial Growth Factor

14.1 Introduction

Mature cells of multicellular organisms have a fixed life span. Death may be natural or due to various other causes such as trauma or infections. These dying cells are replaced by the proliferation and differentiation of appropriate stem cells. This proliferation needs to be strictly regulated for the numbers of any cell type to remain more or less constant; cell renewal and death must be balanced to maintain homeostasis. Growth control mechanisms ensure that most cells grow in a density-dependent manner and cease proliferating after a certain number of cell divisions or after reaching a certain density. However, occasionally, there may be formations of cells that are unresponsive to growth control mechanisms and that refuse to die. ***Neoplasm***[1] **is used to describe such an abnormal mass of tissue whose growth exceeds and is uncoordinated with that of normal tissues and persists even after cessation of the change-invoking stimuli**. The process that converts the cell from a normal to a cancerous form is termed *malignant transformation*.

To explain the point, allow us to compare a cell to a car. There is an appropriate speed at which a car should travel (or a cell should divide). Speeding beyond that invites accidents; too much deceleration causes a halt. The accelerator and the brake determine the speed. In a cell, too, genes that speed up growth are accelerators whereas those that check cell division are brakes. When the brakes fail, cells multiply without restriction. Fortunately, a cell has two copies of every gene, so even if one mutates (i.e., if one set of brakes fails), the other copy controls proliferation. It is only when both brakes fail that unrestricted multiplication results and possibly leads to cancer. Consider what will happen when the accelerator gets jammed to the floor. The car will hurtle at breakneck speed; braking will prove futile. In contrast to the brakes, for genes that accelerate cell growth rate, a single copy is enough to cause malignancy, and having two copies proves to be a disadvantage.

Tumour suppressor genes are recessive genes that code for proteins that regulate, dampen, or have a repressive effect on the cell cycle (the brakes). These genes often promote apoptosis, cell division, and cycling at various points during the cell's life. As these genes are recessive, only the loss of both copies of the normal genes renders the cell cancerous. p53 and the related p63 and p73 proteins belong to this family. Both p63 and p73 have functions in normal development, whereas p53 seems to have evolved in higher animals exclusively to prevent tumour development. The *p53* gene is the most commonly transformed gene, being found in about 50% of human tumours. The p53 protein is induced in response to stress signals encountered during tumour development and malignant progression (such as DNA damage, hypoxia, telomere erosion, and the loss of survival signals; also termed oncogenic stress). It acts in the nucleus to stop the replication of damaged cells. In most cases, the induction of p53 results in irreversible cell-growth inhibition, sometimes by the induction of apoptosis. So long as one allele of *p53* is active, tumour suppression continues. Loss of both alleles of *p53* causes the cell cycle to continue despite oncogenic stress or errors in DNA transcription. **Proto-oncogenes are dominant genes that regulate cell growth, division, and differentiation** (the accelerators). If a proto-oncogene undergoes a mutation that makes it more potent, it can cause cancer. Proto-oncogenes that have undergone such changes are known as oncogenes; oncogenes are derived from proto-oncogenes by mutation, retroviral transduction, and so forth. Genes that normally regulate the expression of growth factors and their receptors, signal transducing proteins, nuclear transcription factors, and cyclins involved in cell-cycle progression are primary candidates for oncogenes.

[1] The term derives from *neo*, meaning new, and *plasia*, meaning growth.

Understanding the Terminology

A tumour is a swelling caused by a mass of cells. Benign tumours are normally slow growing, circumscribed, and encapsulated with well-defined edges. They do not invade surrounding tissues extensively. Malignant tumours are often rapidly growing, aggressive, invasive masses of cells that can spread through the blood or lymph, that is, metastasise. *Cancer* refers specifically to malignant tumours. **Oncology** is the study of malignant tumours.

Metastasis is the transfer of pain, function, or disease from one organ to another through contact, blood vessels, or lymphatics. It is generally used in reference to the transfer of tumour cells. The original tumour is the primary tumour; tumours caused by metastasis are secondary tumours.

Malignant tumours are classified according to the embryonic origin of the tissue from which they are derived.

- **A carcinoma is a malignant tumour arising from ectodermal or endodermal tissues** such as the skin or epithelial lining of internal organs or glands. More than 80% of tumours are carcinomas; a majority of breast, lung, prostrate, or colon tumours are carcinomas (e.g., mammary adenocarcinoma, hepatocarcinoma, and colon carcinoma).
- **A sarcoma arises from mesodermal tissues**, such as those in bone, fat, and cartilage. Examples include fibrosarcoma, osteosarcoma, and liposarcoma.
- **Lymphomas and leukaemias are tumours of haematopoietic cells of the bone marrow**. Lymphomas tend to grow as solid masses. Examples of lymphomas include Hodgkin's disease and multiple myelomas. Carcinomas, sarcomas, and lymphomas are called solid tumours. By contrast, leukaemias are circulating cancerous cells that proliferate as single cells (e.g., acute lymphocytic leukaemia and chronic myelogenous leukaemia).

Mutations or transformations in two other sets of genes can also result in cancer. The first set is of genes involved in DNA repair. To carry the car analogy further, if mutations can be looked upon as damage to the car, DNA repair machinery is a team of mechanics. As the old adage goes — if you lose the mechanic, you lose the car. **The second set of genes that can result in malignancy are those involved in apoptosis regulation**. Apoptosis-regulating genes are highway patrolmen. They keep the roads safe by hauling off speeding cars, thus preventing accidents. Two gene families of apoptosis regulators have been identified (see chapter 11), and mutations in either of them can result in cancer.

- The ***Bcl-2* family** comprises molecules with pro- and anti-apoptotic functions. Antiapoptotic proteins include Bcl-2 and Bcl-XL, whereas the proapoptotic ones are Bax, Bad, Bak, and Bid. Bcl-2 proteins affect apoptosis by decreasing (antiapoptotic) or increasing (proapoptotic) mitochondrial permeability, thus affecting the release of cytochrome *c*[2].
- **IAP** (**I**nhibitors of **Ap**optosis) **proteins** are the other family of apoptosis regulators. They inhibit caspases, the enzymes involved in apoptosis.

Carcinogenesis is the process by which a malignant neoplasm is produced. A random or single mutation cannot cause malignancy. Apart from a nonlethal mutation that imparts the ability to grow indefinitely and resist apoptotic signals (i.e., confers immortality), cancerous cells often show a diminished requirement for growth factors and altered biochemical activities that allow them to survive and multiply (Figure 14.1). Malignant neoplasms grow by the progressive infiltration, invasion, destruction, and penetration of surrounding tissues. For a tumour to grow, it must have an adequate nutrient supply. Most solid tumours secrete angiogenesis factors that result in the formation of new blood vessels. To invade

[2] Mammalian cells use two main pathways to undergo apoptosis. The extrinsic pathway is initiated by the ligation of cell-surface death-receptors, such as TNF-α and Fas (CD95). The intrinsic pathway is centred on the dysregulation of mitochondrial function and the release of cytochrome *c*.

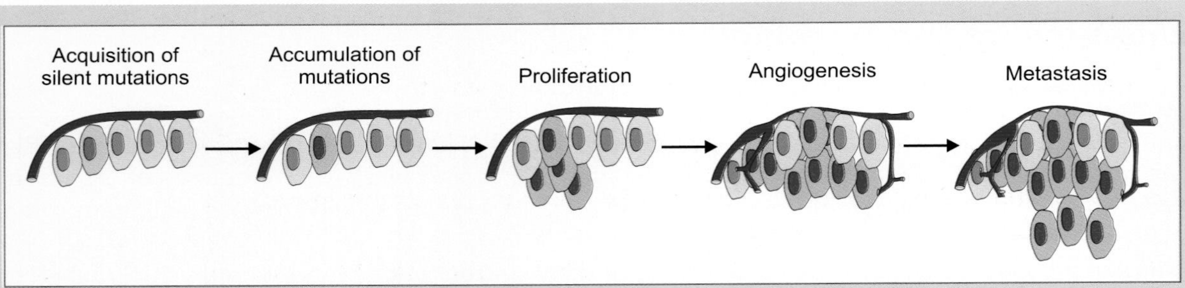

Figure 14.1 Tumour transformation involves the progressive accumulation of multiple mutations that allow tumour cells to survive, proliferate, and resist apoptotic signals. The secretion of angiogenesis factors allows the formation of new blood vessels, securing nutrient supply for the growing neoplasm. Metastasis occurs when cancerous cells detach from the tumour and migrate to other tissues or organs.

other tissues and organs, a cancerous cell must detach from the tumour, adhere to the extracellular matrix, proteolytically degrade the matrix, and then find its way out of the basement membrane. A single mutation cannot endow a cell with all these characteristics that make it malignant. Thus, **tumour formation is a multistep process requiring the accumulation of numerous mutations in the cells**. This multistep process is termed *tumour progression.*

Factors that can induce malignant transformations in normal cells include:

- **Environmental factors** that can increase mutations enhance the chances of malignancies.
  - **Radiation**: UV rays and ionising radiations that induce mutations can cause cancers.
  - **Chemicals**: A variety of chemicals, called carcinogens, can cause malignant transformations. Arsenic, asbestos, benzene, benzo(a)pyrenes in cigarette smoke, vinyl chloride, and aflatoxins are some common carcinogens.
  - **Viruses**: Some viral infections can increase cellular proliferation and lead to cancer. Examples include infections by the human T cell leukaemia virus type 1, mouse mammary tumour virus, the Rous sarcoma virus[3], the human papilloma virus (causes cervical cancer), the hepatitis B virus (causes hepatocellular carcinoma), and the Epstein-Barr virus (linked to Burkitt's lymphoma and Hodgkin's lymphoma).
- **Genetic factors**: Until the 1980s, environmental factors were thought to be predominantly responsible for cancer. However, in the last decade of the 20th century, the importance of genetics came to be appreciated when the first predisposing gene, the *RB1* gene associated with retinoblastoma, was identified. Mutations in genes involved in apoptosis, signalling, or in genes encoding enzymes involved in angiogenesis, DNA repair, and genome stability have been identified as responsible for cancer. Notable examples include the following:
  - Mutations in genes encoding key molecules, such as p53 involved in inducing cell cycle arrest or apoptosis or Ras, that are involved in signal transduction can predispose a person to cancer.
  - A majority of inherited breast cancers are caused by mutations in the *BRCA-1* and *BRCA-2* (**Br**east **Ca**ncer-**1** and -**2**) genes; mutations in the tyrosine receptor kinase gene *ERBB2* can also result in breast cancer.
  - Mutations in the phosphatase PTEN increase the risk of Cowden's syndrome (resulting in gastrointestinal tract, breast, or thyroid cancers).

[3] Studies on v-src encoded by the virus helped prove that oncogenes alone could induce malignant transformations; a group of tyrosine kinases is named after this gene. See the sidetrack *The SAARC family* in chapter 6.

- Xeroderma pigmentosum is caused by a defect in the gene that encodes UV-specific endonuclease (a DNA repair enzyme). The ensuing inability to repair UV-induced mutations results in multiple types of skin cancers.

14.2 Tumour Antigens

Antigens expressed by tumour cells that trigger an immune response in the host are termed tumour antigens. Tumour antigens may be expressed as cytoplasmic constituents or membrane proteins, or they may be secreted in fluids. Secreted tumour antigens are also called neoantigens. Tumour antigens can be classified into two types.

- **Tumour-Specific Antigens (TSAs) are expressed exclusively by tumour cells and are not found on normal body cells**. TSAs can be of two types:
  1. Those encoded by genes exclusively expressed by tumours, such as antigens expressed as a result of viral transformation, and
  2. Antigens expressed by variant (mutated) forms of normal genes.

  Most tumours induced by physical or chemical agents or viruses express neoantigens. By contrast, spontaneously occurring tumours are often non- or weakly immunogenic. TSAs induced by physical or chemical agents are specific for each tumour, as the mutations induced by such agents are likely to be unique to each cell. Thus, multiple tumours induced by these agents in the same animal will have different TSAs. TSAs induced by viral transformation are characteristic of the tumour-inducing virus and are shared by all the tumours induced by that virus. Because TSAs contribute to tumour rejections, they are also referred to as tumour-specific transplantation antigens or tumour-rejection antigens. Examples of TSAs include

  - mutated proteins found in melanomas, such as the cyclin-dependent kinase-4 involved in cell-cycle regulation, and β-catenin involved in signal transduction,
  - mutated caspase-8 found in squamous cell carcinoma, and
  - viral gene products E6 and E7 of the human papilloma virus found in cervical cancer.
- **Tumour-Associated Antigens (TAAs) are found on both normal and cancerous cells**. Some TAAs are antigens expressed at only certain stages of

Tumour Harm: Count the Ways

- At the simplest level, tumour growth starves the host of nutrients.
- It may destroy a vital structure by virtue of its location (e.g., pituitary gland) or obstruct or perforate a hollow organ (bowels, bronchus).
- It may cause the ulceration of intact surfaces (mucosa, epidermis) or the necrosis of tissues.
- Tumours may interfere with normal immune responses, resulting in a predisposition to infections, excessive cytokine production, and wasting.
- If they invade adjacent structures, tumours may cause obstruction, bleeding, thrombosis, or infarction.
- Many tumours cause hormonal dysregulation — increased hormonal secretion or inappropriate hormonal secretion (the wrong hormones produced by the wrong organs).
- Tumour growth may result in pinching of nerves and may interfere with nerve conduction.

differentiation or only by certain differentiation lineages in normal cells, whereas others are antigens overexpressed on cancer cells. Examples include the following:

- Oncofoetal antigens that are normally expressed only in foetal cells. They include α-fetoprotein, secreted in patients with hepatomas and testicular cancers, and carcinoembryonic antigen, expressed on cell membranes and secreted in fluids of patients with colorectal tumours. As these antigens are normally expressed before the immune system gains competence, they can elicit an antitumour response.
- Common acute lymphocytic leukaemia antigen has now been identified as CD10, normally expressed at low levels on B cells.
- Differentiation antigens are overexpressed in tumours and are found in only certain types of normal tissue, such as prostate-specific membrane antigen found in prostate cancer, mucin-1 and ERBB2 (also called her2/neu[4]) found in breast and ovarian cancers, and cancer-testis antigens such as **M**elanoma **A**nti**ge**ns (MAGE). MAGE are a family of proteins expressed only in the testes or in tumour cells, especially in skin cancer. MAGE-1 was the first tumour-specific T cell epitope to be defined. The group also includes a number of differentiation antigens expressed by melanocytes but overexpressed by melanoma cells, e.g., tyrosinase (involved in melanin synthesis), gp75, gp100, MART-1, and Melan-A.
- Telomerase and survivin (an IAP) are TAAs expressed by most tumour-bearing individuals

14.3 Effectors of Antitumour Immunity

Although the role of the immune system in protection against tumours was conceptualised by Ehrlich in the early 1900s, the notion was formalised by Burnet and Thompson in 1967 when they outlined the *immunosurveillance concept*. They suggested that lymphocytes acted as sentinels in recognising and eliminating continuously arising, nascent transformed cells. Experimental evidence for this concept emerged after the advent of knock out mice; mice deficient in certain key features of the immune system (e.g., *RAG* genes, perforin, or IFN-γ) showed an increased susceptibility to tumours. A similar trend emerged from data on transplant patients and immunocompromised individuals. Thus, data obtained from both murine and human studies provide strong support for the existence of and physiologic relevance of cancer immunosurveillance.

Tumours represent a unique problem for the immune system. The immune system is educated to discriminate between self and nonself and to focus its destructive powers on the latter. Thus, the presence of nonself molecules or danger signals is required to activate the immune system. Tumours develop from self-cells, and at least in the initial stages, do not send out distress/danger signals. They are therefore likely to escape immunosurveillance during the initial phases of development. Once solid tumours reach a certain size, they begin to grow invasively and require enhanced blood supply. Tumours often produce angiogenic proteins that induce the growth of blood vessels, ensuring adequate blood supply. However, the invasive growth causes disruptions in surrounding tissue and triggers an immune response. The sequence of events that results in an antitumour response is outlined below (Figure 14.2).

- Invasive growth induces inflammatory signals, such as the expression of stress-induced proteins. These signals induce an influx of leukocytes and the

[4] Her2/neu (human epi dermal growth factor receptor 2) is a cell membrane surface-bound receptor tyrosine kinase that is expressed in trace amounts in normal tissues but in much elevated levels in breast cancer cells. Anti-her2/neu mAbs can therefore be used for the selective elimination of breast cancer cells (see the sidetrack *In Gandhi's Footsteps,* chapter 10).

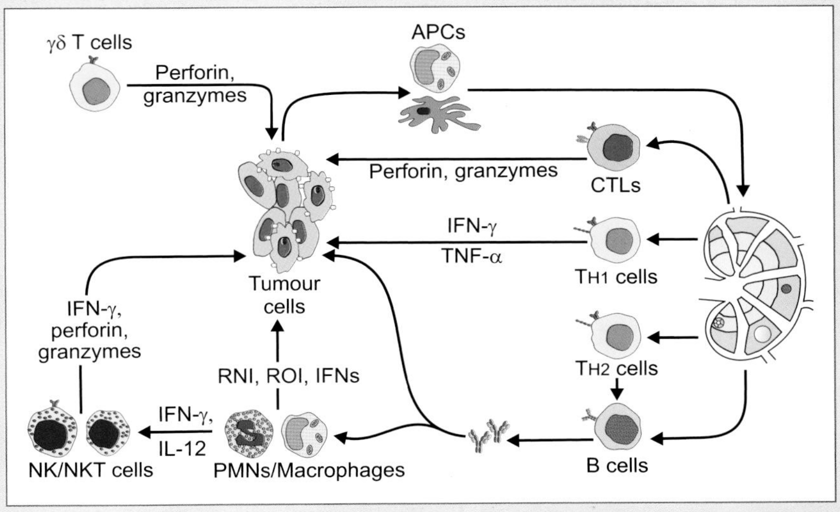

Figure 14.2 Cells of the innate and adaptive immune systems work in tandem to eliminate tumours. Pro-inflammatory signals released by tumour cells recruit cells of the innate immune system, such as γδ T cells, NK cells, and NKT cells. These cytotoxic cells kill tumour cells by inducing apoptosis and secreting pro-inflammatory cytokines such as IFN-γ. NK cells and macrophages activate one another by the reciprocal production of IFN-γ and IL-12. Activated macrophages damage tumour cells by liberating cytotoxic ROI and RNI. Chemokines and cytokines released as a result of the inflammatory response recruit APCs to the tumour site. IFN-γ and tissue debris from dead and dying cells cause maturation of the DCs. APCs transport tumour antigens to the regional lymph node, where they induce the development of tumour specific TH1 cells and CTLs. These effector T cells home to the tumour site and help destroy tumour cells. The TH cells also help B cells in mounting a humoral response.

recruitment of innate immune system cells (NK, NKT, γδ T cells, macrophages, and DCs).

- NK and NKT cells act against tumour initiation, growth, and metastasis. They induce cell death by the perforin–granzyme pathway, as well as by the induction of apoptosis (chapter 11). Additionally, NK cells secrete effector cytokines (such as IFN-γ, TNF-α, and GM-CSF) that are important in recruiting and activating other cell types.
- Structures on the transformed cells are recognised by infiltrating innate immune system cells, and this results in the production of IFN-γ.
- IFN-γ production aids antitumour immunity in multiple ways.
  - It induces limited tumour cell death because of its antiproliferative and apoptotic mechanisms.
  - IFN-γ induces angiostatic chemokines, such as CXCL19, CXCL10, and CXCL11. These chemokines block the formation of new blood vessels within the tumour. The lack of blood supply and the effect of IFN-γ accelerates tumour cell death.
  - It stimulates NO production by phagocytes.
  - IFN-γ increases leukocyte–endothelium interaction and further increases the influx of leukocytes.
  - It upregulates MHC class I and class II molecules on many cell types — including tumour cells.
- Chemokines produced during escalating inflammatory responses recruit more NK cells and macrophages to the site; tumour infiltrating NK cells and macrophages activate one another by the reciprocal production of IFN-γ and

IL-12. Cytotoxic granules, TRAIL (**T**NF-**R**elated **A**poptosis **I**nducing **L**igand), ROI, and RNI produced by NK cells and macrophages are all responsible for killing tumour cells.

- IFN-γ, produced by NK cells, and the tissue debris resulting from the lysis of tumour cells by NK cells are central to DC activation. The debris from tumour cell death is internalised by DCs, which home to the local draining lymph node.
- DCs in draining lymph nodes induce tumour-specific TH1 cells and CTLs. These cells home to the tumour site, where the CTLs destroy tumour cells.
- Memory TH1 and CTL cells formed in the later stages of the response are critical in maintaining protective immunity.
- Humoral responses to tumour antigens are triggered when B lymphocytes internalise tumour antigens and enter into a cognate interaction with activated TH cells. Antibodies against tumour cells or their constituents have been observed in the sera of patients with Burkitt's lymphoma, malignant melanoma, osteosarcoma, lung cancers, and breast cancers. However, antibodies often have difficulties in reaching peripheral tumours and may have a limited role in their elimination. Paradoxically, some antibodies may even promote tumour survival. Such antibodies, termed *enhancing antibodies*, are noncytotoxic antibodies, whose mechanism of action is unclear. They are thought to promote tumour survival by binding cell-surface molecules critical for tumour lysis and inducing their downregulation.

14.4 Tumour Immune Evasion

As explained, tumour formation is a multistep process in which transformed cells undergo a series of changes that allow their survival and proliferation. Although a tumour originates from self-cells, it is a parasite; this host–parasite relationship is as dynamic as the more conventional host–parasite relationship. A newly forming tumour will inevitably express self-antigens; it is also likely to express new or altered proteins. These new or altered proteins trigger the immune response — the tumour eventually becomes a target of immune effector mechanisms. Tumours that survive or flourish through this immune attack are often ones that can withstand the immune system's tumour-suppressing actions. Thus, the immune system exerts selective pressure on the tumours — highly immunogenic tumours are unlikely to escape the immune system's offensive, whereas weakly immunogenic tumours will survive. Aggressive tumours that replace cells faster than those destroyed by the immune attack are also more likely to survive, irrespective of their immunogenicity. Multiple pathways lead to tumour survival.

- **Downregulation of MHC class I antigen processing and presentation pathways**: Decreased or absent HLA class I expression is associated with invasive and metastatic tumours, and a total loss of MHC class I expression is not uncommon in many tumours, including melanomas, colorectal carcinomas, prostate adenocarcinomas, and breast cancers. Some of the mechanisms reported to result in the loss or downregulation of MHC class I expression in cancer patients include
  - loss of or defective β_2-m gene,
  - downregulation of the proteasome multicatalytic complex units LMP-2 and LMP-7 (chapter 7),

- downregulation of the peptide transporters TAP-1 and TAP-2, and
- selective loss of MHC class I haplotype because of loss of transcription factors.

- **Loss/downregulation of the MIC-A and MIC-B molecules**: Loss of MHC class I molecules should make tumours more susceptible to NK cell lysis. However, such tumours also lose or downregulate the MIC-A and MIC-B molecules that engage NKG2D activatory receptors on NK cells. In the absence of these activating signals, NK cells do not attack tumour cells.
- **Lack of costimulatory molecule expression**: Absence of costimulatory molecules allows tumours to escape NK cell and CTL attack. The engagement of MHC class I:peptide complexes in the absence of costimulation will result in CTL anergy and suboptimal NK cell activation, allowing tumour escape.
- **Overexpression of MIF** (macrophage **M**igration **I**nhibitory **F**actor): MIF, overexpressed by many tumour cells, is an immunomodulator associated with tumour progression. It contributes to neoangiogenesis and epithelial cell proliferation. Furthermore, it may contribute to genetic instability in tumours, as it suppresses p53 function.
- **Loss or decreased surface-antigen expression**: Decreased tumour antigen expression often correlates with disease progression although the exact mechanisms of this downregulation are not known.

Cancer and the Immune System

- Tumour formation is a multistep process requiring the accumulation of numerous mutations that allow cells to survive, proliferate, and obtain nutrients from surrounding tissues.
- Genes often implicated in malignant transformation include
  - proto-oncogenes involved in the regulation of cell growth, division, and differentiation,
  - tumour suppressor genes important in halting cell division,
  - apoptosis genes involved in the regulation of cell death, and
  - genes involved in DNA repair.
- $CD4^+$ T_H1 cells and $CD8^+$ CTLs are especially important in destroying tumour cells; Igs play a relatively smaller role in antitumour immunity.
- Strategies that allow tumours to evade host immune responses include
  - downregulation of MHC class I antigen processing and presentation pathways,
  - downregulation of MHC class I-like MIC-A and MIC-B molecules,
  - decrease in costimulatory molecule expression,
  - overexpression of MIF, which contributes to angiogenesis,
  - downregulation of tumour antigen expression,
  - defective apoptosis-inducing pathways or overexpression of FasL,
  - expression of cytokines and chemokines that can negatively affect the maturation and function of immune cells,
  - interference with IFN-γ signalling pathways, and
  - overexpression of protease inhibitor-9.
- Immunotherapeutic strategies for the selective targeting of tumour cells include
  - injecting irradiated tumour cells with adjuvants to stimulate the host immune system,
  - administrating immunostimulatory cytokines,
  - enhancing the expression of costimulatory molecules, and
  - using engineered mAbs that deliver the drug/radionuclide/cytokine directly to tumour cells.

- **Defective apoptosis-inducing pathways**: FasL and TRAIL are important to inducting apoptosis (chapter 11). The engagement of Fas on tumour cells by FasL-expressing cells (e.g., CTLs or NK cells) results in the activation of death-receptors and induces cell death. Mutations at multiple sites in death-receptor pathways can favour tumour escape.
  - Defective death-receptor signalling may contribute to tumour survival and proliferation.
  - The downregulation of Fas expression itself may also contribute to apoptosis resistance. Mutation and loss of the gene encoding Fas has been identified in multiple myelomas, non-Hodgkin's lymphomas, and melanomas.
  - Many tumours show an increased expression of FLIP, an inhibitor of caspase-8. These tumours are resistant to apoptosis induction.
  - Tumour cells may also show a defect in downstream signalling pathways following engagement of the death-receptor, for example, loss or mutation in caspase-8 or caspase-3. Alternatively, they may express lower levels of death-receptors.
- **Overexpression of the protease inhibitor-9**: This enzyme inactivates granzyme B and allows tumours to escape CTL- or NK cell-mediated cytotoxicity (chapter 11).
- **Expression of immunosuppressive factors**: Cytokines can negatively affect the maturation and function of immune cells and create an immunosuppressive environment.
  - Most tumour cells secrete VEGF (**V**ascular **E**ndothelial **G**rowth **F**actor). *In vitro*, VEGF inhibits DC differentiation and maturation through the suppression of the transcription factor NFκB. In patients with lung, head, neck, and breast cancers, a decrease in the function and number of mature DCs was found to be associated with increased plasma concentrations of VEGF.
  - An increased concentration of serum IL-10 is frequently detected in cancer patients. IL-10 inhibits DC differentiation, antigen presentation, IL-12 production, and induction of T_{H1} responses. It enhances spontaneous DC apoptosis and may also protect tumour cells from CTLs through the downregulation of MHC class I, MHC class II, TAP-1, TAP-2, and ICAM-1 expression.
  - PGE2 is also expressed in many tumours. This is often the result of the enhanced expression of cycloxeganse-2, the rate-limiting enzyme in PGE2 synthesis. PGE2 increases the production of IL-10 by lymphocytes and macrophages while inhibiting IL-12 production by macrophages.
  - High concentrations of TGF-β are frequently found in cancer patients and are associated with disease progression and a poor response to immunotherapy. TGF-β is an immunosuppressive cytokine that directly inhibits NK cell activation. It also inhibits activation and proliferation of B and T lymphocytes. Both IL-10 and TGF-β may also induce formation of T_{reg} cells. These cells suppress the activation and proliferation of $CD4^+$ and $CD8^+$ T cells and aid tumour escape.
- **Interference with IFN-γ signalling pathways**: IFN-γ is crucial to antitumour immunity. Many tumour cells have defective IFN-γ signalling pathways, rendering them resistant to its actions.

Chemical Kill: Chemotherapy

Antitumour chemotherapeutic agents in common use kill target cells primarily by causing cellular stress and inducing death by apoptosis. Four broad classes of chemotherapeutic agents are in use.

- **Alkylating agents** generally interact and crosslink DNA nonspecifically. The alkylation of nucleic acids involves a substitution reaction in which a nucleophilic atom on the nucleic acid is replaced by an alkyl group from the alkylating agent. Alkylating agents in clinical use include cyclophosphamide, melphalan, mitomycin C, chlorambucil, and platinum compounds (cisplatin and carboplatin).
- **Antitumour antibiotics** target DNA or other structures, such as tubulin, that are common constituents of eukaryotic cells. Because they affect rapidly proliferating cells, these antibiotics are often more selective for cancerous cells than nonspecific alkylating agents. Some antibiotics introduce protein-associated strand breaks by stabilising DNA–topoisomerase I or II complexes, for example, doxorubicin, etoposide, mitoxantrone, and SN-38. Others inhibit translation (actinomycin D), introduce double-stranded DNA breaks (bleomycin), or crosslink DNA–DNA (mitomycin C).
- **Tubulin-binding agents** prevent microtubule assembly so that the cell is arrested in the G2 phase of growth — examples include vinblastine, vincristine, taxotere, and paclitaxel.
- **Antimetabolites** disrupt or inhibit essential metabolic processes by incorporating themselves into nuclear material or combining irreversibly with vital cellular enzymes. Many anticancer antimetabolites exploit the dividing cells' need for a constant supply of nucleic acid bases required for DNA synthesis.
  - **Methotrexate** inhibits the enzyme dihydrofolate reductase that is essential for the synthesis of purines and pyrimidines.
  - **5-fluorouracil** inhibits thymidylate synthase.
  - **Folic acid analogs**, such as tomudex, act by interfering with thymidylate synthase.
  - **Nucleotide analogues**, whether purine (mercaptopurine and thioguanine) or pyrimidine (floxuridine and fludarabine), are also in clinical use.

14.5 Treatment

Tumours have a greater population of dividing cells than normal tissue and are hence more susceptible to agents that interfere with cell division, for example, γ-radiation and the chemotherapeutic agents listed in the sidetrack *Chemical Kill*. However, these drugs (and radiation) have severe side-effects, as they also kill healthy cells undergoing cell division (such as cells in the bone marrow, gastrointestinal tract, and hair follicles). These side-effects limit the dosage of anticancer drugs, and they often have to be given in suboptimal doses, resulting in the eventual failure of therapy and/or the development of drug resistance and metastatic disease. Newer approaches to cancer treatment include the development of immunotherapeutic strategies that allow the selective targeting of cancer cells.

Immunostimulatory therapies aimed at activating macrophages and NK cells and at promoting pro-inflammatory pathways are also likely to be beneficial, as tumours often escape immunosurveillance by promoting immunosuppressive pathways. Such antigen-nonspecific approaches have the advantage of not requiring an extensive knowledge of the immunogenicity of the tumours being treated. Examples of nonspecific immunostimulation include:

- **Injecting irradiated tumour cells**[5] **or tumour extracts mixed with bacterial adjuvants**, such as BCG or suspension of *Corynebacterium parvum*. The adjuvant is supposed to cause an inflammatory response, stimulate DCs to internalise and present tumour cell debris to T cells, and generate a tumour-

[5] Irradiation ensures that the injected tumour cell is incapable of multiplying in the patient.

specific antigen response. BCG, whether given intralesionally or locally, has resulted in tumour regression in a number of melanoma cases. However, the treatment has failed to live up to its promise with other kinds of tumours.

- **Cytokine administration** is also being investigated for nonspecific immunostimulation. However, the complexity and redundancy of cytokine functions makes their clinical use difficult, and severe side-effects limit their usage.
    - Pro-inflammatory cytokines, such as IFN-γ and TNF-α, have yielded some promising results in clinical trials. Apart from activating innate effector cells, such as NK cells and macrophages, these cytokines also upregulate the expression of MHC molecules on tumour cells. As explained, many tumour cells escape immunosurveillance by downregulating MHC expression. IFN-γ and TNF-α therapy may help in restoring MHC expression, allowing the development of CTL responses. Moreover, IFN-γ may act directly through its antiproliferative effect on tumour cells.
    - IL-2 is a potent lymphocyte activator, and it also induces their proliferation. NK cells obtained from cancer patients can be activated with IL-2 *in vitro*. When re-injected in the patient, these activated cells, called LAK cells, have potent antitumour activity. **T**umour **I**nfiltrating **L**ymphocytes (TILs) can be similarly isolated and activated *in vitro* with IL-2. When administered to the patient, such activated TILs have specific antitumour activity.
- **Increasing the expression of costimulatory signals** has enhanced tumour immunity in animal models. The injection of tumour cells transfected with genes encoding costimulatory molecules CD80/86 results in tumour regression in mice.

The successful development of tumour-specific immunotherapies requires the knowledge of tumour antigens. TSAs are the obvious choice as immunotherapy targets. The past two decades have resulted in the identification of many such targets for antitumour therapy. Antibody-based immunotherapies are of use against membrane-expressed tissue antigens. Differentiation antigens expressed in trace amounts in normal cells but overexpressed in tumour cells (e.g., ERBB2, MART-1) are especially good targets as mAbs against these antigens selectively target tumour cells. The usefulness of secreted tumour antigens (neoantigens) in such therapies is questionable. However, they are used in the diagnosis and monitoring of tumour progression and in gauging the tumour's response to treatment.

Active antigen-specific immunotherapy is often referred to as antitumour vaccination. In contrast to anti-infection vaccines, antitumour vaccines are therapeutic (i.e., curative), not prophylactic (i.e., preventive). The vaccines aim to trigger an antitumour CTL response. Peptides derived from TSAs and TAAs can be good vaccine candidates. Individual differences between patients — both in terms of HLA types and epitopes that are presented and recognised — are a major hurdle in successful vaccination. The identification of T cell epitopes that are most likely to trigger the immune response requires an extensive knowledge of tumour antigens and patient MHC haplotypes. Antitumour vaccines are therefore expensive; they are tailored for the individual and cannot be mass produced. Preclinical and clinical trials of antitumour vaccines demonstrate that they are well tolerated and thus safe. Although some patients report tumour regression, the clinical outcome has not been as successful as anticipated. However, with increasing knowledge of T cell epitopes and a better understanding of the mechanisms of tumour immunoevasion, it is hoped that antitumour vaccinations will someday yield real benefits to cancer patients.

Although inducing antitumour immunity through the use of autologous DCs is complicated and expensive, it is one of the most promising new approaches in cancer therapy. Administration of peptide–adjuvant mixtures has been attempted, but met with less success than did the use of peptide-pulsed DCs. DCs are the most potent APCs of the body and can be used to induce tumour-specific CTLs and TH cells in the patient (see the sidetrack *Licensed to Kill*, chapter 13). Briefly, pre-DC cells, such as monocytes isolated from the patient, are allowed to proliferate and differentiate *in vivo* by exposure to GM-CSF and IL-4. They are then pulsed with tumour antigens. Originally, DCs were pulsed with tumour extracts, but now tumour-derived peptides are used to ensure safety and standardisation. The DCs are also exposed to maturation stimuli, such as IFN-γ. The peptide-loaded mature DCs are then injected into patients. DC vaccination has resulted in efficient CTL responses and tumour regression in several clinical trials for the treatment of melanoma, renal carcinoma, breast cancer, ovarian cancer, and prostate cancer. Moreover, it appears to be safe, with minimal side-effects (e.g., transient fever, local reactions, and autoimmune vitiligo in melanoma patients).

We have discussed the general principle and application of antibody-based therapy in chapter 10. Here, clinical strategies for increasing the efficacy of anticancer therapies will be briefly discussed (Figure 14.3).

- **Combination therapy**: The use of mAbs in conjunction with cytotoxic drugs increases treatment success. For example, herceptin (humanised anti-ERBB2 mAb) is synergistic with chemotherapeutic agents, such as the Pt salt cisplatin, doxorubicin, methotrexate, taxol, and cyclophosphamide. Similarly, the anti-CD20 antibody increases the efficacy of conjugated drugs. For example, tositumomab (an anti-CD20 mAb conjugated to a radioactive iodine isotope), sold in the USA under the brand name Bexxar, is found to be effective in treating lymphomas. On the down side, combination therapies often show increased cytotoxicity.
- **Use of immunoliposomes**: Liposomes are self-assembled lipid bilayers that encapsulate some of the surrounding medium during their formation. Liposomal formulations of chemotherapeutic agents such as doxorubicin have been approved for clinical use. Engineered antibody fragments can be attached to

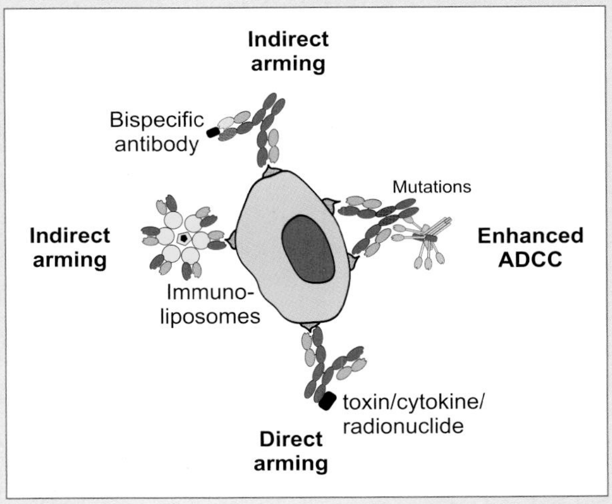

Figure 14.3 Several strategies can increase the efficiency of anticancer immunotherapies. The simplest strategy is to enhance the cytotoxic potential of antibodies by site-directed mutagenesis. Such antibodies are more efficient in complement fixation and ADCC. Alternatively, mAbs can be linked to toxins, cytokines such as IFN-γ or TNF-α, or radionuclides to enhance their killing potential (direct arming). This has the advantage of requiring lower doses of the toxic substance because antibodies deliver their toxic cargo directly to the tumour cell. Engineering mAbs to produce bispecific antibodies capable of binding to two different antigens allows the targeting of cytotoxic molecules directly to the tumour (indirect arming). Immunoliposomes containing an immunotherapeutic agent can be linked to engineered antibody fragments, allowing selective tumour targeting (adapted from Carter, *Nature Reviews in Cancer*, 2001, 1:118).

these liposomes for selective tumour targeting by drugs or toxins or even DNA for gene therapy. Results from studies in animal models are highly encouraging. The large size of the immunoliposomes (usually ~100 nm in diameter) makes liposome extravasation into the tumour difficult. This drawback could be circumvented if the immunoliposomes are targeted to the tumour vasculature rather than to the tumour itself, and this approach is under investigation.

- **Targeting tumour vasculature**: Tumours become vascularised by the proliferation of new blood vessels during angiogenesis. Beyond a certain size, tumours cannot survive without angiogenesis. mAbs that neutralise VEGF or its receptor have potent antitumour activity (e.g. Bevacizumab sold under the brand name Avastin).
- **Targeting minimal residual disease**[6]: Because the difficulties associated with the penetration of mAb in solid tumours are reduced in minimal residual disease, mAbs can be used in their treatment. The anti-epithelial adhesion molecule mAb panorex has recently been approved for the treatment of colorectal cancer minimal residual disease, and herceptin is being tested in clinical trials for breast cancer minimal residual disease treatment.

[6] Minimal residual disease is the term used to describe the small number of cancerous cells that remain in the patient after surgery, chemotherapy, or radiotherapy.

As You Sow so Shall You Reap: Diet and Cancer

The importance of nutrition in cancer development is an area of active study and controversy. Diet is estimated to contribute to about one-third of preventable cancers. The most convincing epidemiological evidence for the role of dietary factors in cancer risk is the inverse relationship between fruit and vegetable consumption and several types of cancers, including lung, oral, oesophageal, gastric, pancreatic, ovarian, and breast cancers. Similarly, whole grain consumption has been associated with a decreased risk of stomach and colorectal cancers. However, a causal link between foods and cancer has not been established. Neither the exact constituents of fruits, vegetables, and grains that are responsible for their protective effect nor the cellular or molecular processes responsible for that protection are understood. Some vitamins and micronutrients whose deficiency increases the risk of cancer are the following:

- Reduced folate intake has been associated with a higher risk of colon cancer, breast cancer, gastric and oesophagal cancer, and pancreatic cancer in smokers. *In vitro* studies have shown that folic acid deficiency causes a dose-dependent increase in uracil incorporation into human lymphocyte DNA and results in increased chromosome breaks.
- Vitamin B_6 and B_{12} deficiencies can increase chromosome breaks by a uracil-misincorporation mechanism that is similar to folate deficiency. Epidemiological studies indicate an association between B_6 deficiency and increased lung and prostate cancers.
- A reduced intake of antioxidants such as vitamins C, D, and E are all associated with an increased risk of cancer. Although both experimental and epidemiological data indicate that vitamin C protects against stomach cancer, especially when consumed in the form of fruits and vegetables, studies on vitamin C supplements have proved largely inconclusive. Increased vitamin C intake reduces DNA strand breaks. Vitamin C intake is especially important in smokers; smoking depletes vitamin C, which is required to protect sperm DNA from oxidative damage. A smoker needs to consume 40% more vitamin C than a nonsmoker to maintain a comparable blood plasma level. Several studies have examined the association between paternal smoking, vitamin C intake, oxidative damage, and childhood cancer in offspring. Available evidence indicates that the risk of cancer is likely to be higher in the offspring of male smokers, especially when antioxidant intake is low.
- Deficiencies in minerals such as Fe, Se, and Zn have also been associated with an increased cancer risk.

15 Autoimmunity

I am the hate you try to hide and I control you
I take you where you want to go
I give you all you need to know
I drag you down I use you up
Mr self destruct

— Nine Inch Nails, *Mr Self Destruct*

Abbreviations

ACh:	Acetylcholine
EAE:	Experimental Autoimmune Encephalomyelitis
HPA: axis	Hypothalamus-Pituitary-Adrenal axis
IDDM:	Insulin-Dependent Diabetes Mellitus
MS:	Multiple Sclerosis
NOD: mice	Non-Obese Diabetic mice
RA:	Rheumatoid Arthritis
SLE:	Systemic Lupus Erythematosus
VLA-4:	Very Late Antigen-4

15.1 Introduction

The paradox of arms possession — the same weapons that protect can also harm — is equally true of the immune system's weapons. The famous immunologist Ehrlich used the term *horror autotoxicus* to describe what he considered the organism's unwillingness to endanger itself through the formation of toxic autoantibodies. He thought an organism would never imperil itself by allowing autoimmune reactions.

In chapter 9, we explained how T cells bearing high-avidity receptors for self-antigens are negatively selected during development. This process is not foolproof — autoreactive T cells[1] and antibodies can be found in the peripheral blood of healthy animals. A number of tolerance mechanisms in the periphery control potentially dangerous self-reactive cells (section 12.3). Most autoreactive antibodies found in healthy individuals are against ubiquitous proteins, such as albumin, transferrin, cytochrome *c*, and nucleic proteins, or against cytoplasmic filaments, such as actin. These are low-affinity IgM antibodies that are often cross-reactive with exoantigens, such as bacterial cell walls. They are usually produced by $CD5^+$ B cells, do not undergo affinity maturation, and are not generally self-damaging. Thus, some physiological autoimmune reactivity against self-antigens is normal. Autoantibodies have been suggested to be regulators of the immune system and, at normal physiological levels, act as components of the body's homeostatic mechanisms. Autoantibodies are also thought to function as 'biological taxis', transporting cellular breakdown products to their site of disposal. For example, after myocardial infarction, apparently harmless autoantibodies to heart tissue are found in patient's blood and are thought to help clear damaged tissue. Thus, transient autoimmune responses are common and do not require intervention.

Autoimmunity can be defined as the specific breakdown of mechanisms responsible for tolerance to self-antigens, and autoimmune diseases result when a specific adaptive immune response against self-antigens is serious enough to cause tissue damage. This covers a large spectrum of diseases. Both humoral immunity and CMI are involved in autoimmunity. A breakdown in peripheral tolerance mechanisms allows the formation of TH1 cells and CTLs and is accompanied by the inappropriate activation of macrophages. Normal immunoregulatory mechanisms are perturbed. The autoreactive T cells can help self-reactive B cell clones produce autoantibodies. These are usually complement-fixing IgG antibodies that can undergo affinity maturation. Together, the autoreactive CTLs and antibodies cause the extensive and indiscriminate damage observed in some autoimmune disorders.

[1] One could argue that autoreactivity is a built-in feature of the immune system. The T cell receptor repertoire is positively selected on MHC:self-peptide complexes in the thymus, and naïve T cells require contact with self-MHC molecules in the periphery for their survival and effector functions. This means that all T cells in the periphery are, by definition, autoreactive! It is only when T cells that recognise self-peptide–self-MHC complexes with moderate to high affinity are allowed to proliferate that autoimmune diseases develop.

15.2 Interplaying Factors

Immune diversity generating mechanisms are active throughout an individual's life, and both B and T cell repertoires are open-ended. Put simply, this implies that cells capable of recognising self-antigens can exist or can be formed throughout an individual's lifetime. Strict regulatory control mechanisms at the central and peripheral level ensure that this self-reactivity does not translate into a threat to the body. In the earlier chapters, we discussed the processes that ensure the absence of autoimmune disorders in healthy individuals. They are summarised again below.

❒ During maturation in the primary lymphoid organs, most self-reactive clones, B as well as T, are deleted, rendering the immune system tolerant to self-antigens.

Figure 15.1 Three factors are important in precipitating autoimmune diseases. The relative importance of the individual factors may vary in individuals and diseases, but the convergence of all three is required for the manifestation of the disease (adapted from Ermann & Fathman, *Nature Immunology*, 2001, 2:759).

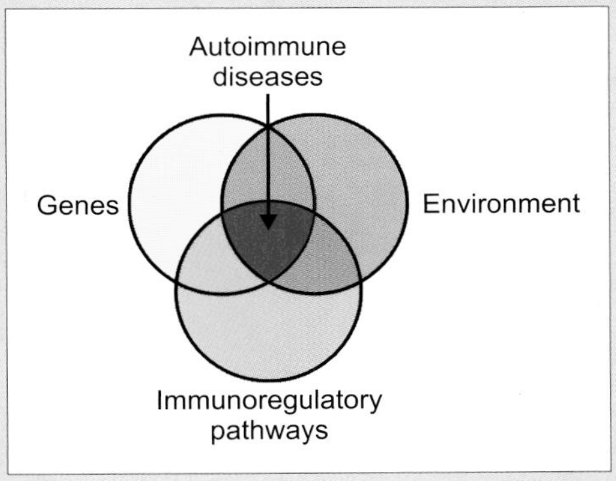

- Most tissue antigens are expressed at levels that are too low to initiate a T cell response; if not, the corresponding T cells are deleted in the thymus.
- The presentation of tissue-specific antigens by nonprofessional APCs (i.e., in the absence of costimulation) induces tolerance to self-antigens through clonal anergy. Surviving self-reactive clones cannot be activated and are said to be in a state of *immunological ignorance*.
- T_{reg} cells have a major role in keeping self-reactive T cell clones in check. Because B cells cannot respond to TD antigens in the absence of T cell help, the control of self-reactive T cell clones automatically results in the control of B cell responses to most antigens.
- Innate immune mechanisms also have a role in preventing autoimmunity by promoting self-tolerance.
  - Self-reactive low-affinity IgM antibodies are thought to be important in the negative selection of autoreactive B cell clones formed in the bone marrow. The crosslinking of BcRs by these IgM–self-antigen complexes is postulated to trigger negative selection.
  - Components of the complement cascade are involved in the opsonisation and clearance of apoptotic cells and immune complexes. This prompt clearance of cell debris helps reduce the incidence of autoimmunity. Interference with these clearance mechanisms leads to chronic inflammation and increased susceptibility to autoimmune diseases in animal models.

The relatively low incidence of autoimmune diseases points to the success of these controls. Currently, no general unifying theory can explain what triggers an autoimmune disease in healthy individuals. The induction and perpetuation of autoimmune diseases seems to depend upon the interplay of three parameters — genetic predisposition, environmental factors, and immune regulation. The importance of each single component may vary for different individuals and diseases; however, the appearance of an autoimmune disease requires the convergence of all three (Figure 15.1).

- **Genetic predisposition**: Autoimmune diseases run in families, which suggests a genetic component in their incidence. Only a few genes involved in the pathogenic mechanisms that underlie autoimmune diseases are known. One of the most important ones implicated is the gene cluster that codes for MHC molecules; certain MHC alleles are associated with specific autoimmune disorders with much greater frequency (Table 15.1). However, autoimmune diseases can develop in the absence of the 'disease-associated' MHC haplotype.

Furthermore, not everyone who has the disease-associated haplotype develops autoimmunity, which argues for the involvement of other genes as well. How MHC molecules affect predisposition to autoimmune diseases is not understood. MHC molecules serve as thymic selection elements to create the repertoire of naïve T cells, and then, in the periphery, present antigenic determinants to the same T cells. Amino acids lining the peptide-binding groove of an MHC molecule dictate which peptides can be loaded on that molecule. Consequently, during thymic development, predisposing MHC gene products could lead to the positive selection of a repertoire of T cells that are specific for certain autoantigens (i.e., a failure of negative selection). It is also possible that in the periphery some MHC alleles are better at presenting foreign peptides that resemble self-peptides. Another possibility is that certain haplotypes may tend to bind and present autoantigenic peptides to T cells in the periphery. Thus, it is not clear whether the role of predisposing MHC gene products is in the thymic selection process, in the presentation of (auto) antigenic peptides in the periphery, or both. The presence of a T cell receptor with a particular Vα or Vβ region has also been linked to some autoimmune pathologies, such as the murine **E**xperimental **A**utoimmune **E**ncephalomyelitis (EAE) and **M**ultiple **S**clerosis (MS; its human counterpart), and myasthenia gravis.

Genes other than the MHC cluster are also important in the development of autoimmune diseases. For example, in humans, an inherited deficiency of the early proteins of the complement pathway (C1, C2, and C4) is very strongly associated with SLE (**S**ystemic **L**upus **E**rythematosus). A point mutation in the Fas–FasL system induces SLE-like disease in murine models. A similar defect results in autoimmune lymphoproliferative syndrome in children, although all children who develop this disorder do not have the defect. An inherited variation in the levels of expression of certain cytokines (e.g., TNF-α) is also thought to increase susceptibility to autoimmune disorders.

Table 15.1 MHC Alleles and Autoimmune Disease Associations

Allele	Disease
HLA-B8 & HLA-DR3	Myasthenia gravis Sjogren's syndrome
HLA-B27	Ankylosing spondylitis Reiter's syndrome Enteropathic arthropathy
HLA-B38	Psoriatic arthritis
HLA-DR2	Goodpasture's syndrome MS
HLA-DR3	Graves' disease SLE IDDM
HLA-DR4	IDDM Pemphigus vulgaris RA
HLA-DR5	Autoimmune pernicious anaemia Hashimoto's autoimmune thyroiditis

- **Environmental factors**: Although autoimmune diseases run in families, both twins in an identical pair do not necessarily suffer from them. This clearly points to the importance of factors other than genes in autoimmune disease development. This low concordance rate within identical twins can only be explained if an environmental factor triggers disease in a genetically predisposed individual. Infectious agents are proposed to be such a trigger. Thus, for SLE, an infection with the Epstein-Barr virus is a trigger. Recent work suggests that autoimmunity may also be triggered by an immune response to highly conserved stress proteins (e.g., heat shock proteins) of human or bacterial origin. Smoking, for example, is a predisposing factor for psoriasis. In some autoimmune pathologies, smoking is linked to certain manifestations of that disease, e.g., in Goodpasture's syndrome, pulmonary haemorrhage is almost exclusively found in smokers. Similarly, smoking is a major factor in the ophthalmopathy associated with Graves' disease.
- **Disruption of immunoregulatory pathways**: In spite of being predisposed and being exposed to predisposing environmental factors, some individuals develop only a low titre of autoreactive antibodies and do not suffer from full-blown autoimmune diseases. This suggests the importance of a third factor — the disruption of immunoregulatory pathways. Though such disruptions are difficult to prove in humans, they have been shown to occur in murine models.
  - **Disruption of apoptotic pathways**: Fas–FasL signalling is involved in apoptosis and several reports establish the importance of apoptotic pathways in autoimmune diseases. For example, *lpr/lpr*[2] mice harbour a disruption of the gene that encodes Fas. These mice spontaneously develop a multiorgan autoimmune disease with symptoms that are similar to SLE. The same phenotype is found in *gld/gld* mice, in whom the gene that encodes FasL is disrupted.
  - **Interference with T_{reg} cell development**: Studies in experimental models of organ-specific autoimmune diseases provide convincing evidence that specialised T_{reg} cells capable of controlling autoimmunity are an integral part of the T cell repertoire in normal animals and that interference with their development leads to autoimmune disorders. For example, IL-2 signalling pathways are known to be important in T_{reg} cell development. Mice with targeted mutations of IL-2 or CD25 (a subunit of its receptor) also develop fatal disease characterised by lymphoproliferation, lymphocytic organ infiltration, colitis, autoantibody formation, and anaemia. Similarly, mice deficient in CTLA-4[3], a key molecule in the development of T_{reg} cells, succumb to severe lymphoproliferative syndrome with organ infiltration within the first 3-4 weeks of life.
  - **Dysregulation of cytokine and neuroendocrine networks**: The integrated immunoregulatory circuit consisting of immune-derived cytokines, the **H**ypothalamus–**P**ituitary–**A**drenal (HPA) axis, and the sympathetic nervous system is thought to be essential in maintaining homeostasis. The disruption of this circuit can predispose to autoimmunity as has been observed in animal models of spontaneous autoimmune thyroiditis, lupus-like disease, and experimental arthritis. Cells of the immune system are responsive to neurotransmitters, such as norepinephrine. Both norepinephrine and epinephrine can inhibit the production of pro-inflammatory cytokines, such as TNF-α, by cells of the immune system. They also stimulate the production

[2] Animal mutations are denoted by simple three letter codes meant to aid recall, although this may not always help outsiders to the field! The *lpr,* in this case, stands for ***l****ympho-***pr***oliferative*, and the *gld*, for ***g****eneralised* ***l****ymphoproliferative* ***d****isease.*

[3] CTLA-4 has been mapped as a susceptibility gene in both human autoimmune thyroid disease and IDDM, whereas IL-2 and CTLA-4 are found in genetic regions linked to disease susceptibility in NOD (**N**on-**O**bese **D**iabetic) mice.

Figure 15.2 An alteration in the balance of TH1:TH2 cytokines may result in autoimmune diseases, and restoring this balance can prevent or reduce the disease severity.

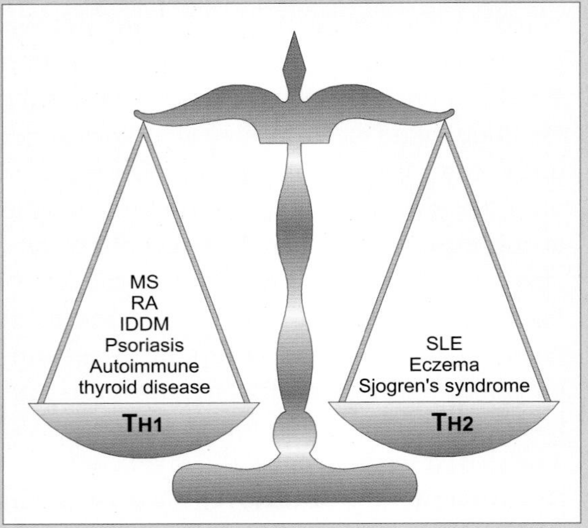

of anti-inflammatory cytokines, such as IL-10 and TGF-β. Several autoimmune diseases are characterised by alterations in the balance of TH1, TH2, or Treg cells. Both **R**heumatoid **A**rthritis (RA) and MS show a bias toward TH1 cytokines, with the overproduction of TNF-α and IL-12 and a deficiency in IL-10. These cytokines appear to be the critical factors that determine the proliferation of autoreactive T cells in these diseases. SLE, on the other hand, shows a shift toward TH2 cells (Figure 15.2). The restoration of this balance through intervention prevents or moderates severity of the disease. Similarly, in experimental models of EAE, animal recovery is dependent upon an increase in endogenous glucocorticoid levels. This endocrine response has recently been shown to be, at least in part, triggered by the immune response to the encephalitogenic antigen and mediated by endogenous IL-1 produced during the disease.

15.3 Triggering Factors

Autoimmune responses are a direct consequence of the breakdown of normal tolerance mechanisms. Factors that can trigger this breakdown are discussed below.

- **There is a strong association between infection and onset of autoimmunity.** Multiple mechanisms could be involved in the infection-associated breakdown of tolerance (Figure 15.3).
  - **Antigen-specific mechanisms** of breakdown of tolerance include the following —
    - **The release of hidden or sequestered antigens caused by infection** could lead to a reversal of anergy, triggering an autoimmune response against these antigens. For example, sperm formed in later stages of development is sequestered from the immune system. Male infertility is caused when cells of the immune system gain access to sperm-forming tissues in infections like mumps. Similarly, an autoimmune reaction against eye-lens protein is observed when this normally sequestered antigen is made accessible to cells of the immune system by trauma.
    - **Cross-reactivity between a foreign and self-antigen** may induce an immune response to self-antigens. This is called *molecular mimicry* and

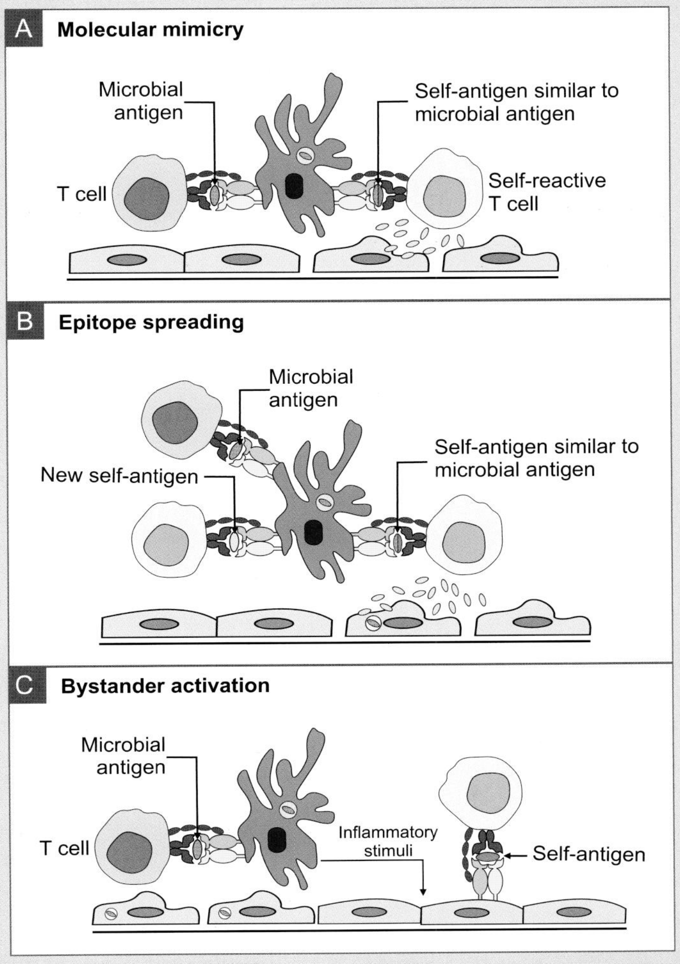

Figure 15.3 Infections can induce autoimmunity by multiple mechanisms. Following infection, macrophages and DCs internalise and process the infecting agent and its products, load fragments derived from these proteins onto MHC molecules, and display them at the cell surface. The recognition of these MHC:peptide complexes, with appropriate costimulatory signals, activates naïve T cells. The activated T cells proliferate and differentiate to effector T cells. The stimulation of effector T cells by MHC:peptide complexes causes them to release pro-inflammatory cytokines that activate macrophages and switch on their destructive potential. If an antigenic determinant on a microbial protein is structurally similar to a host epitope, the process activates cross-reactive T cells and destroys self-tissues (molecular mimicry; panel A). The subsequent release of self-tissue antigens and their uptake by APCs perpetuates the disease. In epitope spreading (panel B), the self-tissue damage caused by a persistent microbial infection causes the release of multiple self-peptides. The pro-inflammatory environment caused by the infection upregulates expression of MHC and costimulatory molecule by APCs. The internalisation and presentation of self-peptides by the APC under these conditions spreads the autoimmune response to multiple self-epitopes. Bystander activation is the activation of self-reactive cells caused by the pro-inflammatory environment in persistent infection (panel C). The activation of T cells specific for microbial antigens results in the influx of more T cells to the site of infection. The pro-inflammatory stimuli can result in the expression of MHC and costimulatory molecules on cells that normally do not express them. Some of these T cells are likely to be self-reactive and get activated. These activated T cells cause self-tissue damage and perpetuate the autoimmune response. For simplicity, the activation of effector T cells is depicted here. The engagement of costimulatory molecules is needed for the activation of naïve T cells.

implies that an antigenic determinant on some microbial protein is structurally similar to a determinant on a host protein. In T cells, this implies a linear peptide of about 8-15 amino acids in length is recognised in the context of a particular MHC haplotype. Most post-infection autoimmune diseases are therefore linked to a particular MHC allele. An initial T cell response to one epitope can lead to B and T cell autoimmune responses to other closely related, diverse determinants and is termed *epitope spreading*. Some examples of infections precipitated autoimmune damage are listed below.

- The cardiac damage observed in the acute rheumatic fever that follows a streptococcal infection is attributed to cross-reactivity between *S. pyogenes* and the myosin in cardiac muscles.
- The coxsackie B virus apparently shares an antigen with myocardium; infection with this virus can trigger lethal myocarditis.
- The outer surface protein of the spirochete *Borrellia burgdorferi* induces the formation of T cells that recognise a peptide derived

from LFA-1 in the context of HLA-DR4. Lyme arthritis that does not resolve after antibiotic treatment is thought to result from this cross-reactivity. Nerve toxicity observed in this disease is attributed to cross-reactivity between the flagellin of *B. burgdorferi* and an axonal protein.

- A recent study established a link between *Klebsiella* infection and ankylosing spondylitis in HLA-B27 individuals.
- An eye infection by the herpes simplex virus-1 can provoke the chronic inflammation of the corneal stroma which is called herpetic stromal keratitis. It is a leading cause of human blindness. Recent studies indicate that the inflammation is the result of cross-reactivity between a viral peptide (UL6) and a peptide derived from the corneal antigen.
- Reiter's syndrome is a crippling arthritis linked to infection with *Chlamydia trachomatis*, and it is associated with HLA-B27. Similarly, infection with *Proteus mirabilis* is linked to reactive arthritis in HLA-DR1 and HLA-DR4 alleles.

➢ For **antigen-nonspecific mechanisms**, no particular microbial determinant is implicated and is often the result of activation of bystander cells. Possible causes of tolerance breakdown include

- the induction of costimulatory molecules on tissue cells because of a local inflammatory response, which may allow nonprofessional APCs to present self-antigens to T cells, triggering a deleterious autoimmune response;
- the binding of infectious agents or their products to self-proteins, which may result in the proteins acting as haptens conjugated to the pathogenic carrier, precipitating an autoimmune response;
- the dysregulation of cytokine networks because of the induction of pro-inflammatory cytokines, such as TNF-α and IL-1 by an infection, which can lead to autoimmune disorders in predisposed individuals; and
- polyclonal lymphocyte activation by a mitogen or superantigen, which could activate autoreactive T or B cells and result in autoimmunity.

Too Much of a Good Thing: Type 1 and Type 2 Diabetes Mellitus

Diabetes mellitus is a chronic disorder of carbohydrate, fat, and protein metabolism, characterised by hyperglycaemia (too much glucose in the blood, not enough in cells) resulting from defects in insulin secretion or signalling. Insulin, a hormone produced by the β cells of the pancreatic islets of Langerhans, is indispensable for glucose metabolism. Glucose entry into β cells triggers insulin secretion. The secreted insulin is then carried through the blood to peripheral tissues. The binding of insulin to its receptors allows glucose uptake by cells. Defects in the β cell–peripheral tissue pathway can result in hyperglycaemia. Diabetes is biology's imitation of a depressing reality — starvation in plenty. Two forms of diabetes are clinically recognised. Type 1 (**I**nsulin-**D**ependent **D**iabetes **M**ellitus (IDDM) or juvenile onset) diabetes is caused by a deficiency in the production of insulin that is due to the immune-mediated destruction of β cells. By contrast, type 2 (non–insulin-dependent, or adult onset) diabetes is caused by inappropriate or inadequate insulin secretion coupled with insulin resistance.

IDDM has a major autoimmune component, with inflammatory infiltrate in pancreatic islets, antibodies to islet cell autoantigens, and a strong MHC allele linkage (Table 15.1). IDDM patients show the presence of at least three different autoantibodies. Antibodies to an isoform of glutamic acid decarboxylase are commonly observed in most patients, as are antibodies to IA-2, a member of the transmembrane protein tyrosine phosphatase family that is thought to play a role in insulin secretion. Anti-insulin antibodies are especially common in young children developing IDDM. The frequency of anti-insulin antibodies is much less in those who develop the disease at an older age. These antibodies appear in the patients months or years before the development of clinical disease. The number (i.e., presence of two or more autoantibodies) rather than the titre of antibodies seems to be of predictive value.

In spite of their predictive value, antibodies are thought to play a minor role in disease pathogenesis. Two processes result in cell-mediated β cell destruction. Antigen-specific destruction occurs because of the recognition of autoantigens in the context of MHC class I molecules on β cells by $CD8^+$ CTLs and culminates in β cell death. β cells do not express MHC class II molecules and are therefore not susceptible to antigen-specific $CD4^+$ T cell destruction. However, APCs in the islets internalise dead or dying cells, degrade autoantigens, and load them onto MHC class II molecules. The engagement of these petide: class II complexes by TH1 cells recruited to the site of insulitis (inflammation of the pancreas) results in the release of cytotoxins by these cells. These cytotoxins cause the death of not only APCs but also bystander β cells in the vicinity. The nature of the autoantigens is not clear, but glutamic acid decarboxylase and insulin are believed to be strong candidates. Childhood viral infections that cause pancreatic inflammation may also have a role in breaking tolerance to these antigens.

Type 2 diabetes begins as a syndrome of insulin resistance — target tissues fail to respond to insulin appropriately. Typically, the disease manifests itself in adulthood, usually around age 40 or more. It is the most common type of diabetes; 85-90% of diagnosed diabetes cases are of this type. The exact cause of the disease is not clear. In some instances, type 2 diabetes can be of autoimmune origin. For example, antibodies to insulin receptors can prevent the binding of insulin to its receptors and cause perturbation of the glucose metabolism. Diabetes is often called a self-induced disease because lifestyle choices (a lack of adequate exercise and obesity) seem to be major predisposing factors, although genes do play a role. People of Native American, Asian, and Hispanic origin show increased susceptibility to the disease. India has earned the dubious distinction of being the diabetes capital of the world, with around 10% of the population suffering from the disease.

- **The inappropriate expression of MHC class II molecules** may stimulate autoreactive clones. Normal tissue cells do not express MHC class II molecules. However, during viral infections, they secrete IFN-γ, a potent inducer of both MHC class II and costimulatory molecules. Mitogens, such as phytohaemagglutinin, can also induce MHC class II expression. This inappropriate expression of MHC class II molecules could allow these nonprofessional APCs to present self-antigens, leading to a breakdown of self-tolerance. Elevated levels of both MHC class I and class II molecules has been observed on the pancreatic β cells of IDDM patients and thyroid acinar cells in individuals with Graves' disease, suggesting a link between the increased MHC expression and occurrence of the disease.
- **Sex hormones**, in addition to their effects on sexual differentiation and reproduction, influence the immune system. This results in a **sexual dimorphism in the immune function**, with females having higher Ig levels and mounting stronger immune responses than males. Women tend to have a TH2 pattern of immune responses, which is further heightened during pregnancy. This greater immune responsiveness in females is also evident in their increased susceptibility to autoimmune diseases. Autoimmune diseases show a clear sex-bias, with women affected at a much greater frequency than men (Table 15.2). Studies in normal mice show that oestrogen treatment induces polyclonal

Table 15.2 Sex-bias and Autoimmune Diseases

Disorder	Female:male ratio
Hashimoto's thyroiditis	50:1
Sjogren's syndrome	9:1
SLE	9:1
RA	4:1
MS	2:1
Myasthenia gravis	2:1
IDDM	2:1
Chronic idiopathic thrombocytopaenic purpura	2:1
Ankylosing spondylitis	1:3
Acute anterior uveitis	1:2

B cell activation, alters their susceptibility to death, and increases the expression of autoantibodies. In addition, recent data indicates that sex hormones influence both apoptosis and the cytokine profile of T cells. In animal studies, androgens change the cytokine profile from predominantly IL-4, IL-5, IL-6, and IL-10 (i.e., T_{H2} type) to TNF-α and IFN-γ (i.e., T_{H1} type). Thus, patients suffering from SLE improve with androgen treatment. Conversely, patients with RA (characterised as a T_{H1} disease) improve with oestrogens (such as in oral contraceptives).

15.4 Mechanisms of Damage

The pathology of an autoimmune disease is determined by the specific antigen(s) against which the autoimmune response is directed and the mechanism of damage to the antigen-bearing tissue. These mechanisms are essentially similar to those of protective immunity, but they cause harm to self-tissues instead of to infectious agents or tumours. Because the offending self-antigen is in constant supply, there is a possibility of constant amplification of response.

- **Damage caused by complement-fixing autoantibodies**: These autoantibodies cause tissue damage through two mechanisms.
  - **Complement-mediated cytolysis**: Autoantibodies against cell-surface antigens cause extensive damage through complement-associated cytolysis, especially in non-nucleated cells. For example, in autoimmune haemolytic anaemia, IgG autoantibodies bind to blood group antigens and cause the destruction of erythrocytes. Likewise, autoantibodies against the fibrinogen receptor expressed on platelets result in the depletion of platelets and cause haemorrhage in autoimmune thrombocytopaenic purpura. Antibodies against basement membrane[4] collagen are responsible for the acute vasculitis and renal failure observed in Goodpasture's syndrome.
  - **Immune complex formation**: This is the underlying mechanism of tissue damage in systemic autoimmune diseases such as SLE. The binding of autoantibodies to soluble antigens deposits immune complexes in tissues and blood vessels. The large quantities of complexes cause the chronic

[4] A basement membrane is a specialised form of an extracellular matrix that consists of laminins, collagen type IV, and glycoproteins. The basement membrane separates the epithelium from the underlying supporting tissue. Basement membranes of different organs have different compositions.

Autoimmune Diseases

- Autoimmune diseases are the result of a breakdown of self-tolerance.
- The interplay of three factors (genetics, environment, and the perturbation of immunoregulatory pathways) results in autoimmunity.
- Triggering factors include
  - infections resulting in tolerance breakdown,
  - inappropriate expression of MHC class II molecules, and
  - hormones.
- Mechanisms of autoimmune damage include
  - cytolysis caused by complement fixation,
  - failure to clear immune complexes,
  - autoantibody-mediated compromised cellular function, and
  - inappropriate CTL activation.
- Treatment includes the alleviation of symptoms by metabolic correction and the administration of anti-inflammatory and/or immunosuppressive agents.

activation of the complement cascade and the release of C3a and C5a. This in turn initiates an inflammatory response with phagocytic cell recruitment that is destructive to tissues. NK cells recruited to the site of immune complex deposition cause further tissue injury through ADCC. The muscle weakening observed in myasthenia gravis is partially attributed to the complement-mediated degradation of **A**cetyl**ch**oline receptors (ACh). In some individuals with RA, the formation of autoreactive IgM antibodies against the Fc region of IgG results in the deposition of IgM–IgG complexes in joints and chronic complement activation. Together, these cause joint inflammation.

- **Compromised cellular function caused by the binding of autoantibodies to cell-surface receptors**: The binding of autoantibodies to cell-surface receptors can mimic the normal function of the receptor, block binding of the ligand to the receptor, or even lead to receptor degradation. In either of these, binding results in a reduction or loss of cellular function. For example, in myasthenia gravis, antibodies to myocyte ACh receptors, located at the neuromuscular junction, interfere with proper neurotransmission. The binding of these antibodies to receptors is believed to promote their internalisation and degradation, diminishing contractility and progressively weakening muscles. By contrast, in Graves' disease, the binding of autoantibodies to the receptor of the thyroid-stimulating hormone mimics the action of the normal hormone, overstimulating the thyroid gland.
- **Damage due to CTL activation**: Cells expressing specific MHC class I:self-peptide complexes are destroyed by CTLs in predominantly cellular autoimmune responses, and this results in extensive damage to specific tissues or organs. For example, in autoimmune thyroiditis, phagocytic and cytolytic cells are found to accumulate in thyroid lesions. Similarly, in IDDM, infiltrating CTLs destroy the β cells located in the islets of Langerhans in the pancreas. Autoreactive T cells also have a role in MS and EAE. In MS, T cells destroy the myelin sheath of nerve fibres, resulting in neurologic dysfunction. EAE is a classic T cell-related autoimmune disease in which autoreactive cells to myelin basic protein cause damage to brain cells.

Table 15.3 Common Autoimmune Disorders

Disorder	Autoantigen	Description
Organ-specific diseases		
Hashimoto's disease (chronic thyroiditis)	Thyroglobulin, thyroid peroxidase	Common in middle-aged women Infiltration of lymphocytes and phagocytes in thyroid causes inflammation, enlargement (goitre), and destruction of thyroid gland The binding of antibodies to the autoantigen causes thyroid malfunction and interferes with iodine uptake, resulting in hypothyroidism
Thyrotoxicosis (Graves' disease)	Thyroid-stimulating hormone receptor	Hyperthyroidism caused by the uncontrolled stimulation of thyroid because of the binding of autoantibodies to the receptor
Autoimmune gastritis, pernicious anaemia	α, β subunits of gastric proton pump (H^+/K^+-ATPase) Intrinsic factor (a membrane protein on gastric parietal cells) involved in the transport of vitamin B_{12} across the small intestine	Perturbation of the gastric proton pump causes loss of acid, gastritis (inflammation of gastric mucosa), and atrophy of the gastric mucosa Destruction of gastric parietal cells by autoantibodies results in intrinsic factor deficiency Autoantibodies also interfere with binding of vitamin B_{12} to intrinsic factor; resultant vitamin B_{12} deficiency affects haemato-poiesis and changes the size, shape, and number of mature erythrocytes
Autoimmune haemolytic anaemia	Rh or Ii blood group antigens	The binding of antibodies to antigens on erythrocyte surface results in complement-mediated lysis of erythrocytes and severe anaemia Anti-Ii antibodies are cold agglutinins that bind RBC-associated Ii antigen, resulting in severe necrosis of body extremities, especially in winter
	Drugs like penicillin or methyldopa (used to treat hypertension); these drugs attach to erythrocytes and act as haptens	An autolytic reaction to the drug-coated erythrocytes induces anaemia
Goodpasture's syndrome	α3 chain of type IV collagen found in the basement membrane of kidney glomeruli and/or lung alveoli	Binding of autoantibodies to collagen causes progressive kidney damage and/or pulmonary haemorrhage
Male infertility	Antigens on spermatozoa	Antibodies cause agglutination of spermatozoa and prevent fertilisation

Thrombocytopaenic purpura	Glycoprotein antigens on cell surface; most common is platelet integrin GpIIb/IIIa	Antiplatelet antibodies cause agglutination and destruction of platelets, resulting in uncontrolled haemorrhage
Myasthenia gravis	Neuronal nicotinic ACh receptor (a neurotransmitter released by somatic motor neurons into the neuromuscular junction) — especially peptides derived from the a subunit involved in muscle contraction	Binding of autoantibody to ACh receptors causes endocytosis of ACh receptors or destruction of the cell by ADCC Signal generated by the binding of ACh to its receptor results in the contraction of skeletal muscle cells; the loss of receptors interferes with muscle contraction and causes progressive muscle weakness
IDDM (type I diabetes)	Antigen on β cells in the islets of Langerhans of the pancreas; β cells produce insulin, a hormone needed for the transport of glucose into cells	Hyperglycaemia is caused by insufficient insulin production because of CTL-mediated destruction of insulin-producing pancreatic β cells; autoantibodies have only a minor role in this destruction
Systemic diseases		
SLE#	Ro and La proteins found in the nucleosome, the spliceosome, and ribonucleoprotein complex* Other ubiquitous self-antigens (DNA, histones, ribosomes, etc.)	Chronic inflammatory multiorgan disorder that may prove fatal if major organs are involved Deposition of immune complexes in renal glomerulii, joints, and other organs results in widespread damage with a spectrum of symptoms
RA	Antigen unknown	Autoantibodies against the Fc domain of IgG/IgM found in the synovial fluid of affected joints; whether cause or effect of disease is unclear Deposition of immune complexes causes joint inflammation and destruction
MS	Components of myelin: myelin basic protein, proteolipid protein, and myelin oligodendrocyte glycoprotein	$CD4^+$ T cell-mediated attack on the myelin sheath of neuronal cells causes paralysis
Pemphigus vulgaris	Desmoglein, the skin glue that attaches adjacent cells	Autoantibodies cause virtual ungluing of skin, resulting in burn-like lesions or blisters that do not heal.

Systemic refers to the multiorgan involvement observed in this disease, *lupus* (Latin for wolf) describes the characteristic butterfly facial rash resembling the colouring of a (European) wolf, and *erythematosus* refers to the redness of the skin rash.

* The spliceosomes are located in the nucleus and splice out intronic nucleotides from pre-mRNA to yield mRNA that is then translated into proteins on ribosomes. Ro and La stand for the first two letters of the surnames of patients in whom these autoantibodies were first discovered.

15.5 Diagnosis and Treatment

Laboratory diagnosis of most autoimmune diseases entails the detection of autoreactive antibodies in tissues or sera of the patients. Immunofluorescence is used to determine the presence of autoantibodies in tissues, whereas RIA or ELISA/EIA remain the techniques of choice for detection of antibodies in sera. The Coombs' test is used to detect autoantibodies in haemolytic anaemias (appendix IV). With diseases such as IDDM, treatment is symptomatic, and diagnosis of the cause of the disease is secondary.

Autoimmune disorders can be divided into two categories (Table 15.3).

- **Organ-specific** autoimmune diseases have localised lesions, such as Hashimoto's disease and pernicious anaemia.
- **Non-organ-specific** diseases, also termed *systemic diseases,* which do not have localised lesions but have lesions distributed in the body. Examples include RA, MS, and SLE.

Metabolic correction can alleviate symptoms of organ-specific diseases, for example, administration of thyroid hormones in thyrotoxicosis or vitamin B_{12} in pernicious anaemia. However, these treatments do not address the underlying cause of the disease. The ideal treatment for autoimmune diseases (whether organ-specific or systemic) is to induce long-lasting, antigen-specific tolerance. Commonly used immunosuppressive and anti-inflammatory drugs have clinical benefits but also have significant side-effects. Conventional immunosuppressive therapy is now being reduced or replaced with new biological agents.

- **The neutralisation of pro-inflammatory cytokines** has proved useful in autoimmune diseases such as Crohn's disease, MS, and RA that have increased levels of the pro-inflammatory cytokine TNF-α. This cytokine acts on multiple target tissues and induces other pro-inflammatory cytokines, such as IL-1, IL-6, and IL-8. Moreover, it potentiates lymphocyte activation and facilitates the recruitment of leukocytes to sites of inflammation by inducing the expression of adhesion molecules and chemokines. Therapies aimed at neutralising TNF-α include
  - the administration of chimeric mouse–human anti-TNF-α mAbs, which has met with encouraging success in clinical trials of RA patients and is found to be especially beneficial with low doses of methotrexate, and
  - the use of modified cytokine receptors that bind the cytokine and inhibit it from activating cellular responses. Engineered TNFR is now used in RA treatment. Etanercept, a dimeric molecule comprising two extracellular domains of TNFR attached to the Fc portion of the human IgG1 antibody, is in clinical use.
- **mAbs against T cell surface molecules**, such as CD3, CD4, CD52, and CD25, have been successful in the treatment of autoimmune diseases.
- **Targeting T cell trafficking pathways** with mAbs against LFA-1, LFA-3, and VLA-4 (**V**ery **L**ate **A**ntigen-**4**) has shown promise in clinical trials. These mAbs have the advantage of interfering with lymphocyte trafficking without decreasing T cell numbers.
- **The administration of antigens by oral or mucosal routes** induces antigen-specific tolerance in animal models. The reasons behind tolerance induction are not clear. The oral or mucosal administration of the antigens is generally believed to lead to the T_{reg} cell formation. T_{reg} cell generation may also cause immune deviation — a shift from a pathogenic ($T_{H}1$ or IFN-γ driven) to a protective ($T_{H}2$ or IL-4/TGF-β) T cell response. Though antigen-specific therapy is logically appealing and demonstrably effective in animal models, the approach has not met with appreciable success in clinical trials.

Jamming the Joints: Rheumatoid Arthritis

RA, an autoimmune disease, is a chronic destructive disease of the joints that is characterised by inflammation, synovial hyperplasia, and abnormal cellular and humoral responses. It is the most common form of inflammatory arthritis that affects primarily the synovium[5] although the integrity, resilience, and water content of the cartilage are also impaired. RA can be distinguished from other forms of arthritis by the location and number of joints involved — joints of the neck, shoulders, elbows, wrists and hands (especially the joints at the base and middle of the finger) as well as hips, knees, ankles, and the joints at the base of the toes. Interestingly, the affected joints tend to be involved in a symmetrical pattern. About 20% of RA patients also develop rheumatoid nodules, which are lumps of tissue that form under the skin, often over bony areas. The disease shows a sex-bias — it is four times more common in women than men. The synovium of an inflamed rheumatoid arthritic joint shows the presence of $CD4^+$ T cells, B cells (especially plasma cells), monocytes, and macrophages. Neutrophils, by contrast, are found almost exclusively in the synovial cavity (fluid) and only rarely in synovial tissue. Also found in the synovial cavity are *rheumatoid factors* — autoantibodies against IgG, and collagen type IV. Although these autoantibodies are characteristic of RA, it is not clear whether they are the cause or the result of the disease.

RA was originally believed to be a T cell-driven disease. The activation of $CD4^+$ T cells by as yet unidentified antigen was postulated to trigger and maintain inflammatory processes in the rheumatoid joints. The large number of $CD4^+$ T cells in the joint, skewed TcR gene usage, and the association of RA with MHC class II haplotypes (e.g., HLA-DR4) lent support to this idea. However, T cells appear to be inactive in the chronic phase of the disease, and the secretion of cytokines, such as IL-2, IL-4, or IFN-γ, that are associated with an activated T cell state is also very low. By contrast, cytokines, such as IL-1, IL-6, IL-8, TNF-α, and GM-CSF, known to be produced primarily by macrophages and connective tissue cells, are expressed in abundance in RA synovium and synovial fluid. This has led to the suggestion that RA is an inflammatory immune complex disease. Thus, although T cells are thought to be important in initiating the disease, chronic inflammation perpetuated by macrophages and fibroblasts in a T cell-independent manner is thought to be the cause of RA. According to this hypothesis, T cell-dependent B cell activation to viral or bacterial antigens results in the formation of autoantibodies and leads to immune complex deposition. The inflammatory mediators released by this deposition cause the migration of monocytes into the synovium and start a self-perpetuating cycle of inflammation and tissue destruction. Thus, secretion of IL-1 and TNF-α by chronically activated macrophages maintains the synovial fibroblasts in an activated state, whereas cytokines, such as IL-6, IL-8, and GM-CSF, secreted by the activated fibroblasts, contribute to further recruitment and maturation of monocytes and the activation of PMNs. Prostaglandins and proteases, secreted by the fibroblasts, erode and destroy nearby connective tissues, such as bone and cartilage, resulting in the characteristically painful joints of RA.

There is no cure for RA at present. The goals of current treatment methods are to relieve pain, reduce inflammation, stop or slow down joint damage, and improve function and patient well-being.

[5] Synovium is the membrane that lines and lubricates a joint; cartilage is normally a very resilient tissue that absorbs considerable impact and stress and is composed primarily of type II collagen and proteoglycans.

16 *Hypersensitivity*

I know, I get cold
Cos I cant leave things well alone
Understand I'm accident prone

— Natalie Imbruglia, *Wishing I was There*

16.1 Introduction

Hypersensitivity or allergy describes any adaptive immune response to innocuous antigens in a presensitised host[1] that is exaggerated or inappropriate, causes tissue damage, and is detrimental to the health of the individual. The term *anaphylaxis* (Greek *ana* — against, *phylaxis* — protection) was introduced to describe the detrimental effect of a second encounter with a foreign substance. Von Pirquet introduced the term *allergy* (once again, Greek *allos* – altered, *ergon* — action) to suggest any alteration in the immune status of an individual after exposure to the antigen. He suggested that *immunity* be used to describe increased resistance, whereas *hypersensitivity* be used to describe its deleterious effects. However, allergy and hypersensitivity (especially type I) are now used synonymously, and the antigen triggering the hypersensitivity reaction is termed the *allergen*.

The current classification of hypersensitivities is based on the one used by Gell and Coombs. They described four types of hypersensitivity reactions (types I to IV); types I, II, and III are antibody-mediated, whereas type IV is a T cell-driven, cell-mediated response (Table 16.1).

Abbreviations

DTH:	Delayed-Type Hypersensitivity
MCP-1:	Macrophage Chemoattractant Protein
MIP-1α:	Macrophage Inflammatory Protein-1α
NCF-A:	Neutrophil Chemotactic Factor-A
PAF:	Platelet Activating Factor
PDGF:	Platelet-Derived Growth Factor
RA:	Rheumatoid Arthritis
SCF:	Stem Cell Factor
VEGF:	Vascular Endothelial Growth Factor

16.2 Hypersensitivity Type I

Also referred to as *immediate*, or *anaphylactic* hypersensitivity, **hypersensitivity type I is characterised by an immediate, inappropriate response to a secondary exposure to allergen**. The primary mediators of type I hypersensitivity reactions are IgE antibodies and cells, such as eosinophils, basophils, and mast cells, that express a high-affinity receptor for IgE. The primary exposure to the allergen that causes IgE production is termed *sensitisation*; the secondary exposure that results in manifestations of type I hypersensitivity is termed *allergenic challenge*. Because allergen-specific IgE antibodies must be developed to precipitate the reaction, at least 14 days must elapse between sensitisation and challenge for the manifestations of type I hypersensitivity to be exhibited. Eosinophils and basophils have already been described in chapter 2. Only mast cells will be discussed here.

Mast cells are sessile cells that do not circulate in the blood but are associated with mucosal epithelial cells and connective tissues. They are heterogeneous in terms of their granule constituents and the patterns of cell-surface proteins that they express. Nevertheless, all of them express FcεRI and can bind and retain IgE for prolonged periods. Large numbers of mast cells are observed in the skin, lungs, and gastrointestinal tract. Their maturation occurs in vascularised peripheral tissues and is tightly regulated by various cytokines. Although mast cells were traditionally classified into *mucosal* and *connective tissue* cells, this classification has been abandoned. Mast cell phenotype has now been established to change markedly in response to its cytokine milieu. Mature mast cells proliferate in response to TH2 cytokines; their numbers increase dramatically during TH2-type responses but return to baseline after the resolution of the process. **S**tem **C**ell **F**actor (SCF) regulates many aspects of mast cell development and survival, although IL-3 can also promote their survival. Mast cells produce an array of mediators and signalling molecules (Table 16.2). Many of these mediators (e.g., histamine, serotonin, proteases such as tryptase, chimase, and kininogenase, TNF-α, NCF-A (**N**eutrophil **C**hemotactic **F**actor-**A**), and ECF-A) are preformed and stored in intracellular granules. Upon activation, mast cells rapidly synthesise bioactive molecules — metabolites of the arachidonic acid pathway, including leukotrienes (LTC4, LTD4,

[1] A host who has previously encountered that antigen.

Table 16.1 Types of Hypersensitivities

Characteristic	Type I (anaphylactic)	Type II (cytotoxic)	Type III (immune complex)	Type IV (delayed-type)
Effectors	IgE antibodies	IgG, IgM antibodies	IgG, IgM antibodies	T cells
Antigen	Exogenous	Cell surface	Soluble	Tissues and organs
Response time	Immediate (15-30 minutes)	Minutes to hours	Hours to days	48-72 hours
Manifestations	Weal and flare reaction, smooth muscle contraction, bronchoconstriction, mucous production	Lysis and necrosis of tissue; erythrocyte damage	Erythema, oedema, and necrosis; kidney damage	Erythema and induration
Pathophysiology	Accumulation of basophils, neutrophils, and eosinophils	Complement-mediated destruction of tissue/cells	Immune complex-mediated complement activation and accumulation of neutrophils and macrophages	Infiltration by monocytes/macrophages and lymphocytes; granuloma formation
Mediators of injury	Bioactive products of mast cells/basophils	Complement components, perforin/granzymes	Complement components, neutrophil exocytosis	Cytokines and chemokines
Examples	Asthma, hay fever, atopic dermatitis, urticaria	Erythroblastosis foetalis, autoimmune haemolytic anaemias, transfusion reactions	Glomerulonephritis, SLE, RA, pigeon fanciers' disease, serum sickness	Granulomatous diseases, such as tuberculosis and leprosy, contact dermatitis

LTE4), prostaglandins (PGD2), and PAF (**P**latelet **A**ctivating **F**actor); cytokines (GM-CSF, IL-1 through IL-8, IL-10, IL-13 through IL-16, neuronal growth factor); chemokines (MIP-1α, or **M**acrophage **I**nflammatory **P**rotein-**1α,** and MCP-1, or **M**acrophage **C**hemoattractant **P**rotein-**1**); and angiogenesis factors such as **V**ascular **E**ndothelial **G**rowth **F**actor (VEGF), and **P**latelet-**D**erived **G**rowth **F**actor (PDGF). The preformed mediators are released immediately upon activation. The synthesis of bioactive molecules occurs after the first immediate hypersensitivity reaction and helps amplify it. Together, the preformed and newly synthesised molecules are responsible for the recruitment and activation of inflammatory cells as well as smooth muscle contraction, bronchoconstriction, mucous production by goblet cells, and the increased gastrointestinal motility observed in anaphylactic shock.

Mast cells, basophils, and eosinophils interact with each other to amplify the hypersensitivity response. The MBP released by eosinophils can result in mast cell exocytosis. Eosinophils also secrete nerve growth factor, which can promote mast cell survival. Mast cells, on the other hand, are responsible for recruiting basophils and eosinophils to the site of allergenic challenge. The mast cell-derived chymase promotes SCF production by eosinophils. As noted earlier, SCF has an important role in mast cell development and survival. In humans, basophils are a major source of IL-4 and IL-13, although mast cells and eosinophils also produce these cytokines, which are crucial to promoting TH2 pathways. Thus the three major cell types are involved in a cross-talk with each other and other cell types to promote pathways that can lead to type I hypersensitivity reactions.

The first step in an immediate hypersensitivity reaction is the preferential production and secretion of IgE antibodies in response to innocuous substances. Most common allergens are small, readily soluble molecules with low MWs. Inhalation or contact with minute doses of an allergen results in a TH2-driven response. B cells undergo an isotype switch from IgM to IgE antibodies when exposed to the TH2-derived cytokines IL-4 and IL-13 (chapter 9). Once the B cells switch to IgE, subsequent exposures result in an IgE antibody response. The type of antigen, dose, and route of administration, as well as the cytokine milieu, are important in this response (chapter 4).

As both mast cells and basophils express FcεRI, the high-affinity IgE receptor[2], they get coated with any IgE present in their vicinity. The crosslinking of FcεRI-bound IgE antibodies by a multivalent antigen results in the degranulation of cells and the release of bioactive mediators responsible for the manifestations and symptoms of type I hypersensitivity[3]. The primary dose of the allergen that results

[2] The strength of FcεRI–IgE binding is ~10^{-9} M.

[3] IgE can also bind other FcRs (FcγRII, FcγRIII) expressed by some mast cell populations. However, the ligation of these FcRs does not result in degranulation.

Table 16.2 Pharmacological Mediators of Basophils and Mast Cells

Mediators	Biological effects
Preformed mediators	
Histamine	Vasodilation, increased capillary permeability, chemokinesis, mucous secretion, bronchoconstriction, cytotoxicity to parasites
Serotonin*	Smooth muscle contraction, increased vascular permeability
Heparin	Anticoagulant, cytotoxicity to parasites
Proteases (tryptase, kininogenase)	Activation of complement cascade, hydrolysis of plasma kininogen to release bradykinin, connective tissue degradation
NCF-A, ECF-A	Recruitment of neutrophils and eosinophils
TNF-α	Pro-inflammatory; activation of vascular endothelium
Synthesised mediators	
PAF	Aggregation and lysis of platelets, resulting in the release of vasoactive amines, heparin, and histamine, that cause smooth muscle contraction, neutrophil chemotaxis, and bronchoconstriction
Leukotrienes** (LTB4, LTC4, LTD4, LTE4)	Prolonged bronchoconstriction and mucous secretion — leukotrienes are a thousand times more potent bronchoconstrictors than histamine; they are vasoactive and cause pain and oedema; LTB4 causes basophil chemotaxis
Prostaglandins (PGD2)	Bronchoconstriction, platelet aggregation, vasodilation
Thromboxanes	Bronchoconstriction
Bradykinin	Bronchoconstriction, vasodilation
Cytokines (GM-CSF, IL-1 to IL-8, IL-10, IL-13 to IL-16, neuronal growth factor, MIP-1α, MCP-1, TNF-α, VEGF, PDGF)	Activation of vascular endothelium, recruitment and activation of neutrophils and macrophages, angiogenesis, increased IgE production, etc.

* Only in mice; platelet-derived in humans

** Originally called slow reacting substance of anaphylaxis as they were released in the later phase of the anaphylactic reaction

One's Food, Another's Poison: Food Allergy

The term *food allergy* is often used inappropriately to encompass disorders related only by their presumed relationship to food ingestion. A more suitable term would be *adverse food reaction*, which is used to describe any abnormal clinical response associated with ingestion of a food or food additive and which may further be differentiated into food allergy and food intolerance. Food allergy is generally immediate and occurs each time the food is ingested, even if it is in small amounts. Such an allergy can be IgE- or cell-mediated. Food intolerance refers to a variety of nonimmunologic reactions that occur after food ingestion and may be caused by inherent properties of the food (such as histamine contamination because of improper storage) or physiologic characteristics of the host (such as lactose intolerance). Food intolerance may be nonreproducible — an adverse reaction may not occur each time the food is consumed — and is often dose dependent.

The most common symptoms associated with IgE-mediated food allergy involve the skin and gastrointestinal tract: urticaria, angioedema, pruritis, nausea and vomiting, abdominal pain or cramping, and diarrhoea. Respiratory and ocular symptoms of IgE-mediated food allergy often accompany skin and gastrointestinal symptoms. Mixed IgE- and cell-mediated food allergy includes allergic eosinophilic gastrointestinal disorders, atopic dermatitis, and asthma. Cell-mediated food allergy represents a heterogeneous group of disorders caused by the ingestion of common food proteins that do not initiate mast cell or basophil degranulation.

Manifestations of adverse food allergies are almost as varied as the foods that cause them. The same food (e.g., wheat) can induce in the same individual atopic dermatitis, urticaria, anaphylaxis, asthma when wheat antigens are inhaled, or coeliac disease caused by gluten when ingested. IgE-mediated food allergies exhibit acute onset of symptoms after ingestion of the offending food and affect one or more target organs — skin (urticaria or angioedema), respiratory tract (rhinitis or asthma), gastrointestinal tract (pain, emesis, diarrhoea), and the cardiovascular system (anaphylactic shock). Allergies can be triggered by direct exposure of the involved organ to the food or by the systemic distribution of proteins after digestion. Those mediated by T cells are subacute or chronic and of delayed onset (occurring after a few hours of ingestion). They typically affect the gastrointestinal tract or skin, with a propensity to affect infants and children. Sensitised T cells can home to different organs or tissues (e.g., skin) and result in food-responsive atopic dermatitis. Coeliac disease and the related skin disorder dermatitis herpetiformis are rather serious manifestations of T cell-mediated food allergies to gluten.

Food hypersensitivity is caused by a malfunction of the normal immune responses to dietary antigens. In healthy nonallergic individuals, oral tolerance ensures that the gastrointestinal immune system does not respond to dietary antigens or commensal bacteria colonising the gut. Furthermore, humoral immune responses in the gut are of the complement nonfixing, noninflammatory IgA type. The exact cause of the breakdown of tolerance and a switch from IgA to IgE antibody response is not clear. An alteration in intestinal permeability is often observed in patients with allergies; however, it is not clear if it is a cause or effect of the disease. Cytokines, such as IL-4, superantigens, and dietary lectins are all known to affect mucosal membrane permeability. In both animals and humans with food allergy, increases in the number of IELs, mucosal mast cells, and eosinophils are observed in intestinal biopsies. Alterations in TH1:TH2 balance or a reduction in T_{reg} cells have therefore been proposed to result in food hypersensitivity.

Potent food allergens are usually water-soluble glycoproteins of about 10-60 KD that are stable at a low pH. Cooking can affect allergenicity variably — it can reduce allergenicity because of the destruction of allergenic epitopes (as is the case with eggs and fish) but increase the allergenicity of other foods because of covalent modifications that result in the formation of new epitopes or that improve stability (e.g., roasting of peanuts results in greater resistance to digestion and heightened allergenicity). Patient history and physical examination are very important to the diagnosis of food allergies. The information important for diagnosis is — food suspected of provoking the adverse reaction, quantity of food required to provoke symptoms, length of time from ingestion of food to onset of symptoms, and reproducibility of symptoms. IgE-mediated food allergies can be diagnosed through the skin-prick test or the detection of antigen-specific IgE antibodies. Antihistamines can be used for symptomatic treatment. However, epinephrine injections remain the mainstay of treatment for anaphylaxis, and avoiding the culprit foodstuff is the best way to prevent allergic episodes.

Table 16.3 Classical Anaphylactic Reactions

Type of reaction	Characteristics
Hay fever (allergic rhinoconjunctivitis)	Characterised by catarrh of conjunctiva and respiratory tract Precipitated when inhaled allergens, such as pollen, (house) dust, or mould spores, come in contact with exposed mucous membranes
Asthma	Characterised by airway hyper-reactivity and the obstruction of small bronchioles, which together result in difficulty in breathing Caused mainly by inhaled allergens; although ingested allergens may also trigger asthmatic attacks
Urticaria	Characterised by the appearance of skin rashes — crops of intensely itching weals with raised white centres surrounded by erythema Precipitated by contact with or inhalation of a wide variety of sensitisers — sea food, egg whites, wheat, tomatoes, chocolates, etc.
Angioedema	Characterised by skin rashes similar to urticaria but with larger weals restricted mainly to the face and neck Caused by contact with or inhalation of the sensitising substance

in the IgE response is known as the *sensitising dose*, and the later dose that precipitates the reaction is the *shocking dose*. The ligation of FcεRI-bound IgE by the antigen induces the activation of protein tyrosine kinases associated with the receptor and results in phosphorylation of ITAMs that are present on receptor subunits. The activation of downstream signalling pathways leads to the degranulation of cells and the synthesis and release of lipid mediators, cytokines, chemokines, and growth factors that are responsible for the symptoms of anaphylaxis.

Mast cell degranulation is preceded by the massive influx of Ca^{2+} ions into cells. This is a crucial step in the process, as ionophores that cause an increase in cytoplasmic Ca^{2+} can cause degranulation even in the absence of IgE-ligation, and agents that deplete cytoplasmic Ca^{2+} suppress degranulation[4]. **The immediate result of IgE-mediated degranulation is an inflammatory response that starts within seconds and which is followed by a late-phase response that takes 6-12 hours to develop**. The immediate inflammation is due to the release of preformed pharmacological mediators, such as histamine, and the rapid synthesis and release of prostaglandins, leukotrienes, and other metabolites following degranulation. Receptors for histamine are expressed on a variety of cells. The binding of histamine to these receptors mediates different effects — intestinal and bronchial smooth muscles contract, endothelial cells in blood vessels contract (increasing vasopermeability), and mucous secretion by goblet cells increases. Leukotrienes and prostaglandins also cause a further rapid increase in vascular permeability, smooth muscle contraction, and mucous production. Thus although the initial (within 60 seconds) smooth muscle contraction occurs because of histamine, later bronchoconstriction is mediated by prostaglandins and leukotrienes. The cytokines and chemokines released as a consequence of degranulation also contribute to the clinical manifestations of type I hypersensitivity and often induce a localised inflammatory response termed the *late-phase reaction*. This reaction develops 4-6 hours after the initial reaction and can persist for upto two

[4] Degranulation can also be triggered by other stimuli, such as exercise, emotional stress, chemicals, anaphylatoxins such as C3a and C5a, and MBP released by eosinophils. Although the resulting symptoms are similar to anaphylaxis, these are not hypersensitivity reactions as IgE antibodies are not involved.

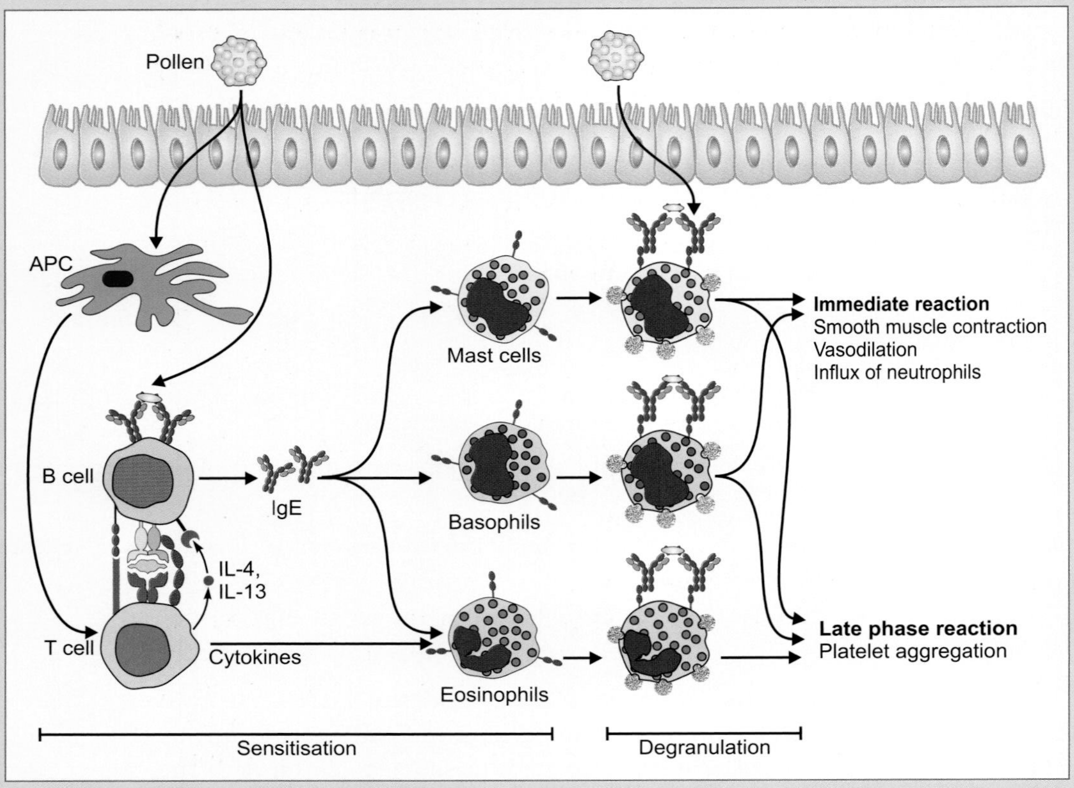

Figure 16.1 The allergen-mediated crosslinking of FcεR-bound IgE antibodies leads to the degranulation of mast cells, basophils, and eosinophils and results in manifestations of hypersensitivity type I reactions. Most allergic reactions occur at mucosal surfaces in response to allergens that enter by inhalation or ingestion. In the sensitisation phase, antigen presentation by APCs results in the differentiation of naïve T cells into effector T_{H2} cells. The effector T_{H2} cells help antigen-activated B cells switch to an IgE isotype by producing cytokines, such as IL-4 and IL-13. IgE produced by B cells binds to the FcεRI expressed on mast cells, basophils, and eosinophils. Upon subsequent challenge, the allergen binds to the FcεRI-bound IgE molecules on mast cells and basophils and triggers their degranulation. The release of preformed mediators present in the granules is responsible for the immediate effects of type I hypersensitivity. Degranulation also leads to the synthesis and release of new mediators, resulting in platelet aggregation and the influx of T cells and eosinophils. Cytokines produced by T cells induce the proliferation and degranulation of eosinophils, resulting in the late-phase reaction.

days. ECF-A, IL-4, IL-5, LTC4, and histamine cause an influx of eosinophils, whereas MIP-1α, MCP-1, NCF-A, and IL-8 are responsible for the recruitment of phagocytic monocytes and neutrophils. PAF, released in the later phase of the response, causes platelet aggregation and further release of histamine, heparin, and vasoactive amines. The combined effect of these mediators is an influx of leukocytes to the site of allergenic challenge and a sustained late-phase reaction (Figure 16.1). Thus, **even though the initial cells involved in anaphylactic reactions are mast cells and basophils, platelets, neutrophils, T_{H2} cells, and eosinophils also get involved in the later phases**. Eosinophils play a critical role in this late-phase reaction. They express FcRs for IgG and IgE isotypes and degranulate upon binding of antigen–antibody complexes. The release of preformed and newly synthesised mediators from eosinophils causes extensive tissue damage in this phase of the type I response. Although most of these responses to allergen challenge are transient and reversible, repeated challenges can lead to chronic inflammation with a progressively increasing severity of symptoms and may prove fatal.

The binding of IgE antibodies to FcεRI has biologically important consequences beyond the sensitisation and eventual degranulation of mast cells and basophils. IgE-binding upregulates the surface expression of FcεRI by mast cells and basophils. Following FcεRI aggregation, these IgE-bound cells have an enhanced ability to secrete histamine, MIP-1α, and lipid mediators. MIP-1α (like IL-4 and IL-13 also produced by mast cells) has been reported to increase IgE production by B cells, suggesting a potential positive feedback mechanism — increased IgE levels lead to increased FcεRI expression, which in turn increases the probability of degranulation and increases the release of IL-4, IL-13, and MIP-1α — themselves responsible for increased IgE production. *In vitro*, monomeric IgE increase mast cell survival by an unknown mechanism; it may have the same effect *in vivo*. Monocytes and DCs are known to express a trimeric version of FcεRI. Crosslinking of this receptor prevents apoptosis and triggers IL-10 production. These data suggest that **IgE antibodies may contribute directly and indirectly to chronic allergic inflammation and the persistence of symptoms associated with it**.

Manifestations of type I hypersensitivity vary according to the site of response (Table 16.3).

- A local reaction is limited to the target organ or tissue and is often observed when the allergen comes in contact with epithelial surfaces or respiratory mucous membranes — the involvement of the skin results in urticaria or eczema, a reaction in the eyes results in conjunctivitis, a reaction in the nasopharynx results in rhinitis, bronchopulmonary tissue involvement results in hay fever and/or asthma, and a reaction in the gastrointestinal tract causes gastritis.
- Systemic hypersensitivity reaction is often fatal and develops when the allergen is given parenterally. Termed *anaphylactic shock*, it occurs within minutes of allergen administration. Disseminated mast cell activation results in a widespread increase in vascular permeability (causing disastrous capillary leakage and a precipitous fall in blood pressure), constriction of airways (causing difficulty in breathing), laryngeal oedema, etc. The syndrome can be controlled by the immediate injection of epinephrine. A wide range of substances can trigger systemic anaphylactic shock, including insect venom, antitoxins, seafood, and peanuts. The administration of drugs, such as penicillin, vitamin B_6, or insulin, or exposure to anaesthetics, such as benzocaine and lidocaine, can also result in anaphylactic shock in sensitised individuals[5].
- If the allergen enters the body *via* the mucous membranes of the gut, a mixed reaction may develop. It is characterised by gastrointestinal symptoms (vomiting, diarrhoea, etc., caused by the activation of intestinal mast cells), as well as skin rashes, asthma, and so forth.

16.2.1 Interplaying Factors

The factors responsible for the onset and maintenance of type I hypersensitivities are still unclear. As in autoimmune diseases, a number of interplaying factors are thought to contribute to their occurrence.

- **Genetic factors**: It has been recognised since the 1920s that allergic parents have a higher proportion of allergic children, i.e., type I hypersensitivity tends to run in families. The word *atopy* is used to describe this propensity to develop allergies which is strongly linked to TH2-type responses and raised IgE antibody levels. Three atopic diseases in particular are inherently connected within families — asthma, atopic eczema, and rhinoconjunctivitis. There is almost a

[5] Although drug induced, this reaction is different from type II hypersensitivity as the damage is caused by the IgE-mediated degranulation of mast cells and is not the result of complement activation by antidrug IgG antibodies.

Taking Your Breath Away: Asthma

Asthma is a pulmonary disorder characterised by a generalised reversible obstruction of airflow that results in symptoms, such as wheezing, shortness of breath, chest tightness, and airway inflammation. Airway hyper-responsiveness (an exaggerated bronchospastic response to nonspecific agents such as cold air or specific antigens) is a cardinal feature of asthma. It is thought to be a result of complex interactions between genes and the environment. The pathology of asthma is characterised by changes in airways — mucous plugging, shedding of epithelial cells, thickening of the basement membrane, the engorgement of vessels, angiogenesis, inflammatory cell infiltration, and smooth muscle hypertrophy and hyperplasia.

Increased IgE levels are characteristic of allergic asthma. DCs seem to be the key APCs that cause a TH2 type response that results in IgE production after antigen exposure. The crosslinking of FcεRI on mast cells by IgE–allergen complexes activates and degranulates these cells. The bioactive mediators liberated by mast cells are responsible for the acute bronchospasm that occurs within 15-30 minutes of exposure. These mediators are also responsible for an inflammatory response in the airways. The chemokines released by mast cells result in neutrophil recruitment (caused by IL-8), eosinophil recruitment (caused by RANTES, eotaxin, IL-5), and a late-phase response that peaks 4-6 hours after exposure and causes prolonged symptoms. The presence of activated eosinophils is a defining feature of asthma. Chemokines, such as eotaxin, cause their activation, and TH2 cytokines, such as IL-5, increase the production of eosinophils from the bone marrow. Once activated, eosinophils contribute to airway epithelial damage through the release of products, such as MBP. A type IV hypersensitivity caused by TH2-dependent IL-4 and IL-13 may also contribute to the pathology of asthma. Both these cytokines, especially IL-13, are thought to contribute to airway hyper-responsiveness and goblet cell metaplasia[6].

Epithelial damage and the loss of its protective barrier function because of repeated asthmatic episodes may be responsible for smooth muscle proliferation in the airway, the thickening of basement membranes, and fibrosis. These changes lead to a reduced airway diameter and increased bronchial responsiveness. Thus, chronic asthma is thought to be the result of a disturbance in repair and remodelling. Although allergens and endotoxins have long been identified as modifiers of asthma, the genes involved have proven hard to identify. Genetic alterations in two proteinases — ADAM-33 and MMP9 — have been linked to both asthma and chronic obstructive pulmonary disease. Expressed by lung fibroblasts and bronchial smooth muscles, these metalloproteinases are thought to be important in small airway remodelling and airway hyper-reactivity. They may thus be (at least partially) responsible for turning an allergic runny nose into a wheezy asthmatic response.

Asthma is commonly provoked by allergens, upper respiratory tract infections, exercise, perfumes, fumes, changes in temperature and humidity, drugs (e.g., aspirin and other nonsteroidal anti-inflammatory drugs), food additives (metabisulphite or tartrazine), and food and drink (peanuts, alcohol, cola). Chemicals in the workplace, such as ink, paints, adhesives, Pt salts, or epoxy resin hardening agents can also cause asthma. Conventional asthma treatment includes inhaled corticosteroids, β-agonists, and agents that interfere with Ca^{2+} influx or increase cytoplasmic cAMP levels. New therapies involving humanised mAb to IgE, IL-4 and IL-5, and soluble IL-4R are currently being investigated for their efficacy.

50% chance of children suffering from these if both parents are allergic. The figure reduces to 30% if one parent is allergic. Several loci for specific atopy-associated genes have been determined on various chromosomes (Table 16.4). Not surprisingly, these include genes for cytokines, TcRs, and MHC class II molecules. A particular HLA-DR locus seems to increase the risk of allergy to a specific allergen. Thus, the HLA-DR2 allele is found to be associated with ragweed pollen allergy, whereas the HLA-DR3 allele is associated with allergy to grasses.

[6] Goblet cells are found in the epithelium of the intestines and respiratory tracts. They secrete *mucous* a viscous fluid composed primarily of highly glycosylated proteins called *mucins*.

Table 16.4 Genetic Loci Associated with Allergic Diseases

Chromosome	Gene product
5q31-35	T_H2 cytokines: IL-3, IL-4, IL-5, IL-9, IL-13 CD14 β-adrenergic receptor Glucocorticoid receptor
6p21	MHC class II molecules MHC class III molecules (TNF-α)
7	TcR α chain
11q13	β chain of FcεRI

However, genetic background alone cannot explain the incidence of allergies, particularly the increase that has been observed over the last few decades.

- **Environmental factors**: The environment clearly influences the incidence of allergies. Allergies are more prevalent in urban areas, whether in Ethiopia or in Germany. This has led to the suggestion that the traditional lifestyles of indigenous peoples reduce the occurrence of allergic disease, and progressive westernisation/urbanisation places them at a greater risk of developing these conditions. This hypothesis is supported by studies that compare the incidence of asthma in immigrant groups in industrialised countries with that of the populations in the native developing country; the prevalence of asthma in the immigrant groups was significantly greater than that of their compatriots in the native country. The observed increase of allergies in urban areas has given credence to the idea that increased environmental pollution results in increased type I hypersensitivity. However, different pollutants seem to affect humans differently. Type I pollutants are characterised by SO_2, large particles, and dust emitted from predominantly outdoor sources, and they have adverse health effects such as upper respiratory tract inflammation and infection. Type II air pollutants are emitted from outdoor and indoor sources and include volatile organic acids, ozone, NO_x, and cigarette smoke. These have been found to lead to allergic sensitisation of the airways and are considered risk factors for atopic disease.

  Exposure to allergens is a prerequisite for allergies. The increased use of indoor furnishings such as carpets and curtains tends to concentrate house-dust mites (*Dermatophagoides pteronyssinus*) in the immediate environment of an individual; this has been recorded to increase the incidence of adult asthma in case studies. House-dust mites, and especially their faecal pellets, are major allergens in the house environment, and the removal of fixtures that house these allergens has often been shown to decrease the severity of asthma attacks in patients. Pollen grains are common outdoor allergens, and their counts seem to have increased with increased pollution. Also, the release of allergens from pollen grains seems to have altered in urban areas. The interaction of pollen grains with type II air pollutants has been shown to lead to agglomeration and changes in surface structure with a concomitant increase in allergen release.
- **Lifestyle choices**: Lifestyle seems to influence the incidence of allergies. Children raised in big families and on farms tend to show a lower incidence of

Too Clean for Comfort: The Hygiene Hypothesis

There has been a remarkable increase in allergy-linked diseases such as asthma, atopic dermatitis, and hay fever in developed (read 'Westernised') countries. Even after accounting for factors such as better diagnosis and a hereditary component, there is little doubt that this increase in the incidence of allergies is real. The best documented amongst these has been asthma — the prevalence of asthma has increased 75%, in the USA alone from 1980 to 1994. By contrast, the low baseline prevalence of allergic disease has not changed appreciably over the same period in the developing world. The rapidity of the epidemiological shift suggests a change in environment as a possible culprit. Interestingly, a similar trend has been observed in the incidence of autoimmune diseases as well. While the incidence of childhood infections has shown a continuous downward trend since the 1950s in Western Europe and USA, there is an almost parallel upward trend in the incidence of autoimmune diseases. The *Hygiene Hypothesis*, put forward to explain this phenomenon, is increasingly gaining acceptance in the scientific community. It is derived from a number of studies across the developed world. These studies noted an inverse relation between the risk of allergy and family size (bigger the family, the lower the chances of developing allergies), exposure to animals (allergies are less frequent in children raised on farms), and a direct correlation between parental economic status (the higher the income group, the more prone to allergies), educational background (the more educated the parents, the higher the chances of allergy), and even birth order (only or first-born children seem more prone to allergies). Thus, childhood infections show an overwhelming and consistent negative association with allergic disorders.

The Hygiene Hypothesis postulates that infectious stressors guide the development of the immune system and that decreased exposure to infectious agents early in life increases susceptibility to allergy (and perhaps autoimmune diseases) by limiting immune system development. The core of the Hygiene Hypothesis is the notion that the microbial environment interfaces with the innate immune system and modulates its ability to impart instructions to adaptive immune responses, particularly when such interactions occur *in utero* and/or in early life. Children with a Westernised lifestyle who are protected from childhood infections and/or exposure to a variety of infectious agents fail to get this 'education' and have an increased risk of developing allergic and/or autoimmune diseases. Thus, exposure to food and orofecal pathogens, such as the hepatitis A virus, *Toxoplasma gondii*, or *Helicobacter pylori*, reduces the risk of atopy by >60%. Studies have also revealed a difference in the rate and quality of bacterial colonisation in children with and without a predisposition to allergy. An early colonisation with enterobacteria or lactobacilli has a protective effect, whereas early exposure to antibiotics changes the bacterial flora and is a risk factor. A shift in the balance from TH1 to TH2 was initially thought to be responsible for allergy development. Worldwide, however, helminth infections and allergic diseases do not overlap, despite both conditions being accompanied by strong TH2-type responses. Recent work on DCs and the induction of T_{reg} cells has resulted in a plausible explanation for this phenomenon and provided an immunological framework for the Hygiene Hypothesis. Chronic parasitic infections are known to cause T cell hyporesponsiveness in humans. Downregulatory molecules, such as IL-10, TGF-β, and NO, are implicated in this suppression, and they seem to be effective against both TH1- and TH2-type responses. A high overall infection turnover (such as observed in developing countries) is therefore proposed to be instrumental in the development of an immunoregulatory network. When uncontrolled, strong TH1 or TH2 responses can lead to autoimmunity and allergy. High pathogen burden is thought to endow DCs with the ability to induce T_{reg} cells. These T_{reg} cells produce immunosuppressive cytokines, such as IL-10 and TGF-β, and thus ensure that inflammatory responses (either of the TH1 or TH2 type) with their negative health repercussions, are kept under control. It is rather ironic that just as we were about to breathe a sigh of relief for winning the war against childhood infections, we have to find a way to increase our exposure to them in order to avoid increasing our risks of allergies.

allergies than those raised in nuclear families. This has given rise to the Hygiene Hypothesis (see the sidetrack *Too Clean for Comfort*). Cigarette smoking and exposure to cigarette smoke is also known to predispose an individual to allergies. Other lifestyle choices, such as increased psychological stress, lack of exercise — linked to increased TV viewing and greater exposure to indoor allergens — and changed food habits with early introduction to exotic foods, have all been implicated in increased allergies. However, in the absence of controlled studies, it is difficult to draw firm conclusions about the effects of these factors on the incidence of allergies.

16.2.2 Diagnosis

Prausnitz and Kustner first demonstrated the passive transfer of allergy *via* serum. Kustner was allergic to fish; Prausnitz was not. Prausnitz injected Kustner's serum intracutaneously into his forearm (life, or for that matter science, was easier in the pre-HIV/AIDS days). An injection of fish extract in the area elicited a weal and flare (oedema and erythema) reaction characterised by inflammation and reddening, indicating the successful passive transfer of hypersensitivity. The reaction is called the P–K reaction in their honour. Ovary's passive cutaneous anaphylaxis test is essentially similar to the P–K test and is used to demonstrate the presence of sensitising IgE antibodies in serum, whereas an individual's allergy status can be diagnosed with a skin prick, a patch test, or through the demonstration of allergen-specific IgE antibodies in the patient's serum with ELISA or RIA (see appendix IV).

16.2.3 Treatment

Although it may not always be feasible, the best strategy for allergy management is to avoid the allergen. In case of contact with the allergen, immediate treatment

Table 16.5 Drugs Used to Treat Type I Hypersensitivities

Drug	Mechanism of action
Antihistamines (e.g., diphenylhydramine)	Prevent mast cell degranulation, relieve early-phase symptoms
Cortisone	Blocks conversion of histidine to histamine; stimulates cAMP production in mast cells
Epinephrine (adrenaline)	Stimulates cAMP production by binding to β-adrenergic receptors; mainstay of anaphylactic shock treatment*
Isoproterenol derivatives (isoprenaline or salbutamol)	Bind to β-adrenergic receptors and stimulate cAMP production
Leukotriene-receptor blockers, inhibitors of cyclooxygenase pathways	Inhibit late-onset symptoms mediated by leukotrienes
Phenoxybenzamine	Elevates cAMP by blocking β-adrenergic receptors
Sodium chromoglycate	Inhibits mast-cell degranulation by inhibiting Ca^{2+} influx
Theophylline	Increases intracellular cAMP levels by inhibiting cAMP-phosphodiesterase, which cleaves cAMP

* Individuals who have experienced anaphylactic shock or who have scored very high on skin-prick allergy tests are advised to carry a single dose of epinephrine in the form of an autoinjector with them at all times to avoid life-threatening episodes.

is needed to alleviate symptoms. Most of the drugs used to treat type I hypersensitivities interfere with mast cell degranulation, either by inhibiting Ca^{2+} influx into the cells or increasing intracellular cAMP levels (see Table 16.5)

Immunotherapy (desensitisation or hyposensitisation) aimed at the elicitation of a non-IgE antibody response to allergens can be successful, especially for insect venom and pollen allergy. It consists of repeated injections of small but increasing amounts of adjuvant-bound antigen extracts. Although the mechanism of desensitisation is unclear, the treatment results in an increase in allergen-specific serum IgG levels and a decrease in IgE levels. Because of their ability to inhibit immediate skin reactions to allergen provocation, these IgG antibodies are termed *blocking antibodies*. However, the immunotherapeutic administration of allergens is itself likely to cause local or systemic side-effects, such as urticaria or anaphylactic shock. The allergen is therefore chemically modified or adsorbed to an adjuvant to delay systemic release. As most extracts are not chemically pure, there is also the risk that the therapy may induce new allergic reactions towards other extract components.

16.3 Hypersensitivity Type II

Multiple factors are brought into play when IgG or IgM antibodies bind the antigens expressed on cell surfaces of pathogens or transformed cells. The process is briefly listed here.

- Complement is activated, resulting in lysis of the target cell (chapter 3).
- C3a and C5a, released during complement activation, cause vasodilation and chemotaxis of PMNs to the site of the reaction. The released anaphylatoxins also cause the degranulation of mast cells so that the cytotoxic and pro-inflammatory arsenal of these cells is available for target cell destruction.
- Neutrophils, recruited to the site of interaction, phagocytose dead and dying cells. They also liberate an array of microbicidal agents.
- K cells may be recruited to the site of infection and are responsible for complement-independent ADCC.
- Antigen–antibody complexes bind to platelets *via* their FcRs and cause platelet aggregation and microthrombus formation; aggregation leads to the release of vasoactive amines, further increasing PMN influx.
- If the antigen–antibody complex is embedded in tissue or is too large for engulfment, phagocytosis fails; neutrophils release their lysosomal contents in the external milieu by exocytosis, thereby releasing a corrosive cocktail of enzymes/acids in the surrounding tissue.
- The end result of these events is an inflammatory response that helps get rid of intruders. However, other cells in the vicinity (known as bystanders) may also be killed in the process — the protective process is not without collateral damage. Healing and tissue remodelling begin immediately, and eventually, normalcy returns.

The protective response outlined above can become destructive if the antibodies are against cell-surface or tissue antigens, and the reaction is then called a cytotoxic reaction or type II hypersensitivity. The antigen could be endogenous (molecules expressed by cells) or exogenous (e.g., haptens adsorbed to tissue cells). The kind of tissue or organ affected and the intensity of the reaction depend upon the location and concentration of the antigen.

Figure 16.2 The adsorption of drugs to erythrocytes can trigger an immune response that results in the complement-mediated lysis of RBCs and their accelerated removal by macrophages. Multiple mechanisms can result in drug-induced anaemias. The primary event common to all of them is the induction of an immune response to the drug. Drugs that get adsorbed to erythrocytes or tissue cells appear as new epitopes to the immune system, and this results in an immune response and production of antidrug antibodies. The recognition of the drug-epitopes on RBCs by antibodies results in the complement-mediated lysis of the erythrocytes (top panel). A similar mechanism is responsible for the lysis of incompatible donor RBCs in transfusion reactions. Complement-mediated erythrocyte lysis can also occur if drug–Ig complexes bind to RBCs *via* FcRs or complement receptors. Antibody- and complement-coating of the RBCs also accelerates their disposal by macrophages (lower panel), further increasing the severity of anaemia.

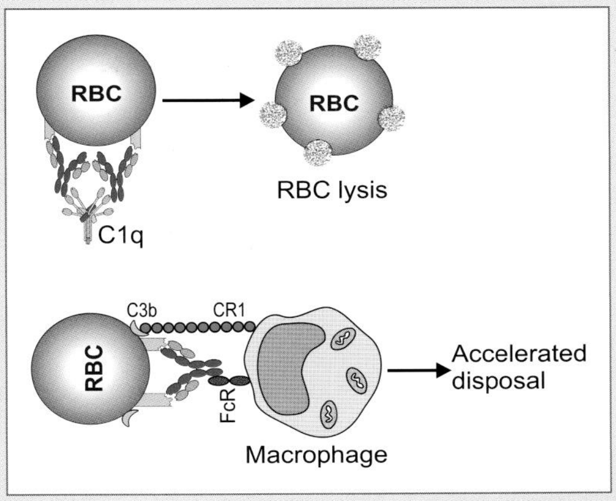

Manifestations of type II hypersensitivities include the following:

- **Autoimmune diseases**: Type II hypersensitivity can lead to a variety of autoimmune diseases, including myasthenia gravies, Goodpasture's syndrome, autoimmune gastritis, autoimmune haemolytic anaemia, and pemphigus vulgaris.
- **Transfusion reactions**: In chapter 17, we have explained that erythrocytes express multiple membrane antigens and that the transfer of incompatible blood results in severe transfusion reactions (Table 17.2). Other components of blood, including leukocytes and platelets, may also give rise to transfusion reactions, though they are often of lesser severity. The transfusion reaction is acute and immediate if the recipient has been previously exposed to donor antigens. If a considerable number of antibodies are present in the blood at the time of transfusion, donor RBCs are rapidly lysed either intravascularly (ABO incompatibility) or extravascularly (Rh incompatibility). Symptoms include fever, chills, nausea, and vomiting. Circulatory shock, as well as acute necrosis of the kidneys, is a complication associated with intravascular lysis. A previously unsensitised transfusion recipient may develop antibodies to antigens on donor erythrocytes over a period of time, and this may give rise to delayed jaundice and anaemia.
- **Haemolytic disease of the newborn (erythroblastosis foetalis)**: This is observed when blood groups of the foetus and the mother are mismatched and when maternal IgG antibodies to foetal erythrocytic antigens cross the placenta. The antibodies cause the destruction of foetal erythrocytes, resulting in anaemia and jaundice within the first 24 hours of life[7]. Rh incompatibility — an Rh^- mother carrying an Rh^+ foetus — is one of the most common causes of this disease. Because the mother has to be sensitised to the antigen, the disease is not manifested during the birth of the first Rh^+ child. Sensitisation may occur during this first childbirth, when foetal cord blood can enter the mother's circulation, and the risk of the disease increases exponentially with each new Rh^+ foetus. Sensitisation can be avoided by the administration of anti-RhD serum within 72 hours of the delivery of the first Rh^+ child. The anti-RhD antibodies will also opsonise any foetal erythrocytes that may have entered the

[7] Detection of erythroblastosis foetalis is done by Coombs' test described in appendix IV.

mother's circulation and facilitate their removal before the mother's immune system is activated. As explained in chapter 8, memory B cells suppress naïve B cell activation. The administration of anti-Rh antibodies exploits this phenomenon. The antibodies must be administered before the mother is sensitised to the Rh^+ antigen as activation of naïve but not memory B cells will be inhibited by the antibodies. Severe haemolytic disease during pregnancy can be treated with intrauterine blood exchange, in which foetal Rh^+ blood is replaced with Rh^- blood. Haemolytic disease can also be caused by ABO incompatibility; however, the disease is rarely severe enough to require treatment.

- **Drug-induced anaemia**: A variety of drugs (e.g., quinines, sulphonamides, and penicillin) can get adsorbed to erythrocytes, platelets, etc., and provoke hypersensitivity type II reactions. Autoimmune anaemias (whether drug-induced or otherwise) can also result from phagocytosis of autoantibody-coated erythrocytes (Figure 16.2).

16.4 Hypersensitivity Type III

Type III hypersensitivity (also known as *immune complex disease*) is caused by the persistence of immune complexes in circulation. Although both type II and type III reactions are at least partially the results of complement activation, unlike in a type II reaction, **the antigen in a type III reaction is soluble and not attached to the tissue**. As seen in chapter 3, complement is activated as a consequence of antigen–IgG (or IgM) antibody interaction and results in cytolysis and PMN and mast cell recruitment. Scavenger cells, such as macrophages and monocytes, remove these immune complexes by FcR-mediated endocytosis. However, moderate antigen excess can result in the formation of small complexes (Ag.Ab2 or Ag3.Ab2) that are not easily removed by the scavengers. Consequently, these small complexes remain in circulation, depositing on tissues and initiating complement-mediated and neutrophil exocytosis-mediated tissue injury and inflammation (Figure 16.3). As small antigen–antibody complexes are barely soluble and are often filtered out in capillary beds or may be driven into vessel walls at places of especially turbulent flow, the kidneys and lungs are often the sites of tissue damage in immune complex disease. Nephrotoxicity is therefore often an

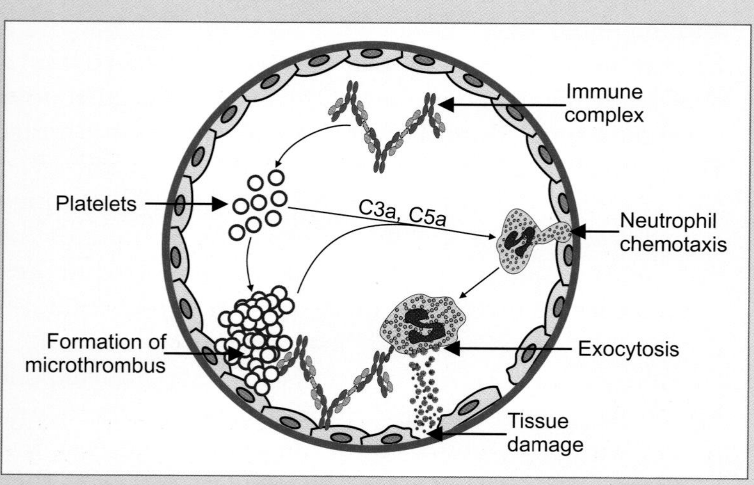

Figure 16.3 Type III hypersensitivity is caused by the persistence of large quantities of small, soluble immune complexes in circulation. Small antigen–antibody complexes are barely soluble and are often filtered out in capillary beds or may be driven into vessel walls at places of turbulent flow, such as the kidneys. Immune complexes also adhere to platelets *via* FcRs, leading to microthrombus formation. The C3a and C5a released as a result of immune complex formation and the chemokines released by platelets attract PMNs and monocytes to the site of the reaction. Unable to endocytose the complexes, the scavengers release their intracellular contents by exocytosis, which results in tissue injury and inflammation.

identifying feature of type III hypersensitivity. Experiments in knock out mice suggest that mast cells may have a major role in the pathology of immune complex disease; mast cell degranulation caused by the binding of IgG–antigen complexes to FcγRIII may contribute significantly to the type III reaction.

Many human diseases are associated with circulating immune complexes, and the spectrum of symptoms depends upon the site of immune complex deposition. Chronic immune complex formation is often observed in chronic and persistent infections that elicit a weak antibody response (e.g., malaria and viral hepatitis). Complex deposition may occur in the small blood vessels, kidneys, joints, and skin, resulting in vasculitis, glomerulonephritis, scar tissue formation, and urticaria, respectively. By contrast, the repeated inhalation of an antigen (spores of fungi, actinomycetes, tissue (or hair) of animals or plants) may result in the deposition of complexes in the lungs, as in farmer's lung and pigeon fancier's disease. Immune complexes may also be formed by autoantibodies and may result in autoimmune diseases, such as SLE or rheumatoid arthritis (RA).

Hypersensitivity Type I

- IgE antibodies play a central role in this hypersensitivity.
- It is triggered by the degranulation of mast cells, basophils, or eosinophils by crosslinking of IgE-bound FcεRI.
- The early-phase reaction begins within seconds to minutes after allergenic challenge and is caused by the release of preformed mediators.
- The late-phase reaction sets in 6-12 hours later and is caused by the leukotrienes and cytokines released by mast cells.
- Manifestations include urticaria, hay fever, rhinitis, asthma, and anaphylactic shock.
- Treatment includes
  - Medications, such as antihistamines, sodium chromoglycate, epinephrine, isoprenaline, salbutamol, theophylline, phenoxybenzamine, and corticosteroids, that alleviate symptoms and
  - immunotherapy aimed at the elicitation of a non-IgE antibody response.

Hypersensitivity Type II

- Type II hypersensitivity is a cytolytic reaction caused by antibodies to cell-surface antigens.
- Complement-mediated cytolysis is responsible for some of the damage observed.
- Manifestations are seen in blood transfusion reactions, erythroblastosis foetalis, and drug-induced anaemias.

Hypersensitivity Type III

- Persistence of small complexes consisting of a soluble antigen and low-affinity antibodies is responsible for type III hypersensitivity.
- Tissue damage is often seen in areas of turbulence and high blood pressure, such as the kidneys, alveoli, and joints.
- Examples of type III hypersensitivity include farmer's lung, pigeon fancier's disease, serum sickness, and autoimmune disorders, such as RA and SLE.

Hypersensitivity Type IV

- Type IV reaction is of the delayed-type, with symptoms observed a minimum 24 hours after antigenic challenge.
- It is observed in diseases in which pathogens tend to persist intracellularly.
- Examples of the type IV reaction include leprosy, tuberculosis, leishmaniasis, blastomycosis, histoplasmosis, and schistosomiasis.

Manifestations of type III hypersensitivity include the following:

- **Serum sickness**: This disease was originally observed in the beginning of the 20th century when passive immunisation with massive doses of horse or rabbit antisera was attempted in children with acute bacterial infections such as diphtheria. After 8-12 days of therapy, the children developed a self-limited syndrome characterised by fever, urticaria, muscle and joint pains, lymphadenopathy, and proteinuria. Serum sickness has become rare since the use of heterologous serum has been discontinued. However, similar symptoms are sometimes observed in hypersensitivity reactions to drugs such as penicillin.
- **Arthus reaction**: This is the name given to a local type III reaction that can be demonstrated experimentally by a subcutaneous injection of a soluble antigen in pre-immunised animals having a high titre of IgG antibodies. Oedema and haemorrhage develop at the site of the injection within 4-10 hours and subside within 48 hours. The reaction is slower than the type I reaction because the FcγRIII receptor is a low-affinity receptor, and the threshold of activation of mast cells *via* this receptor is considerably higher than for the FcεRI receptor. A passive cutaneous form of the arthus reaction can be demonstrated by injecting a high-titre antiserum intravenously in an unsensitised animal and challenging it later with an intradermal injection of the antigen.
- **Generalised type III reactions**: Persistent soluble complexes are found in post-infection complications and may result in conditions, such as glomerulonephritis and arthritis. Type III reactions are also observed in a variety of autoimmune diseases, such as RA, SLE, and Sjogren's syndrome.

16.5 Hypersensitivity Type IV

Although type IV hypersensitivity was first described by Koch in 1882, it was not until the 1940s that Landsteiner and Chase proved that it was caused by cellular, and not humoral, components of immunity. For the first time, they experimentally demonstrated that this type of immunity can be transferred solely by transferring cells. Unlike the antibody-mediated type I, type II, and type III responses that are manifested within minutes or hours after antigen exposure, type IV hypersensitivity takes between 24-72 hours to develop and is therefore called DTH (**D**elayed-**T**ype **H**ypersensitivity). **DTH is mediated by CD4⁺ T cells in conjunction with CD8⁺ CTLs**. The lag between antigenic challenge and manifestation of DTH thus reflects the slower nature of cellular responses that require the induction of cytokines/chemokines and an influx of cells. Thus, if type I, II, and III hypersensitivity responses are humoral responses gone awry, DTH responses are considered to be CMI gone awry. Unfortunately, the exact relation between the protective and destructive functions of CMI is not clear. DTH responses are characterised by erythema and induration[8], with a large influx of macrophages/monocytes and lymphocytes at the site of antigenic challenge.

As in other hypersensitivity reactions, type IV hypersensitivity also requires sensitisation — it is a secondary response to the sensitising antigen. The sequence of events in the development of DTH responses is similar to that of CMI responses. Upon primary exposure to the antigen, APCs, such as Langerhans[9] cells and macrophages, internalise the antigen, transport it to the draining lymph node, and present it to CD4⁺ T cells. The resulting activation and differentiation gives rise to TH cells and CTLs capable of responding to epitopes on that antigen. A subsequent

[8] Derived from the Latin *indurare*, meaning hard, induration is the hardness or lumpiness observed at the site of physical trauma or antigen challenge.

[9] Langerhans cells are a type of skin DC; see chapter 5.

exposure to antigen induces the effector phase of the response, which is a result of complex interactions between the cells of innate and adaptive arms of the immune system. The secondary antigenic challenge leads to the release of pro-inflammatory cytokines such as TNF-α and IFN-γ by DCs. Antigen-specific T_H1 cells that recognise the MHC class II:peptide complexes on the DCs also secrete chemokines and cytokines. The result is increased vascular permeability and influx of monocytes to the site of injection; monocytes and neutrophils begin to infiltrate the site within hours of antigenic challenge. The infiltrating monocytes differentiate to activated macrophages during this process. The IL-3 and GM-CSF produced by the activated T_H1 cells stimulate monocyte production by bone marrow stem cells, further enhancing their influx into the site of inflammation. Each of these events can take several hours to develop. As explained in chapter 2, activated macrophages have a potent arsenal of antimicrobial substances that are crucial in host defence against intracellular parasites or pathogens that are not accessible to circulating antibodies. The cocktail of enzymes and microbicidal factors released by the macrophages can damage cells in the immediate vicinity; this is a small price to pay for rapid pathogen clearance.

Unfortunately, prolonged macrophage activation becomes destructive to the host. The persistence of the antigen/pathogen results in an intense inflammatory response and chronic macrophage activation. The high concentrations of the lytic enzymes liberated by these cells result in the destruction of surrounding tissues and may even lead to tissue necrosis. Continuous activation causes the macrophages to adhere closely and assume an epithelioid shape (Figure 16.4). They may even fuse together to form multinucleated giant cells that displace normal tissue cells and form palpable nodules termed *granulomas*. The granuloma consists of an outer kernel of infected macrophages, surrounded by foamy macrophages and other mononuclear phagocytes, with a mantle of lymphocytes in association with a fibrous cuff of collagen and other extracellular matrix components. Such granulomatous immune responses are observed both in protective immune responses and disease pathology. Only 5% of the participating cells in a fully developed DTH reaction are antigen-specific T cells. A variety of intracellular pathogens (e.g., *Schistosoma* spp., *Leishmania* spp., the herpes simplex virus, the mumps virus, *Blastomyces dermatitidis*, and *Histoplasma capsulatum*) or their products can give rise to deleterious DTH responses.

The prototypical type IV response is observed during the tuberculin test, which is used to determine the immune status of an individual to *M. tuberculosis*. Other

Figure 16.4 A granuloma is a characteristic localised inflammatory response that is the result of persistent macrophage activation coupled with a failure to eliminate intracellular pathogens or their constituents. Failure to get rid of the intracellular pathogen causes the macrophage to remain in a continuously activated state. The macrophages assume an epithelioid shape and closely adhere to each other. Some of them fuse to form multinucleated giant cells. This central macrophage-derived core is surrounded by T cells. T cells provide TNF-α, which is crucial for the initiation and maintenance of granulomas. An extracellular matrix shields this cellular conglomerate from healthy tissue. The pathogen may survive in the granuloma, but its proliferation and spread are inhibited.

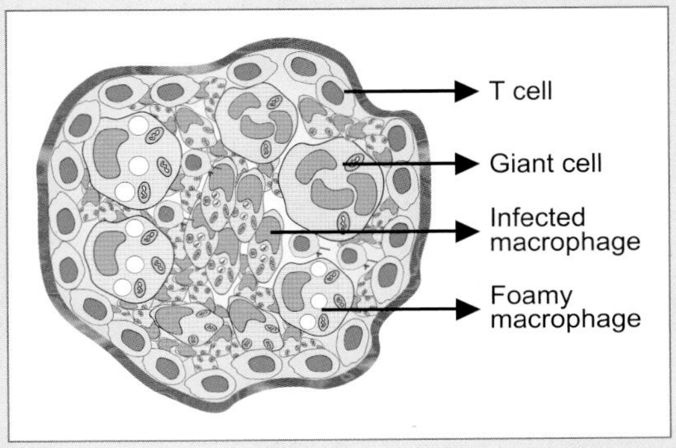

substances that induce DTH responses include insect venom, poison ivy, and Ni or Cr salts. Some of these give rise to painful cutaneous (or contact) hypersensitivity responses. Substances like picryl chloride or the lipid soluble pentadecacatechol, present in the leaves of poison ivy, interact and modify cellular proteins upon contact with the skin to elicit a deleterious T cell response to the modified proteins. Although DTH was originally proposed to be an exclusively TH1 phenomenon, later experiments have proven that both TH1 and TH2 cells can participate in DTH reactions. IL-4 seems to be crucial for systemic DTH responses, although IFN-γ and TNF-α appear to be key cytokines in the initial phase of the response. The fact that IL-4 knock out mice cannot mount contact hypersensitivity responses seems to support this idea. Paradoxically, TH2 cytokines, such as IL-4 and IL-10, may also be important in DTH suppression and lesion healing. Different types of DTH responses are listed in Table 16.6.

Table 16.6 Type IV Hypersensitivity Reactions

Type	Time (days)	Features
Jones–Mote reaction	1	Induced by soluble antigens Characterised by infiltration of basophils in the epidermis; basophils constitute 50% of cellular exudate
Contact hypersensitivity	2-3	Induced by contact with small haptens conjugated to tissue antigens, e.g., salts of nickel, chromates, and poison ivy Characterised by a local epidermal reaction (eczema) at the site of contact
Tuberculin type	1-2	Induced by injection of soluble antigens of *M. tuberculosis*, *M. leprae*, etc., although certain nonmicrobial antigens may also induce a similar reaction. Characterised by fever, generalised sickness, induration, and swelling at the site of injection
Granulomatous type	Minimum 14	Induced by the persistence of antigen that cannot be cleared by phagocytes. Characterised by a nodular mass consisting of a core of multinucleated giant cells and epithelioid cells surrounded by fibrotic scar tissue; T cells are instrumental in the formation and maintenance of granulomas

Transplantation and Transfusion Immunology 17

You think we look pretty good together
You think my shoes are made of leather

But I'm a substitute for another guy
I look pretty tall but my heels are high
The simple things you see are all complicated
I look pretty young, but I'm just back-dated, yeah

— The Who, *Substitute*

Abbreviations

ATG:	Antithymocyte Globulin
AZT:	Azathioprine
CsA:	Cyclosporin A
FKBP:	FK506 Binding Protein
GVHD:	Graft *Versus* Host Disease
IDO:	Indoleamine 2,3-Dioxygenase
MMF:	Mycophenolate Mofetil
MNA:	Malononitrilamide
TRALI:	Transfusion-Related Acute Lung Injury
vCJD:	Variant Creuzfeldt Jacob Disease

17.1 Introduction

The replacement of defective or damaged body parts has been a long cherished human goal. Sushruta, an ancient Indian surgeon, has been credited with using self soft tissues in reconstructive surgery of the nose and ears as far back as 700 BC. However, the transfer of tissues between individuals was never successful until recently. The problems associated with tissue transfer became apparent when experiments in tissue and tumour transfers were undertaken in the later decades of the nineteenth century. **The transfer of cells or tissues from one individual (animal) to another was termed *transplantation*, and the transferred tissue was called a *graft*.** A graft was said to be *rejected* if the recipient's immune system mounted a vigorous response against the graft and damaged the grafted tissue. Rejection of the first graft from a particular donor is the *first-set rejection*, and rejections of second or additional grafts are termed *second-set rejections* (in parallel with primary and secondary immune responses). In 1912, the German scientist Georg Schöne coined the term *transplantation immunity* and formulated the rules of transplantation, paraphrased below.

- Heterografts (xenografts) — the tissues transplanted from one species to another — invariably fail.
- Allografts (homografts) — transplants between genetically nonidentical individuals of the same species — usually fail; they may initially seem to survive but are subsequently rejected.
- Autografts — self-tissues transplanted from one region of the body to another — and isografts (syngrafts; tissues transplanted from one genetically identical individual to another) are almost always successful.
- Second grafts undergo accelerated rejection if the recipient has previously rejected a graft from the same donor. Grafts are also rapidly rejected if the recipient has been preimmunised with material from the donor.
- A close blood relationship between the donor and recipient is more likely to ensure graft success.

Eyes of an Eagle, Heart of a Lion: Xenotransplantation

Xenotransplantation is the transfer of cells, tissues, or organs between disparate species. Perceived to be an experiment in biological curiosity, xenotransplantation has been made possible by the advent of genetic engineering, and a severe shortage of human organs and tissues for transplantation has given added impetus to research. The idea of animals raised to express genes of therapeutic value or engineered expressly to reduce the risk of rejection does not seem like a fantasy anymore. Nonetheless, several hurdles need to be crossed before xenotransplantation becomes a reality, not the least among them being the fear of unwittingly transferring infectious agents or aiding the emergence of new infectious agents.

- The first hurdle is physiological — whether or not the xenogeneic organ or tissue will function adequately in humans. Porcine liver has been shown to function adequately in baboons, suggesting that physiological barriers may not be insurmountable. Such experiments with pig organs suggest that their kidneys, lungs, and hearts may function satisfactorily in humans.
- A major barrier to xenograft acceptance is nonimmunological. The survivability of the xenograft will depend upon its ability to sustain angiogenesis, that is, its ability to stimulate the growth of blood vessels. Incompatibility of the donor and recipient growth factors may impair the growth of new blood vessels, and the graft cells will quickly die because of impaired blood supply.

- If the xenograft survives this barrier, it has to face the immune system.
  - The complement system provides the most potent threat to xenograft acceptance; it can cause hyperacute rejection that begins within minutes of engraftment. Both classical and alternative pathways have a role in xenograft rejection. Xenograft cells do not express complement regulatory proteins, such as CD59 and DAF, and are particularly susceptible to complement-mediated injury. Xenoreactive natural antibodies produced by B1 B cells recognise a variety of structures on foreign cell surfaces and activate the classical pathway. Especially important are antibodies specific for Galα1,3Gal — a saccharide epitope expressed on the cells of lower mammals but not humans or apes. These anti-carbohydrate antibodies seem to arise during the first few years of life through the interaction of the immune system with gut bacteria; they belong to the same class of antibodies as anti-blood group A antibodies.
  - Evidence suggests that neutrophils and macrophages can be activated by direct interaction with xenogeneic cells. They can initiate xenograft tissue destruction without the involvement of inflammatory mediators.
  - Porcine endothelial cells seem to generate thrombin spontaneously; human platelets get recruited to the site of thrombin generation and trigger an inflammatory response.
  - NK cells are particularly cytotoxic to xenogeneic cells. The absence of MHC class I expression, the stimulation of lectin receptors on NK cells by saccharides (e.g., Galα1,3Gal) on the xenografted cells, and ADCC mediated by xenoreactive antibodies are all thought to contribute to this susceptibility.
  - As in allografts, T cells can also be involved in xenograft rejection. Both the direct and indirect pathways of MHC recognition may operate in this T cell activation.

17.2 Antigens Involved in Graft Rejection

Genetic differences between the donor and the recipient are responsible for graft rejection; alloantigens[1] are the root cause of graft rejection.

- **ABO blood group antigen** incompatibility results in the hyperacute rejection of vascularised organ transplants. Preformed anti-A or anti-B antibodies in host circulation bind to antigens expressed on the donor vascular endothelium and trigger complement-mediated lysis of the graft vasculature (see section 17.7.1).
- Experiments by Peter Gorer, George Snell, and Peter Medawar identified the role of **MHC molecules**, that is, the classical MHC class I molecules (HLA-A, HLA-B, and HLA-C) and MHC class II molecules (HLA-DR, HLA-DP, and HLA-DQ) in graft rejection.
- **Minor histocompatibility antigen** mismatch can result in graft rejection, even when the donor and recipient MHC are matched. Minor histocompatibility antigens are polymorphic proteins encoded by alleles of genes on sex chromosomes, autosomes, mitochondrial DNA, etc. Minor histocompatibility antigens may be present only in the donor (e.g., antigens encoded by the Y chromosome), or may be present in both the donor and the recipient but may differ in a single or a few amino acids. These antigens were originally termed *minor* because of their weaker potential to effect rejection as compared to the HLA antigens. Subsequent studies established that multiple differences in these antigens in MHC-matched donors and recipients can elicit graft rejection with a speed comparable to MHC mismatch. Minor H antigens are presented by the indirect pathway (section 17.3). Examples of murine minor histocompatibility antigens include the autosome-encoded β_2-m, mitochondrial DNA-encoded and maternally transmitted factor-α, and H–Y antigens encoded by the Y chromosome.

[1] An antigen with the potential to elicit an immune response when transferred from one individual to another of the same species is called an alloantigen. The immunogenicity of the alloantigen is due to allelic variance of the gene encoding that antigen. *Allogeneic* refers to genetic differences between individuals of the same species, whereas *syngeneic* implies genetic identity (e.g., monozygotic twins or inbred laboratory animals).

17.3 Allorecognition

MHC molecules are involved not only in graft rejection but also in the initiation of immune responses to antigens — whether self or foreign. As explained in chapter 9, self-MHC restriction of T cells is a consequence of selection during thymic development. These self-restricted T cells are activated through the engagement of their TcRs by peptide:self-MHC complexes in conjunction with appropriate costimulatory signals, and this activation culminates in an immune response. Graft rejection is a consequence of a system that cannot discriminate between the deliberate introduction of a beneficial foreign immunogen and the accidental introduction of a potentially harmful one. Transplantation immunity is different from the general model of T cell activation outlined above. It is a seeming contradiction of the imposed rule of self-restriction, as nonself MHC molecules appear to trigger the T cells involved in graft rejection. Thus, although the vigorous nature of T cell responses to allogeneic MHC molecules has been known for more than half a century, the details of the mechanism by which these antigens are recognised remain controversial. Two pathways could result in allorecognition.

- **Direct allorecognition** involves the stimulation of recipient T cells by donor APCs. In this, *the direct pathway,* recipient TcRs react directly with intact MHC molecules expressed on the surface of donor APCs. About 1-7% of T cells are estimated to show alloreactivity (i.e., they recognise nonself-MHC), and this seems to fly in the face of self-MHC restriction. Much evidence points to direct allorecognition. First, matching the donor and recipient for MHC antigens improves graft survival. Second, depleting MHC-mismatched grafts of potential APCs (e.g., DCs) improves graft survival, whereas restoring the APCs to the graft accelerates rejection. When tested *in vitro*, recipient T cells respond with vigorous proliferation in an MLR[2] with donor APCs, suggesting a similar reaction between the cells *in vivo*. The structural analysis of alloreactive TcRs indicates that the recognition of allogeneic MHC occurs because of cross-reactivity. Thus, some of the TcRs specific for self-MHC cross-react with nonself-MHC, and T cells proliferate in response to this recognition. It is not clear if the alloreactive TcRs recognise the allogeneic MHC itself (i.e., irrespective of the peptide bound to it) or only a particular peptide:allo-MHC complex. These possibilities are likely to represent two extreme scenarios, and both MHC molecules and peptides may contribute in varying degrees to the overall binding energy between the TcR and its ligand. The density of allogeneic MHC molecules on donor APCs can be in the neighbourhood of 10^5 or more molecules per cell. Recipient T cells bearing even weakly cross-reactive TcRs for the allogeneic MHC are therefore likely to be stimulated by the high density of MHC molecules present, explaining, at least in part, why the frequency of alloreactive T cells is much higher than that of antigen-specific T cells.
- **The indirect pathway of allorecognition** is triggered when peptides derived from allogeneic MHC are processed and presented in the context of self- (i.e., recipient) MHC molecules. Originally, the direct pathway was thought to be the only major pathway of allorecognition, and the importance of the indirect pathway has only recently been recognised. Thus, self-APCs internalise and degrade allogeneic MHC molecules, load the peptides derived from them on self-MHC molecules, and present them to self-T cells, triggering an immune response against graft tissues and cells expressing the allogeneic MHC molecules (Figure 17.1).

[2] MLR (**M**ixed **L**eukocyte **R**eaction) is used to determine a recipient T cell proliferative response to alloantigens on donor APCs *in vitro*. Donor leukocytes are irradiated to ensure that only recipient, and not donor, T cells proliferate (see appendix IV).

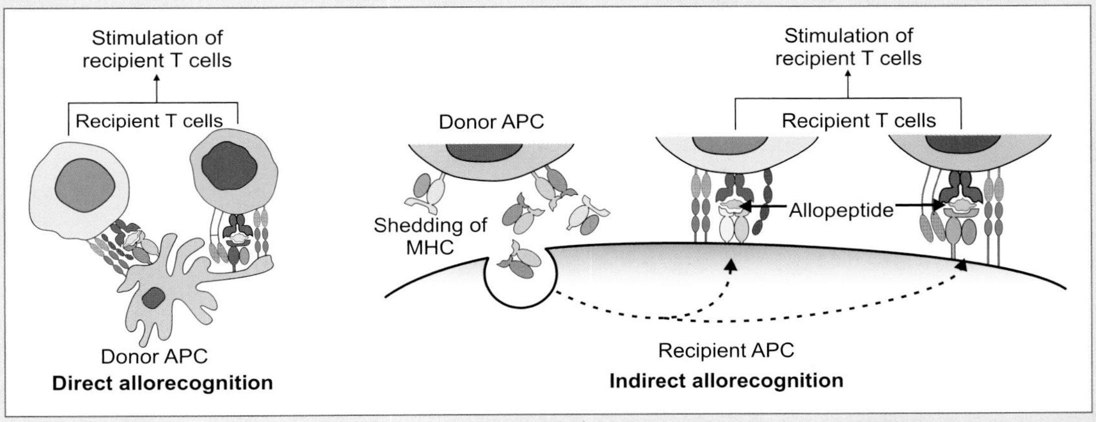

Figure 17.1 Graft antigens can activate recipient T cells through two pathways. Direct allorecognition occurs because of cross-reactivity between donor and recipient MHC molecules. Recipient T cells recognise and bind the nonself-MHC molecules on donor APCs and induce an immune response against donor antigens (left panel). In the indirect pathway, donor alloantigens (MHC molecules) are internalised and processed by recipient APCs. Peptides derived from these molecules are loaded onto recipient MHC and presented to recipient T cells, activating the cells (right panel).

17.4 Graft Rejection

Clinically, allografts are usually obtained from cadavers or brain-dead individuals; bone marrow transplants and sometimes kidney transplants are notable exceptions. The organ to be transplanted is removed from the cadaver and treated before being packed in ice for transport. The process results in ischemic/reperfusion damage[3] to the organ, characterised by complement deposition, upregulation of adhesion molecules, inflammatory cell infiltration, and cytokine release. This injury is compounded by emergency interventions that precede organ harvest. Recipients are also subjected to treatments before and during transplantation. Moreover, the transplant surgery itself causes trauma, which is compounded if the transplant site gets infected. All these factors act as immunological stimuli and play a role in graft rejection. Although the relative contributions of the direct and indirect pathways of allorecognition to graft rejection are not established, the mechanism of graft rejection and the effector pathways leading to it are becoming clearer. The process of graft rejection has two stages — the sensitisation stage and the effector stage. During the sensitisation stage, recipient CD4$^+$ T cells and CTLs proliferate in response to the alloantigens expressed on the graft tissue, whereas in the effector stage, a variety of effector mechanisms participate in graft rejection.

Three phases of graft rejection are recognised.

- The **hyperacute** response occurs within 48 hours of engraftment. It is induced by preformed recipient antibodies that mediate graft rejection by binding to antigens (e.g., blood group antigens), which are expressed by the vascular endothelium of the graft. This binding of recipient antibodies to antigens on the graft endothelium activates complement by the classical pathway, resulting in vasoconstriction, an influx of PMNs and monocytes, formation of platelet thrombi, ischemic damage, and infarction.
- **Acute rejection** typically occurs 1-2 weeks after transplantation and may be prevented or reversed with immunosuppressive regimens.

[3] Ischemia is the insufficient supply of blood to an organ; re-establishing the blood supply causes damage to the organ, and this is called ischemia/reperfusion damage.

- **Chronic rejection** is a more insidious rejection and may take months or years to develop. The process is characterised by luminal narrowing, occlusion of arteries and arterioles, and fibrosis of the graft parenchyma, resulting in organ failure. The relation between acute and chronic rejection is not clear, but early acute rejection appears to be the most important antigen-dependent risk factor in chronic rejection.

17.4.1 Role of APCs

The immune system treats the graft as a potentially dangerous intruder, and the immunological events that follow engraftment are essentially similar to those occurring after a major infection or trauma. The pro-inflammatory signals that result from the stress, ischemia, and perfusion injury that are caused by the surgery lead to the maturation of donor DCs, which leave the graft and move to the draining lymph node. There, donor APCs evoke a direct alloresponse involving both $CD4^+$ and $CD8^+$ effector T cells. This response is thought to be due to cross-reactive (recipient) memory T cells primed against various environmental antigens in the context of self-MHC molecules. The activated effector T cells migrate to the graft. The ischemic injury and resultant microvascular stress help this trafficking of leukocytes into the graft. Chemokines play a major role in this recruitment and may further increase the damage by recruiting other effector cells, such as NK cells, macrophages, and B cells, to the site of the graft. The site becomes a seat of intense inflammation, and necrosis and eventual graft rejection follow. Thus, the direct pathway is thought to be responsible for acute graft rejection. If the graft survives this phase, the direct pathway may also contribute to tolerance induction once the professional APCs in the graft die. The recipient alloreactive T cells entering the graft will recognise MHC molecules on nonprofessional antigen-presenting graft cells in the absence of costimulatory signals. Such nonprofessional antigen presentation will result in anergy, inducing tolerance to the graft.

The death of donor APCs starts in the draining lymph nodes and can set the stage for the activation of the indirect pathway of allorecognition and chronic graft rejection. The dying APCs become a source of donor MHC molecules to recipient APCs present in the lymph node. Migrant recipient APCs trafficking through the lymph nodes can also capture alloantigens from dying donor APCs. The alloantigens will then be processed and presented to recipient T cells — predominantly to $CD4^+$ T cells. Because of cross-presentation[4], some of the alloantigen-derived peptides will also be loaded onto MHC class I molecules, resulting in the activation of $CD8^+$ T cells. Thus, both naïve $CD4^+$ and $CD8^+$ T cells are primed by recipient DCs by the indirect pathway. This pathway is thought to be important in chronic graft rejection, as the perpetual trafficking of recipient DCs through the transplanted tissue is likely to provide a continuous stimulus for the indirect alloresponse. The indirect pathway therefore poses a threat to long-term graft survival.

17.4.2 Role of Effector Cells

The central role of $CD4^+$ T cells in graft rejection is demonstrated by the inability of CD4-deficient mice to reject allografts. As explained in chapter 9, effector $CD4^+$ T cells can be of two types and both have a role in graft rejection. T_H1 cells produce IFN-γ and IL-2. Both these cytokines activate $CD8^+$ CTLs and NK cells, whereas IFN-γ acts on macrophages. CTLs and NK cells cause the graft cells to apoptose (chapter 11), whereas chronically activated macrophages result in a DTH reaction

[4] Cross-presentation is the loading of peptides derived from exogenous antigens onto MHC class I molecules. Cross-presentation and the role of DCs in this process are discussed in chapter 7.

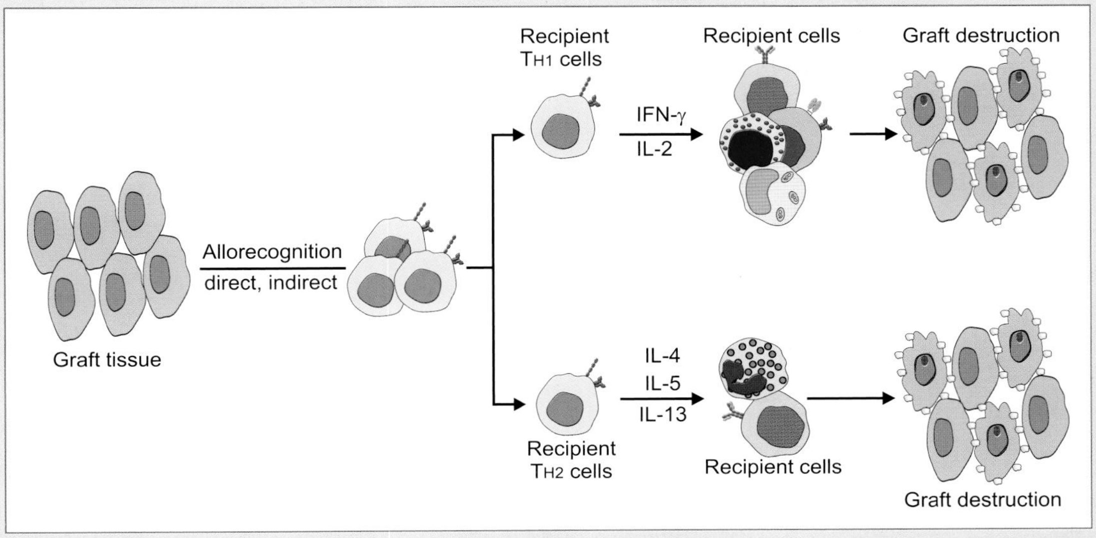

Figure 17.2 Both recipient TH1 and TH2 cells are involved in graft rejection. Upon first exposure, graft antigens activate naïve host $CD4^+$ T cells by both the direct and indirect pathways. The activated T cells eventually differentiate to TH1 and TH2 effector cells. Cytokines, such as IFN-γ, secreted by recipient TH1, help in the activation of $CD8^+$ T cells, NK cells, and macrophages; the cytotoxic substances released by these cells cause extensive graft tissue damage. TH1 cells also induce apoptosis in graft cells expressing Fas. Cytokines such as IL-4, IL-5, and IL-13 released by TH2 cells, activate host eosinophils. The activated eosinophils participate in graft destruction. Additionally, the TH cells also help antigen-activated B cells proliferate and differentiate to plasma cells. Igs secreted by the plasma cells may also participate in graft rejection.

as outlined in chapter 16. Furthermore, TH1 cells express FasL, and can therefore cause the apoptosis of Fas-expressing graft cells. TH1 cells also promote the synthesis of complement-fixing antibody isotypes by B cells (IgG2a and IgG2b in mice, IgG1 and IgG3 in humans). Such antibodies are capable of causing lysis of graft cells. By contrast, TH2 cells trigger eosinophil activation. Eosinophils are recruited and activated within the allograft through the combined action of IL-4, IL-5, and IL-13 produced by alloreactive TH2 cells. Activated eosinophils have been shown to mediate graft rejection in experimental models without the participation of CTLs, macrophages, and NK cells. This pathway may have a greater role in rejection if the TH1 pathway is inhibited. Hypereosinophilia preceding graft rejection has been reported in a number of cases, and the presence of activated eosinophils has been demonstrated in liver, kidney, or heart allografts undergoing acute rejection. The newly found TH17 cells may also be involved in acute graft rejection; several studies report IL-17 production in acute rejection.

Although not as important as T cells, B cells do play a role in graft rejection. Antibodies are more important in the rejection of tissues that are either directly connected to the host's blood supply, such as the kidney and the heart, or in recipients sensitised to donor antigens. A history of transplant rejections is associated with an increase in the risk of acute humoral rejection. Patients that have undergone multiple pregnancies, blood transfusions, or previous transplantations (termed *hyperimmunised patients*) often have a higher frequency of anti-alloantigenic B cells. These B cells present alloantigens to recipient T cells by the indirect pathway and enter into cognate interaction with them. This results in an antibody response against graft cell constituents and the opsonised graft cells are subject to complement-mediated lysis or ADCC (Figure 17.2).

17.5 Graft *Versus* Host Disease

GVHD or **G**raft ***Versus*** **H**ost **D**isease is observed after allogeneic bone marrow transplantation, and it is a significant cause of morbidity in patients. It occurs when donor immunocompetent T cells react to and attack the genetically disparate host. In contrast to graft rejection, the host tissue is under attack in GVHD, and the *donor T cells* are responsible for the damage. The clinical manifestations of acute GVHD include weight loss and solid organ toxicity affecting the lungs, liver, skin, and gut. GVHD is a multistep process (Figure 17.3).

- **The induction phase follows a typical T cell recognition and activation cascade**. The stage for GVHD is set by the activation of recipient APCs for any number of reasons including infections suffered by the patient that caused the organ failure to begin with, transplant procedure, ischemia, and reperfusion injury. The systemic vasculature, including the capillary beds, is potentially the first extensive area of contact between donor T cells and alloantigens on recipient APCs. MHC mismatch therefore results in the immediate and extensive activation of these T cells. If the major MHC antigens have been matched, minor MHC antigens can still activate donor T cells and lead to GVHD. Direct allogeneic recognition (i.e., the recognition of recipient MHC molecules with or without peptides, expressed on recipient APCs) is thought to be responsible for the activation of donor T cells. $CD4^+$ T cells from the donors are the first to be activated, although $CD8^+$ donor T cells do get involved later.
- **In the expansion phase, activated donor T cells migrate to lymphoid tissues, proliferate, differentiate to effector cells, and produce cytokines and chemokines**. This release of cytokines is termed a *cytokine storm*; the term seems to accurately describe the clinical picture. Although TH1 cytokines

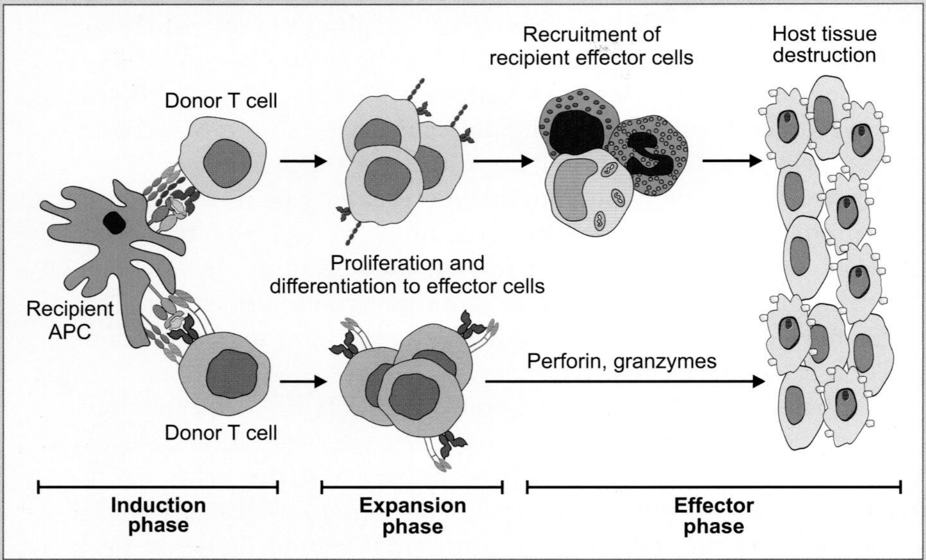

Figure 17.3 GVHD occurs in three phases. In the induction phase, donor T cells are activated by contact with recipient APCs. Donor $CD4^+$ T cells are the first to be activated, although $CD8^+$ T cells may also be involved at a later stage. The activated T cells undergo proliferation and differentiation in the expansion phase. In the effector phase, the cytokines and chemokines produced by activated donor T cells recruit recipient effector cells such as macrophages, PMNs, and NK cells to the graft site. Cytokines and cytotoxins released by these cells are responsible for the extensive host tissue damage observed in GVHD.

(IFN-γ, IL-2, and TNF-α), have been implicated in the pathophysiology of acute GVHD, the role of T cell subsets remains unclear. Experimental administration of various cytokines has failed to give an unambiguous picture of the type of T cells (TH1, TH2, NKT, and TH17) involved. The experimental data so far collectively suggest that the timing of the administration of cytokines, the local production of different cytokines, the type of conditioning regimen before transplantation, and the donor–recipient combination may all be critical to the eventual outcome of acute GVHD.

- **In the effector phase, the recruited recipient cells act in tandem with donor T cells to damage recipient organs.** The release of cytokines by donor T cells leads to the recruitment of recipient cells such as NK cells, macrophages, and PMNs. Host tissue damage occurs by multiple pathways — release of cytotoxic molecules by NK cells and CTLs, damage due to ROI and RNI released by macrophages, induction of apoptosis by the Fas pathway, and damage caused by pro-inflammatory cytokines.

GVHD can be divided into acute or chronic, based on the time of onset, pathophysiological processes, and clinical presentations. Acute GVHD involves mostly the skin, gastrointestinal tract, and liver, and it occurs within the first 100 days of transplant. It is directed against multiple host cells, including the epithelial cells and mucosa of the skin, hair follicles, bile duct, cryptic cells of the intestines, airways, mucous membranes, the bone marrow, and cells of the immune system. Between 9-50% of HLA-matched bone marrow transplants with intensive immunosuppressive therapy report clinically significant GVHD; the figure is 100% without such therapy. GVHD is said to be chronic if it appears within 100 days of engraftment. However, much remains to be elucidated on the mechanisms underlying chronic GVHD. As donor T cells are responsible for GVHD, methods aimed at depleting them from the graft, in conjunction with immunosuppressive therapy, are the only means of prophylaxis. mAbs against T cell-surface proteins, such as CD2, CD3, and CD5, have been used successfully in experimental models and clinical situations. Treating GVHD once it has occurred is much more difficult, and use of steroids with immunosuppressive therapy remains the treatment of choice.

Patience of a Saint, Tolerance of a Mother: Why the Foetus Is not Rejected

Mother and foetus are never genetically identical in an outbred population because the foetus inherits a set of polymorphic genes from each parent. Thus, the foetus differs from the mother in multiple tissue antigens. The pregnant mother's immune system must protect both, the mother and the foetus, against invading pathogens or altered cells while allowing the survival and growth of the allogeneic foetoplacental unit. The mother's immune system is exposed to paternal alloantigens repeatedly during multiple pregnancies. This results in the formation of memory cells and increases the chances of secondary responses to the alloantigens. Nevertheless, the immune system has to be subverted enough to allow these pregnancies to succeed. How the allogeneic foetus avoids immune rejection during pregnancy is one of the enduring enigmas of transplantation immunology. Elucidating the mechanisms involved in the maternal tolerance of the foetus have proved difficult, as most experiments can only be carried out in animal models and then merely extrapolated to humans. Experimental data suggests that multiple mechanisms are responsible for the acceptance of foetal allograft.

- **Lack of MHC expression**: Similar to cells of other immunoprivileged sites, the cells of the trophoblastic surface that envelops the embryo and constitutes the foetomaternal interface lack MHC class I and class II molecules. They also have an increased expression of FasL. The extravillous trophoblast cells, which come in direct contact with maternal blood, express the classical MHC class I molecule HLA-C and the nonclassical MHC class I molecules HLA-E and HLA-G; this allows them to escape both CTL and NK cell attack.
- **Infiltration by NK cells**: The predominant lymphocytes at the foetal implantation site are $CD56^{bright}$ NK cells. The infiltration of the site by NK cells is due to the influence of progesterone, endometrial IL-15, and prolactin; the cells disappear at 20 weeks of gestation. These uterine NK cells express high levels of inhibitory receptors CD94/NKG2 and are thought to influence both the cytokine milieu and innate immune responses at the foetomaternal interface.
- **Cytokine milieu**: A balance of TH1 and TH2 cytokines is also thought to be critical to the survival of the foetus and a successful pregnancy. The local dominance of TH2 cytokines is thought to protect the immunologically foreign foetoplacental unit against CMI and nonspecific innate and inflammatory phagocytic responses. Pro-inflammatory TH1 cytokines, such as IFN-γ and TNF-α, have been shown to be embryotoxic. IL-1 and TNF-α have been reported to regulate trophoblastic cell apoptosis, protease production, and angiogenesis. By contrast, anti-inflammatory TH2 cytokines, such as IL-4, IL-10, and TGF-β, are thought to play a role in preventing the maternal rejection of the foetal allograft. They have been shown to deactivate macrophages and downregulate their cytokine production.
- **Treg cells**: Tregs are postulated to produce immunosuppressive cytokines, such as TGF-β and IL-10, that are responsible for the induction of tolerance.
- **Complement regulation**: Murine trophoblastic embryo cells express the complement inhibitor Crry. This protein is structurally related to DAF and MCP and inhibits the deposition of activated complement components C3 and C4 on the surface of autologous cells. It therefore prevents the inappropriate and potentially destructive effect of the activation of the complement cascade. Human placental cells and trophoblasts do not express Crry but express DAF and MCP instead. Hence, these molecules are thought to play a similar role in protecting the foetus.
- **Tryptophan starvation**: Activated T cells are sensitive to tryptophan concentration; tryptophan starvation induces T cell-cycle arrest and accelerated AICD. **I**ndoleamine 2,3-**dio**xygenase (IDO) is a haem-containing enzyme that catalyses the first step in the oxidative degradation of tryptophan. Maternal cells expressing IDO are found to surround the murine embryo soon after implantation. IDO is also expressed in human placental tissue. IDO-expressing cells are therefore postulated to help in tolerance induction or maintenance by locally decreasing the amount of tryptophan available to activated T cells. Thus, the IDO expressing cells may act as potent immunosuppressive cells.

17.6 Immunosuppressive Therapies

The ultimate aim of any transplantation procedure is the induction of tolerance to the transplant, allowing the graft to survive indefinitely in the absence of immunosuppression. Although this has proven difficult to achieve, the introduction of a number of drugs and strategies has allowed transplantation to become a standard option in the case of organ failure (Figure 17.4). Currently, immunosuppressive drugs are used clinically for three purposes.

- **Induction therapy** is given at the time of transplantation to reduce the likelihood of immediate rejection. Drugs used in induction therapy are broad-spectrum immunosuppressive agents and are often discontinued when the patient goes home from the hospital.
- **Maintenance therapy** reduces the immune system's ability to recognise and reject foreign tissue. Often, combinations of synergistic drugs are used in maintenance. The drugs are chosen on the basis of their ability to interfere with different aspects of the immune response. This allows the use of each

drug at low dosages and reduces drug-related toxicity. This approach allows the immune system to remain functional at levels sufficient to protect the patient against infections and malignancies.

- **Specific treatments** are aimed at treating episodes of acute rejection.

Immunosuppression jeopardises the well-being of transplant patients in two ways. Firstly, there is an increased risk of malignancies. Between 1-5% transplant recipients develop malignancies within a few years of receiving their allografts. This represents an almost 100% increase in risk compared to the general population, and underscores the importance of a functional immune system in controlling neoplasms. Transplant patients are particularly prone to developing B cell lymphomas. Interestingly, there does not seem to be an increased risk of developing other types of cancers (breast, lung, colon, prostrate) that are common in the rest of the population. Moreover, the patient is at an increased risk of infections. Infections are a major cause of morbidity in solid organ transplant patients. The type of infection is often dependent upon the type of transplant — pyelonephritis and cystitis is common in renal transplants and bronchitis is common in lung transplants. The administration of pre- and post-operative antibiotics is therefore necessary to combat infections. Immunosuppressive agents currently used in clinical practice include the following:

- **Calcineurin inhibitors**: **C**yclo**s**porin **A** (CsA) and tacrolimus (FK506) are the cornerstones of successful long-term immunosuppressive regimens. Calcineurin is a phosphatase crucial for IL-2 gene transcription in T cells (chapter 6). Both FK506 and CsA are called *prodrugs* because they must form complexes with cellular proteins, known as *immunophilins*, to exert their effects. FK506 binds FKBP (**FK**506 **B**inding **P**roteins), whereas CsA binds cyclophilin. Both these drugs are highly specific for T cells, and their effects are reversed upon discontinuation. They inhibit the activation of mature $CD4^+$ and $CD8^+$ T cells after they have received activation signals through APCs. The haematopoiesis and maturation of lymphoid stem cells is not affected. Both drugs have been shown to prolong the survival of transplanted hearts, kidneys, and livers in clinical trials, and they have similar renal and hepatic toxicities.
- **IL-2R antagonists**: IL-2R is a complex of noncovalently linked polypeptide chains ($\alpha\beta\gamma$). The α chain (CD25) is expressed on activated T cells. IL-2 activates T cells in an autocrine manner; the binding of the IL-2 produced by an activated T cell to the IL-2R expressed on its surface results in proliferation. Anti-IL2R mAbs bind IL-2R and prevent IL-2 binding. IL-2R antagonists therefore inhibit the proliferation of T cells. Two antagonists — basiliximab and daclizumab — have been successfully used in the induction phase in clinical trials; neither was found to increase the incidence of malignancy or opportunistic infections.
- **Lymphocyte-depleting agents**: These can be used in the induction phase of rejection and to treat steroid-resistant rejection. They are useful in depleting T cells from donor bone marrow prior to engraftment. Being powerful immunosuppressants, they increase the risk of infections and malignancy. Consequently, the use of these agents is limited to less than 21 days, and they are used in conjunction with steroids to reduce adverse reactions.
    - **Anti-CD3 antibody** (OKT3) is a murine mAb against human CD3 that interferes with antigen-binding and signal transduction. Its action is very rapid, with peripheral T cell depletion occurring within minutes of administration. Its effect is reversible; T cell numbers return to normal

after its discontinuation. OKT3 has been used to prevent the acute rejection of heart, kidney, and liver transplants, as well as to prevent GVHD. The administration of the first dose of OKT3 can result in a sudden and massive release of cytokines, leading to manifestations ranging from mild flu-like symptoms to life-threatening shock. Other toxic side-effects include encephalopathy, nephropathy, and hypotension. Repeated usage can result in the development of human anti-mouse antibodies. Although not yet in routine use, anti-CD2 and anti-CD154 antibodies have been demonstrated to selectively deplete T cells as well.

- **ATG** (**A**nti**t**hymocyte **g**lobulin) is a purified Ig preparation of hyperimmune sera of horses or rabbits immunised with human thymocytes. ATG administration results in the depletion of peripheral T cells. Major side-effects include anaphylactic reactions, serum sickness, leucopenia, thrombocytopenia, and nephritis.

- **Antiproliferative agents**: Purine and pyrimidine analogues are powerful antiproliferative agents that interfere with DNA synthesis.
  - Azathioprine (AZT), a purine analogue**,** interferes with the proliferative cycles of B and T cells and prevents effector cell formation. It is thus a powerful inhibitor of primary, but not secondary, immune responses. AZT also suppresses neutrophil generation and macrophage activation. Its use has become less frequent because of potent adverse effects on all rapidly growing cells and because of its hepatotoxicity.
  - **M**ycophenolate **M**o**f**etil (MMF), also a purine analogue, inhibits the enzyme inosine monophosphate dehydrogenase involved in the *de novo* pathway of DNA synthesis. This is the sole pathway of DNA synthesis in both B and T cells. Hence, unlike AZT, MMF is a selective inhibitor of T and B cell proliferation; other cells such as bone marrow cells and parenchymal cells

Transplantation

- Transplantation is the transfer of cells, tissues, or organs from one site to another.
- Autografts and isografts are generally accepted; allografts and xenografts are generally rejected.
- Blood group antigens and major and minor histocompatibility molecules are involved in graft rejection.
- T cells involved in graft rejection are alloreactive; they can recognise peptide:nonself-MHC complexes.
- Two pathways can lead to transplant rejection:
  - direct allorecognition, involving the stimulation of recipient T cells by the MHC (loaded with or without peptides) on donor APCs, and
  - indirect allorecognition, involving the recognition of peptides derived from the donor's MHC in the context of self-MHC molecules.
- $CD4^+$ T cells have a central role in graft rejection.
- CMI is the major effector mechanism of rejection; eosinophils and antibodies may also play a role in the process.
- GVHD is the result of donor T cells attacking recipient tissue cells.
- Immunosuppressive therapies used to increase transplant acceptance include the following:
  - Calcineurin inhibitors, such as CsA and FK506
  - IL-2R antagonists, such as basiliximab and daclizumab
  - Lymphocyte depleting agents, such as anti-CD3 mAb and ATG
  - Antiproliferative agents, such as AZA, MMF and MNA
  - Gene transcription inhibitors, such as corticosteroids and sirolimus

that use the alternative salvage pathway of purine synthesis are spared. MMF is therefore replacing AZT in the treatment of acute rejection. Preliminary results indicate that it may also be of use in the treatment of chronic rejection and may promote tolerance to donor antigens. Its adverse effects are less severe than those of AZT and include diarrhoea and gastritis.

- **M**alono**n**itril**a**mide (MNA) inhibits both B and T cell proliferation by interfering with pyrimidine biosynthesis. It is also thought to interfere with T cell tyrosine kinases. Clinical trials are underway to assess its suitability for use in the treatment of acute rejection.

- **Agents interfering with cytokine gene expression**: Two types of agents belong to this group — those that interfere with the expression of a number of cytokine genes and those that interfere with downstream IL-2 signalling.

Islets in the Right Portals: Successfully Transplanting Islet Cells

Diabetes mellitus and associated complications affect more than 6% of people worldwide (chapter 15). Insulin administration has been the only real treatment for these patients to date. Although islet transplantation had been tried in the past, this approach has not resulted in exogenous insulin-free life for the patient. Less than 10% of patients achieved insulin-independence for periods of a year or more. One of the major problems with this approach was that treatment used to induce immunosuppression often resulted in damage to the insulin-producing β cells or induced peripheral insulin resistance. In 1999, researchers at the University of Alberta in Edmonton, Canada, developed a method of transplanting islet cells that allowed >80% of patients to remain insulin-independent for more than a year. Known as the *Edmonton protocol*, this method has since been used in international multicentre trials in several patients. It involves transplanting well-characterised pancreatic islet cells into portal circulation. The protocol uses islets from two donors; for as yet unexplained reasons, the use of islets from a single donor has failed to result in insulin-independence. The donor and recipient are matched for the blood group antigens and cross-matched to ensure absence of lymphocytotoxic antibodies. More important (and surprising), the protocol does not call for HLA-matching of the donor and the recipient. Islet cells are transplanted as quickly as possible to minimise cold ischemic damage. The protocol uses a xenoprotein-free medium for preparing the islet cells. Normally, media used in tissue culture contain between 5-10% foetal bovine serum. The researchers argued that the use of bovine serum could result in the graft cells being coated nonspecifically with bovine proteins and could target the cells for immediate destruction. Therefore, they used 25% human serum albumin instead. A key element of the Edmonton protocol is avoiding corticosteroids. The protocol uses a combination of sirolimus and low-dose tacrolimus for immunosuppression, instead. Patients are administered an inductive course of daclizumab (anti-CD25 mAb) to enhance graft survival. The combined therapy prevents the activation of T cells and the triggering of an immune response.

Despite the success of the protocol, islet transplantation remains restricted to patients with severe hypoglycaemia and is presently unsuitable for the majority of patients with type I diabetes mellitus. Most patients require two to three procedures before they can achieve complete insulin-independence. Although the risk of malignancies and life-threatening sepsis has been low in the patients treated to date, fears of these complications limit the broader application of this technique (e.g., in patients with a less severe form of diabetes and in children). Also, even though they are less toxic than earlier drugs, the medications used do result in (amongst other things) mouth ulcerations, weight loss, anaemia, and elevated cholesterol. Thus, although the Edmonton protocol is an important step forward in diabetes management, further refinements are needed to improve its safety and applicability.

- **Corticosteroids** are the most commonly used immunosuppressive and anti-inflammatory agents in clinical practice, although because of their adverse side-effects, most regimens try to minimise their use. Corticosteroids inhibit the gene transcription of a number of cytokines (such as IL-1, IL-2, IL-6, IFN-γ, and TNF-α). They also induce lymphocytopaenia by causing the redistribution of lymphocytes from the intravascular space to the lymphoid space, although the reason for this redistribution is not clear. Predinisone, the prototypic agent, is analogous to the major endogenous corticosteroid cortisol (hydrocortisone), except that it is four times more potent than cortisol. Low doses of corticosteroids are used in combination with drugs, such as CsA and MMF, for maintenance therapy, whereas high doses may be used for short terms to treat acute rejections. They may also be used to minimise hypersensitivity reactions to ATG or mAbs. High doses can cause severe side-effects, including hypertension, diabetes, weight gain, osteoporosis, gastrointestinal bleeding, opportunistic infections, cataracts, and poor wound healing.
- **Sirolimus** (rapamycin) is a macrolide antibiotic that is structurally similar to FK506. It binds to FKBP but causes the arrest of T cells in the G1 phase by binding to a unique cellular target, termed mammalian target of rapamycin. The binding of sirolimus to this protein inhibits the activation of $p70^{s6}$ kinase and thus arrests the synthesis of the proteins required for cell-cycle progression. In addition, it also blocks the signals delivered by IL-2, IL-4, and IL-6 to T cells. It thus interferes with late events in the signalling cascade. Although they bind to the same protein, FK506 and sirolimus are found to be synergistic *in vivo*. The use of sirolimus in combination therapy with FK506 and CsA has allowed a reduction in the dosages of all the drugs involved. It also has relatively fewer side-effects, although it can cause hypercholesterolaemia and increase the risk of heart disease. It is often used in maintenance regimens and in treating chronic rejection.

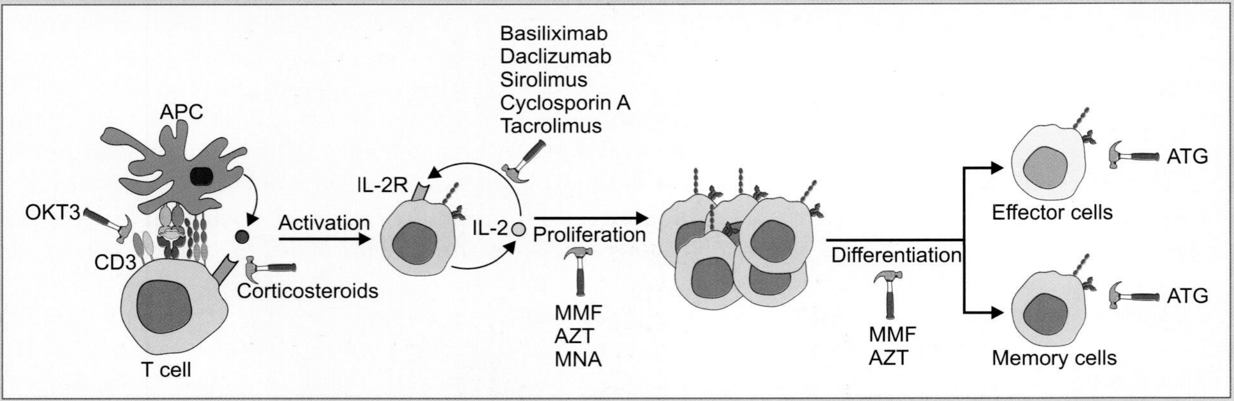

Figure 17.4 Therapeutic agents can be used to target the various stages of a developing anti-graft immune response. OKT3, a murine anti-human CD3 mAb interferes with antigen-binding and signal transduction. Corticosteroids inhibit the gene transcription of a number of cytokines. Various agents can be used to target T cell proliferation. Basiliximab, daclizumab, and IL-2R antagonists block the binding of IL-2 to its receptor, whereas cyclosporin A, tacrolimus, and sirolimus interfere with IL-2 gene expression. Sirolimus also causes the arrest of T cells in the G1 phase. AZT, MMF, and MNA are nucleotide analogues that interfere with T cell proliferation and differentiation. ATG is a purified Ig preparation. Its administration results in the depletion of peripheral T cells.

17.7 Blood Transfusion

Jean-Baptiste Denis in France and Richard Lower in England separately reported the transfusion of blood from lambs to humans in 1667, but it was Karl Landsteiner who opened the door to routine transfusions with his discovery of antigens on human erythrocytes. He described the presence of the A and B blood groups and called the third group O (lacking both A and B). His colleagues Descatello and Sturli reported the fourth (AB) blood group. To date, the international society of blood transfusion recognises 270 blood group determinants ascribed to 26 blood group systems. Most erythrocytic surface antigens are located on integral membrane polypeptides and glycoproteins or, in some cases, membrane glycolipids. Most blood group antigens are proteinic in nature, with the blood group specificity determined primarily by the amino acid sequence. The more well-known ABO blood group antigens are, however, carbohydrate in nature. The membrane antigens are divided into four types on the basis of their integration into the lipid layer. The first two categories consist of antigens that pass the membrane once and are anchored in it by either their NH_2- or -COOH termini (the Indian and Kell group antigens, respectively). A third group consists of antigens that span the membrane several times (e.g., Rh and Duffy group antigens). The last type consists of glycoproteins that are anchored in the membrane by a glycosylphosphatidylinositol anchor. This is a fatty acid anchor that is inserted into the membrane and is attached to the protein through a carbohydrate (Figure 17.5).

Despite the cloning and sequencing of most of the genes involved in erythrocyte antigenicity, relatively little is known about their functions and/or the biological significance of their polymorphisms.

- Some blood group antigens are membrane transporters involved in the transport of molecules across the erythrocyte membrane.
  - The Diego blood antigen is an anion transporter.
  - The Kidd glycoprotein is a urea transporter.
  - The Colton glycoprotein is a water channel.
  - The Rh proteins and Rh-associated glycoproteins have a structure characteristic of transporters. They belong to a family of proteins that facilitate ammonium transport in lower organisms. This has led to the suggestion that Rh proteins could be involved in ammonium transport and that erythrocytes may protect against ammonium toxicity to the brain by transporting ammonium ions to the liver or kidney for metabolism or excretion.

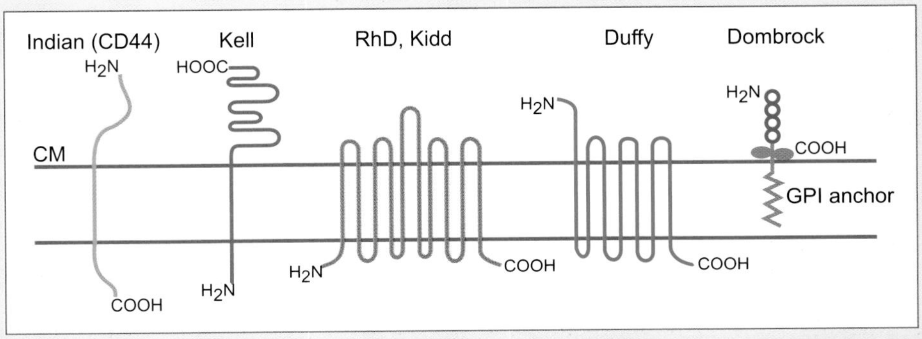

Figure 17.5 Different types of integral cell-membrane proteins confer antigenicity to erythrocytes. The proteins are anchored either at their amino- or carboxy-termini or anchored through a glycosylphosphatidylinositol anchor (adapted from Daniels, *Wiener Klinische Wochenschrift*, 2001, 113:781).

O Bombay, No *Kidd*ing: Blood groups in the Indian subcontinent

The O blood group is the most common in the world, but the incidence of type B is unusually high in Asia; the incidence of the type B group reaches a whopping 21% in North India. The recognition that the O group has the H antigen and that the phenotype is due to *Hh* alleles was a result of a report by Bhende et al. in 1952. They reported the first three individuals who completely lacked A and B antigens but did not belong to the O group. This null phenotype was named *Bombay* after the city where it was first reported. This phenotype, extremely rare in most populations, is relatively easier to find in India; its reported frequency is 1:7600. The Bombay phenotype results from a recessive gene at the H locus (labelled *h*). Persons with the Bombay phenotype do not have the *H* gene and hence cannot produce α (1,2) fucosyl transferase. Without the basic H structure, the A and B enzymes cannot function as they use UDP-N-acetylgalactose or UDP-galactose to convert the H antigen to A and B antigens, respectively. Individuals with the Bombay phenotype have severe transfusion reactions when transfused with the normal (*Hh* genotype) O blood group because of the presence of anti-H antibodies in their serum.

The first occurrence of the Kidd antibody was reported in 1951. It is named after Mrs. Kidd, who had antibodies that caused haemolytic disease in her newborn son. Many individuals of Asian or Polynesian backgrounds, including Indians and Japanese, completely lack the Kidd antigen. The antigen is on the urea transporter, which is found not only on erythrocytes but also on kidney cells. RBCs from the Kidd$^-$ phenotype (called Jk^{a-b-}) have been shown to be resistant to lysis in high concentrations of urea as opposed to normal cells that lyse within a minute of urea exposure.

- The Duffy glycoprotein is a member of the G-protein-coupled superfamily of chemokine receptors.
- The Kell glycoprotein is an endopeptidase and may be involved in the formation of endothelin, a vasoconstrictor.
- The Cromer antigen is the complement-regulatory protein DAF (CD55).

17.7.1 ABO and Rh Blood Groups

Of the myriad of blood groups recognised, only the ABO and Rh groups are matched for routine transfusions. Only those will be covered here.

- The **ABO** (and the related Lewis or Le) blood group antigens are different from other groups described earlier in that these are not proteinic but carbohydrate antigens, that is, the epitopes are the result of oligosaccharides attached to glycoproteins or glycolipids. The genes that govern their polymorphisms do not encode the antigen directly but encode glycosyl transferases that catalyse the addition of the defining monosaccharide to an oligosaccharide substrate. The ABO antigens are attached to a variety of glycoproteins on erythrocyte surfaces. Most are attached to antigens, such as the Diego antigen, glycophorin A, and MN, or to glucose transporters, though they may also be present on any of the other glycoproteins and glycolipids. These carbohydrates at the erythrocyte surface comprise the glycocalyx (or cell coat), and this extracellular matrix of carbohydrates protects the cell from mechanical damage and microbial attack. It may also prevent cellular aggregation.

  Many of the original assumptions about these antigens have now proven erroneous. Originally, four groups had been recognised on the basis of the presence or absence of two epitopes, A and B. The A group consists of two subgroups, A_1 and A_2. Less prevalent than A_1, A_2 was missed in earlier

Table 17.1 ABO Blood Group System

Group	Genotype	Enzyme(s) present		Structure of minimal epitope	Serum antibodies
A	*AA, AH*	A (N-acetylgalactosaminyl transferase) H (α(1,2) fucosyl transferase)	A H	GalNAc-α1-3 ↘ Gal-β1-R ↗ Fuc-α1-2	Anti-B
B	*BB, BH*	B (galactosyl transferase) H (α(1,2) fucosyl transferase)	B H	Gal-α1-3 ↘ Gal-β1-R ↗ Fuc-α1-2	Anti-A
AB	*AB*	A (N-acetylgalactosaminyl transferase) H (α(1,2) fucosyl transferase) B (galactosyl transferase)	A H B H	GalNAc-α1-3 ↘ Gal-β1-R ↗ Fuc-α1-2 Gal-α1-3 ↘ Gal-β1-R ↗ Fuc-α1-2	None
O	*HH, Hh*	H (α(1,2) fucosyl transferase)	H	Fuc-α1-2-Gal-β1-R	Anti-A, anti-B
Bombay	*hh*	α(1,2) fucosyl transferase *absent*	–		Anti-A, anti-B, anti-H

investigations[5]. The group O was originally thought to completely lack antigen but is now known to possess the H antigen. The Bombay group lacks this H antigen and is thus truly O (see the sidetrack *O Bombay, No Kidding*). Structures similar to ABO antigens are found on some plants and microorganisms. Consequently, individuals who do not express these antigens produce anti-A, anti-B, or anti-O antibodies. The transfusion of ABO incompatible blood to a recipient therefore results in severe reactions.

The structure and synthesis of ABO antigens is well established. A core polysaccharide consisting of N-acetylgalactose-galactose-N-acetylglucose-galactose attached to proteins or lipids that are anchored in the erythrocyte cell membrane forms the backbone of the ABH and Le antigens. H[6] antigen is the result of the addition of a fucose to this core polysaccharide by a fucosyl transferase enzyme (FUT1, H glycosyl transferase, or H enzyme). Hexoses attached to the branched polysaccharide H antigen give the A, B, or Le epitopes (Table 17.1). The *A* allele of the *ABH* gene codes for the A enzyme, and the *B* allele codes for the B enzyme. Being codominant, both the A and B epitopes, when present together, combine to yield the AB blood group. The lack of both the enzymes results in an inability to modify the H antigen and generates the O blood group (Table 17.1). The structurally related Lewis groups are also coded by allelic genes that result in four groups (Le^a, Le^b, Le^-, and Le^{a+b+}). Interestingly, the synthesis of these glycolipids does not occur in erythroid tissue, but the epitopes are acquired by erythrocyte membranes.

❐ **The Rh system** is named after the Rhesus monkey. Individuals with the antigen were called Rh^+, whereas those who lacked it were termed Rh^-. The **Rh antigen was later realised not to be a single antigen but a group of alloantigens.** It is one of the most complex blood group systems, with 45 determinants. The D antigen is the most immunogenic of all the protein blood groups. D^- individuals lack the whole RhD protein and therefore mount an immune response to a variety of epitopes on the extracellular domains of the RhD protein.

[5] Type A_1 appears to exist to the exclusion of type A_2 among Australian aborigines and Eskimos and amongst inhabitants of parts of Indonesia, the Pacific islands, Canada, India, and northern USA. The highest incidence is amongst the Blackfoot and Blood Indians in Canada and the USA. Its highest frequency is 50% among the Lapps.

[6] Why H? Because it is a **H**eterophile antigen shared with many plants and microbes.

Sweet Disposition, Sweet Advantages: Secretors and Nonsecretors

ABH and Le alloantigens are present not only on erythrocytes but also on other tissue cells, such as those of the kidneys, liver, and sperm. About 80% of the human population also produces blood group alloantigens in a soluble form, and these are found in saliva, sweat, gastric juices, and so forth. People belonging to this majority are called secretors. The remaining 20% do not produce soluble blood group antigens and constitute the nonsecretors. Alleles of a single gene (*Se/se*) control this trait. The presence of these secreted blood group mucins is thought to influence the type of bacteria that take up residence in the gut — some gut bacteria produce enzymes that allow them to degrade the terminal sugar of ABH antigens and use it as a source of energy. The secreted antigens may also bind and neutralise dietary lectins. Such lectins that can bind ABH antigens could otherwise have a disruptive effect on cells expressing these antigens. Secretors are found to have less dental cavities than nonsecretors, giving rise to the suggestion that salivary ABH antigens may aggregate some oral bacteria, preventing them from colonising dental surfaces. Overall, nonsecretors seem to be at higher risk of infections, including those caused by *Helicobacter pylori*, *Neisseria* spp., and *Candida* spp., and to bacterial urinary tract infections.

Rh incompatibility is found to cause erythroblastosis foetalis, a severe haemolytic disease in newborns. The disease results from Rh incompatibility between the foetus and the mother — when the mother is Rh^- and the foetus Rh^+ (because of an Rh^+ father). If the mother has been previously exposed to Rh antigens (as during previous pregnancies), IgG antibodies from the sensitised mother cross the placenta and coat the foetal erythrocytes. Such coated erythrocytes are destroyed by liver phagocytic cells, causing haemolytic anaemia in the foetus and the newborn infant (see chapter 16).

17.7.2 Potential Transfusion Hazards

Two major hazards have to be considered when transfusing blood or blood products.

- **Transfusion reactions**: The identification of alloantigens on erythrocyte surfaces is termed *blood typing* or *blood grouping* (appendix IV). The matching of donor and recipient blood groups for ABO and Rh antigens is essential for the success of solid organ transplants and of blood transfusion. The transfusion of improperly matched blood can result in severe immunological reactions that may prove fatal (Table 17.2). Although transfusion reactions are often caused by the activation of the recipient's humoral response against antigens on donor erythrocytes, antibodies in the donor blood may cause the lysis of recipient erythrocytes. This reaction is akin to GVHD and results in acute respiratory distress and lung injury or TRALI (**T**ransfusion-**R**elated **A**cute **L**ung **I**njury). Cross-matching of donor and recipient blood ensures the absence of a high titre of anti-recipient antibodies in the donor's blood and helps in avoiding TRALI.
- **Infections**: Asymptomatic blood donors suffering from infections such as AIDS, hepatitis, and malaria can transmit the infection to unsuspecting recipients. Transfusion-related AIDS cases have been reported in many countries in the early 1980s. A number of tests developed since then have dramatically reduced the risks of transfusion-related infections. Proper collection and storage procedures are also essential, as blood may get contaminated with skin flora during venupuncture and/or collection.

Table 17.2 Transfusion Reactions

Type of reaction	Signs and symptoms	Comments
Febrile (nonhaemolytic)	Rise in temperature, rigours, headache, malaise, vomiting	Usually occurs in patients with a history of previous transfusions or pregnancies and is caused by HLA-mismatch
Acute haemolytic reaction (immediate)	Fever, chills, haemoglobinuria, renal failure, hypotension, oozing from the intravenous site, back pain, and pain along infusion vein; symptoms occur within 15 minutes of transfusion	Caused by the administration of ABO- or Rh-incompatible blood; bacterial contamination of the blood can also result in similar symptoms; treatment is aimed primarily at prevention of renal failure
Delayed haemolytic reaction	Weakness, unexplained fall in post-transfusion haemoglobin, elevated serum bilirubin	Usually caused by previous sensitisation to red cell antigens; antibody titre usually too low to be detected in initial screening
Anaphylactic shock	Urticaria, erythema, respiratory distress, hypotension, laryngeal or pharyngeal oedema, bronchospasm	Caused by antibodies to donor plasma proteins or ingested substances (e.g., drugs, foods, chemicals) in donor plasma; treatment consists of administration of antihistamines and epinephrine
Allergic reaction	Rash, urticaria, flushing	Caused by antibodies to plasma proteins or ingested substances; treated with antihistamines
TRALI	Shortness of breath, hypoxaemia, chills, fever, cyanosis, hypotension, pulmonary oedema	Caused by donor antibodies to recipient HLA antigens; treatment aimed at reducing respiratory distress

Donor procedures that help protect blood supplies from infectious agents are listed here.

- Collection of blood only from volunteers with nothing to gain from the procedure
- Extensive screening to ensure that the blood donation will not harm the donor or the recipient; donor health history interpreted strictly to assure highest recipient safety
- Donor physical examination to ensure good health
- Examination of donor's arms to confirm the donor is not a drug user
- Extensive laboratory testing of each unit of blood — blood group and antibody titre is established; blood is also tested for haemoglobin content, presence of hepatitis B antigen, and antibodies to infections such as hepatitis C, HIV-1 and -2, syphilis, and HTLV-I and –II; nonetheless, blood cannot be tested for all possible infectious agents, including ones that can prove fatal (e.g., bovine spongiform encephalitis). There is thus a certain risk inherent in any transfusion

17.7.3 Transfusion Alternatives

The advent of the AIDS epidemic led to the emergence of alternatives to blood transfusions. The emergence of vCJD (**v**ariant **C**reuzfeldt **J**acob **D**isease: the human variant of the mad cow disease) has given a further impetus to their development.

One alternative is to stimulate RBC production by the administration of recombinant human erythropoietin or altered erythropoietin. A second approach is the development of artificial oxygen carriers. Such carriers have multiple advantages. They are nonantigenic, sterilisable, and have an extended shelf-life and unlimited availability. However, many carriers in clinical trials seem to have a short intravascular life (~1-2 hours) and have shown renal toxicity. In addition, they have been shown to lead to pulmonary and systemic hypertension and immune suppression. Perfluorocarbon emulsions that can dissolve any gas (including O_2 and CO_2), raffinose-crosslinked and polymerised human haemoglobin, and recombinant human haemoglobin are being tested in clinical trials. The two principal applications for these artificial oxygen carriers are in patients with trauma and those undergoing surgery, as well as individuals belonging to religious sects or groups (e.g., Jehovah's witnesses) that object to blood transfusions.

Appendix I

CDs Mentioned in This Book

CD (*alternate name*)	Cellular expression	Known functions
CD1a-e	Cortical thymocytes, Langerhans cells, DCs, B cells, cells in the intestinal epithelium, smooth muscles, blood vessels	MHC class I-like molecules associated with β2-microglobulin Present nonpeptide and glycolipid antigens
CD2 (*T11, Leukocyte Function associated Antigen-2; LFA-2*)	T cells, thymocytes, NK cells	Adhesion molecule, binds CD58 (LFA-3) Involved in T cell and NK cell activation
CD3 α, δ, ε, ζ (*T3*)	Thymocytes, T cells, NKT cells	Signal transduction molecules associated with TcR; essential for signal transduction by and cell surface expression of TcR
CD4 (*L3T4, T4*)	Thymocyte subsets, MHC II restricted T cells, monocytes, macrophages, granulocytes	Coreceptor for MHC class II molecules; cytoplasmic domain binds Lck Receptor for HIV gp120
CD5 (*T1, Ly1, Leu1*)	Thymocytes, T cells, majority of B1 B cells	Modulates signalling through TcR and BcR A phenotypic marker for some B cell lymphoproliferative disorders (B cell lymphocytic leukaemia, mantle zone lymphoma, hairy cell leukaemia, etc)
CD7	Pleuripotent haematopoietic stem cells, thymocytes, T cells, NK cells, pre-B cells	Function unclear Marker for T cell acute lymphatic leukaemia and pleuripotent stem cell leukaemias
CD8 (*T8, Leu2*)	Thymocyte subsets, MHC I restricted T cells, DC subsets	Coreceptor for MHC class I molecules; cytoplasmic domain binds Lck
CD9	Pre-B cells and a subset of B1 cells, eosinophils, basophils, monocytes, platelets, activated B and T cells, cells in the brain and peripheral nerves, vascular smooth muscle cells	Mediates platelet aggregation and activation *via* FcγRIIa May play a role in cell adhesion and migration
CD10 (*Common Acute Lymphocytic Leukaemia Antigen; CALLA*)	B and T cell precursors, bone marrow stromal cells	Zinc metalloproteinase that cleaves a variety of pro-inflammatory and vasoactive peptides Marker for pre-B acute lymphatic leukaemia
CD11a (*Leukocyte Function-associated Antigen-1 α chain; LFA-1*)	Lymphocytes, granulocytes, monocytes, and macrophages	Subunit of integrin LFA-1 (associated with CD18) Intercellular adhesion and costimulation Binds to CD54 (ICAM-1), CD102 (ICAM-2), and CD50 (ICAM-3)

CD11b (*Mac-1 α chain*)	Myeloid cells and NK cells, subsets of B and T cells	Subunit of integrin MAC-1/CR3 (associated with CD18) Binds CD54, CD102, complement component iC3b, extracellular matrix proteins, fibrinogen Promotes phagocytosis of iC3b or IgG coated particles
CD11c (*integrin/CR4 α chain*)	Myeloid cells, subsets of B and T cells	Subunit of integrin/CR4 (associated with CD18) Binds fibrinogen Similar in function to CD11b/CD18 with which it acts cooperatively
CD14	Myelomonocytic cells, subset of DCs	Receptor for complex of LPS and LBP; may bind other pathogen-associated molecules
CD16a,b (*FcγRIII*)	Neutrophils, NK cells, macrophages, mast cells	Component of low affinity Fc receptor — FcγRIII Mediates phagocytosis and ADCC
CD18 (*β2 integrin chain*)	Leukocytes	Integrin β-2 subunit, associates with CD11a, b, c Adhesion and signalling
CD19	B cells (except plasma cells), FDCs	Component of the B cell coreceptor complex along with CD21 (CR2) and CD81 (TAPA-1) Cytoplasmic domain binds tyrosine kinases and PI 3-kinase
CD20	B cells (except pro-B cells and plasma cells)	Function unclear; may act as a Ca^{2+} channel Often expressed in B cell lymphomas, hairy cell lymphomas and B cell chronic lymphocytic leukaemia; target of many monoclonal antibodies used in cancer therapy
CD21 (*CR2, Epstein-Barr virus receptor*)	Mature B cells, FDCs	Component of the B cell coreceptor complex along with CD19 and CD81 Binds C3d
CD22	Mature B cells	Inhibitory receptor of BCR signalling; binds sialoconjugates
CD23a, b (*FcεRII*)	B cells, activated macrophages, eosinophils, FDCs, platelets	Low affinity receptor for IgE Regulates IgE synthesis
CD25 (*IL-2R α chain; Tac*)	Activated T cells, B cells, monocytes, thymocyte subset, macrophages, NK cells, a subset of DCs	Low affinity IL-2 receptor, associates with β and γ chains to form the high affinity receptor Activation marker
CD28	T cell subsets, activated B cells, plasma cells	Binds CD80 (B7.1) and CD86 (B7.2) and delivers a costimulatory signal required for activation of naïve T cells

CD31 (*Platelet Endothelial Cell Adhesion Molecule-1; PECAM-1*)	Endothelial cells, platelets, monocytes, granulocytes, NK cells, T cell subsets	Homotypic and heterotypic adhesion molecule Mediates both leukocyte–endothelial and endothelial–endothelial interactions; important in transendothelial migration of leukocytes
CD32 (*FcγRII*)	Monocytes, macrophages, granulocytes, B cells, Langerhans cells, platelets	Low affinity IgG receptor Regulates B cell function Has a major role in immune complex-induced tissue damage
CD33	Myeloid progenitors, monocytes, macrophages	Sialoadhesin
CD34	Haematopoietic precursors, capillary endothelial cells	Adhesion molecule Ligand for CD62L (L-selectin) expressed on T cells; required for T cell entry into lymph nodes
CD35 (*CR1*)	Erythrocytes, B cells, monocytes, neutrophils, eosinophils, FDCs, a subset of T cells	Binds C3b and C4b coated particles, facilitating their phagocytosis Negative regulator of complement cascade
CD40	B cells (except plasma cells), macrophages, DCs, FDCs, activated monocytes, $CD34^+$ haematopoietic cells	Binds CD154 (CD40L) Promotes growth, differentiation, and isotype switching of B cells; promotes cytokine production by macrophages and DCs Rescues germinal centre B cells from apoptosis
CD44	Surface of most cell types, erythrocytes, H isoform (CD44H) expressed on lymphocytes	Cell–cell interaction, adhesion, migration Binds hyaluronic acid and other molecules such as collagen, osteopontin, and matrix metalloproteinases Involved in lymphocyte activation, recirculation and homing, haematopoiesis, and tissue metastasis CD44H binds E-selectin and L-selectin
CD45 (*Leukocyte Common Antigen, B220; T220*)	All haematopoietic cells, except erythrocytes; especially high on lymphocytes	Tyrosine phosphatase that regulates a variety of cellular processes — cell growth, differentiation, oncogenic transformation Essential role in BcR and TcR signalling; augments signalling through these receptors Multiple isoforms formed by the alternative splicing of three exons (A, B, C).
CD45RO	B and T cell subsets, monocytes, macrophages	Presence of the RO isoform is used to distinguish activated/memory T cells from naïve T cells Isoform does not contain either of the three A, B, or C exons

CD46 (*Membrane Cofactor Protein; MCP*)	Haematopoietic and non-haematopoietic nucleated cells	Binds C3b and C4b; allows their degradation by Factor I Receptor for the measles virus and *S. pyogenes*
CD49d (*Very Late Antigen-4; VLA-4*)	A variety of cells including B and T cells, NK cells, monocytes, DCs, granulocytes	Adhesion molecule; binds VCAM-1, MAdCAM-1, fibronectin, thrombospondin Role in lymphocyte homing
CD50 (*Intercellular Cell Adhesion Molecule-3; ICAM-3*)	All leukocytes, Langerhans cells, endothelial cells, APCs	Adhesion molecule Binds integrin CD11a/CD18 (LFA-1), DC-SIGN (CD209) Important in the initial DC–T cell interaction
CD52	Thymocytes, T cells, B cells (except plasma cells), monocytes, macrophages, granulocytes, spermatozoa, epithelial cells lining the male reproductive tract	Unclear
CD54 (*Intercellular Cell Adhesion Molecule-1; ICAM-1*)	Haematopoietic and non-haematopoietic cells	Adhesion molecule Binds CD11a/CD18 integrin (LFA-1) and CD11b/CD18 integrin (Mac-1) Important in the DC–T cell interaction Receptor for rhinoviruses; ligand for RBCs infected with the malarial parasite
CD55 (*Decay accelerating factor; DAF*)	Haematopoietic and non-haematopoietic cells	Interacts with complement components including C3b; inhibits formation of C3 convertase and disassembles C3/C5 convertase Receptor for coxsackie B virus
CD56 (*Neural Cell Adhesion Molecule; NCAM*)	NK cells, subset of T cells, brain, neuromuscular junctions	Adhesion molecule
CD57	NK cells, subset of B and T cells	Adhesion molecule
CD58 (*Leukocyte Function-associated Antigen-3; LFA-3*)	Haematopoietic and non-haematopoietic cells	Adhesion molecule; binds CD2
CD59	Haematopoietic and non-haematopoietic cells	Binds complement components C8 and C9, and blocks Membrane Attack Complex (MAC) assembly
CD62L (*L-selectin, Leukocyte adhesion molecule; LAM*)	B cells, T cells, thymocytes, monocytes, granulocytes, subset of NK cells	Adhesion molecule; binds CD34 Mediates leukocyte rolling interactions with the endothelium

CD69	Activated leukocytes (B and T cells, macrophages, granulocytes, NK cells), Langerhans cells	Involved in early events following leukocyte and platelet activation Earliest inducible cell surface glycoprotein acquired during lymphoid activation
CD72	B cells (except plasma cells)	Unknown; may play a role in regulating signal threshold in B cells
CD79a,b (*Ig-α, Ig-β*)	B cells	Components of BcR complex Required for BcR cell surface expression and signal transduction
CD80 (*B7.1*)	Activated B cells, T cells, macrophages, DCs	Costimulatory molecule Ligand for CD28 and CTLA-4 (CD152)
CD81 (*Target of Antiproliferative Antibody-1; TAPA-1*)	Haematopoietic cells, endothelial and epithelial cells	Component of the BcR coreceptor complex (with CD19 and CD21) Receptor for hepatitis C virus
CD86 (*B7.2*)	Monocytes, activated B cells, DCs	Costimulatory molecule Ligand for CD28 and CTLA-4 (CD152)
CD90 (*CD90.1, thy-1*)	Haematopoietic stem cells, neurons, thymocytes and peripheral T cells	Unclear, may contribute to inhibition of proliferation and differentiation of haematopoietic stem cells
CD91 (*α2-macroglobulin receptor*)	Monocytes, many nonhaematopoietic cells	Endocytosis-mediating receptor
CD94	T cell subsets, NK cells	Associates with NKG2 receptors Depending upon the NKG2 molecule associated with may inhibit or activate NK cells
CD95 (*Fas*)	Activated B and T cells, monocytes, fibroblasts, neutrophils	Binds FasL; induces apoptosis
CD100	Haematopoietic cells including RBCs and platelets, activated T cells, germinal centre but not MZ B cells	Monocyte migration, T–B and T–DC interaction Increases CD3 and CD2 induced T cell proliferation
CD102 (*Intercellular Cell Adhesion Molecule-2; ICAM-2*)	Vascular endothelial cells, resting lymphocytes, monocytes	Adhesion molecule Binds CD11a/CD18 (LFA-1)
CD117 (*c-kit*)	Developmental marker for most haematopoietic cells	Stem Cell Factor (SCF) receptor A proto-oncogene; mutations associated with cancers such as mast cell disease and chronic myelogenous leukaemia

CD122 (*IL-2 receptor β chain*)	T cells, B cells, NK cells, monocytes, macrophages	A critical subunit of IL-2R and IL-15R
CD127 (*IL-7R*)	Lymphoid progenitors, pro-B cells, mature T cells, monocytes	IL-7 receptor α chain, associates with CD132 (IL-2Rγ) to form the high affinity receptor for IL-7
CD134 (*OX40*)	Activated T cells	Costimulatory molecule expressed 24 to 72 hours following T cell activation; binds OX40L (CD252) Member of the TNFR family Enhances T cell survival
CD152 (*Cytotoxic T Lymphocyte Antigen-4; CTLA-4*)	Activated T cells and T_{reg} cells	Receptor for CD80 and CD86 Negative regulator of T cell activation Mutations in this gene are associated with autoimmune diseases such as insulin-dependent diabetes mellitus, Graves' disease, Hashimoto's thyroiditis, and SLE
CD154 (*CD40L, TRAP*)	Activated $CD4^+$ T cells	Binds CD40, α5β1 integrin and αIIbβ3 Induces activation of APCs Induces B cell proliferation; a defect in this gene results in an inability to undergo CSR and it is associated with hyper-IgM syndrome
CD161 (*NKR-P1*)	NK cells, T cells	Unclear; may regulate NK cell cytotoxicity
CD178 (*CD95L; FasL*)	Most T cells, NK cells, neutrophils, breast epithelial cells, microglia, a subset of DCs	Expression upregulated upon activation Induces the trimerisation of surface Fas on target cells initiating apoptotic pathways In mice, defective Fas–FasL system results in generalised lymphadenopathy
CD209 (*DC-SIGN*)	Macrophages, DCs	Adhesion molecules that binds ICAM-3 Enables TcR engagement by stabilization of the DC–T cell contact zone Binds to mannose type carbohydrates commonly found on bacteria, viruses, and fungi; initiates their phagocytosis Receptor for HIV and Hepatitis C virus
CD252 (*OX40L*)	Activated B cells, DCs, vascular endothelial cells	Costimulatory molecule Enhances T_H2 responses

— I —

Appendix II

Cytokines Mentioned in This Book

Cytokine *(Alternate names)*	Major source	Function
Erythropoietin (*EPO, Haematopoietin*)	Renal fibroblasts, liver	Promotes RBC survival Cooperates with other factors in stimulating erythroid progenitors
G-CSF (*Granulocyte CSF**)	Macrophages, bone marrow stromal cells, endothelial cells	Essential for the growth and differentiation of granulocytes especially neutrophils
GM-CSF (*Granulocyte Macrophage CSF*)	Monocytes and macrophages, T cells, mast cells, endothelial cells, fibroblasts	Induces growth and differentiation of granulocytes, monocytes, and DCs
IFN-α, β (*Interferon-α, β, Type I Interferons*)	T and B cells, monocytes and macrophages, fibroblasts, epithelial cells	Critical in antiviral and antitumour defence Increase MHC class I expression on nucleated cells Activate NK cells
IFN-γ (*Interferon-γ, Type II Interferon*)	NK cells, NKT cells, T cells, $T_{H}1$ cells	Pro-inflammatory; critical for innate and adaptive immune responses Activates macrophages, affects activation, growth and differentiation of T cells, B cells, macrophages, and NK cells Promotes $T_{H}1$ pathways
IL[#]-1α, β (*Lymphocyte Activating Factor; LAF, Mononuclear Cell Factor; MCF, Endogenous Pyrogen; EP*)	Monocytes and macrophages, DCs, B and T cells, NK cells, epithelial and endothelial cells	Pro-inflammatory Pleiotropic; involved in various immune responses, inflammatory processes, and haematopoiesis Growth and stimulatory factor for macrophages, B, and T cells Mediator of fever, hypertension; induces acute-phase response
IL-2 (*T Cell Growth Factor; TCGF*)	T cells	Stimulates growth and differentiation of T cells, B cells, and NK cells
IL-3 (*Multi-CSF, Mast Cell Growth Factor; MCGF, Haematopoietic Cell Growth Factor; HCGF*)	Activated T cells, thymic epithelial cells, mast cells, eosinophils	Haematopoietic cell and lymphocyte growth factor; stimulates proliferation of cells of the myeloid lineage (granulocytic, erythroid and monocytic lineages)

IL-4 (*IgE Inducing Factor, B cells SF*[$]*-1; BSF-1*)	NKT cells, TH2 cells, mast cells, bone marrow stromal cells	Promotes growth and development of B cells Induces isotype switching to IgG4 and IgE in humans or IgG1 and IgE in mice Promotes differentiation of antigen-activated naïve $CD4^+$ T cells to TH2 type
IL-5 (*Eosinophil CSF; EoCSF, Eosinophil Differentiation Factor; EDF*)	TH2 cells, mast cells, eosinophils	Induces proliferation and differentiation of eosinophils
IL-6 (*B cell Stimulating Factor-2; BSF-2, Hepatocyte SF; HSF*)	B and T cells, monocytes and macrophages, fibroblasts, endothelial cells, bone marrow cells, stromal cells, astrocytes, adipocytes	Regulates B and T cell growth and function Mediator of fever and acute-phase response
IL-7 (*Lymphopoietin-1*)	Stromal cells of thymus, bone marrow and spleen	Stimulates the differentiation of haematopoietic stem cells into lymphoid progenitor cells Growth factor for B and T cell progenitors
IL-8 (*Neutrophil Activating Factor; NAF; Granulocyte Chemotactic Protein; GCAP*)	Variety of immune and non-immune system cell types including monocytes, lymphocytes, granulocytes, fibroblasts, epithelial and endothelial cells, hepatocytes	Chemokine; chemotactic for and activator of neutrophils Promotes angiogenesis
IL-9 (*T Cell Growth Factor-III; TCGF-III*)	TH cells	Induces erythropoiesis Stimulates proliferation of T cells
IL-10 (*Cytokine synthesis inhibitor factor*)	Monocytes and macrophages, T cells, B cells	Anti-inflammatory Inhibits generation of and cytokine synthesis by TH1 cells Suppresses macrophage and NK cell functions Stimulates proliferation of B cells, thymocytes, and mast cells Enhances IgA synthesis in association with TGF-β
IL-12 (*NK cell stimulatory factor; NKSF, Cytotoxic Lymphocyte Maturation Factor; CLMF*)	Monocytes, macrophages, DCs, B cells	Induces IFN-γ synthesis in T cells and NK cells Important for development of TH1 pathways Enhances NK cell activity
IL-13	TH2 cells, Mast cells, NK cells	Inhibits pro-inflammatory cytokine production by macrophages Induces B cell growth and differentiation Synergizes with IL-4 in isotype switching to IgE

IL-15	Variety of cell types; especially DCs and cells of the monocytic lineage	Promotes NK cell and T cell activation and proliferation and survival of memory CTLs Promotes NK cell and CTL cytotoxicity
IL-17A-D, F	$CD4^+$ T cells, γδ T cells	Pro-inflammatory Induces the production of cytokines (such as IL-6, G-CSF, GM-CSF, IL-1β, TGF-β, TNF-α), chemokines (including IL-8 and MCP-1) and prostaglandins from many cell types (fibroblasts, endothelial cells, epithelial cells, keratinocytes and macrophages) Essential for TH17 pathway
IL-18 (*Interferon-γ Inducing Factor; IGIF*)	Macrophages, Kupffer cells, DCs, B cells, intestinal and airway epithelial cells	Induces IFN-γ production Accelerates differentiation to TH1 type Activates cytotoxic activity of NK cells and CTLs Inhibits angiogenesis
IL-21	T cells	Induces proliferation and cytokine production by activated CTLs and NK cells
IL-22 (*T cell derived Inducible Factor; TIF*)	Activated $CD4^+$ T cells	Unclear; could alter TH1/TH2 balance Induces acute-phase response
IL-23	Activated DCs	Pro-inflammatory, functionally similar to IL-12 Stimulates differentiation to TH17 cells in conjunction with IL-6 and TGF-β
IL-25 (*IL-17E*)	Bone marrow stromal cells, TH2 cells, mast cells	Induces IL-4, IL-5, IL-13, and eotaxin
IL-27	APCs	Regulates B and T cell activity Synergizes with IL-12 to enhance IFN-γ production by T cells
M-CSF (*CSF -1*)	Monocytes, granulocytes, endothelial cells, and fibroblasts, activated B and T cells	Induces proliferation and differentiation of haematopoietic stem cells into macrophages Enhances growth, survival, and differentiation of monocytes
MIF (*Macrophage migration Inhibitory Factor*)	Variety of cell types including activated T cells, hepatocytes, monocytes, macrophages, and epithelial cells	Activates macrophages and inhibits their migration
SCF (*Stem Cell Factor*)	Bone marrow stromal cells, thymic stromal cells, fibroblasts and endothelial cells	Important in haematopoiesis, spermatogenesis, and melanogenesis

TGF-β (*Transforming growth factor-β, TGF-β1*)	Most nucleated cell types and platelets	Anti-inflammatory Inhibits growth of several cell types and induces apoptosis Induces T_{reg} cells Induces isotype switching to IgA In mice, promotes $T_{H}17$ pathways in conjunction with IL-6
TNF-α (*Catchectin*)	Macrophages, lymphoid cells, endothelial cells, fibroblasts, adipocytes, neuronal cells	Pro-inflammatory Induces apoptosis Induces acute-phase reaction Regulates growth and differentiation of a wide variety of cell types
TNF-β (*Lymphotoxin*)	Activated B and T cells, fibroblasts, endothelial and epithelial cells	Promotes fibroblast proliferation Induces terminal differentiation of monocytes Enhances phagocytosis and ROI production by neutrophils

*CSF — colony stimulating factor, #IL — Interleukin, $SF — Stimulating factor

— II —

Appendix III

'Nobel' Immunologists

Year	Recipient(s)	*Citation* and the Prize-winning Discovery
1996	Peter C. Doherty Rolf M. Zinkernagel	*'For their discoveries concerning the specificity of cell-mediated immune defence'* They established that cytotoxic T lymphocytes' coordinate-recognition of self H-2:viral antigen defined the basis of MHC restriction of immune responses in 1974
1987	Susumu Tonegawa	*'For his discovery of genetic principle for generation of antibody diversity'* He established that single immunoglobulin proteins were encoded by separate rearranging genes in 1976
1984	Niels K. Jerne Georges J.F. Kohler Cesar Milstein	*'For theories concerning the specificity in development and control of the immune system and the discovery of the principle of production of monoclonal antibodies'* Jerne was cited for the influence of his theories concerning the development and control of the immune system Kohler and Milstein were awarded the prize for their 1975 discovery of the principle for production of monoclonal antibodies
1980	Baruj Benacerraf Jean Dausset George D. Snell	*'For their discoveries concerning genetically determined structures on the cell surface that regulate immunological reactions'* They were recognized for their separate work over three decades to define genetically the H-2 and HLA molecules that regulate immunological reactions
1977	Rosalyn Yalow	*'For the development of radio-immunoassays of peptide hormones'* She shared the prize with R. Guillemin and A.V. Schally who were awarded the prize for their work concerning peptide hormone production in the brain
1972	Gerald M. Edelman Rodney R. Porter	*'For their discoveries concerning the chemical structure of antibodies'* They received the prize for their work done during the late 1950s and early 1960s that showed that immunoglobulins were composed of covalently bonded two heavy and two light chains
1960	Sir Frank MacFarlane Burnet Peter Brian Medawar	*'For discovery of acquired immunological tolerance'* Burnet was cited for developing the theories of clonal selection of antibody production and for applying this to the concept of acquired immunological tolerance Medawar was awarded the prize for experimentally confirming Burnet's theory, showing that graft rejection is due to an immunological reaction and that tolerance can be built up by injections into embryos

1951	Max Theiler	*'For his discoveries concerning yellow fever and how to combat it'* He was awarded the prize for his contribution to the 1938 development of the universally successful vaccine for yellow fever
1930	Karl Landsteiner	*'For his discovery of human blood groups'* He was cited for serologically defining the ABO blood group system
1919	Jules Bordet	*'For his discoveries relating to immunity'* He was awarded the prize for developing the basis of immune haemolysis of foreign erythrocytes including the involvement of separate heat labile (complement) and heat stable (antibody) components.
1913	Charles Robert Richet	*'In recognition of his on work on anaphylaxis'* He showed that injection of dead or attenuated microbes not only led to specific immunity but that subsequent re-exposure could provoke severe illness or death due to anaphylactic shock
1908	Ilya Ilyich Mechnikov Paul Ehrlich	*'In recognition of their work in immunity'* Mechnikov was cited for developing the cellular theory of immunity, emphasizing a key role for phagocytes Ehrlich was awarded the prize for developing the first general theories of specific immunity and natural self-tolerance. His side-chain receptor theory anticipated Burnet's clonal selection theory by half a century
1901	Emil Adolf von Behring	*'For his work in serum therapy, especially its application against diphtheria'* He was awarded the prize for defining the concept of serum therapy by showing that diphtheria and tetanus exotoxins could be used to raise antitoxins that could be passively transferred to protect against disease

— III —

Appendix IV

Tools of the Trade

Science asks questions and then designs experiments to answer them. The earlier chapters are intended to help in identifying the questions that need to be asked. To be able to design experiments for answering those questions, a basic understanding of available techniques becomes necessary. This appendix is meant as a general aid to understanding common immunological techniques — the tools needed to practice the trade. Igs, with their high specificity and their ability to specifically recognise and bind antigens even in the presence of high concentrations of other molecules, are excellent tools for the detection and/or measurement of virtually any biological molecule. They are almost indispensable in immunological investigations.

Kinetics of Antigen–Antibody Interaction

The basic thermodynamic principles of the monovalent antigen–antibody interaction are the same as those for any reversible bimolecular chemical binding. The reaction between an antigen and its homologous antibody is essentially a reaction between the antigenic epitope and the paratope (or antigen-binding site) of the antibody. Antigen–antibody binding involves multiple noncovalent bonds, such as hydrogen bonds, electrostatic bonds, hydrophobic bonds, Van der Waal's bonds, and salt bridges (Table 4.1, Figure 4.1). Although each single bond is weak, together, the large numbers of bonds formed yield considerable binding energy[1]. Nevertheless, the interacting groups must be closely associated to allow formation of these bonds — the forces of attraction that form these bonds rapidly decline with increasing distance.

Consider the reaction between epitopes at concentration $[E]$ and paratopes at concentration $[P]$. Both the reactants will be in thermal motion and will collide with each other. The rate of complex formation will be governed by the Law of Mass Action.

Epitope–paratope complexes $[x]$ will be formed at a rate proportional to

— the concentrations of the free reagents, that is, $[P-x]$ and $[E-x]$, and

— the rate of effective collisions, k.

The rate of complex formation with respect to time can be given by the equation

$$\frac{d[x]}{dt} = k_1[E-x][P-x] \qquad \text{(I)}$$

The formed complexes will dissociate at a rate (k_2) proportional to the concentration of the complex, and this decrease in concentration with respect to time can be defined by the equation

$$-\frac{dx}{[dt]} = k_2[x] \qquad \text{(II)}$$

The rate of change of concentration of the complex with respect to time can be obtained by combining equations I and II

$$\frac{dx}{[dt]} = k_1[E-x][P-x] - k_2[x] \qquad \text{(III)}$$

When the reaction reaches equilibrium, the rate of the forward reaction (complex formation) is equal to the rate of the backward reaction (complex dissociation), and therefore, $dx/[dt] = 0$.

That is,

$$k_1[E-x][P-x] = k_2[x] \text{ or}$$

$$\frac{[E-x][P-x]}{[x]} = \frac{k_1}{k_2}\frac{[\text{free epitopes}][\text{free paratopes}]}{[\text{complexes}]} \qquad \text{(IV)}$$

1 This is akin to the Lilliputians using numerous ropes, nailed into the ground, to immobilize Gulliver. Each rope was not strong enough alone, but the collective strength of all the ropes kept him bound.

This ratio k_1/k_2 is the equilibrium constant, denoted by K_D. It has the dimensions of concentration and is conventionally expressed in moles (M). It may also be expressed in as K_A, which is the inverse of K_D and has the peculiar dimension of M^{-1} (or litres/Mole). The equilibrium constant (whether K_D or K_A) is a measure of the antibody's affinity for the antigen (*affinity* — liking or attraction). The smaller the K_D, the greater the forward reaction, the stronger the bonds between antigen and antibody, and the more stable the complex. As with any chemical reaction, antigen–antibody binding is affected by temperature, pH, and ionic strength.

Equation III describes an interaction of monovalent reactants (e.g., haptens and Fab fragments). Thus, K_D is a summation of all the attractive and repulsive forces of a reaction between monovalent reactants. In practice, however, the situation is more complex since a multivalent antigen reacts with a bivalent or multivalent antibody. When a multivalent antigen combines with more than one of the antibody's combining sites, the binding strength is considerably more than just the sum of the individual site's binding energies because all antigen–antibody bonds must be simultaneously broken before the reactants can dissociate. The strength with which a multivalent antibody binds to a multivalent antigen is its avidity.

Determination of K_D

As explained, information regarding the antigen–antibody reaction is contained in its equilibrium constant K_D. To estimate the equilibrium constant, at least one of the three entities needs to be separated from the equilibrium mixture (free epitopes, free paratopes, and epitope–paratope complexes), without disturbing the equilibrium. Equation IV (above) can be transformed to its linear form ($y = mx + C$) to estimate the K_D.

Thus,

$$\frac{[x]}{[E-x]} = \frac{[P-x]}{K_D} = \frac{[P]}{K_D} - \frac{[x]}{K_D} \qquad \text{(V)}$$

This equation requires the determination of the amount of bound $[x]$ and free fraction $[E - x]$ of one of the reactants at equilibrium over a range of concentrations, keeping the concentration of the other reactant $[P]$ in large excess. The values can then be fitted in a graph of $[x]$ *vs* $[x]/[E - x]$; the inverse of the slope of this graph is K_D.

- **Equilibrium dialysis** is one of the most unequivocal methods of K_D determination. It is particularly suited to determining the K_D of antibodies to small dialyzable haptens. In this technique, antibodies are retained on one side of a semipermeable membrane, whereas the antigen (hapten) can freely pass through it. K_D is computed by fitting the initial and bound concentrations of the hapten into equation V.
- **Labelled reactant techniques** can be used to determine K_D of large reactants. The major drawback of equilibrium dialysis is that it is only applicable to small haptens (MW<3000 daltons) that can pass through the membrane. To determine the reaction between large antigens and antibodies, one of the reactants (usually the antigen) is labelled, and the complexes formed at equilibrium are separated by various methods (e.g., centrifugation, filtration, precipitation, and adsorption to a solid surface), to obtain data needed for equation V. These techniques presume that both the labelling and the separation of one of the reactants at equilibrium do not alter binding kinetics.
- **Optical biosensors** allow the measurement of biomolecular interactions with minimal distortions because they enable interaction monitoring in real-time (i.e., as they occur). In this method, K_D is measured by immobilising one of the reactants to a nonreactive gel. The kinetics of the interaction at the gel surface is measured by adding the second reactant. The measurements are taken as the reaction proceeds, that is, before it reaches equilibrium, and this allows use of equations I, II, and III.

Table A.1 Precipitin Tests

Test	Application	Method
Interfacial ring test	Antigen identification	Antigen and antibody react in a capillary tube; a ring of precipitate is formed at the zone of equivalence
Precipitation in gels	Determination of homogeneity and identity of reactants	Reactants are added to wells bored in gel and a band of precipitate is formed at the zone of equivalence; the band position depends upon the rate of diffusion and concentrations of the reactants
Immunoelectrophoresis	Identification of antigen components	Components of a complex antigen are separated electrophoretically before addition of the antiserum; the number and intensity of precipitin bands give information about the antigen components
Flocculation tests	Diagnosis of syphilis	A colloidal suspension of an alcoholic extract of beef-heart is used as the antigen; infection-elicited anticardiolipin response is detected in the VDRL and Kahn tests

A. Antibodies as Diagnostic Tools

Antibody binding to its homologous antigen can alter the physical state of the antigen; this property is exploited in a variety of ways in immunodiagnosis.

A.1 The Precipitin Reaction

The complexing of antibodies with soluble antigen results in the formation of insoluble aggregates that precipitate from the liquid. The quantitative precipitin assay was described by Heidelberger in 1897 and is used as a qualitative and semiquantitative procedure for serum antibody detection and measurement. The highest dilution of a serum that yields a positive reaction is its titre. When fixed aliquots of antiserum are added to increasing concentrations of the homologous soluble antigen, first, the amount of the precipitate obtained increases, then a maximum is reached, and eventually, the quantity of precipitate decreases. The curve obtained by plotting the amount of precipitate formed against the concentration of antigen added is called the precipitin curve (Figure A.1). When the unused reactant in the supernatant is correlated with the precipitate formed, three distinct zones are observed:

- the initial antibody excess zone in the ascending limb of the precipitin curve,
- a zone of equivalence at the plateau (when neither excess antigen nor antibody is left in the supernatant; the concentrations of reactants that yield maximum precipitate are *equivalent concentrations*), and
- a zone of antigen excess in the descending limb of the curve.

The Lattice hypothesis explains the nature of the precipitin curve. It assumes that precipitation is a consequence of the growth of aggregates, with each antigen/antibody molecule displaying multiple linkages to other molecules. As the aggregate size reaches critical volume, the complex becomes too large to remain suspended, and spontaneously precipitates. Such large linear aggregates are formed in the zone of equivalence (Figure A.1). By contrast, small complexes that fail to precipitate are formed in the antigen or antibody excess zones. *In vivo* formation of such small antigen–antibody aggregates (termed *immune complexes*) may cause significant pathology (see chapters 15 and 16). With the advent of accurate and easy techniques such as RIA and ELISA, precipitin reaction-based assays are now more of historical interest than of practical use and are summarised in Table A.1.

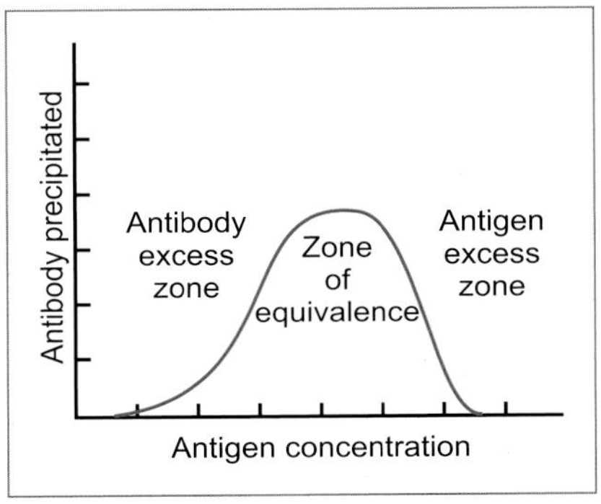

Figure A.1 The precipitin curve.

Table A.2 Diagnostic Applications of Agglutination

Infection (sample)	Antigen	Diagnostic titre
Typhoid (serum)	*S. typhi* H/O	1:80/1:80 active infection 1:80/<1:80 carriers, previous infection <1:80/1:80 usually infection with a related organism A four-fold increase in paired sera taken 10 days apart is confirmatory
	S. typhi Vi	>10 suggestive of carrier state
Paratyphoid (serum)	*S. paratyphi A* H/O *S. paratyphi B* H/O	As above
	S. paratyphi A Vi *S. paratyphi B* Vi	As above
Undulating fever (serum, milk)	*B. abortus* strain 456 O	>1:80 of diagnostic value
Typhus (serum)	*Proteus* spp OXK/OX2/OX19	1:80 diagnostic
Atypical pneumonia (serum)	Human O RBCs	>40 diagnostic High titres may be found in other diseases
Infectious Mononucleosis (serum)	Sheep RBCs	1:10 or more; differential adsorption of antibody by bovine RBCs but not guinea pig kidney extract is confirmatory

A.2 Agglutination

When a particulate antigen reacts with its homologous antibodies, the antigen molecules clump or agglutinate because they are crosslinked by antibody molecules. Agglutination tests have multiple applications.

- **Diagnosis**: Antibodies against infecting organisms generally appear in peripheral blood a week after onset of infection, and they can be demonstrated using the agglutination reaction (Table A.2). The test is therefore used for presumptive diagnosis of the infection. Agglutination is normally carried out in a physiological salt solution (0.8% or 0.15 M NaCl, pH 7.4). Although agglutination tests are easy to perform, their interpretation is complicated by many factors:
  - Low or moderate titres can be observed in healthy individuals for endemic or widely prevalent infections (e.g., cholera or typhoid/paratyphoid in parts of India).
  - Anamnestic responses can result in false positive tests with moderate titres (see section 10.3).
  - Vaccinated individuals may have a high titre of antibodies. To avoid false positive diagnoses, two consecutive tests are performed a week apart — an active infection yields a rising titre.
  - Some antibiotics are immunosuppressive; early antibiotic therapy can lead to false negative results.
  - An anomalous phenomenon, called the prozone phenomenon, is sometimes observed in certain infections (e.g., brucellosis); higher concentrations of the antiserum fail to agglutinate the antigen but lower concentrations do so. The cause of this phenomenon, however, is not clear.
- **Serotyping**: The classification of strains on the basis of their antigenic make-up is termed *serotyping*. The identification and typing of bacteria is a major application of the agglutination reaction.
- **Haemagglutination**: This test is used to determine blood groups (see chapter 17). A drop of the patient's blood is mixed with standard high titre sera and observed for the pattern of agglutination (Table A.3). The donor's serum or plasma is also tested to ensure the absence of antibodies to recipient erythrocytes and to avoid TRALI. If a 1:10 dilution of the donor serum agglutinates recipient

Table A.3 ABO Blood Typing

Blood type	Antiserum A	Antiserum B
A	+	−
B	−	
AB	+	+
O	−	−

erythrocytes, the donor blood is deemed unsuitable for transfusion to that recipient. For anti-D antibodies, the acceptable titre of donor serum is 0.5 IU/mL or lower.

- **Haemagglutination inhibition test**: This is an interesting variation of the haemagglutination test and is used in the diagnosis of some viral infections. Many viruses — picorna to pox — can bind to receptors on human RBCs and agglutinate them. Antiviral antibodies inhibit this agglutination. In the test, dilutions of patients' sera are mixed with aliquots of viral suspension. The mixture is then added to washed test erythrocytes. The highest dilution of the serum that can inhibit haemagglutination is the antibody titre of the serum.
- **The Coombs' test**: Developed by Robin Coombs, this test is used to detect anti-Rh antibodies. Rh incompatibility (an Rh^- mother carrying an Rh^+ foetus) can cause erythroblastosis foetalis, a potentially fatal complication (chapter 17). Detecting these anti-Rh antibodies is difficult. For unexplained reasons, anti-Rh antibodies in maternal serum (termed incomplete antibodies) do not agglutinate erythrocytes and hence cannot be detected by standard agglutination tests. Even if foetal RBCs are coated with such antibodies, Rh antigens are so widely spaced on the erythrocytic surface that the doublet of antigen-bound IgG molecules necessary for C1q activation is not formed (see chapter 3). As a result standard tests cannot demonstrate the presence of these antibodies. To circumvent these problems, Coombs suggested the use of anti-IgG antibodies. Two types of tests are used. The direct Coombs' test allows detection of sensitised foetal erythrocytes. In the test, anti-IgG serum is added to washed foetal RBCs. Anti-IgG antibodies crosslink any maternal antibody coating the foetal erythrocytes and agglutinate them (Figure A.2, left panel). The indirect Coombs' test is used to detect nonagglutinating anti-Rh antibodies in maternal serum. In this test, maternal serum is incubated with Rh^+ RBCs. After thoroughly washing the cells, anti-IgG antibodies are added to the erythrocytes and checked for agglutination (Figure A.2, right panel). The Coombs' test can also be used in the diagnosis of autoimmune haemolytic anaemia (chapter 16).
- **Passive agglutination**: One of the most versatile applications of the agglutination test is to passively adsorb (or conjugate) the antigen to latex or sepharose beads, and use these beads in immunodiagnosis. For example, in the agglutination test for rheumatoid arthritis, IgG-coated latex beads are used to detect anti-IgG antibodies in the patient's serum. The latex pregnancy test uses latex beads coated with the antibodies to human choriogonadotropin. This hormone is found in the urine of pregnant women, and hence, bead agglutination confirms pregnancy.

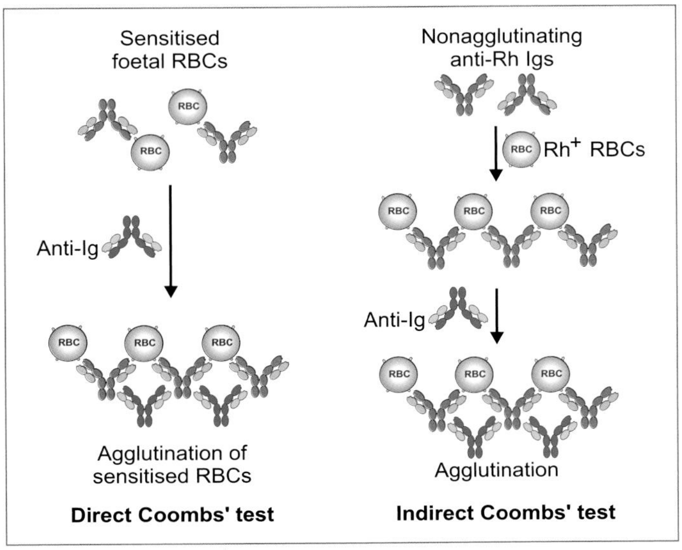

Figure A.2 Direct and Indirect Coombs' test.

A.3 Complement Fixation Test

Although no longer in common use, this test was routinely used in the diagnosis of many viral infections (such as those caused by enteroviruses, myxoviruses, varicella, rubella, and variola viruses) and mycoplasmal infections. The complement cascade is activated when IgG or IgM antibodies react with their homologous antigen (chapter 3). The amount of complement used up in the antigen–antibody interaction is a function of the amount of the two reactants; hence, the extent of complement activation (or *fixation*) can be used to quantify the amount of antigen in the mixture. It is difficult to directly quantify the amount of complement used up in the test, instead, the quantity of leftover complement is determined. A mixture of sheep RBCs (SRBC) and anti-sheep RBCs (αSRBC) is used for this purpose; the sensitised SRBC lyse in the presence of complement. The extent of the SRBC lysis is a measure of the unused complement in the primary reaction, which is in turn a function of the concentration of primary reactants. Historically, the test was used to diagnose any infection that elicited an IgM or IgG antibody response. Because IgE or IgA antibodies cannot fix complement, the test could not be used to demonstrate their presence.

In the test, dilutions of the patient's serum are added to aliquots of the antigen in the presence of a fixed amount of complement (complement in the patient's serum is inactivated by heating to 56°C for 30 minutes, and standardised guinea pig serum is the source of complement). The indicator system (SRBC–αSRBC) is next added to the tubes. If SRBC are not lysed, it indicates that there is no free complement in the system and that all of it was utilised in the primary antigen–antibody interaction. By contrast, SRBC lysis implies the presence of free complement in the system, indicating that the primary reaction was negative. The highest dilution of the patient's serum that fails to give a predetermined degree of lysis is its titre.

A.4 Labelled Antibody Techniques

Most of the early immunodiagnostic techniques that we have described so far used polyclonal antisera raised by repeated injections of whole cell extracts or crude antigen preparations with an adjuvant in animals (e.g., rabbits or mice). As the antisera contained a mixture of antibodies with different specificities and affinities, the preparations had limited applications. Most of these techniques also had low sensitivities (Table A.4).

Kohler and Milstein's development of mAb (monoclonal antibody) technology changed everything. It allowed the large-scale production of antibodies with a defined specificity, and it increased the number of

Table A.4 Relative Sensitivities of Serodiagnostic Assays

Assay	Sensitivity
Immunoelectrophoresis	20-50 mg/mL
Precipitation	1-20 mg/mL
Double diffusion in agar	1-5 mg/mL
Complement fixation	0.5-1 mg/mL
Radial immunodiffusion	0.5-0.05 mg/mL
Agglutination	0.1-0.01 mg/mL
Haemagglutination inhibition	0.001-0.005 mg/mL
RIA	picogram/mL achievable
ELISA	picogram/mL achievable

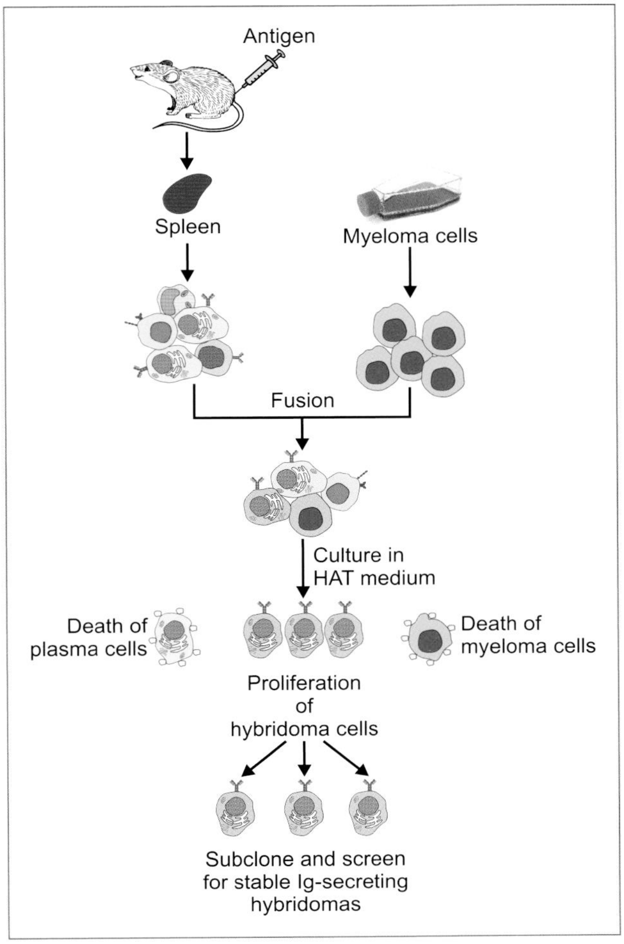

Figure A.3 Major steps in mAb production.

applications significantly. A number of laboratories now routinely produce mAbs. In this technique, the animal (usually a mouse) is injected with an antigen repeatedly to elicit a strong immune response and ensure a large population of plasma cells specific to the antigen. The animal is sacrificed, and the harvested splenocytes are fused with a myeloma cell line that lacks the enzyme hypoxanthine guanine phosphoribosyltransferase. Myeloma cells lacking this enzyme cannot use exogenous hypoxanthine to synthesise purines and die when placed in a purine-deficient medium. The fused cells are grown in a medium containing Hypoxanthine Aminopterin Thymidine (HAT) for a few weeks. Plasma cells, being short lived, die within a few days. Unfused myeloma cells die because they cannot use HAT as a purine source. Hybridoma cells with an unlimited capacity for growth of the myeloma cell and the plasma cell's ability to utilise HAT survive. The survivors are tested for Ig secretion, and the Ig secreting hybridomas are further cultured and expanded for use (Figure A.3). T cell hybridomas can be obtained in a similar manner. For this, it is necessary to establish a clone of T cells from the splenocytes of immunised animals. The T cell clone is fused with a malignant T cell lymphoma line. Such T cell hybridomas recognise a particular peptide loaded on a specific haplotype of MHC class I, class II, or CD1 molecule.

Labelled mAbs are used as probes in the detection of specific molecules in or on cells, tissues, or biological fluids. Antibodies are usually labelled at their Fc region or at the CH1 domain of the Fab region. Two types of assays are in common use. In direct assays, a labelled antibody is allowed to react with an unlabelled ligand, and the residue is examined after washing to remove excess of the labelled antibody. In indirect assays, both the antigen and the antibody are unlabelled, and the bound antibody is detected by labelled anti-Ig preparation. An advantage of the indirect technique is that the same labelled reagent can be used to detect a variety of antigens.

- **Immunofluorescence**: The use of antibodies labelled with fluorescent dyes is one of the most popular methods of detecting antigens in tissues or cells. The dyes chosen for immunofluorescence are usually excited by light of one wavelength (usually blue or green), and they emit a light of a different wavelength in the visible spectrum. The difference in between the two wavelengths is called the Stokes shift (Figure A.4). Greater the Stokes shift, the better suited the dye for use in immunofluorescence. Appropriate filters allow the detection of only the emitted light from the dye. The most commonly used dyes include green light emitting fluorescein, red light emitting phycoerythrin, and texas red. In immunofluorescent microscopy, tagged antibodies are used in diagnosis (e.g., to detect viruses or rickettsiae) or in research (e.g., to detect expression of particular molecules) in tissue sections

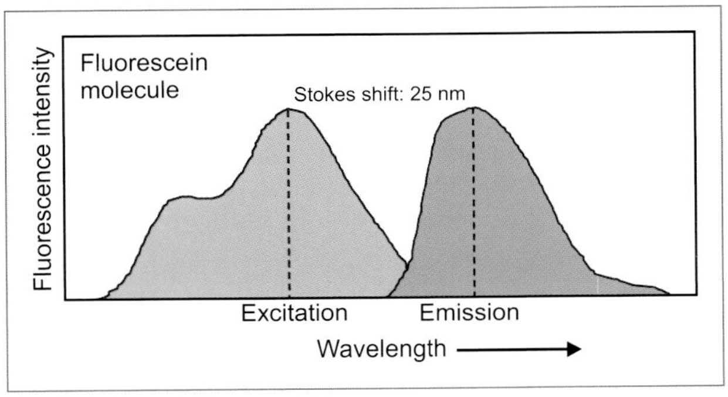

Figure A.4 Immunofluorescent dyes.

or cells. The confocal microscope takes this a step further. It uses computer-aided techniques to produce ultrathin optical sections of a cell or tissue, allowing the study of the localisation of molecules in different organelles of the cell without elaborate sample preparation.

A major application of fluorescent-labelled antibodies is in flow cytometry and FACS (Fluorescent Assisted Cell Sorter) analysis. Flow cytometry allows the study of particular cell subset in a mixed population. Individual cells in the population are tagged with fluorescent-labelled antibodies. The tagging can be direct or indirect (i.e., by the use of anti-Ig antiserum). The labelled and unlabelled cells are mixed with a large volume of saline (called sheath fluid) and forced through a nozzle to create a fine stream of liquid containing cells that are singly spaced. The stream of cells passes through a laser beam. As each cell passes through the laser beam, it scatters the light. Fluorescent antibody-tagged cells fluoresce and emit a light in the visible range. A sensitive photomultiplier tube detects the intensity and polarisation of scattered light as well as fluorescence emission. Both the forward light scatter (FSC; generally over a range of 2° to 15°) and side light scatter (SSC; at 90°) of the light is detected separately and conveys details about the size and granularity of the cell, respectively. By contrast, the fluorescence emission provides information about the extent of the binding of the tagged antibody and, hence, the extent of the expression of molecule of interest. The data from the flow cytometer is usually displayed as a histogram of fluorescence intensity *vs* cell numbers if cells are labelled with a single dye. If two dyes are used for labelling, the data is usually in the form of a scatter plot, where the fluorescence from one dye is plotted against the other (Figure A.5). Each dot in the plot represents a single cell. FACS analysis allows the separation and purification of a subpopulation of cells on the basis of the degree of the binding of labelled antibodies, and flow cytometry instruments

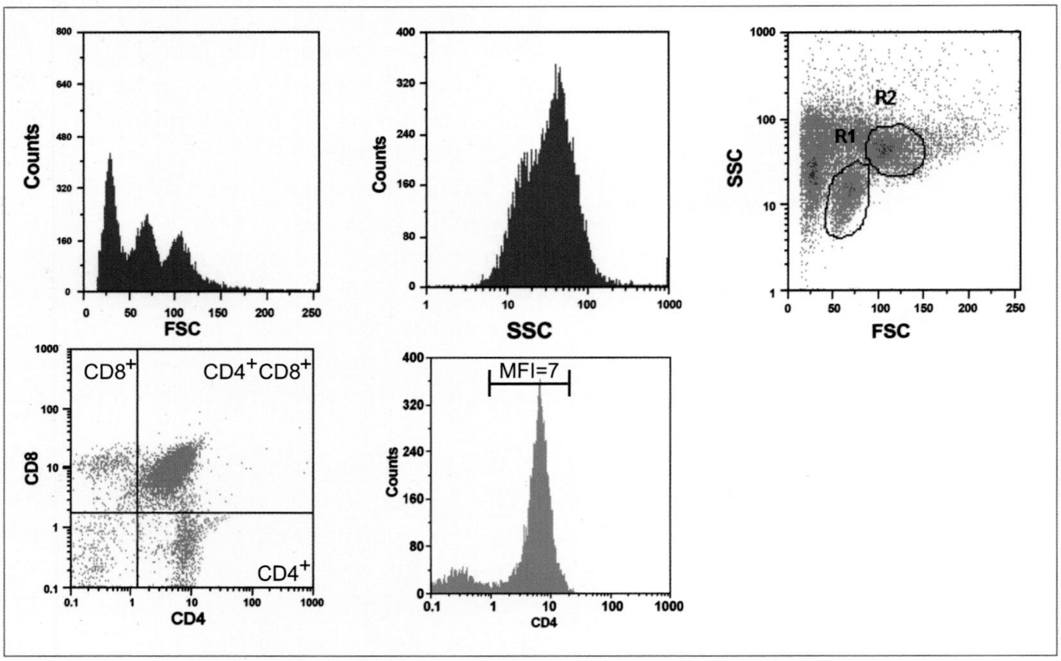

Figure A.5 Flowcytometric analysis. The top panels show histogram (top left and central panels) and dot plots (top right panel) of murine splenocytes. Cell types can be separated by their size and granularity using a FSC and SSC plot (top panels). Each dot in the plot represents a cell. Macrophages (R2), being larger and more granular than lymphocytes, fall in a region separate from lymphocytes (R1) in a FSC *vs* SSC plot. The lower left panel is a dot plot of CD4 and CD8 staining of thymic T cells. Based on the CD4 and CD8 expression, thymocytes can be differentiated into CD8 positive (upper left), CD4 positive (lower right), $CD4^+$ CD^+ double positive (upper right) or $CD4^-CD8^-$ (lower left) cells. The lower right panel is a histogram of the CD4 staining of the cells from the CD4 *vs* CD8 dot plot. The CD4 negative cells fall in the minor peak in the left of the graph. Greater the CD4 expression, more shifted the peak to the right. The extent of CD4 expression is given in terms of MFI (Mean Fluorescence Index).

IMMUNODIAGNOSTIC ASSAY PARAMETERS

- The specificity of an assay is its ability to give a positive reaction with only the ligand, and it is related to the quality of the detecting reagents.
- The detection limit of a test is defined as the lowest concentration of the ligand that gives a significantly different response from the zero concentration response (i.e., the negative control).
- The sensitivity of an assay is the change in response per unit of reactant.
- The precision and reproducibility of an assay are defined by the standard deviation (SD) obtained with multiple readings of the same concentration of reactants; the smaller the SD, the greater the precision and reproducibility.
- The practicability of the assay refers to the speed and ease of use, possibility of automation, and other such factors related to its usage.

have become standard equipment in research laboratories. A latest innovation in flow cytometry is the use of cytometric bead arrays. Each type of bead in a bead array has its own spectral emission, which can be used to capture and quantitate soluble analytes. These bead arrays therefore allow the quantitation of multiple analytes (e.g., cytokines) in the pictogram levels in very small volumes (μL) of the samples.

- **Radioimmunoassay (RIA)**: This highly sensitive method was introduced by Barsson and Yalow to measure insulin concentrations in serum. Although RIA has multiple applications, the most common is in the measurement of peptide hormones in blood and tissue fluids (Figure A.6). RIA is based on radioactive isotope labelling, and both direct and indirect assays are in use. Iodine isotopes are most commonly used for labelling. Several methods are used to separate the labelled complexes from the mixture. Of these, the most common is to conjugate or adsorb the unlabelled antigen to a solid support — plastic tubes, cellulose beads, and so on. The excess labelled ligand can be easily separated from the complexes through simple washing or centrifugation. The degree of radioactivity on the solid support is directly related to the labelled complexes formed and can be used as a measure of the labelled complexes formed.

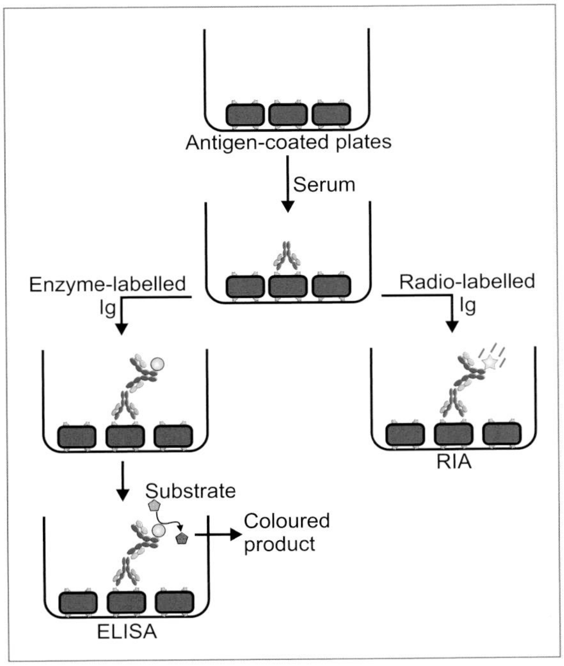

Figure A.6 Schematic representation of Direct ELISA and RIA.

- **Enzyme-Linked Immunosorbent Assay (ELISA)**: Since its introduction by Engvall and Perlman in 1971, the ELISA test has become the most widely used of all immunological tests. It is a quick, sensitive, and specific assay that allows the detection and quantification of antigens (or antibodies) in the picogram range, even in the presence of numerous background proteins (Figure A.6). One of the reactants (usually the antigen) is adsorbed to a solid support. Polyvinylchloride or polystyrene trays (microtitre plates) are most commonly used, although beads or tubes can also be employed. Direct ELISAs are simple to perform but may be limited in their sensitivity. Sandwich ELISAs (indirect ELISAs) are much more sensitive. A high-affinity antigen-specific antibody (called the catching antibody) is adsorbed to the solid support. The solid-bound antibodies bind any antigen present in the solution and concentrate it on the plate's surface. This technique allows the detection of antigen even when it is present at very low concentrations (e.g., cytokines or hormones in secretions).

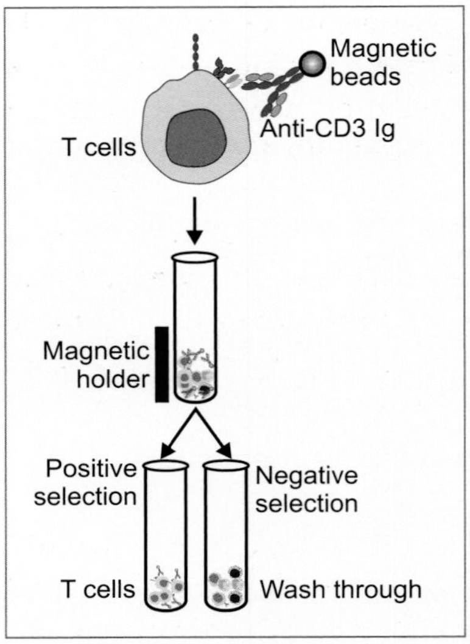

Figure A.7 MACS separation.

A secondary antibody that recognises a separate epitope on the antigen is linked to an enzyme and used as the detecting reagent. The two most commonly used enzymes are alkaline phosphatase and horseradish peroxidase. The amount of the soluble coloured product that is formed is a direct readout of the amount of antigen present.

❒ **Immunohistochemistry**: This technique is essentially similar to ELISA, in that it uses enzyme-linked Igs. The difference is that the reaction occurs in tissues instead of in plates, and the localised deposition of the insoluble coloured product is observed under a light microscope. Tissue-fixation techniques need to be very gentle for successful staining. To avoid problems caused by fixing, the antibodies can be added to frozen tissue sections and developed before fixing.

❒ **Magnetic-Activated Cell Sorting (MACS)**: Antibodies to cell-surface molecules coupled with paramagnetic beads can be used for quick and easy cell separation. In this technique, cells are mixed with antibody-coated paramagnetic beads and run through columns containing materials that attract these beads in a strong magnetic field (Figure A.7). Cells coated with antibody-coupled paramagnetic beads are retained on the column (i.e., they are positively selected) while those not expressing that molecule run through it (i.e., they are negatively selected).

❒ **Immunoblotting and immunoprecipitation**: Immunoblotting allows the detection of one protein in a complex mixture. It also gives information about the molecular size and quantity of the protein. The most common technique of separation is called Western blot[2]. It combines separation of proteins by SDS-PAGE with immunodetection. In SDS-PAGE, cells are lysed with non-ionic detergents such as Triton X-100 or NP40 that disrupt the cell membranes but do not interfere with antigen–antibody interactions. The proteins are dissolved in a strong ionic detergent Sodium Dodecyl Sulphate (SDS). SDS binds to the proteins relatively homogenously and confers a charge that allows their electrophoretic separation. The proteins are loaded on a Polyacrylamide Gel (PAGE) and subjected to an electrical field. Molecular weight markers are normally loaded on the gel to aid in protein identification. The SDS-bound proteins migrate at different rates depending upon their molecular size and can be visualised using protein-staining dyes such as Coomassie blue. The separated protein bands are driven (blotted) into a nitrocellulose membrane by the application of a second electrical field. For unknown reasons, the proteins stick to nitrocellulose and retain their relative positions even when flooded with fluids in the following steps. The blot is then developed using enzyme-linked antibody preparations. The antibodies bind specifically to the protein of interest and upon addition of a colourless substrate yield specific bands of coloured product (Figure A.8).

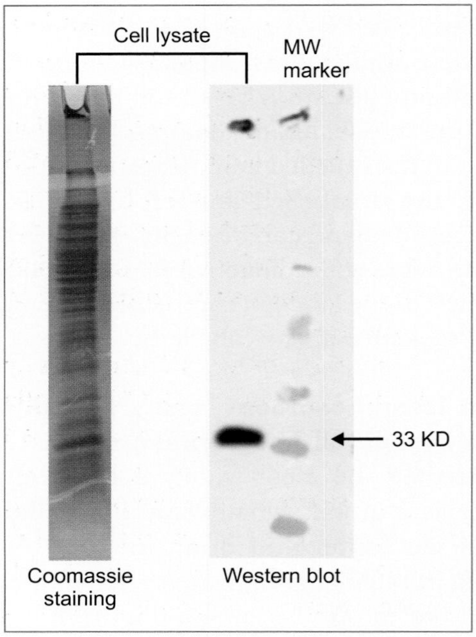

Figure A.8 Western blot. The Western blot technique allows for the detection and quantitation of a protein of interest in the presence of several other proteins. The left panel shows Coomassie staining of a reducing SDS-PAGE gel loaded with the whole cell lysate of a fibroblast line transfected with HLA-DR4 (human MHC class II molecule). The right panel shows a Western blot of a parallel gel that was probed by an anti-HLA-DR4α antibody.

[2] A comparable technique for the detection of specific DNA sequences, the *Southern blot*, was developed by E. M. Southern. This gave rise to the term *Northern blot* for the size separation of RNA and *Western blot* for separation of proteins and proves that scientists can have a sense of humour ☺.

In immunoprecipitation, antibodies linked to solid support such as agarose or sepharose, beads can be used to isolate a radiolabelled protein from a complex mixture. Protein radiolabelling can be done through two principal methods. All the proteins in a cell can be labelled by growing the cells in a medium containing radioactive amino acids. Alternatively, only surface (membrane) proteins can be labelled by radioiodination using a method that does not allow the radioactive iodine to cross the cell membrane. The cells are then lysed and the proteins separated by SDS-PAGE. The size and concentration of the protein can be determined by exposing the gel to an X-ray film. Apart from basic research (e.g., to assess the intracellular concentration and distribution of a protein or to determine whether it undergoes changes in MW because of intracellular processing), Western blotting and immunoprecipitation are frequently used to ascertain if sera contain antibodies to specific proteins.

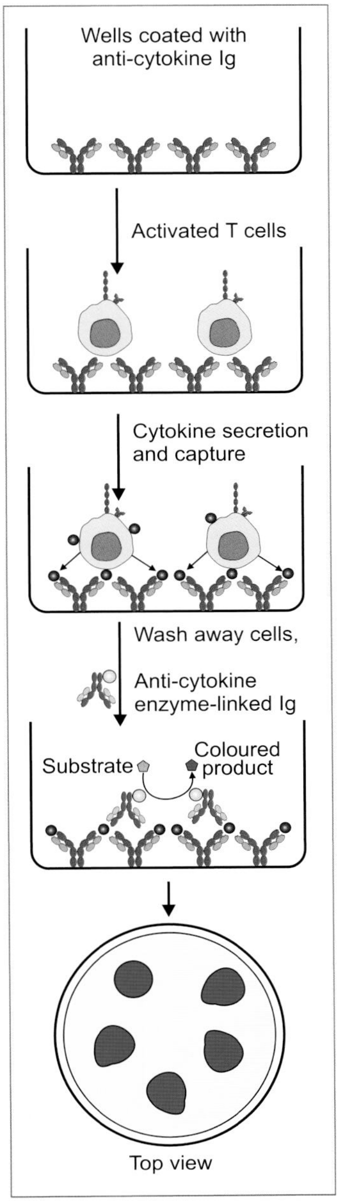

Figure A.9 Schematic representation of a cytokine ELISPOT assay and a top view of the developed ELISPOT well.

B. Cell-based Immunoassays

To perform cell-based immunoassays, it is necessary to separate lymphocytes from other cells. Human lymphocytes can be easily isolated from peripheral blood using density gradient centrifugation. A mixture of the carbohydrate polymer Ficoll and a dense compound, such as metrizamide or sodium diatrizoate (called Ficoll Hypaque® or Ficoll–Histopaque®, respectively), is generally used, and 1:1 diluted blood is layered on the gradient. The heavier erythrocytes and granulocytes sediment at the bottom while a mixture of monocytes and lymphocytes (PBMCs or Peripheral Blood Mononuclear Cells) band at the gradient:plasma interface. Unwanted cell populations can be eliminated by complement-mediated lysis or a population of cells can be positively isolated using antibody-linked magnetic beads, FACS (described in section A.4), or antibodies bound to solid surfaces (panning). Cells can also be differentially eluted from columns containing antibody-coated nylon wool. A 95-99% pure population can be obtained using FACS or magnetic bead cell separation. The isolated populations can then be used to obtain stable cell lines (or clones) for laboratory use.

❒ **ELISPOT**: A modification of ELISA, this technique measures the frequency of cells secreting a particular cytokine or Ig. In cytokine ELISPOT, T cells stimulated with a mitogen are allowed to settle on a plate coated with an anti-cytokine antibody. Any cytokine secreted by a T cell is captured by antibodies in the cell's neighbourhood. The cells are eventually removed, and a second anti-cytokine antibody linked to an enzyme is added to the plate. After addition of the substrate, the presence of the cytokine-secreting T cell is revealed by a spot of coloured product (hence the name; Figure A.9). Given the number of T cells originally added to the plate, the frequency of T cells secreting that particular cytokine can be easily calculated. The frequency of B cells secreting a particular isotype can similarly be determined using anti-isotype antibodies. To determine the frequency of antigen-specific B cells, the plate is coated with the antigen, and the assay is developed using enzyme-linked anti-Ig antiserum.

❒ **Mixed Lymphocyte (or Leukocyte) Reaction (MLR)**: In this technique, used to determine histocompatibility, PBMCs are isolated from the peripheral blood. Donor PBMCs are either irradiated or treated with Mitomycin C to ensure that they will not proliferate but will be able to present antigen. The donor cells are mixed with recipient PBMCs and incubated for 5-7 days. The culture is assessed for either T cell proliferation (usually by ^{3}H thymidine incorporation) or CTL response (by ^{51}Cr release). If the two individuals are

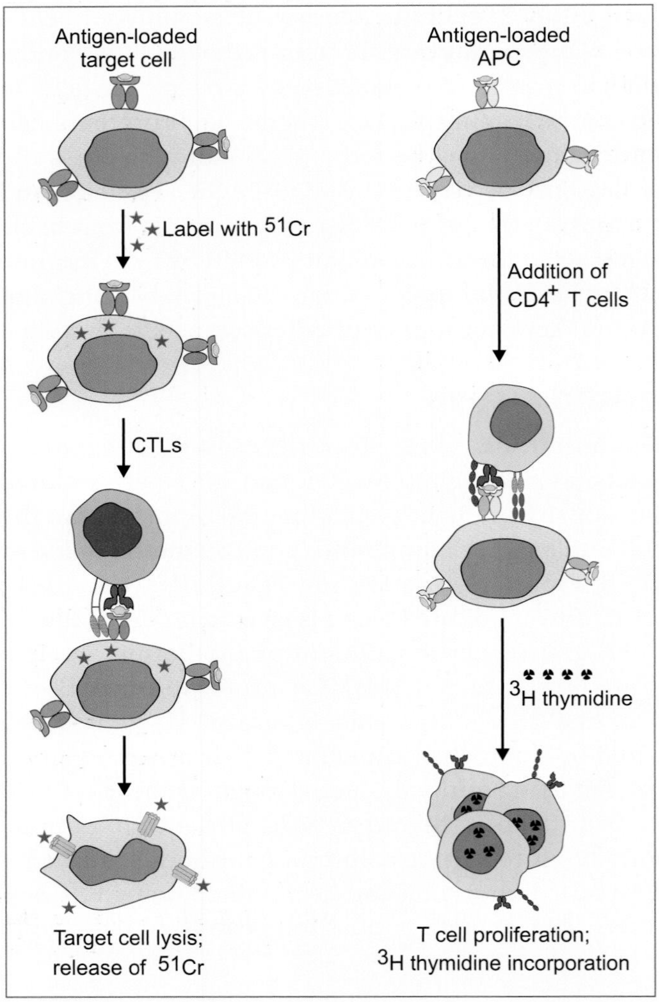

Figure A.10 Schematic representation of CTL (left panel) and CD4⁺ T cell assays (right panel).

incompatible, recipient $CD4^+$ T cells will proliferate in response to the irradiated donor MHC class II alleles. By contrast, $CD8^+$ T cells will proliferate and differentiate to effector CTLs in response to MHC class I alleles. These effector $CD8^+$ T cells cause the lysis of the ^{51}Cr labelled target cells added to the culture (see below).

- **Polyclonal activation assay**: Lymphocytes, whether B or T, proliferate rapidly in response to polyclonal mitogens; phytohaemagglutinin, obtained from red kidney beans, or concanavalin A, obtained from Jack beans, induce T cell proliferation whereas LPS is a B cell mitogen. These mitogens induce lymphocyte proliferation, independent of their antigenic specificity, and they can test the proliferative ability of human (or animal) lymphocytes and determine whether the individual is immunodeficient or immunosuppressed. The extent of proliferation is measured by ^3H thymidine incorporation into the DNA.
- **Assays for CTLs**: Effector $CD8^+$ cytotoxic T lymphocytes kill APCs expressing MHC class I: peptide complexes that they recognise. Therefore, CTL assays are designed to determine target cell lysis and use the propensity of live cells to take up, but not spontaneously release, Na_2CrO_3. The chromate taken up by cells is released only upon cell death. In the assay, the antigen is incubated with autologous (i.e., expressing the same haplotype of MHC class I molecules) target cells to allow MHC class I loading. The cells are labelled with radioactive $Na_2{}^{51}CrO_3$ and incubated with $CD8^+$ T cells. The amount of released radioactive chromate in the supernatant is measured (Figure A.10, left panel).
- **$CD4^+$ T cell assays**: Upon recognition of specific MHC class II:peptide complexes, $CD4^+$ T cells proliferate rapidly and release cytokines such as IL-2 or IFN-γ. Hence, the quantity of cytokine released can be used to measure the extent of $CD4^+$ T cell activation. In the assay, APCs are incubated with the antigen to allow MHC class II loading. Excess antigen is washed off, and the APCs are incubated with autologous $CD4^+$ T cells. The amount of a particular cytokine (whether IL-2 or IFN-γ) released by the T cells can be quantitated by ELISAs. Alternatively, a bioassay using a cytokine-dependent cell line can also be used. The supernatant obtained from the $CD4^+$ T cell assay is added to the cytokine-dependent cells in a 96-well plate. After incubation, ^3H thymidine is added to the wells. The rapidly proliferating test cells incorporate the radioactive thymidine in their DNA. The plate is harvested on a glass filter mat, and the extent of radioactivity is measured with a scintillation counter (Figure A.10, right panel).

C. Use of Animal Models

Animal models are required for understanding a disease process in its entirety. Samples of tissues taken from diseased patients (whether human or animal) cannot adequately explain a disease process, as tissues and cells are influenced by their environment, and tissues may behave differently in isolation than in animals. Animal models allow the testing of new drugs and the design of novel therapies before clinical trials in human patients. This is not to say that alternatives do not exist. In fact, the law requires alternatives (mathematical or computer models, tissue cultures, etc.) to be used before animal models. Virtually all the medical advances of the last century have been made through animal studies. The development of transgenic and knock out mice has furthered the understanding of a variety of diseases, and examples of their use include the following:

- the analysis of the functioning of genes involved in receptor signalling (e.g., TcR/BcR signalling);
- understanding of cellular development and migration through markers such as bacterial β-galactosidase or versions of the jelly-fish green fluorescent protein (GFP) under the control of ubiquitous or highly specific promoter elements;
- insights into tumour development through the introduction of specific oncogenes under the control of tissue-specific promoters; and
- the use of knock out technology to knock out virtually any gene (from cytokines to MHC molecules to *RAG* genes), which helps to study the role of the knocked out gene in immune processes.

In transgenic mice, the genes responsible for a trait or disease susceptibility are chosen, extracted, and injected into fertilized mouse eggs. The embryos are harvested and implanted into the uterus of a surrogate mother. Less than one-third of the embryos develop into healthy pups. The DNA from the pups is tested to ensure the presence of a transgene. The pups will be heterozygous for that gene even if they have the transgene. The pups are then mated with each other. One in four pups will be homozygous for the transgene.

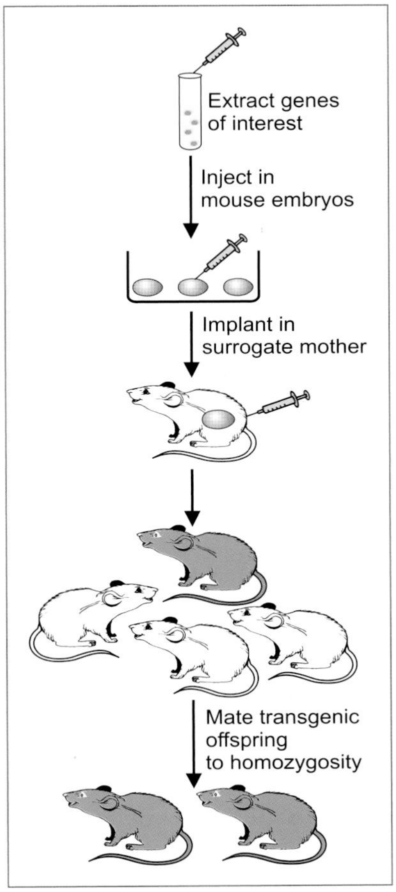

Figure A.11 Major steps in the production of transgenic animals.

To obtain knock out mice, specific genes are targeted for gene disruption by homologous recombination (also called gene targeting). In this technique, an original gene sequence in the mouse genome is replaced by another related, but mutated, sequence. The replacement occurs through homologous recombination. Gene targeting is carried out in mouse embryonic stem cells (ES cells). These cells are derived from a very early (usually male) mouse embryo and can therefore differentiate into all types of cell when introduced into another embryo. The aim is to get the modified ES cells to contribute to the germ line, giving rise to sperm. Some of the spermatozoa produced will carry the desired mutation, and when they fertilize a normal egg, the resultant progeny will be heterozygous for that gene — every cell in that mouse will have a copy of the mutated gene. The mice are mated to obtain homozygous knock out mice. In traditional, constitutive knock out mice, the mutation is present throughout development and in all cells of the adult. Conversely, in conditional knock out mice, other genetic strategies are incorporated to allow mutations to be induced at different developmental stages or in selected cell types. Examples of knock out mice include the SCID mice and lpr mice described in chapters 6 and 15 respectively.

D. Assessment of Immune Status

The immune status of an individual can be determined by various means; the simplest of these is testing serum for the presence of specific antibodies through agglutination tests, ELISA or RIA. For autoimmune diseases, the patient's serum is allowed to react with tissue sections, and the bound antibody is examined by immunofluorescence using labelled anti-human Ig antiserum. The presence of activated T cells can be determined using CTL assays or $CD4^+$ T cell assays. Such an analysis of serum or cells is sometimes inconclusive, and some tests may need to be performed directly on the patient. These tests involve injections of minute quantities of the antigen into the patient. The antigen remains localised at the site of injection and elicits a local response without causing a systemic reaction. Nevertheless, the risk of a systemic reaction needs to be considered before testing.

- **The tuberculin test** or Mantoux test is a skin-prick test that provides information about exposure to the tubercle bacillus. It tests for a DTH response (see chapter 16) to a Purified Protein Derivative (PPD; hence, the test is also known as the PPD test) obtained from *M. tuberculosis*. A small quantity of PPD is injected into the individual's forearm, and the area is observed for up to 72 hours. An area of induration and reddening of ≥10 mm is considered significant. The test is likely to be positive for people in countries with a high incidence of tuberculosis, and a confirmatory chest X-ray is recommended before commencing treatment.
- **The skin test for allergic responses** detects substance(s) responsible for an immediate-type hypersensitivity reaction. A local intracutaneous injection of minute doses of the antigen is administered by pricking the patient's skin, usually that of the forearm, with a small test needle through a minute drop of fluid containing the suspected allergen. A negative control comprises a drop of saline. A positive test may develop within minutes and consists of a weal and flare reaction.
- **The passive cutaneous anaphylaxis test** developed by Ovary is essentially similar to the P–K test described in chapter 16, and is used to demonstrate the presence of sensitising IgE antibodies in serum. In the test, antiserum or purified antibodies are injected intradermally in the test animal (usually a guinea pig). The area is then challenged with a subcutaneous injection of allergen a few hours later. A dye such as Evan's Blue is injected with the antigen to facilitate observation. An irregular circular stained area appears at the site of challenge; the size of the area is an indication of the concentration of antibodies in the preparation.
- **The patch test** is used to determine contact hypersensitivity to drugs, cosmetics, surfactants, botanicals, etc. The test is performed by placing antigen extract in contact with the skin of the arm and then occluding the area for up to 48 hours. A negative control (without the antigen extract) is placed on the other arm to eliminate false-positive reactions. A positive test is characterised by a typical weal and flare reaction at the site of contact with the allergen.

— IV —

Glossary

ABO blood groups — Grouping of human blood according to antigens expressed on red blood cells. Determining the blood group is vital for blood for transfusions; individuals who do not express A or B antigens on their erythrocytes form antibodies against them and this can result in severe disease.

Accessory cells — Cells that aid in the immune response but do not directly mediate specific responses. These cells are often involved in antigen presentation to T cells.

Acute lymphoblastic leukaemia — A highly aggressive, undifferentiated form of lymphoid malignancy.

Acute-phase proteins — Serum proteins that rapidly increase in concentration after an infection. They are part of the first line of host defence.

Adaptive (or acquired) immune response — An immune response generated by the antigen-mediated activation of antigen-specific lymphocytes and characterised by the development of immunological memory.

Adaptor proteins — Small linker proteins that connect cell-surface receptors and downstream members of signalling pathways. All adaptor proteins have SH2 and SH3 domains. They bind to phosphotyrosine residues generated by receptor-associated tyrosine kinases by the SH2 domain and other proteins by the SH3 domain, bringing the two entities in close proximation.

ADCC (antibody-dependent cell-mediated cytotoxicity) — A cytotoxic reaction mediated by effector cells bearing receptors for the Fc region of IgG antibodies. The antibodies bind to the effector cell through their Fc region and link this cell to the target cell by binding to epitopes on the target cell *via* their Fab region.

Adhesion molecules — Molecules that mediate the binding of one cell to another cell or to extracellular matrix proteins; examples include selectins, integrins, and ICAMs.

Adjuvant — A substance that can nonspecifically enhance the immune response to an immunogen when administered along with the immunogen.

Adoptive immunity — Immunity conferred by the transfer of lymphoid cells from an actively immunised donor to a naïve recipient.

Adoptive transfer — The transfer of the ability to participate in an immune response through the transfer of cells of the immune system.

Affinity — The measure of the binding strength between an antigenic determinant and antibody-binding site of the Ig.

Affinity maturation — The increase in average affinity (or avidity) of antigen-specific antibodies with time that is observed in an immune response to TD antigens. Affinity maturation is the result of mutations in the rearranged V region genes of antibodies.

Agglutination — The crosslinking of particulate antigen by antibodies that leads to the separation of the complex (called the agglutinate) from the suspending liquid. The agglutination of RBCs is referred to as haemaglutination.

Agonist peptide — Peptide antigen that activates its specific T cell when presented in context of appropriate MHC haplotype and induces the T cell to synthesise cytokines and proliferate.

AICD (activation-induced cell death) — The phenomenon wherein antigen-activated lymphocytes upregulate the expression of apoptosis-inducing proteins such as Fas or FasL and die unless they receive specific survival signals.

AID (activation-induced deaminase) — A B cell specific enzyme that is essential for the twin processes of somatic hypermutation and class-switch recombination. It increases the diversity of the antibody repertoire by introducing mutations in rearranged V region genes.

AIDS (acquired immune deficiency syndrome) — A disease caused by infection with the human immunodeficiency virus (HIV). Although the patient may harbour HIV for years, full-blown AIDS occurs only after the loss of most $CD4^+$ T cells.

AIRE (autoimmune regulator) — A gene that regulates autoimmunity by promoting the expression of tissue-specific genes in the thymus.

Allele — The intraspecies variation observed in a particular gene locus.

Allelic exclusion — The expression, in heterozygous individuals, of only one allele of the C region of TcR or BcR, in spite of the presence of the alternate allele.

Allergens — Substances that elicit hypersensitivity type I (or allergic) responses.

Allergy — An exaggerated or inappropriate immune response to innocuous environmental substances that causes tissue damage. Most common allergies, such as allergic rhinitis, hay fever, or allergic asthma, are the result of mast cell degranulation because of the allergen-mediated crosslinking of FcεRI-bound IgE antibodies.

Alloantigens — Polymorphisms at the MHC locus that stimulate intense reactions to allografted tissues. Individuals who differ in their alloantigens are allogeneic.

Allograft — A graft of tissue from an allogeneic or nonself donor of the same species. Allografts are invariably rejected unless the recipient is immunosuppressed.

Alloreactive T cells — T cells that can respond to nonself MHC molecules.

Allotype — The protein product of an allele that may be recognised as an antigen by another individual of the same species.

Altered peptide ligand — A peptide that, when loaded on the appropriate MHC molecule, induces only a partial response from T cells specific for the agonist peptide. It is usually closely related to the agonist peptide in its amino acid sequence.

Alternative pathway — One of the pathways of complement activation that is triggered by the binding of complement protein C3b to the surface of a pathogen. Being independent of Ig, it is a feature of innate immunity.

Alveolar macrophage — A type of macrophage found in the alveoli of the lungs.

Anamnestic immune responses — See *Secondary immune responses.*

Anaphylactic shock or systemic anaphylaxis — An allergic reaction to systemically administered antigen that causes circulatory collapse and suffocation due to tracheal swelling. It is the result of the binding of antigen to FcεRI-bound IgE antibodies on mast cells, leading to the degranulation of connective tissue mast cells throughout the body and the disseminated release of inflammatory mediators.

Anaphylatoxins — By-products of the complement cascade (C3a and C5a); they cause mast cell degranulation and smooth muscle constriction.

Anaphylaxis — Classical anaphylaxis is an IgE-mediated response that causes the degranulation of mast cells, vasodilation, and smooth muscle constriction and has varied manifestations including urticaria, asthma, rhinoconjunctivitis, and gastrointestinal symptoms.

Anergy — The state of unresponsiveness of a cell to its homologous antigen. The cells remain in circulation, but are unable to respond to antigenic stimulation. Also see *Clonal anergy.*

Angiogenesis — The process of new blood vessel formation.

Antagonist peptides — Variants of agonist peptides that deliver a negative signal to specific T cells (when presented in the context of appropriate MHC haplotype) and inhibit responses to agonist peptides.

Antibody — A glycoprotein produced by B lymphocytes in response to an antigen that can specifically and reversibly combine with the molecule that induced its formation. Also see *Immunoglobulin molecules.*

Antibody-Dependent Cell-mediated Cytotoxicity — See *ADCC.*

Antigen — A substance (molecule) capable of stimulating the immune system of an animal and reacting with the product of such stimulation. Also see *Immunogen.*

Antigen processing — The degradation of proteins or (glyco)lipids into fragments that can be loaded onto MHC or CD1 molecules for presentation to T cells.

Antigenic determinant — The part of the antigen molecule that is recognised by the antibody or the T cell receptor. Also see *Epitope.*

Antigen presentation — The display of antigen fragments loaded onto MHC or CD1 molecules at the APC cell surface for T cell recognition.

Anti-Ig antibodies — Antibodies that are produced against the constant domains of immunoglobulins from a different species.

Antiserum — Serum from an immune individual that contains antibodies against the molecule used for immunisation.

APCs (antigen-presenting cells) — Cells that internalise and process antigen, load the antigen-derived fragments onto MHC or CD1 molecules and present them to T cells. Professional APCs trigger immune responses by presenting antigen and delivering costimulatory signals to T cells (e.g., DCs, macrophages, eosinophils, and B cells). Nonprofessional APCs present antigen in the absence of costimulatory signals, which causes T cell anergy.

Apoptosis — Cell death, brought about by the activation of an internal death programme. It is characterised by nuclear DNA degradation, nuclear degeneration, and condensation.

Artemis — A protein related to the DNA repair enzyme DNA-PKcs, Artemis is thought to be involved in opening of RAG-generated DNA hairpins during *V*(*D*)*J* recombination.

Asthma — A respiratory disorder, usually of allergic origin, characterised by wheezing that is caused by the constriction of the bronchial tree.

Atopy — The propensity to produce IgE antibodies and develop type I hypersensitivity reactions to innocuous environmental antigens.

Attenuate — Treatment of a pathogen to decrease its virulence and render it incapable of causing disease. Attenuated organisms are used in vaccine preparations.

Autoantibody — An antibody specific for a self-antigen.

Autografts — The transfer of self-tissues from one site to another.

Autoimmune diseases — Diseases caused by an adaptive immune response to self-antigens.

Autologous — That which is part of the same individual.

Avidity — The strength with which a multivalent antibody binds to a multivalent antigen.

Azathioprine — A potent immunosuppressive drug that is converted to its active form *in vivo* and then kills rapidly proliferating cells.

B1 B cells — Type of B cells found in pleural and peritoneal cavities that are self-renewing. B1 B cells respond predominantly to T-independent antigens and do not undergo affinity maturation in response to antigenic stimulation.

B2 B cells — Type of B cells that do not express CD5 or CD9. These are the major type of B cells in the periphery. They respond predominantly to T-dependent antigens. These cells undergo affinity maturation and isotype switching in response to antigenic stimulation.

B cell antigen receptor (BcR) — A complex of a transmembrane immunoglobulin molecule associated with Ig-α–Ig-β (CD79a and b) molecules found on the surface of B lymphocytes. The immunoglobulin molecule confers antigen specificity; Ig-α–Ig-β chains are the signalling subunits. The receptor is associated with a coreceptor complex.

B cell coreceptor complex — A complex of three proteins associated with the B cell receptor (CD21, CD19, and CD81) that is associated with the B cell receptor and that can modulate signal transduction through the receptor.

B lymphocyte — A type of lymphocyte that expresses surface immunoglobulin molecules and, in mammals, matures in the bone marrow. It is the only cell that can produce immunoglobulins.

B1 B cells — A subset of B cells; they generally (but not necessarily) express the CD5 molecule. These cells can be activated to proliferate and differentiate independent of T cell help. About 5% total B cells are of the B1 type.

B2 B cells — The majority of B cells in the body. These cells do not express CD5, require T cell help for proliferation and differentiation, and can undergo affinity maturation.

BAFF (B cell activating factor of the TNF family) — A pro-survival factor secreted by macrophages and DCs that promotes the survival of B cells.

BALT — Bronchial-associated lymphoid tissue.

Bare lymphocyte syndrome — An immunodeficiency disease in which, as a result of one of several regulatory gene defects, MHC class II molecules are not expressed on cells.

Basophils — White blood cells that contain granules that stain with basic dyes. They are thought to be functionally similar to mast cells.

BCG (Bacille of Calmette and Guerin) — An attenuated strain of *Mycobacterium tuberculosis* used in vaccines. It is also used as an adjuvant in animal experiments.

Bcl-2 family of proteins — A group of proteins with pro- and antiapoptotic functions. Members of this family affect apoptosis by decreasing (antiapoptotic) or increasing (proapoptotic) mitochondrial permeability.

Bence Jones Proteins — Free immunoglobulin light-chain dimers found in the serum and urine of patients with multiple myelomas.

BLNK (B cell linker protein) — An adaptor protein found in B cells that recruits signalling molecules to lipid rafts.

Blood group antigens — Surface molecules on red blood cells that are detectable with antibodies from other individuals. The ABO and Rh antigens are routinely tested in blood transfusions.

Blood typing — Determination of the compatibility of the blood groups (usually ABO and Rh) of the donor and the recipient.

Bone marrow — The site of haematopoiesis, the source of B and T cell progenitors, and the site of B cell development in mammals.

Bradykinin — A vasoactive peptide that is produced as a result of tissue damage and that acts as an inflammatory mediator.

Bruton's X-linked agammaglobulinaemia — See *X-linked agammaglobulinaemia.*

Btk — Bruton's tyrosine kinase, an enzyme indispensable for B cell receptor signalling.

Bursa of Fabricus — Found at the junction of the hind gut and cloacae, it is a primary lymphoepithelial organ in birds and is the site of B cell maturation.

C regions (constant regions) — Parts of B and T cell antigen receptors that are distal from the ligand binding site and show minimal variability in amino acid sequences.

C3a/C5a — See *Anaphylatoxins.*

C3b — A product of the breakdown of the complement component C3. It is the principal effector molecule of the complement system.

Calnexin — A chaperone protein found in the endoplasmic reticulum that binds to partly folded members of the immunoglobulin superfamily of proteins and retains them in the endoplasmic reticulum until folding is completed.

CALT — Cutaneous-associated lymphoid tissue.

CAMs (cell-adhesion molecules) — Cell surface proteins involved in binding cells together in tissues as well as in less permanent cell–cell interactions.

Carcinogenesis — The process by which a malignant tumour is formed. Malignant tumour formation requires the accumulation of multiple mutations that allow the growing mass of cells to resist apoptotic signals and survive, proliferate, and eventually invade other tissues.

Carcinoma — A malignant tumour of ectodermal or endodermal tissues such as the skin or epithelial lining of internal organs or glands. More than 80% of all tumours are carcinomas.

Carrier — An immunogenic molecule to which other non-immunogenic antigens or haptens can be conjugated to render them immunogenic. The carrier elicits a T cell response, whereas the hapten elicits an antibody response.

Caspases — A family of closely related cysteine proteases that cleave proteins at the aspartic acid residues. Caspases have a central role in apoptosis.

Cathelicidins — Antimicrobial peptides that are part of the innate immune system.

CD (cluster of differentiation) — A prefix used to identify markers on cells of the immune system. These are groups of monoclonal antibodies that identify the same cell surface molecule. See *Appendix I.*

CD3 complex — A multicomponent constituent of the T cell antigen receptor complex that is responsible for signal transduction. It is composed of five chains — γ, δ, ε, η, and ζ.

CD4 — A coreceptor expressed on a subset of T cells that binds the lateral face of MHC class II molecules. See *Appendix I.*

CD8 — A coreceptor expressed on a subset of T cells that binds the lateral face of MHC class I molecules. See *Appendix I.*

CDRs (complementarity-determining regions; hypervariable regions) — Short segments of about 10 amino acid residues found within the variable region of antigen receptors (or free immunoglobulins) that are in contact with the epitope. CDRs are the most variable part of receptors, and they determine its specificity.

Cell-mediated immunity — See *CMI.*

Centroblasts — Large, rapidly dividing cells that arise from antigen-activated B cells in germinal centres.

Centrocytes — Small B cells that arise from centroblasts in germinal centres.

Chemokines — A family of structurally homologous, chemoattractant cytokines that stimulate the migration and activation of cells, especially phagocytic cells and lymphocytes.

Chemotaxis — The increased directional migration of cells in response to concentration gradients of chemotactic substances.

Chimera — An organism consisting of two or more tissues of different genetic composition, produced as a result of organ transplant, grafting, or genetic engineering. A chimeric protein is produced by splicing together the genetic sequences of two different proteins.

c-Kit (CD117) — See *Appendix I.*

Class switching (class switch recombination) — See *Isotype switching.*

Classical complement pathway — The pathway of the activation of complement that is initiated by antigen–antibody complexes and that proceeds to terminal pathway.

CLIP (class II-associated invariant chain peptide) — A fragment of invariant chain that remains in the peptide-binding groove of the class II molecule, stabilises it, and protects it from premature loading.

Clonal anergy — Functional inactivation of T or B lymphocytes rendering them incapable of responding to their homologous antigen. The mechanisms of clonal anergy induction differ in the two kinds of lymphocytes. Clonal anergy induction is an important mechanism of maintaining self-tolerance.

Clonal deletion — The elimination, in the primary lymphoid organs, of immature lymphocytes that bind self-antigens.

Clonal selection theory — A central paradigm of adaptive immunity, which states that adaptive immune responses derive from individual, self-tolerant, antigen-specific lymphocytes. The specific lymphocytes proliferate in response to antigen and differentiate into antigen-specific effector cells and memory cells.

Clone — Progeny of a single cell that are genetically identical to the parent cell. Clonotypic features are features that are unique to individual cells or members of a clone.

Cluster of Differentiation — See *CD.*

CMI (cell-mediated immunity) — Adaptive immunity conferred by presence of cells, not antibodies.

Coding joint — The joint formed by the imprecise joining of a *V* gene segment to a *(D)J* gene segment in immunoglobulin or T cell receptor genes.

Codominant alleles — Alleles at one locus that are expressed in roughly equal amounts in heterozygotes.

Cold agglutinins — Antibodies that agglutinate at temperatures below 37°C.

Collectins — A family of calcium-dependent sugar-binding proteins (C-type lectins) containing collagen-like sequences that are important in innate immunity, e.g., mannose binding protein.

Combinatorial diversity — The diversity generated by combining separate units of genetic information.

Common lymphoid progenitors — The stem cells that give rise to all lymphocytes. They are derived from pleuripotent haematopoietic stem cells.

Complement — Heat labile serum component consisting of a highly complex group of serum proteins that can attack and bring about the lysis of a variety of pathogens. Complement can be activated by three pathways — the classical, alternative, and MBL pathways.

Complement receptors (CRs) — Cell surface proteins present on various cells that specifically recognise and bind the products of the complement cascade, such as C1q, C3a, C3b, C4a, and C5a.

Complementarity-determining regions — See *CDRs.*

Conformational (discontinuous) epitopes — The epitopes formed from several regions in the primary sequence of a protein that are brought together by protein folding.

Constant regions — See *C regions.*

Contact hypersensitivity — Manifestation of hypersensitivity type IV (delayed-type hypersensitivity) caused by sensitised T cells responding to antigens that are introduced by contact with the skin.

Continuous (linear) epitopes — Antigenic determinants on proteins that are contiguous in the amino acid sequence.

Coombs' test — A test for antibody binding to red blood cells. See *Appendix IV.*

Coreceptor — A cell surface protein that increases the sensitivity of the binding of the antigen to its receptor by binding to associated ligands, e.g., CD4 and CD8 coreceptors that are expressed on T cells and CD19 coreceptor on B cells.

Corticosteroids — A family of drugs related to steroids, such as cortisone, that are naturally produced in the adrenal cortex.

Costimulatory molecules — Molecules expressed on the surfaces of antigen-presenting cells that bind ligands on T or B cells. The engagement of costimulatory molecules along with antigen recognition results in lymphocyte activation.

Costimulatory signal — The additional signal required to induce the proliferation of antigen primed cells, obtained through the engagement of costimulatory molecules.

CpG nucleotides — Unmethylated cytidine–guanine sequences found in bacterial DNA that stimulate the mammalian immune system and are used as adjuvants.

C-reactive protein — An acute-phase protein that binds to the phosphatidylcholine moiety of the cell wall polysaccharide of *Streptococcus pneumoniae.*

Cross-matching — A blood typing and histocompatibility test to ascertain that both the donor and recipient do not have antibodies against each other's cellular antigens.

Cross-presentation — Antigen processing and presentation of exogenous antigens in the MHC class I pathway.

Cross-priming — The activation of cytotoxic T cells by exogenous antigens.

Cross-reactivity — The binding of an antibody to an antigen which is not used to elicit its production.

CTLs — See *Cytotoxic T lymphocytes.*

CTLA-4 — CD152. See *Appendix I.*

Cyclophosphamide — An immunosuppressive drug. It is an alkylating agent that kills all rapidly dividing cells.

Cyclosporin A — An immunosuppressive drug that interferes with T cell antigen receptor signalling. It prevents T cell activation and effector functions.

Cytokines — Small glycoproteins, secreted by cells, that can affect the activity of other cells. Cytokines exert their effect through specific receptors present on target cells. Cytokines have varied roles in immune responses — they can be pro- or anti-inflammatory, and they can boost or suppress an immune response. They are the principal mediators of communication between cells of the immune system.

Cytotoxic granules — Lytic vesicles containing perforin and granzymes. They are a defining feature of cytotoxic T cells and NK cells.

Cytotoxic T lymphocytes (CTLs) — Generally $CD8^+$ MHC class I restricted T cells that can kill infected or transformed host cells.

Cytotoxins — Proteins such as perforin and granzymes that are produced by cytotoxic T cells and NK cells and cause the destruction of target cells.

***D* (diversity) gene segments** — Short DNA segments that join the *V* and *J* gene segments in rearranged immunoglobulin heavy chains and the T cell receptor β and δ chains.

Death by neglect — A mode of apoptosis induction caused by the failure to obtain the necessary survival signals, such as growth factors or costimulatory signals. Thymocytes that fail to bind self-MHC molecules undergo death by neglect.

Decay-Accelerating Factor (DAF) — CD55. See *Appendix I.*

Defensins — Cysteine-rich, cationic, non-enzyme, antimicrobial proteins found on the skin and in neutrophil granules.

Degranulation — In the context of esoinophils or mast cells, the release of pharmacological mediators contained in intracytoplasmic vesicles following appropriate stimulus (e.g., the coligation of Fc receptors expressed on the eosinophil or mast cells surface).

Delayed-type hypersensitivity — See *DTH.*

Dendritic cells (DCs) — Bone marrow derived accessory cells found in the epithelial and lymphoid tissue and characterised by their thin membranous projections. These are the major antigen-presenting cells of the body and are important in initiating an adaptive immune response.

Dermicidin — An antimicrobial protein specifically and constitutively expressed in sweat glands.

Desensitisation — The process of exposing an allergic individual to increasing doses of allergen so as to elicit an IgG (instead of IgE) antibody response.

Diacylglycerol (DAG) — An intracellular signalling molecule generated by phospholipase Cγ-mediated hydrolysis of the membrane phospholipid phosphatidylinositol 4,5-bisphosphae (PIP_2) during the antigen-mediated activation of lymphocytes. It activates cytosolic protein kinase C, which participates in the generation of active transcription factors.

Diapedesis — The movement of blood cells (especially leukocytes) from blood to tissues across blood vessel walls.

Differentiation antigens — Antigens that are produced at a particular stage of cell differentiation. Many differentiation antigens have important functional roles.

Double negative (DN) thymocytes — Immature thymic T cells that lack both CD4 and CD8 expression. They represent ~5% of all thymocytes.

Double positive (DP) thymocytes — T cells at an intermediate stage of development in the thymus that are characterised by the expression of both the CD4 and the CD8 molecules. Approximately 80% of thymocytes are DP.

Draining lymph node — Any lymph node that is downstream of a site of infection and thus receives antigens and microbes from the site *via* the lymphatic system.

DTH (delayed-type hypersensitivity) — Type IV hypersensitivity mediated by $CD4^+$ T cells. The reaction is elicited hours or days after antigenic challenge.

Effector mechanisms — The processes by which pathogens are destroyed and cleared from the body.

Effector lymphocytes — Effector cells that are formed upon the differentiation of naïve lymphocytes and that can mediate the removal of pathogens from the body without the need for further differentiation.

Electrophoresis — The movement of molecules in a charged field.

ELISA (enzyme-linked immunosorbent assay) — An immunoassay in which an antigen that is bound to a solid support is detected by an enzyme that is linked to an antibody. The enzyme converts a colourless substrate to a coloured product. See *Appendix IV.*

ELISPOT — An adaptation of ELISA that is used to detect cells that secrete a particular protein (e.g., cytokines or immunoglobulins). See *Appendix IV.*

Embryonic stem cells — Embryonic cells that have the potential to grow continuously in culture and that retain the ability to differentiate to all cell lineages.

Endogenous — Originating from the organism.

Endogenous pyrogens — Cytokines that can induce a rise in body temperature. Exogenous pyrogens (e.g., LPS) trigger the release of endogenous pyrogens.

Endosomes — Vesicles of the endocytic pathway in which the antigen internalised by antigen-presenting cells undergoes progressive degradation to peptide fragments.

Endothelium — The inner layer of capillaries, blood vessels, etc.

Enhancers — Sequences within genomic DNA that act as cell-specific enhancers of RNA transcription.

Enzyme-linked immunosorbent assay — See *ELISA.*

Eosinophils — Polymorphonuclear leukocytes with azurophilic granules containing pharmacological mediators and vasoactive amines. Crosslinking of the FcεR expressed by these cells leads to their degranulation.

Epithelium — A diverse group of tissues that cover or line nearly all body surfaces, cavities, and tubes. Epithelial layers provide physical protection and containment; all of the body's internal epithelial organs are lined with mucous-coated epithelium.

Epitope spreading — The phenomenon in which responses to autoantigens tend to become more diverse as the response persists; also called determinant spreading.

Epitopes — Regions on a macromolecule (antigen) recognised by an antigen receptor. A B cell epitope is the part of the antigen that is recognised by an antibody. A T cell epitope consists of a small fragment of the antigen that is loaded on an autologous MHC molecule; the fragment:MHC complex is recognised by the T cell antigen receptor.

Epstein-Barr virus — A herpes virus that selectively infects human B cells by binding to complement receptor 2 (CD21). The virus causes infectious mononucleosis and can establish a permanent latent infection. It is associated with some B cell malignancies and nasopharyngeal carcinoma.

Equilibrium dialysis — A technique of antibody affinity determination in which the two reactants are separated by a semipermeable membrane. See *Appendix IV.*

ERBB2 — The protein encoded by the gene *c-erbB2*. ERBB2 is an epidermal growth factor receptor tyrosine kinase found on some breast and ovarian cancer tumours. It is also known as Her2/neu.

Erythroblastosis foetalis — The severe haemolytic disease caused in an Rh^+ foetus by the crossing over the placenta of maternal anti-Rh IgG antibodies.

Exon — Gene segment that encodes a protein.

Experimental allergic encephalomyelitis (EAE) – An inflammatory disease of the murine central nervous system. It is the animal model of the human disease multiple sclerosis.

Extravasation — The movement of cells or fluid from within blood vessels to surrounding tissues.

Fab (fragment, antigen-binding) — A proteolytic fragment of an immunoglobulin molecule that contains the antigen-binding site. It consists of one complete light chain and a part of the heavy chain.

$F(ab')_2$ fragment — Two Fab molecules linked by disulphide bonds between the heavy chains; obtained by the digestion of an antibody with pepsin.

FACS (fluorescence-activated cell sorter) — A machine used to separate cell populations by tagging them with fluorescent antibodies. See *Appendix IV.*

Factors B, D — Serine proteases. The Bb fragment of Factor B triggers the alternative pathway of complement activation by binding C3 and allowing it to be cleaved by Factor D.

Factors H, I — Regulators of the complement cascade.

Fas molecule (CD95) — A member of the TNF receptor family. Ligation of Fas by its ligand can trigger apoptosis in Fas-expressing cells.

Fc (fragment, crystallizable) — The portion of the immunoglobulin molecule responsible for complement fixation and binding to cells. It consists of disulphide-linked carboxy-terminal regions of heavy chains.

Fc receptors (FcRs) — Receptors found on a variety of cell types, e.g., macrophages, B and T cells, platelets, eosinophils, mast cells, and basophils that bind the Fc regions of immunoglobulin molecules. The cytoplasmic domains of most FcRs have signalling motifs. FcRs mediate many of the cell-dependent effector functions of antibodies.

FcεRI — A type of FcR expressed on eosinophils, basophils, and mast cells that binds IgE with high affinity. Antigen-mediated coligation of FcεRI-bound IgE results in the degranulation and release of pharmacological mediators that are responsible for some of the symptoms of immediate-type hypersensitivity.

FcγR — A type of FcR that binds IgG antibodies. FcγRI are high affinity receptors expressed on neutrophils and macrophages that mediate phagocytosis; FcγRII are low affinity receptors that mediate various functions.

FDCs (follicular dendritic cells) — Cells of uncertain origin, found in the lymphoid follicles, that are characterised by long branching processes that make intimate contact with many B cells. They are crucial for the process of affinity maturation.

Ficolins — Activators of alternative pathway of complement activation that are characterised by the presence of both, a collagen-like domain and a fibrinogen-like domain.

First-set rejection — Allograft rejection by a recipient who has not previously received a graft or has not been exposed to alloantigens from the same donor.

FK506 (tacrolimus) — An immunosuppressive drug that is similar in activity to cyclosporin A.

FLICE (FADD-like interleukin-1 beta-converting enzyme) — A protease component of the death-inducing signalling complex that is initiated by Fas binding.

FLIP (FLICE-inhibitory proteins) — Potent inhibitors of Fas-mediated apoptosis that are expressed in lymphoid tissues. Cells overexpressing viral FLIPs are protected from the apoptotic cell death.

Flow cytometry — A technique for the analysis of cell population phenotypes through the use of a flow cytometer. See *Appendix IV.*

Freund's adjuvant — An oil-in-water emulsion used as an adjuvant in animals. Complete Freund's adjuvant (CFA) contains cells or cell walls of *Mycobacterium tuberculosis*, whereas the incomplete adjuvant is only an oil-in-water emulsion, without mycobacterial cell walls.

FRs (framework regions) — Relatively invariant sequences of the variable regions of antigen receptors. The FRs are responsible for the molecular architecture of the molecule, whereas the CDRs (hypervariable regions) are responsible for antigen contact.

Fyn — A type of tyrosine kinase.

G proteins — Intracellular proteins that bind GTP and convert it to GDP in the process of cell signal transduction. Heterotrimeric G proteins are receptor-associated; small G proteins (e.g., Ras) act downstream of transmembrane signalling events.

GALT (gut-associated lymphoid tissues) — Lymphoid tissues that are closely associated with the gastrointestinal tract; they include the tonsils, Peyer's patches, and intraepithelial lymphocytes.

GATA-3 — A TH2-cell specific transcription factor that is the master regulator of TH2 differentiation. GATA-3 induces heritable remodelling of the IL-4 locus and promotes the expression of several TH2 cytokines, thereby committing the cell to the TH2 lineage.

Gene therapy — The correction of a genetic defect through the introduction of a normal gene into bone marrow or other cell types.

Germline hypothesis — A theory of antibody diversity that proposes that all the genes needed to generate the antibody repertoire are found in the fertilised ovum.

Germinal centres — The sites of B cell isotype switching, somatic hypermutation, antigen-driven selection, and differentiation. Found in secondary lymphoid tissue during B cell responses to TD antigens, germinal centres are formed around follicular dendritic cell networks when activated B cells migrate into lymphoid tissue.

Germline diversity — In the context of antigen receptors, receptor diversity caused by the inheritance of multiple gene segments that encode V domains. Antigen receptor diversity that is generated during gene rearrangement or after receptor gene expression is termed somatically generated.

Goodpasture's syndrome — A type of autoimmune disease in which antibodies are produced to the type IV collagen of the basement membrane.

Graft — A tissue or organ removed from one site and placed at another. A graft is considered rejected when the grafted tissue is destroyed by recipient's adaptive immune system.

Graft *versus* host disease — See *GVHD.*

Granulocyte-macrophage colony-stimulating factor (GM-CSF) — A cytokine involved in the growth and differentiation of myeloid and monocytic lineage cells, including dendritic cells, monocytes and tissue macrophages, and cells of the granulocyte lineage.

Granulomas — Palpable nodules formed at sites of chronic inflammation as a result of delayed-type hypersensitivity responses. Granulomas can be triggered by persistent infectious agents. They consist of a central area with macrophages fused into multinucleate giant cells surrounded by T lymphocytes.

Granulysin — An antimicrobial protein found in the cytotoxic granules of cytotoxic lymphocytes (e.g., CTLs and NK cells).

Granzymes — Serine esterases found in the granules of cytotoxic T cells and NK cells. They induce nuclear fragmentation.

Graves' disease — An autoimmune disease in which antibodies against the thyroid stimulating hormone receptor cause overproduction of the thyroid hormone, resulting in hyperthyroidism.

Gut-associated lymphoid tissues — See *GALT.*

GVHD (Graft *versus* host disease) — A disease resulting from an attack on recipient tissue by the donor T cells present in a graft.

H chain — See *Immunoglobulin molecule.*

Haematopoiesis — The generation of cellular elements of blood (erythrocytes, leukocytes, platelets) from haematopoietic stem cells.

Haplotype — A set of genetic determinants located on a single chromosome.

Hapten — A molecule that can bind to an immunoglobulin molecule without eliciting an immune response. Haptens have to be conjugated to carrier molecules to elicit an immune response.

Helper T cells — See *TH cells.*

Heterophile antigens — Antigens or epitopes shared by unrelated (widely divergent) species.

HEVs (High endothelial venules) — Specialised venules in lymphoid tissues through which lymphocytes migrate. HEV cells express special receptors that allow transmigration of lymphocytes.

Hinge region — The flexible region of the immunoglobulin molecule that joins the Fab and Fc fragments. The flexibility of this region allows the molecule's Fab region to adopt a wide range of angles for efficient interaction with epitopes on the antigen.

Histamine — A vasoactive amine stored in mast cell granules. It causes vasodilation and smooth muscle contraction.

Histocompatibility antigens — MHC molecules expressed on the surfaces of tissue cells. They mediate the rejection of allogeneic grafts.

HIV (human immunodeficiency virus) — A retrovirus of the lentivirus family; it causes AIDS.

HLA (human leukocyte antigen) — The genetic designation for human MHC antigens.

HLA-DM and HLA-DO — In humans, molecules involved in loading peptides onto MHC class II molecules. The murine equivalents are termed H2-M and H2-O.

Hodgkin's lymphoma — An immune system tumour derived from mutated B lineage cells.

Homeostasis — A generic term used to describe the status of physiological normality.

Homing receptor — A receptor that directs various populations of lymphocytes to particular lymphoid or inflammatory tissues.

HPA (hypothalamus–pituitary–adrenal) axis — The classical neuroendocrine system, comprising the paraventricular nucleus in the hypothalamus, the anterior pituitary gland, and the adrenal glands. It responds to stress, and its final products, corticosteroids, which target components of the limbic system, particularly the hippocampus. The HPA axis functions through the interaction of the nervous and endocrine systems; the nervous system regulates the endocrine system, and endocrine activity modulates the activity of the central nervous system.

Humanised antibodies — Designer monoclonal antibodies produced by genetic engineering. These are human antibodies containing murine hypervariable regions of a desired specificity.

Humoral immunity — Humoral immunity is the immunity conferred by the presence of antibodies in body fluids.

Hybridomas — *In vitro* cell lines created by fusing two cell types, one of which is a tumour cell. The fusion of myeloma cells with plasma cells yields B cell hybridomas, which are the source of monoclonal antibodies. The fusion of myeloma cells with T cell clones yields T cell hybridomas, which are used to detect specific peptide:MHC haplotype complexes.

Hyperacute graft rejection — An immediate (minutes to hours) reaction against allogeneic grafts caused by preformed antibodies that react with antigens on the graft.

Hypereosinophilia — The presence in blood of an abnormally large number of eosinophils.

Hypersensitivity — An exaggerated and inappropriate immune response to innocuous antigens; it is usually detrimental to the health of the individual. Hypersensitivity type I reactions are due to the IgE-mediated degranulation of mast cells, type II reactions are due to the binding of IgG antibodies to tissue antigens, type III reactions are the result of the formation of small antigen–antibody complexes, and type IV reactions are $CD4^+$ T cell mediated.

Hypervariable regions — See *CDRs.*

ICAMs (intercellular adhesion molecules) — The cell surface ligands for leukocyte integrins. The ICAMs are crucial to the binding of lymphocytes and other leukocytes to cells, including antigen-presenting cells and endothelial cells.

Iccosomes — Antigen-coated liposome-like particles consisting of antigen, C3b or its fragments, and immunoglobulins derived from membranes of follicular dendritic cells of lymphoid follicles early in a secondary or subsequent antibody response.

ICOS (inducible T cell costimulator) — A costimulatory molecule expressed on activated T cells. The binding of ICOS to its ligand, known as ICOSL, enhances T cell responses.

IDDM (insulin-dependent diabetes mellitus) — An autoimmune disease caused by the destruction of β cells of the pancreatic islets of Langerhans. It results in a reduced or total lack of insulin production. IDDM is characterised by an inability to metabolise glucose and by increased glucose levels in blood and tissue fluids.

Idiotopes — Antigenic determinants on the variable region, usually the CDRs, of immunoglobulins or T cell receptors. A collection of idiotopes yield the idiotype of the antibody.

IFNs — See *Interferons.*

Ii (invariant chain) — A chaperone molecule synthesised in the endoplasmic reticulum, along with MHC class II molecules. Ii promotes the folding and assembly of the class II molecules and helps in their transport to peptide-loading compartments. It also protects the peptide-binding groove of the class II molecule during its synthesis and transport.

ILs — See *Interleukins.*

Immature B cells — Those B cells that have had their heavy and light chain V region genes rearranged and that express surface IgM but not surface IgD.

Immediate hypersensitivity — See *Allergy.*

Immune complex — A multimolecular complex of antibody molecules bound to antigen molecules. These complexes can greatly vary in size. The deposition of immune complexes on blood vessel walls or in glomeruli can lead to disease.

Immune deviation — A term used to describe the conversion of an immune response from one dominated by $T_{H}1$ to that dominated by $T_{H}2$ type (or *vice versa*).

Immune modulation — A general term encompassing various alterations in an immune response.

Immune-privileged sites — Sites of the body where immune responses are limited or constitutively suppressed, e.g., the brain, the testes, or the eyes. Immune responses at such sites are often *deviant*, i.e., they are non-inflammatory, as opposed to the pro-inflammatory responses observed at non-immunoprivileged sites.

Immune response — The response mounted by a host organism to defend itself against a pathogen.

Immune response genes — See *Ir genes.*

Immune system — A system comprising the tissues, cells, and molecules involved in adaptive immunity. The term is sometimes used to describe the totality of host defence mechanisms.

Immune tolerance — See *Tolerance.*

Immunity — The ability to resist infections. It is derived from the Latin *immunis* — exempt from duty to the state.

Immunisation (active immunisation) — The deliberate provocation of an adaptive immune response through the introduction of an immunogen. By contrast, passive immunisation is the administration of immunoglobulins or immune serum.

Immunoblotting (Western blotting) — A technique for protein identification which uses gel electrophoresis, followed by transferring (blotting) of the separated proteins to nitrocellulose membranes and then the detection of particular proteins by specific antibodies. See *Appendix IV.*

Immunodeficiencies — Inherited or acquired disorders in which some aspect(s) of host defence are absent or functionally defective.

Immunodiffusion — A technique for the detection of antigen or antibodies which allows them to interact in clear agar gel. See *Appendix IV.*

Immunoelectrophoresis — The identification of antigens through separation on the basis of their electrophoretic mobility and then detection by precipitation in gel.

Immunofluorescence — A technique for detecting molecules expressed by cells that uses antibodies labelled with fluorescent dyes. See *Appendix IV.*

Immunogen — Any molecule that can elicit an adaptive immune response upon introduction (usually by injection) into a person or an animal.

Immunoglobulin (Ig) molecule — An antibody molecule. Each Ig molecule is composed of a pair of H chains (~50 KD) and L chains (~25 KD) that associate through disulphide and noncovalent interactions to yield a Y-shaped molecule. Each H chain consists of three (or four) constant domains and one variable domain; the L chain consists of one each of a constant and a variable domain. Each immunoglobulin monomer can combine with two antigenic determinants. The membrane immunoglobulin (mIg), expressed on the surface of B cells, contains an additional transmembrane domain.

Immunoglobulin domain — A three-dimensional structure found in many proteins of the immune system. It is about 110 amino acids in length and consists of two layers of β-pleated sheets and an internal disulphide bond.

Immunoglobulin superfamily — A family of proteins having at least one immunoglobulin-like or immunoglobulin domain.

Immunohistochemistry — A technique for the detection of tissue antigens that uses antibody-linked enzymes. See *Appendix IV.*

Immunological ignorance — The mechanism of nonresponsiveness to a self-antigen, even when the antigen is expressed in organs or tissues and when T cells capable of responding to the antigen exist in the body. It could be due to levels of antigen expression that are too low for T cell activation or to the inability of lymphocytes to gain access to the site of antigen expression.

Immunological memory — The phenomenon of second or subsequent antigen encounters eliciting speedier and more effective protective immune responses.

Immunology — The study of all the aspects of host defence against infectious challenge and of the adverse consequences of such challenge.

Immunoprecipitation — The detection of soluble proteins through the use of specific antibodies. See *Appendix IV.*

Immunoreceptor tyrosine-based activation/inhibitory motifs — See *ITAMs/ITIMs.*

Immunoregulation — The ability of the immune system to sense and regulate its own responses.

Immunosuppression — The inhibition of one or more components of the adaptive or innate systems, either as a result of disease or of drug administration.

Immunotoxins — Cytotoxic agents generated by conjugating an antibody with a toxin.

Inflammation — The local accumulation of fluid, plasma proteins, and leukocytes at a site of injury, infection, or local immune response.

Inhibitors of apoptosis (IAPs) — Members of a family of antiapoptotic proteins that are conserved across several species. IAPs are the only known endogenous inhibitors of caspases, and several of them are regulated *via* the transcription factor NF-κB.

Innate immunity — A form of immune defence that is active in the early phases of the host's response to an infectious challenge. It is an intrinsic system of host defence that discriminates between groups of related pathogens. It responds similarly and with equal intensity to a pathogen, even upon repeat exposures. Unlike the adaptive immune response, it is not directed against any particular epitopes of an antigen.

Inositol-(1,4,5)-trisphosphate (IP_3) — One of the cleavage products of phospholipase Cγ-mediated hydrolysis of phosphatidylinositol-(4,5)-bisphosphate) (PIP_2) during lymphocyte activation. It releases calcium ions from intracellular stores in the endoplasmic reticulum.

Insulin-dependent diabetes mellitus — See *IDDM.*

Integrins — Heterodimeric cell surface proteins involved in cell–cell and cell–matrix interactions. Some integrins are also involved in pathogen recognition and phagocyte activation.

Intercellular adhesion molecules — See *ICAMs.*

Interferons (IFNs) — Cytokines produced by the cells of the immune systems of most animals in response to viruses, bacteria, parasites, and tumour cells. They are important in innate and adaptive immune responses.

Interleukins (ILs) — Originally used to describe cytokines that are produced by or act on leukocytes. The term is now used generically, irrespective of the source or target.

Introns — The gene segments between exons; they do not encode a protein.

Invariant chain — See *Ii.*

Ir (Immune response) genes — The genes that determine the intensity of the immune response to a particular antigen. They encode MHC class II molecules.

ISCOMs — Adjuvants consisting of immune stimulatory complexes of antigen held within a lipid matrix.

Isotype switching — The phenomenon wherein the class of antibodies produced by a B cell in the early phases of the immune response is changed in the later phases of the response (from IgM to IgG/IgA/IgE) without affecting the antigen-binding specificity of the antibodies produced. Class switch recombination is the molecular mechanism that results in isotype switching. In this, the rearranged immunoglobulin V gene segment recombines with one of the downstream C region genes, and the intervening C genes are deleted.

Isotypes — Classes of immunoglobulins, recognised on the basis of the constant regions of the heavy or light chains. Five major isotypes (IgM, IgD, IgG, IgE, and IgA) are recognised on the basis of heavy chain constant regions and two (κ and λ) on the basis of the light chain constant region. Each isotype is encoded by a distinct constant region gene.

ITAMs (immunoreceptor tyrosine-based activation motifs) — A pair of tyrosine-containing motifs (tyrosine–X–X–leucine/isoleucine, where X is any amino acid) that are found in the cytoplasmic tails of activatory receptors. These are the sites of tyrosine phosphorylation and the association of other phosphotyrosine-binding moieties involved in receptor signalling.

ITIMs (immunoreceptor tyrosine-based inhibitory motifs) — Tyrosine-containing motifs (isoleucine/valine–X–tyrosine–X–X–leucine) found in the cytoplasmic tails of inhibitory receptors. These motifs recruit phosphatases to the receptor site; the phosphatases remove the phosphate groups added by tyrosine kinases.

IVIg (intravenous Ig) — A pooled immunoglobulin preparation obtained from the plasma of donors and used in passive immunisation.

J chain — A small, disulphide-bonded peptide found in the tail pieces of multimeric immunoglobulins (IgM and IgG).

J region — Short coding sequences between the variable and constant gene segments in all immunoglobulin and T cell receptor loci. They are combined (with or without the *D* segment) with the variable gene segments during lymphocyte development.

Jaks — See *Janus kinases.*

Janus kinases (Jaks) — A family of intracellular tyrosine kinases involved in the signalling cascade of cytokines. These kinases phosphorylate proteins known as STATs and the signalling pathway is often referred to as Jak/STAT pathway.

Junctional diversity — The diversity created in antigen-specific receptors during the process of the joining of *V*, *D*, and *J* gene segments.

K cells — A subset of NK cells; K cells express the low-affinity receptor for IgG.

Kappa chain — See *κ chain.*

KIRs (killer cell Ig-like receptors) — Inhibitory receptors expressed by NK cells that recognise MHC class I molecules.

Knock out mice — Mice with targeted disruption of one or more genes created by homologous recombination. The technique enables the study of the function of specific gene product(s).

Kupffer cells — Phagocytic cells that line the hepatic sinusoids.

L chain — See *Immunoglobulin molecule.*

Lactoferrin — An iron-chelating protein found in milk and in mucosal secretions such as tears.

LAK (lymphokine-activated killer) cells — *In vitro* activated NK cells that are used in cancer therapy. LAK cells are generated by exposing patients' NK cells to high doses of IL-2.

Lambda chain — See *λ chain.*

Langerhans cells — Phagocytic immature dendritic cells found in the epidermis.

LAT (linker of activation in T cells) — An adaptor protein with several tyrosines that become phosphorylated by the tyrosine kinase ZAP-70. LAT becomes associated with membrane lipid rafts and coordinates downstream signalling events in T cell activation.

Latency — In the context of viral infection, the state in which the virus enters a cell but does not replicate; when reactivated, the virus can replicate and cause disease.

LBP (LPS binding protein) — An acute-phase protein with the ability to bind and transfer bacterial LPS to CD14.

Lck (lymphocyte-specific protein tyrosine kinase) — A Src family tyrosine kinase associated with the cytoplasmic tails of CD4 and CD8 coreceptors in T cells.

Lectin pathway — See *MBL pathway.*

Lentiviruses — A group of retroviruses that include the human immunodeficiency virus, HIV. They cause disease after a long incubation period; infection can take years to become apparent.

Leukaemia — A malignancy of bone marrow blood cell precursors in which a large number of malignant cells are found in the bone marrow and the blood. Leukaemia can be lymphocytic, myelocytic, or monocytic.

Leukocyte common antigen — CD45. See *Appendix I.*

Leukocytes — A general term used to describe white blood cells including lymphocytes, polymorphonuclear leukocytes, and monocytes.

Leukocytosis — The presence of an increased number of leukocytes in the blood. It is commonly observed in acute infection.

Leukotrienes (LTB4, LTC4, etc.) — Lipid inflammatory mediators produced by the lipoxygenase pathway. They have powerful pharmacological effects.

LFAs (leukocyte functional antigens) — Cell adhesion molecules. LFA-1 is a β2 integrin; LFA-2 and LFA-3 are members of the immunoglobulin superfamily. LFA-1 is important in T cell adhesion to endothelial cells and to antigen-presenting cells.

Ligand — A linking or binding molecule.

Lipid rafts — Specialised cell membrane domains enriched in cholesterol and glycosphingolipids. They act as assembly points and platforms that facilitate the interaction of particular signalling components.

Lipoxins — Endogenous lipid anti-inflammatory mediators.

LPS (lipopolysaccharide) — A component of Gram-negative bacterial cells walls. It is also known as endotoxin. It stimulates multiple innate immune responses and is also a powerful B cell mitogen.

L-selectin — An adhesion molecule of the selectin family that is found on lymphocytes.

Lymph —The extracellular fluid that accumulates in tissues and is carried by lymphatic vessels back to the blood through the lymphatic system which empties into the thoracic duct.

Lymph nodes — Secondary lymphoid organs found at locations of lymphatic vessel convergence. They are the sites of the initiation of adaptive immune responses.

Lymphatic system — The system of lymphoid channels and tissues that drains extracellular fluid from the periphery *via* the thoracic duct to the blood. It includes the lymph nodes, Peyer's patches, and other organised lymphoid elements except the spleen, which communicates directly with the blood.

Lymphatic vessels or Lymphatics — Thin-walled vessels that carry lymph through the lymphatic system. Afferent lymphatics drain fluid from the tissues and help transport antigen and antigen-presenting cells from sites of infection to the lymph nodes. Lymphocytes leave the lymph node through the efferent lymphatic vessel.

Lymphoblast — A lymphocyte that has enlarged and increased its rate of RNA and protein synthesis, usually in preparation for accelerated proliferation.

Lymphocyte homing — The directed migration of subsets of circulating lymphocytes into particular tissue sites that are regulated by the selective expression of the adhesion molecules called homing receptors.

Lymphocytes — A class of white blood cells. They consist of T, B, NK, and NKT cells. T and B lymphocytes bear variable cell surface receptors for antigen and participate in adaptive immune responses. NK cells express invariant surface receptors and are part of the innate immune system. NKT cells express NK and T cell receptors.

Lymphoid follicles — B cell rich regions consisting of clusters of B cells organised around follicular dendritic cells. They are the sites of antigen-induced B cell proliferation and differentiation. Primary follicles contain resting B cells and give rise to secondary follicles or germinal centres when antigen-activated B cells enter them.

Lymphoid organs — Organised tissues that are characterised by very large number of lymphocytes interacting with a non-lymphoid stroma. The central or primary lymphoid organs (the thymus and bone marrow) are sites of lymphocyte generation. The peripheral or secondary lymphoid organs (lymph nodes, spleen, and mucosal-associated lymphoid tissues such as the tonsils and Peyer's patches) are sites of the initiation of adaptive immune responses.

Lymphokines — A term originally used to describe the cytokines produced by lymphocytes.

Lymphomas — Tumours of lymphocytes that grow in lymphoid and other tissues. They do not enter the blood in large numbers.

Lymphotoxin — See *TNF.*

Lysosomes — Acidified organelles of the endocytic pathway that contain many degradative hydrolytic enzymes.

M cells — Delicate membranous cells found in the dome epithelium of Peyer's patches. M cells endocytose and transport antigens, without lysosomal degradation.

MAC (membrane attack complex) — A complex formed in the final stages of complement-associated lysis. It comprises terminal components of the complement cascade and multiple molecules of C9. The assembled complex generates a membrane spanning hydrophilic pore in the target cell membrane, which causes lethal ionic and osmotic changes in the cell.

Macrophages — Large, mononuclear, phagocytic, and migratory cells found in most body tissues. They are derived from blood monocytes and have a critical role in host defence. They are important to innate immunity, as antigen-presenting cells, and as effector cells in humoral and cell-mediated immunity.

MAGE — A family of proteins that are normally expressed only in the testes but are often expressed by melanoma (skin cancer) cells.

MALT (mucosal-associated lymphoid tissue) — It comprises all lymphoid cells in the epithelia and in the lamina propria lying below the body's mucosal surfaces.

Mannan-binding lectin (MBL) or mannose-binding protein (MBP) — A plasma protein that binds to mannose residues found on procaryotic cell walls and opsonises them. It can activate the complement cascade.

Mannose receptor — An endocytic pattern recognition receptor that binds mannoses and fucoses on microbial cell walls and mediates their phagocytosis.

MAP (mitogen-activated protein) kinases — Kinases that are present downstream of a variety of receptor-mediated signalling pathways. The phosphorylation of MAP kinases results in new gene expression because of the phosphorylation of key transcription factors.

MASP-1 & MASP-2 — Serine proteases that are components of the MBL pathway of complement activation. They cleave C4.

Mast cells — Large cells found in connective tissues throughout the body, most abundantly in the submucosal tissues and the dermis. They contain granules rich in pharmacological mediators and vasoactive amines, and they are the major effectors of immediate-type hypersensitivity. Antigen-mediated crosslinking of FcεRl receptors expressed on mast cells results in their degranulation.

MBL pathway (mannose-binding lectin pathway) — The pathway of complement activation initiated by the binding of MBL to the mannose-containing component of microbial cell walls.

MBP (major basic protein) — An antiparasitic protein found in eosinophils. It can also induce tissue injury in allergic and inflammatory diseases, and, because it induces injury to the bronchial epithelium, it is linked to asthma.

Medulla — Generally, the central or collecting point of an organ.

Memory cells — Progeny of antigen-activated naïve lymphocytes that mediate immunological memory. Due to a reprogramming of antigen receptors and signalling molecules, memory cells are more sensitive to antigen and respond more rapidly to it than naïve lymphocytes.

Metastasis — The transfer of pain, function, or disease from one organ to another through contact, blood vessels, or lymphatics. In the context of tumours, it is the formation of secondary tumours at sites away from the primary malignancy.

MHC (major histocompatibility complex) — A cluster of genes that encode the highly polymorphic proteins serving as peptide display molecules for T cells. MHC class I molecules present processed antigens to $CD8^+$ T cells, whereas MHC class II molecules present processed antigen to $CD4^+$ T cells. The cluster also encodes nonpolymorphic MHC class III proteins that have varying roles in host defence.

MHC restriction — A term used to describe the fact that T cells recognise a peptide antigen only when it is bound to a particular MHC allele (usually a self-MHC molecule).

Minor histocompatibility antigens (minor H antigens) — Polymorphic cellular proteins. Peptides derived from these proteins, when loaded onto MHC molecules, can cause graft rejection.

Mitogen — A substance that can induce cell division. B or T cell mitogens can induce B or T cell proliferation irrespective of their antigen specificity and are hence called polyclonal activators.

mIg (membrane immunoglobulin) — See *Immunoglobulin*.

MLR (mixed lymphocyte reaction) — A technique used to test histocompatibility in which lymphocytes from two unrelated individuals are cultured together. T cells in the culture proliferate in response to allogeneic MHC molecules on the donor's cells. See A*ppendix IV*.

Molecular mimicry — A postulated mechanism of autoimmunity induction as a result of cross-reactivity between the epitopes present on infectious agents and self-antigens.

Monoclonal antibodies (mAbs) — Antibodies of a single specificity produced by a single clone of B cells. See *Appendix IV*.

Monocytes — Bone marrow-derived circulating blood cells that are precursors of tissue macrophages.

Mucins — Highly glycosylated cell-surface proteins that are a major component of mucous.

Mucosal-associated lymphoid tissue — See *MALT*.

Multiple sclerosis — An autoimmune neurological disease characterised by focal demyelination in the central nervous system, lymphocytic infiltration in the brain, and a chronic, progressive course.

Myasthenia gravis — An autoimmune disease in which autoantibodies against the acetylcholine receptor on skeletal muscle cells cause a block in neuromuscular junctions, leading to progressive muscular weakness and eventual death.

Myeloid progenitors — Bone marrow cells that give rise to the granulocytes and macrophages of the immune system.

Myeloma — A tumour of plasma cells. Myeloma proteins are incomplete or complete immunoglobulins secreted by myeloma tumours.

Myeloperoxidase (MPO) — A critical enzyme in the conversion of hydrogen peroxide (H_2O_2) to the hypochlorous acid found in neutrophil granules. Along with H_2O_2 and a halide cofactor, MPO forms the most effective microbicidal and cytotoxic mechanism of leukocytes.

MZ (marginal zone) — Lymphoid tissue of the spleen that borders the white pulp. MZ contains a unique population of B cells that do not circulate, express a distinct set of surface proteins (such as CD5 and CD9), and are a part of the innate immune system.

Naïve lymphocytes — Lymphocytes that have never encountered their specific antigen and are not a progeny of antigen-stimulated mature lymphocytes.

NALT — Nasopharyngeal-associated lymphoid tissue.

Natural killer cells — See *NK cells*.

Necrosis — Death of cells or tissues caused by physical or chemical injury.

Negative selection — The removal of self-reactive lymphocytes during their development.

Neonatal FcR (FcRn) — An IgG-specific Fc receptor that mediates the transport of maternal IgG across the placenta and neonatal intestinal epithelium. It is also thought to regulate plasma IgG antibody catabolism in adults.

Neoplasm — An abnormal mass of tissue whose growth exceeds and is uncoordinated with that of normal tissues, and which persists in the same excessive manner after the cessation of the stimuli that evoked changes.

Neutropaenia — The occurrence of fewer than normal neutrophils in the blood.

Neutrophils — Phagocytic cells having multilobed nuclei. Neutrophils are one of earliest cells recruited to sites of inflammation.

NFκB — A family of transcription factors important in the transcription of many genes involved in innate and adaptive immune responses.

NK (natural killer) cells — Non-T, non-B cytotoxic lymphocytes. They are important in innate immune responses and anti-tumour responses.

NKT cells — Lymphocytes that express a NK cell receptor and a T cell receptor of limited diversity. NKT cells recognise antigen loaded on CD1 molecules.

Non-homologous DNA end-joining (NHEJ) pathway — A DNA repair pathway that repairs DNA strand breaks without relying on marked homology.

Nude mouse — A strain of mice having a gene defect that causes hairlessness and the defective formation of thymic stroma. These mice lack mature T cells.

Oedema — The swelling caused by the entry of fluid and cells from the blood into tissues. It is one of the characteristics of inflammation.

Opsonin — A macromolecule that becomes attached to microbial surfaces and can increase the efficiency of phagocytosis by being recognised by receptors on phagocytic cells, e.g., complement components and immunoglobulin molecules.

Opsonisation — The alteration of the surface of a pathogen/particle by the deposition of opsonins in order to facilitate phagocytosis.

Oral tolerance — Tolerance induced by the oral administration of antigen.

Original antigenic sin — A phenomenon which results in the generation of antibody responses to epitopes shared between the original infecting strain of a virus and subsequent infections by related viruses while ignoring other highly immunogenic epitopes on the second and subsequent viruses.

PALS (periarteriolar lymphoid sheath) — A part of the inner region of the white pulp of the spleen that contains mainly T cells.

PAMPs (pathogen-associated molecular patterns) — Molecules present on the cell envelope of microorganisms that form a unique geometric pattern recognised by special receptors on host cells.

Paraproteins — See *Bence Jones proteins.*

Passive immunity — Adaptive immunity conferred by the transfer of immune products such as antibodies or sensitised T cells from an immune to a nonimmune individual.

Pattern recognition receptors (PRRs) — Receptors of the innate immune system that recognise common molecular patterns on pathogen surfaces. PRRs include molecules such as TLRs, collections, ficolins, C-reactive proteins, and MBP, that can recognise distinctive ligands found on the cell walls of certain pathogens. PRRs are an integral part of the innate defence system.

PCA (passive cutaneous anaphylaxis) — A skin test used to detect antigen-specific IgE antibodies.

PD-1 (programmed death gene-1) — An inhibitory molecule expressed by activated T cells, B cells, and myeloid cells. The engagement of PD-1 by its ligands is thought to control the proliferation and cytokine expression of these cells.

PECAM-1 — CD31. See A*ppendix I.*

Perforin — A protein found in the granules of cytotoxic T cells and NK cells that polymerises to form membrane pores in the target cell membrane.

Periarteriolar lymphoid sheath — See *PALS.*

Peripheral tolerance — See *Tolerance.*

Peyer's patches — Organised lymphoid tissue in the lamina propria of the small intestines, especially the ileum.

Phagocytosis — A process by which large particulate matter (>0.5 μm) is internalised by cells. The ingested material is contained in a phagosome, which fuses with one or more lysosomes to form a phagolysosome where the ingested material undergoes degradation.

Phospholipase Cγ (PLCγ) — A key signal transducing enzyme that gets activated upon antigen recognition by lymphocytes. It catalyzes hydrolysis of phosphatidylinositol 4,5-bisphosphate (PIP_2) to Inositol-(1,4,5)-triphosphate (IP_3) and diacylglycerol (DAG).

pIgR (poly-Ig receptor) — An Fc receptor expressed by mucosal epithelial cells that mediates the transport of polymeric immunoglobulins (IgA and IgM) through the epithelial cells and into the intestinal lumen.

Plasma — The whitish coloured fluid that is left behind after all cells (leukocytes, erythrocytes, and platelets) are removed from blood. It contains water, electrolytes, and plasma proteins.

Plasma cell — An antibody-secreting, terminally differentiated B cell.

Plasmablast — A B cell in a lymph node that is on the path to becoming a plasma cell and shows some of its features.

Platelet activating factor (PAF) — A lipid mediator that activates the blood clotting cascade. It can cause bronchoconstriction and vascular dilation.

Platelets — Small, cell-like fragments of bone marrow derived megakaryocytes, found in the blood, that are crucial to blood clotting.

Pleuripotent stem cells — Cells capable of continuously dividing and differentiating into the progenitors of multiple lineages.

Polyclonal antibodies — Antibodies produced by many clones of cells, e.g., those found in serum.

Polygenic — Proteins, such as MHC molecules, that have identical functions but are encoded by several genetic loci.

Polymorphism — Literally, to exist in a variety of different shapes; when used in the context of genes (i.e., genetic polymorphism) it means variability at a gene locus in which the variants occur at a frequency of greater than 1%. Each common variant is called a polymorphic gene. The major histocompatibility complex is the most polymorphic gene cluster known in humans.

Polymorphonuclear granulocytes (PMNs) — Leukocytes having multilobed nuclei and large numbers of granules in the cytoplasm; PMNs are classified into neutrophils, basophils, and eosinophils on the basis of their staining reaction.

Positive selection — A thymic process by which only those developing T cells that have receptors recognising self-MHC molecules are allowed to survive and reach maturity.

Pre-B cell — A developing B cell present only in haematopoietic tissues; it is characterised by the expression of the μ heavy chain and the surrogate light chain.

Precipitation — The separation of the complex of a soluble antigen and its homologous antibody from the suspending fluid. See *Appendix IV*.

Primary immune response — An adaptive immune response to an initial exposure to antigen. Primary immunisation generates both the primary immune response and immunological memory.

Pro-B cells — The earliest haematopoietic cells committed to the B cell lineage; they express B lineage-specific markers, such as CD19 and CD10.

Progenitors — The more differentiated progeny of stem cells that give rise to distinct subsets of mature blood cells. They lack the capacity for self-renewal that is possessed by true stem cells.

Promoters — Relatively short nucleotide sequences extending up to 200 basepairs upstream (i.e., 5') from the transcription initiation site, where the proteins that initiate transcription bind.

Properdin — A positive regulator of the alternative pathway of complement activation.

Prostaglandins (PG) — Lipid by-products of the arachidonic acid pathway that have a variety of effects on several tissues.

Proteasome — Large, multiprotein enzyme complex with a broad range of activity that is the main proteolytic system of all eukaryotic cells. Proteasome-generated peptides are presented by MHC class I molecules.

Protein kinase C (PKC) — A serine/threonine kinase activated by diacylglycerol. It is crucial to many receptor-mediated signal transduction pathways.

Protein kinases — Enzymes that add phosphate groups to proteins; those adding phosphate groups to tyrosine residues are called protein tyrosine kinases, or PTKs.

Proto-oncogenes — The genes involved in regulating cell growth. Mutations in these genes may result in tumours; the mutated genes are called oncogenes.

RAG (recombination activating genes) — Genes encoding the proteins RAG-1 and RAG-2. These proteins are critical to gene rearrangements of T and B cell antigen receptors.

RANTES — A chemoattractant for peripheral blood monocytes.

Rapamycin (sirolimus) — An immunosuppressive drug that acts on a protein kinase involved in several cell growth and regulation pathways.

Reagins — Old term for IgE antibodies.

Receptor — A cell surface molecule that can specifically bind to a particular protein (ligand) in the fluid phase.

Receptor editing — The process of further rearranging of the light or heavy chains of a self-reactive antigen receptor on immature B cells to ablate autoreactivity.

Receptor-mediated endocytosis — The internalisation of the molecules bound to cell surface receptors.

Recombination signal sequences (RSSs) — Flanking gene segments found in *V*, *D*, and *J* gene segments that consist of conserved heptamer and nonamer sequences separated by 12 or 23 nonconserved nucleotide sequences. RSSs are targets for the site-specific RAG-1/RAG-2 recombinases that join the gene segments.

Regulatory T cells (T_{reg} cells) — T cells that can modulate (usually, inhibit) immune responses.

Respiratory burst — A process mediated by phagocyte oxidase and by which reactive oxygen intermediates are produced by phagocytes. It is usually triggered by bacterial products or pro-inflammatory mediators.

Rhesus (Rh) antigen — A blood group antigen that is also found on the red blood cells of rhesus monkeys.

Rheumatoid arthritis — An autoimmune inflammatory joint disease.

RIA (radioimmunoassay) — A technique of detection and quantitation of antigens or antibodies in which one of the reactants is labelled with a radioactive isotope, while the unlabelled reactant is attached to a solid support. See *Appendix IV*.

RNI (reactive nitrogen intermediates) — Potent antimicrobial molecules produced by the cells of the immune system and other cells involved in the immune response.

ROI (reactive oxygen intermediates) — Highly reactive metabolites of oxygen, such as superoxide anions and H_2O_2, that are produced by activated phagocytes.

SALT — Skin-associated lymphoid tissue.

Sarcoma — A malignant tumour arising from mesodermal tissue such as in the bone, cartilage, and fat.

Scavenger receptors — Types of pattern recognition receptors found on cells of the innate immune system. They bind to numerous ligands and remove them from the blood.

SCID — See *Severe combined immune deficiency.*

SDS-PAGE — A common abbreviation for polyacrylamide gel electrophoresis. See *Appendix IV.*

Secondary immune response — An immune response induced by a second or subsequent exposure to the antigen.

Secretory component — A fragment of the extracellular domain of poly-Ig receptor that is left attached to the IgA after transport across epithelial cells.

Selectins — A family of carbohydrate-binding, cell-surface adhesion molecules of leukocytes and endothelial cells that include L-selectin, P-selectin, and E-selectin, which is found on activated endothelium.

Sensitisation — In the context of an allergic reaction, prior exposure to the allergen that results in an IgE antibody response.

Sequence motif — A pattern of nucleotides or amino acids shared by different genes or proteins that often have related functions.

Sequestered antigens — Cellular constituents of tissue (e.g., lens of the eye) that are hidden or anatomically sequestered from the immune system during embryonic development.

Seroconversion — The phase of an infection when antibodies against the infecting agent are first detectable in the blood.

Serology — The study of serum antibodies and their reaction with antigens. The term is often used to describe the diagnosis of infections based on the detection of microbe-specific serum antibodies.

Serotonin (5-hydroxytryptamine) — The principal vasoactive amine released by mast cells. It is also a neurotransmitter, synthesised in the central nervous system, that is believed to play an important role in depression, bipolar disorder, and anxiety.

Serpins — A family of protease inhibitors.

Serum — The fluid component of clotted blood; a straw-coloured liquid obtained after the removal of fibrin from plasma.

Serum sickness — A disease caused by the injection of large doses of proteinic antigens, foreign serum, or serum proteins. Characterised by fever, joint pains, and nephritis, the disease is the result of the formation of immune complexes between the injected protein and the antibodies formed against it.

Severe combined immune deficiency (SCID) — An immune deficiency disease in which both antibody- and T cell-mediated immune responses are absent. It is usually the result of T cell deficiencies. The SCID mutation results in the loss of an enzyme needed for DNA-repair (DNA-PK), and such mice are used extensively in immunological research.

Signal joint — A joint formed by the precise joining of recognition signal sequences in the process of somatic recombination that generates lymphocyte antigen receptors.

Signal transduction — The process of converting a signal from one form to another. The binding of a ligand to its receptor causes receptor clustering. This original clustering signal is transduced into chemical signals in the cytoplasm by the activation of receptor-associated protein kinases.

Silencers — Nucleotide sequences that downregulate transcription in both directions over a distance.

Sirolimus — See *Rapamycin.*

SLE (systemic lupus erythematosus) — An autoimmune disease in which autoantibodies against DNA, RNA, and proteins associated with nucleic acids form immune complexes that damage small blood vessels, especially of the kidney.

Somatic hypermutation — High frequency point mutations in the V region genes of the B cell antigen receptor that occur in germinal centre B cells.

Somatic recombination — A process of DNA recombination that occurs in developing lymphocytes by which functional antigen receptor V region genes are assembled from separate *V(D)J* gene segments.

Somatic mutation theory — A theory of antibody diversity postulating that relatively few genes for the V locus are inherited and that the V regions of cells destined to be lymphocytes mutate at a rate higher than the rest of the DNA, yielding large numbers of clones of immunologically competent cells.

Spleen — An organ in the upper left side of the peritoneal cavity which contains a red pulp that is involved in removing senescent blood cells and a white pulp consisting of lymphoid cells. It is the major site of adaptive immune responses to blood-borne antigens.

Src-family tyrosine kinases — Receptor-associated protein tyrosine kinases that have several domains, called Src-homology (SH) domains 1, 2, and 3. The SH1 domain contains the active site of the kinase, the SH2 domain can bind to phosphotyrosine residues, and the SH3 domain is involved in interactions with proline-rich regions in other proteins.

STATs (signal transducers and activators of transcription) — A family of cytoplasmic proteins that function as both signal transducers and transcription activators. STATs are inactive as monomers, and activation involves phosphorylation and dimerisation, after which the STATs translocate to the nucleus.

Stem cells — Undifferentiated cells that divide and give rise to additional stem cells and to cells of multiple lineages.

Superantigens — Molecules that stimulate lymphocytes by binding to antigen receptors outside the antigen-binding site. T cell superantigens glue T cell receptors to MHC class II molecules, whereas B cell superantigens coligate adjacent membrane immunoglobulin molecules.

Suppressor T cells — T cells that were postulated to suppress the activity of other T and B cells.

Surrogate light chain — A complex of two invariant proteins, V pre-B and λ5, that associate with the μ heavy chain in pre-B cells to form the pre-B cell antigen receptor.

Syk — A key tyrosine kinase involved in B cell antigen receptor pathways.

Syndrome — Association of several clinically recognizable features, spectrum of signs and symptoms, behaviour, and phenomena or characteristics that often occur together and are feature of a disease.

Syngraft (syngeneic graft) — A graft between two genetically identical individuals that is usually not rejected.

Systemic anaphylaxis — The degranulation of mast cells all over the body. It results in widespread vasodilation, tissue fluid accumulation, and epiglottal swelling, and it may result in death.

Systemic lupus erythematosus — See *SLE*.

T cell antigen receptor — A clonally distributed, disulphide-linked heterodimer that consists of αβ or γδ chains and recognises peptides (or glycolipids) loaded onto MHC (or CD1) molecules. It includes a membrane proximal constant (C) region as well as a membrane distal variable (V) region that takes part in antigen binding. It is associated with the CD3 complex involved in signal transduction.

T lymphocytes — A subset of lymphocytes. They are defined by their development in the thymus and by their heterodimeric antigen receptors that are associated with the proteins of the CD3 complex.

TAAs — See *Tumour-associated antigens.*

Tacrolimus — See *FK506.*

TAP (transporters associated with antigen processing) — A heterodimeric protein that is involved in the transportation of short peptides from the cytosol into the lumen of the endoplasmic reticulum, allowing for the loading of the peptides onto MHC class I molecules.

Tapasin — TAP-associated protein. It is thought to promote the stability and peptide transport activity of TAP and to hold the MHC class I molecule in its peptide-receptive conformation.

T-bet — A T-box family transcription factor that is expressed in developing and committed TH1 cells. It has a central role in TH1 lineage development.

TD (thymus-dependent) antigens — Antigens that require the involvement of both, antigen-specific B and T cells, to elicit a humoral response.

TdT (terminal deoxynucleotidyl transferase) — An enzyme expressed in developing lymphocytes that inserts non-templated or N-nucleotides into the junctions between gene segments in T cell receptor and immunoglobulin V region genes.

TGF-β (transforming growth factor-β) — An immunosuppresive cytokine that inhibits the proliferation and differentiation of T cells and activation of macrophages and which counteracts the effects of pro-inflammatory cytokines.

TH cells (helper T cells) — $CD4^+$ T cells that help B cells or $CD8^+$ T cells respond to an antigen. Three subsets of TH cells are recognised, based on their secretory cytokine pattern. TH1 cells are mainly involved in activating macrophages and cytotoxic T lymphocytes, TH2 cells stimulate B cells, whereas TH17 cells are involved in inflammatory reactions.

Thymocytes — A precursor of T lymphocytes that are found in the thymus.

Thymus — A bilobed lymphoepithelial organ overlying the heart that is the site of T cell maturation.

TI (thymus-independent) antigens — Antigens that can mediate the activation and proliferation of antigen-specific B cells without the involvement of antigen-specific T cells.

Tangible body macrophages — The phagocytic cells in the germinal centres that engulf apoptotic B cells.

TNFs (tumour necrosis factors) — Pro-inflammatory cytokines. TNF-α is produced by macrophages and T cells and has multiple functions in immunity. TNF-β, also called lymphotoxin, is cytotoxic. It is also critical for the development of lymphoid organs.

TNFRs (tumour necrosis factor receptors) — Trimeric cell surface proteins having a varied role in immunity.

Tolerance — Antigen-specific unresponsiveness of the immune system induced by previous exposure to the antigen. Substances that induce tolerance are called tolerogens. Central tolerance is established in lymphocytes developing in central lymphoid organs. Peripheral tolerance is acquired by mature lymphocytes in the peripheral tissues.

Tolerogens — Substances that induce tolerance.

Toll-like receptors (TLRs) — Members of a family of pattern recognition receptors that are important in innate immunity. All members are homologues of the Toll originally described in *Drosophila*.

Tonsils — Lymphoepithelial structures found at the openings of the respiratory and digestive tracts.

Toxoids — Inactivated toxins that have lost their toxicity but retained their immunogenicity.

Transcytosis — The active transport of molecules across epithelial cells.

Transplantation — The grafting or transfer of tissue or cells from one individual to another.

Transfusion — The transfer of blood or blood products from one individual to another. An immunological reaction to transfused products is termed a transfusion reaction.

Transgenic animals — Animals that express an exogenous gene. See *Appendix IV*.

TSAs — See *Tumour-specific antigens*.

TSS (toxic shock syndrome) — A systemic toxic reaction caused by the massive production of cytokines by $CD4^+$ T cells that are activated by the bacterial superantigen toxic shock syndrome toxin-1 (TSST-1).

Tumour — A swelling caused by a mass of cells. Benign tumours are normally slow growing, circumscribed, encapsulated, have well-defined edges, and do not invade surrounding tissue. Malignant tumours are often rapidly growing, aggressive, and invasive masses of cells.

Tumour-associated antigens (TAAs) — Antigens found on both normal and cancerous cells.

Tumour-specific antigens (TSAs) — Antigens expressed exclusively by tumour cells and that are not found on normal body cells.

Tyrosine kinases (protein tyrosine kinases or PTKs) — Enzymes that specifically phosphorylate tyrosine residues in proteins. The tyrosine kinases critical for B cell activation are Blk, Fyn, Lyn, and Syk, whereas those critical for T cell activation are Lck, Fyn, and ZAP-70.

Urticaria — Red itchy skin welts (hives) that are the results of an allergic reaction.

V (Variable) region — The N-terminal domain of the light or heavy chain of a T cell receptor or immunoglobulin molecule; it contains the antigen-binding site and shows a high degree of sequence variability.

V(D)J recombination — A process found exclusively in lymphocytes of jawed vertebrates that allows the recombination of different gene segments into sequences encoding complete protein chains of immunoglobulins and T cell receptors.

Vaccination — The deliberate introduction of a dead or attenuated (nonpathogenic) form of the pathogen or its products in an individual with the intention of eliciting an adaptive immune response.

Valency — In the context of an antibody or antigen, the number of different molecules of the ligand that the antigen or antibody can combine with at one time.

Vasoactive amines — Amines such as histamine, and serotonin that causes vasodilatation and increases small vessel permeability. They also act on the endothelium of smooth muscles, causing its retraction. Granules of basophils, mast cells, platelets, and eosinophils contain vasoactive amines.

Vesicles — Small membrane-bound compartments within the cytosol.

Western blot — A technique of protein detection. It involves the separation of proteins by gel electrophoresis followed by transferring (blotting) to a nitrocellulose membrane and then probing with specific antibodies. See *Appendix IV*.

Wheal and flare reaction — The local swelling and redness of the skin at a site of an immediate hypersensitivity reaction.

Xenotransplantation — The transfer of cells, tissues, or organs between disparate species.

X-linked agammaglobulinaemia (Bruton's agammaglobulinaemia) — A genetic disorder in which B cell development is arrested at the pre-B-cell stage, and mature B cells or antibodies are not formed. The disease is due to a defect in the gene that encodes the protein tyrosine kinase Btk.

X-linked hyper IgM syndrome — A disease characterised by high serum IgM levels but deficient IgG, IgE, or IgA antibody formation. It is due to a defect in the gene that encodes the CD40 ligand (CD154).

ZAP-70 — A tyrosine kinase that is indispensable for T cell receptor signalling.

β_2-microglobulin — A polypeptide that is a constituent of some membrane proteins. It is noncovalently associated with, and is an integral part of, MHC class I molecules.

αβ T cells — T cells expressing an antigen receptor made of α and β subunits. These are conventional T cells.

γδ T cells — T cells bearing antigen receptor that consists of γ and δ chains. They recognise antigen independent of MHC molecules.

κ chain — One of two types of light chains of an immunoglobulin molecule.

λ chain — One of two types of light chains of an immunoglobulin molecule.

Bibliography

Online Resources

http://en.wikipedia.org/wiki/Main_Page
http://encyclopedia.thefreedictionary.com/
http://games.slashdot.org/article.pl?sid=08/06/29/198238
http://mcb.harvard.edu/BioLinks/Immunology.html
http://ocw.mit.edu/OcwWeb/index.htm
http://pathmicro.med.sc.edu/book/welcome.htm
http://stemcells.nih.gov
http://users.rcn.com/jkimball.ma.ultranet
http://www.aai.org
http://www.abcam.com/index.html?pageconfig=resources
http://www.biocarta.com
http://www.bioscience.org
http://www.britannica.com/
http://www.cellsalive.com/toc_immun.htm
http://www.copewithcytokines.de/
http://www.els.net/
http://www.expasy.ch
http://www.immunology.klimov.tom.ru/
http://www.med.sc.edu:85/links/immunol-link.htm
http://www.ncbi.nlm.nih.gov/omim
http://www.ncbi.nlm.nih.gov/prow
http://www.ncbi.nlm.nih.gov/PubMed/
http://www.ncbi.nlm.nih.gov/structure

Chapter 1

Matzinger P (2002) The danger model: a renewed sense of self. *Science* 296:301.

Medzhitov R, Janeway CA Jr (2002) Decoding the patterns of self and nonself by the innate immune system. *Science* 296:298.

Chapter 2

Beutler B (2004) Innate immunity: an overview. *Molecular Immunology* 40:845.

Born WK, Reardon CL, O'Brien RL (2006) The function of $\gamma\delta$ T cells in innate immunity. *Current Opinion in Immunology* 18:31.

Brown KL, Hancock REW (2006) Cationic host defense (antimicrobial) peptides. *Current Opinion in Immunology* 18:24.

Bryceson YT, March ME, Ljunggren H-G, Long EO (2006) Activation, coactivation, and costimulation of resting human natural killer cells. *Immunological Reviews* 214:73.

Cheent K, Khakoo SI (2009) Natural killer cells: integrating diversity with function. *Immunology* 126:449.

Dale DC, Boxer L, Liles WC (2008) The phagocytes: neutrophils and monocytes. *Blood* 112:195.

Ganz T (2009) Iron in innate immunity: starve the invaders. *Current Opinion in Immunology* 21:63.

Gibbs BF (2005) Human basophils as effectors and immunomodulators of allergic inflammation and innate immunity. *Clinical and Experimental Medicine* 5:43.

Gruys E, Toussaint MJM, Niewold TA, Koopmans SJ (2005) Acute-phase reaction and acute-phase proteins. *Journal of Zhejiang University Sci B.* 6:1045.

Hoyles L, Vulevic J (2008) Diet, immunity and functional foods. *Advances in Experimental Medicine and Biology* 635:79.

Jacobsen EA, Taranova AG, Lee NA, Lee JJ (2007) Eosinophils: singularly destructive effector cells or purveyors of immunoregulation? *Journal of Allergy and Clinical Immunology* 119:1313.

Kohchi C, Inagawa H, Nishizawa T, Soma G (2009) ROS and innate immunity. *Anticancer Research* 29:817.

Kumar H, Kawai T, Akira S (2009) Toll-like receptors and innate immunity. *Biochemical and Biophysical Research Communications* 388:621.

Lee HK, Iwasaki A (2007) Innate control of adaptive immunity: Dendritic cells and beyond. *Seminars in Immunology* 19:48.

Ley K, Laudanna C, Cybulsky MI, Nourshargh S (2007) Getting to the site of inflammation: the leukocyte adhesion cascade updated. *Nature Reviews Immunology* 7:678.

Maggini S, Wintergerst ES, Beveridge S, Hornig DH (2007) Selected vitamins and trace elements support immune function by strengthening epithelial barriers and cellular and humoral immune responses. *British Journal of Nutrition* 98:S29-35.

Martin F, Kearney JF (2001) B1 cells: similarities and differences with other B cell subsets. *Current Opinion in Immunology* 13:195.

Mogensen TH (2009) Pathogen recognition and inflammatory signaling in innate immune defenses. *Clinical Microbiology Reviews 22:*240.

Montecino-Rodriguez E, Dorshkind K (2006) New perspectives in B-1 B cell development and function. *Trends in Immunology* 27:428.

Nathan C (2002) Points of control in inflammation. *Nature* 420:826.

O'Neill LAJ (2005) Immunity's early-warning system. *Scientific American* 292:38.

Ravichandran KS, Lorenz U (2007) Engulfment of apoptotic cells: signals for a good meal. *Nature Reviews Immunology* 7:964.

Rock KL, Kono H (2008) The inflammatory response to cell death. *Annual Review of Pathology: Mechanisms of Disease* 3:99.

Sarhan CN, Sevill J (2005) Resolution of inflammation: the beginning programs the end. *Nature Immunology* 6:1191.

Silva MT (2010) When two is better than one: macrophages and neutrophils work in concert in innate immunity as complementary and cooperative partners of a myeloid phagocyte system. *Journal of Leukocyte Biology* 87:93.

Stämpfli MR, Anderson GP (2009) How cigarette smoke skews immune responses to promote infection, lung disease and cancer. *Nature Reviews Immunology* 9:377.

Steinman L (2004) Elaborate interactions between the immune and nervous systems. *Nature Immunology* 5:575.

Sternberg EM (2006) Neural regulation of innate immunity: a coordinated nonspecific host response to pathogens. *Nature Reviews Immunology* 6: 318.

Voehringer D (2009) The role of basophils in helminth infection. *Trends in Parasitology* 25:551.

Woods JA, Vieira VJ, Keylock KT (2009) Exercise, inflammation, and innate immunity. *Immunology and Allergy Clinics of North America* 29:381.

Chapter 3

Arumugam TV, Magnus T, Woodruff TM, Proctor LM, Shiels IA, Taylor SM (2006) Complement mediators in ischemia-reperfusion injury. *Clinca Chimic Acta* 374:33.

Bohlson SS, Fraser DA, Tenner AJ (2007) Complement proteins C1q and MBL are pattern recognition molecules that signal immediate and long-term protective immune functions. *Molecular Immunology* 44:33.

Diepenhorst GM, van Gulik TM, Hack CE (2009) Complement-mediated ischemia-reperfusion injury: lessons learned from animal and clinical studies. *Annals of Surgery* 249:889.

Dommett RM, Klein N, Turner MW (2006) Mannose-binding lectin in innate immunity: past, present and future. *Tissue Antigens* 68:193.

Dunkelberger JR, Song WC (2010) Complement and its role in innate and adaptive immune responses. *Cell Research* 20:34.

Gros P, Milder FJ, Janssen BJC (2008) Complement driven by conformational changes. *Nature Reviews Immunology* 8:48.

Ip EWK, Takahashi K, Ezekowitz AR, Stuart LM (2009) Mannose-binding lectin and innate immunity. *Immunological Reviews* 230:9.

Kim DD, Song W-C (2006) Membrane complement regulatory proteins. *Clinical Immunology* 118:127.

Klos A, Tenner AJ, Johswich KO, Ager RR, Reis ES, Köhl J (2009) The role of the anaphylatoxins in health and disease. *Molecular Immunology* 46:2753.

Perry VH, O'Connor V (2008) C1q: the perfect complement for a synaptic feast? *Natural Reviews Neuroscience* 9:807.

Rooijakkers SHM, van Strijp JAG (2007) Bacterial complement evasion. *Molecular Immunology* 44:23.

Tedesco F (2008) Inherited complement deficiencies and bacterial infections. *Vaccine* 26 Suppl 8:I3

Chapter 4

Casetti R, Martino A (2008) The Plasticity of γδ T Cells: Innate Immunity, Antigen Presentation and New Immunotherapy. *Cellular & Molecular Immunology* 5:161.

González-Fernández A, Faro J, Fernández C (2008) Immune responses to polysaccharides: lessons from humans and mice. *Vaccine* 26:292.

Klein Klouwenberg P, Bont L (2008) Neonatal and infantile immune responses to encapsulated bacteria and conjugate vaccines. *Clinical & Developmental Immunology* 2008:628963.

Llewelyn M, Cohen J (2002) Superantigens: microbial agents that corrupt immunity. *The Lancet Infectious Diseases* 2:156.

Mond JJ, Kokai-Kun JF (2008) The multifunctional role of antibodies in the protective response to bacterial T cell-independent antigens. *Current Topics Microbiology Immunology* 319:17.

Reed SG, Bertholet S, Coler RN, Friede M (2009) New horizons in adjuvants for vaccine development. *Trends Immunology* 30:23.

Silverman GJ, Goodyear CS (2006) Confounding B-cell defences: lessons from a staphylococcal superantigen. *Nature Reviews Immunology* 6:465.

Sundberg EJ, Deng L, Mariuzza RA (2007) TCR recognition of peptide/MHC class II complexes and superantigens. *Seminars in Immunology* 19:262.

Zinkernagel RM (2000) Localization dose and time of antigens determine immune reactivity. *Seminars in Immunology* 12:163.

Chapter 5

Cesta MF (2006) Normal structure, function, and histology of the spleen. *Toxicologic Pathology* 34:455.

Corr SC, Gahan CC, Hill C (2008) M-cells: origin, morphology and role in mucosal immunity and microbial pathogenesis. *FEMS Immunology and Medical Microbiology* 52:2.

Junt T, Scandella E, Ludewig B (2008) Form follows function: lymphoid tissue microarchitecture in antimicrobial immune defence. *Nature Reviews Immunology* 8:764.

Nagler-Anderson C (2001) Man the barrier! Strategic defences in the intestinal mucosa. *Nature Reviews Immunology* 1:59.

Ruddle NH, Akirav EM (2009) Secondary lymphoid organs: responding to genetic and environmental cues in ontogeny and the immune response. *Journal of Immunology* 183:2205.

Takahama Y (2006) Journey through the thymus: stromal guides for T-cell development and selection. *Nature Reviews Immunology* 6:127.

von Andrian UH, Mempel TR (2003) Homing and cellular traffic in lymph nodes. *Nature Reviews Immunology* 3:867.

http://www.scid.net

Chapter 6

Brezski RJ, Monroe JG (2008) B-cell receptor. *Advances in Experimental Medicine and Biology* 640:12.

Chaudhuri J, Alt FW (2004) Class-switch recombination: interplay of transcription, DNA deamination and DNA repair. *Nature Reviews Immunology* 4:541.

Di Noia JM, Neuberger MS (2007) Molecular mechanisms of antibody somatic hypermutation. *Annual Review of Biochemistry* 76:1.

Dustin ML (2009) The cellular context of T cell signaling. *Immunity* 30:482.

Gauld SB, Cambier JC (2004) Src-family kinases in B-cell development and signaling. *Oncogene* 23:8001.

Gearhart PJ (2002) Immunology: the roots of antibody diversity. *Nature* 419:29.

Geier JK, Schlissel MS (2006) Pre-BCR signals and the control of Ig gene rearrangements. *Seminars in Immunology* 18:31.

Gellert M (2002) *V(D)J* recombination: RAG proteins, repair factors, and regulation. *Annual Reviews of Biochemistry* 71:101.

Gupta N, DeFranco AL (2007) Lipid rafts and B cell signaling. *Seminars in Cell Development and Biology* 18:616.

Guy CS, Vignali DA (2009) Organization of proximal signal initiation at the TCR:CD3 complex. *Immunological Reviews* 232:7.

Herzog S, Reth M, Jumaa H (2009) Regulation of B-cell proliferation and differentiation by pre-B-cell receptor signalling. *Nature Reviews Immunology* 9:195.

Jolly CJ, Cook AJ, Manis JP (2008) Fixing DNA breaks during class switch recombination. *Journal of Experimental Medicine* 205:509.

Krangel MS (2009) Mechanics of T cell receptor gene rearrangement. *Current Opinion in Immunology* 21:133.

Le Deist F, Poinsignon C, Moshous D, Fischer A, de Villartay JP (2004) Artemis sheds new light on *V(D)J* recombination. *Immunological Reviews* 200:142.

Nemazee D (2006) Receptor editing in lymphocyte development and central tolerance. *Nature Reviews Immunology* 6:728.

Neuberger MS (2008) Antibody diversification by somatic mutation: from Burnet onwards. *Immunology Cell Biology* 86:124.

Palacios EH, Weiss A (2004) Function of the Src-family kinases, Lck and Fyn, in T-cell development and activation. *Oncogene* 23:7990.

Rickert RC (2005) Regulation of B lymphocyte activation by complement C3 and the B cell coreceptor complex. *Current Opinion in Immunology* 17:237.

Rojo JM, Bello R, Portolés P (2008) T-cell receptor. *Advances in Experimental Medicine and Biology* 640:1.

Rooney S, Chaudhuri J, Alt FW (2004) The role of the non-homologous end-joining pathway in lymphocyte development. *Immunological Reviews* 200:115.

Saunders AE, Johnson P (2010) Modulation of immune cell signalling by the leukocyte common tyrosine phosphatase, CD45. *Cellular Signaling* 22:339.

Smith-Garvin JE, Koretzky GA, Jordan MS (2009) T cell activation. *Annual Review of Immunology* 27:591.

Stavnezer J, Guikema JEJ, Schrader CE (2008) Mechanism and Regulation of Class Switch Recombination. *Annual Review of Immunology* 26:261.

Tonegawa S (1983) Somatic generation of antibody diversity. *Nature* 302:575.

http://www.antibodyresource.com/educational.html

http://www.bioinf.org.uk/abs/

Chapter 7

Amigorena S, Savina A (2010) Intracellular mechanisms of antigen cross presentation in dendritic cells. *Current Opinion in Immunology* 22:109.

Barral DC, Brenner MB (2007) CD1 antigen presentation: how it works. *Nature Reviews Immunology* 7:929.

Berger AC, Roche PA (2009) MHC class II transport at a glance. *Journal of Cell Science* 122:1.

Bielekova B, Martin R (2001) Antigen-specific immunomodulation via altered peptide ligands. *Journal of Molecular Medicine* 79:552.

Blanchard N, Shastri N (2010) Cross-presentation of peptides from intracellular pathogens by MHC class I molecules. *Annals of New York Academy of Sciences* 1183:237.

Donaldson JG, Williams DB (2009) Intracellular assembly and trafficking of MHC class I molecules. *Traffic* 10:1745.

Geissmann F, Manz MG, Jung S, Sieweke MH, Merad M, Ley K (2010) Development of monocytes, macrophages, and dendritic cells. *Science* 327:656.

Gruen JR, Weissman SM (2001) Human MHC class III and IV genes and disease associations. *Frontiers of Bioscience* 6:D960.

Hansen TH, Bouvier M (2009) MHC class I antigen presentation: learning from viral evasion strategies. *Nature Reviews Immunology* 9:503.

Klein J, Sato A (2000) The HLA system. Second of two parts. *New England Journal of Medicine* 343:782.

León B, Ardavín C (2008) Monocyte-derived dendritic cells in innate and adaptive immunity. *Immunology Cell Biology* (2008) 86:320.

Lipscomb MF, Masten BJ (2002) Dendritic cells: immune regulators in health and disease. *Physiological Reviews* 82:97.

Ramachandra L, Simmons D, Harding CV (2009) MHC molecules and microbial antigen processing in phagosomes. *Current Opinion in Immunology* 21:98.

Rivett AJ, Hearn AR (2004) Proteasome function in antigen presentation: immunoproteasome complexes, peptide production, and interactions with viral proteins. *Current Protein and Peptide Science* 5:153.

Rocha N, Neefjes J (2008) MHC class II molecules on the move for successful antigen presentation. *The EMBO Journal* 27:1.

Rock KL, Farfán-Arribas DJ, Shen L (2010) Proteases in MHC class I presentation and cross-presentation. *Journal of Immunology* 184:9.

Salio M, Silk JD, Cerundolo V (2010) Recent advances in processing and presentation of CD1 bound lipid antigens. *Current Opinion in Immunology* 22:81.

Shatz CJ (2009) MHC Class I: An unexpected role in neuronal plasticity. *Neuron* 64:40.

Trombetta ES, Mellman I (2005) Cell biology of antigen processing in vitro and in vivo. *Annual Review of Immunology* 23:975.

Ueno H, Klechevsky E, Morita R, Aspord C, Cao T, Matsui T, Di Pucchio T, Connolly J, Fay JW, Pascual V, Palucka AK, Banchereau J (2007) Dendritic cell subsets in health and disease. *Immunological Reviews* 219:118.

Yamazaki K, Beauchamp GK (2007) Genetic basis for MHC-dependent mate choice. *Advances in Genetics* 59:129.

http://www.ebi.ac.uk/imgt/hla/

http://www.bshi.org.uk

Chapter 8

Allman D, Pillai S (2008) Peripheral B cell subsets. *Current Opinion in Immunology* 20:149.

Baumgarth (2000) A two-phase model of B-cell activation. *Immunological Reviews* 176:181.

Carsetti R, Rosado MM, Wardmann H (2004) Peripheral development of B cells in mouse and man. *Immunological Reviews* 197:179.

Chan KF, Siegel MR, Lenardo JM (2000) Signaling by the TNF receptor superfamily and T cell homeostasis. *Immunity* 13:419.

El Shikh ME, El Sayed RM, Sukumar S, Szakal AK, Tew JG (2010) Activation of B cells by antigens on follicular dendritic cells. *Trends in Immunology* 31:205.

Elgueta R, Benson MJ, de Vries VC, Wasiuk A, Guo Y, Noelle RJ (2009) Molecular mechanism and function of CD40/CD40L engagement in the immune system. *Imunological Reviews* 229:152.

Fairfax KA, Kallies A, Nutt SL, Tarlinton DM (2008) Plasma cell development: from B-cell subsets to long-term survival niches. *Seminars in Immunology* 20:49.

Hardy RR (2006) B-1 B cells: development, selection, natural autoantibody and leukemia. *Current Opinion in Immunology* 18:547.

Harwood NE, Batista FD (2008) New insights into the early molecular events underlying B cell activation. *Immunity* 28:609.
Klein U, Dalla-Favera R (2008) Germinal centres: role in B-cell physiology and malignancy. *Nature Reviews Immunology* 8:22.
LeBien TW, Tedder TF (2008) B lymphocytes: how they develop and function. *Blood* 112:1570.
Mårtensson IL, Almqvist N, Grimsholm O, Bernardi AI (2010) The pre-B cell receptor checkpoint. *FEBS Letters* 584:2572.
Martin F, Kearney JF (2002) Marginal-zone B cells. *Nature Reviews Immunology* 2:323
McHeyzer-Williams LJ, McHeyzer-Williams MG (2005) Antigen-specific memory B cell development. *Annual Review of Immunology* 23:487.
Notarangelo LD (2010) Primary immunodeficiencies. *Journal of Allergy Clinical Immunology* 125:S 182.
Tew JG, Wu J, Fakher M, Szakal AK, Qin D (2001) Follicular dendritic cells: beyond the necessity of T-cell help. *Trends in Immunology* 22:361.

Chapter 9

Alcami A (2003) Viral mimicry of cytokines, chemokines and their receptors. *Nature Reviews Immunology* 3:36.
Amsen D, Spilianakis CG, Flavell RA (2009) How are TH1 and TH2 effector cells made? *Current Opinion in Immunology* 21:153.
Andersen MH, Schrama D, Thor Straten P, Becker JC (2006) Cytotoxic T cells. *Journal of Investigative Dermatology* 126:32.
Bendelac A, Savage PB, Teyton L (2007) The biology of NKT cells. *Annual Review of Immunology* 25:297.
Castellino F, Germain RN (2006) Cooperation between CD4+ and CD8+ T cells: when, where, and how. *Annual Review of Immunology* 24:519.
Commins SP, Borish L, Steinke JW (2010) Immunologic messenger molecules: cytokines, interferons, and chemokines. *Journal of Allergy & Clinical Immunology* 125 S2:S53-72.
De Libero G, Mori L (2005) Recognition of lipid antigens by T cells. *Nature Reviews Immunology* 5:485.
Dong C (2008) TH17 cells in development: an updated view of their molecular identity and genetic programming. *Nature Reviews Immunology* 8:337.
Gardner JM, Fletcher AL, Anderson MS, Turley SJ (2009) AIRE in the thymus and beyond. *Current Opinion in Immunology* 2:582.
Goncharova LB, Tarakanov AO (2008) Why chemokines are cytokines while their receptors are not cytokine ones? *Current Medicinal Chemistry* 15:1297.
Greenwald RJ, Freeman GJ, Sharpe AH (2005) The B7 family revisited. *Annual Review of Immunology* 23:515.
Haddad JJ (2002) Cytokines and related receptor-mediated signaling pathways. *Biochemical and Biophysical Research Communications* 297:700.
JT Opferman (2008) Apoptosis in the development of the immune system. *Cell Death and Differentiation* 15:234.
Kaech SM, Wherry EJ, Ahmed R (2002) Effector and memory T-cell differentiation: implications for vaccine development. *Nature Reviews Immunology* 2:251.
Kaer LV (2007) NKT cells: T lymphocytes with innate effector functions. *Current Opinion in Immunology* 19:354.
Kaiko GE, Horvat JC, Beagley KW, Hansbro PM (2007) Immunological decision-making: how does the immune system decide to mount a helper T-cell response? *Immunology* 123:326.
Krammer PH, Arnold R, Lavrik IN (2007) Life and death in peripheral T cells. *Nature Reviews Immunology* 7:532.
Ladi E, Yin X, Chtanova T, Robey EA (2006) Thymic microenvironments for T cell differentiation and selection. *Nature Immunology* 7:338.
Lefrançois L, Masopust D (2002) T cell immunity in lymphoid and non-lymphoid tissues. *Current Opinion in Immunology* 14:503.
Leitner J, Grabmeier-Pfistershammer K, Steinberger P (2010) Receptors and ligands implicated in human T cell costimulatory processes. *Immunology Letters* 128:89.
Marelli-Berg FM, Cannella L, Dazzi F, Mirenda V (2008) The highway code of T cell trafficking. *Journal of Pathology* 214:179.
Prlic M, Williams MA, Bevan MJ (2007) Requirements for CD8 T-cell priming, memory generation and maintenance. *Current Opinion in Immunology* 19:315.
Reiner SL (2007) Development in Motion: Helper T Cells at Work. *Cell* 129:33.
Schindler C, Plumlee C (2008) Inteferons pen the JAK-STAT pathway. *Seminars in Cell Development and Biology* 19:311.
Singer A, Adoro S, Park JH (2008) Lineage fate and intense debate: myths, models and mechanisms of CD4- versus CD8-lineage choice. *Nature Reviews Immunology* 8:788.
Starr TK, Jameson SC, Hogquist KA (2003) Positive and negative selection of T cells. *Annual Review of Immunology* 21:139.
Stein JV, Nombela-Arrieta C (2005) Chemokine control of lymphocyte trafficking: a general overview. *Immunology* 116:1.
Stockinger B, Bourgeois C, Kassiotis G (2006) CD4+ memory T cells: functional differentiation and homeostasis. *Immunological Reviews* 211:39.
Stockinger B, Veldhoen M (2007) Differentiation and function of TH17 T cells. *Current Opinion in Immunology* 19:281.
Tang Q, Bluestone JA (2008) The Foxp3+ regulatory T cell: a jack of all trades, master of regulation. *Nature Immunology* 9:239.

van Leeuwen EM, Sprent J, Surh CD (2009) Generation and maintenance of memory CD4+ T Cells. *Current Opinion in Immunology* 21:167
Vignali DA, Collison LW, Workman CJ (2008) How regulatory T cells work. *Nature Reviews Immunology* 8:523.
Williams MA, Bevan MJ (2007) Effector and memory CTL differentiation. *Annual Review of Immunology* 25:171.
Zhu J, Paul WE (2010) Heterogeneity and plasticity of T helper cells. *Cell Research* 20:4.
Zhu J, Yamane H, Paul WE (2010) Differentiation of Effector CD4 T Cell Populations. *Annual Review of Immunology* 28:445.

Chapter 10

Aalberse RC, Stapel SO, Schuurman J, Rispens T (2009) Immunoglobulin G4: an odd antibody. *Clinical & Experimental Allergy* 39:469.
Behn U (2007) Idiotypic networks: toward a renaissance? *Immunological Reviews* 216:142.
Chan AC, Carter PJ (2010) Therapeutic antibodies for autoimmunity and inflammation. *Nature Reviews Immunology* 10:301.
Cohn M, Mitchison NA, Paul WE, Silverstein AM, Talmage DW, Weigert M (2007) Reflections on the clonal-selection theory. *Nature Reviews Immunology* 7:823.
Johansen FE, Braathen R, Brandtzaeg P (2000) Role of J chain in secretory immunoglobulin formation. *Scandinavian Journal of Immunology* 52:240.
Jordan SC, Toyoda M, Vo AA (2009) Intravenous immunoglobulin a natural regulator of immunity and inflammation. *Transplantation* 88:1.
Kaetzel CS (2005) The polymeric immunoglobulin receptor: bridging innate and adaptive immune responses at mucosal surfaces. *Immunological Reviews* 206:83-99.
Knop E, Knop N (2007) Anatomy and immunology of the ocular surface. *Chemical Immunology and Allergy* 92:36.
Kraft S, Kinet J-P (2007) New developments in FcεRI regulation, function and inhibition. *Nature Reviews Immunology* 7:365.
Lee SJ, Chinen J, Kavanaugh A (2010) Immunomodulator therapy: monoclonal antibodies, fusion proteins, cytokines, and immunoglobulins. *Journal of Allergy & Clinical Immunology* 125 S2:S314.
Macpherson AJ, McCoy KD, Johansen FE, Brandtzaeg P (2008) The immune geography of IgA induction and function. *Mucosal Immunology* 1:11.
Muyldermans S, Cambillau C, Wyns L (2001) Recognition of antigens by single domain antibody fragments: the superfluous luxury of paired domains. *Trends in Biochemical Sciences* 26: 230.
Neuberger MS (2008) Antibody diversification by somatic mutation: from Burnet onwards. *Immunology & Cell Biology* 86:124.
Nimmerjahn F, Ravetch JV (2008) Fcγ receptors as regulators of immune responses. *Nature Reviews Immunology* 8:34.
Phalipon A, Corthésy B (2003) Novel functions of the polymeric Ig receptor: well beyond transport of immunoglobulins. *Trends in Immunology* 24:55.
Pre'homme J, Petit I, Barra A, Morel F, Lecron J, Lelie'vre E (2000) Structural and functional properties of membrane and secreted IgD. *Molecular Immunology* 37: 871.
Roopenian DC, Akilesh S (2007) FcRn: the neonatal Fc receptor comes of age. *Nature Reviews Immunology* 7:715.
Saha NR, Smith J, Amemiya CT (2010) Evolution of adaptive immune recognition in jawless vertebrates. *Seminars in Immunology* 22:25.
Schroeder HW Jr, Cavacini L (2010) Structure and function of immunoglobulins. *Journal of Allergy & Clinical Immunology* 125 S2:S41.
Shmagel KV, Chereshnev VA (2009) Molecular bases of immune complex pathology. *Biochemistry (Moscow)* 74:469.
Silverstein AM (2003) Splitting the difference: the germline-somatic mutation debate on generating antibody diversity. *Nature Reviews Immunology* 4:829.
Snoeck V, Peters IR, Cox E (2006) The IgA system: a comparison of structure and function in different species. *Veterinary Research* 37:455.
Walker A (2010) Breast milk as the gold standard for protective nutrients. *The Journal of Pediatrics* 156:S3-7.

Chapter 11

Benedict CA, Norris PS, Ware CF (2002) To kill or be killed: viral evasion of apoptosis. *Nature Immunology* 3:1013.
Brenner D, Mak TW (2009) Mitochondrial cell death effectors. *Current Opinion in Cell Biology* 21:871.
Brown DM (2010) Cytolytic CD4 cells: Direct mediators in infectious disease and malignancy. *Cellular Immunology* 262:89.
Chávez-Galán L, Arenas-Del Angel MC, Zenteno E, Chávez R, Lascurain R (2009) Cell death mechanisms induced by cytotoxic lymphocytes. *Cellular & Molecular Immunology* 6:15-25.
Cullen SP, Martin SJ (2008) Mechanisms of granule-dependent killing. *Cell Death and Differentiation* 15:251.
Hoves S, Trapani JA, Voskoboinik I (2010) The battlefield of perforin/granzyme cell death pathways. *Journal Leukocyte Biology* 87:237.
Jonjiã S, Babiã M, Poliã B, Krmpotiã A (2008) Immune evasion of natural killer cells by viruses. *Current Opinion in Immunology* 20:30.
Krensky AM, Clayberger C (2009) Biology and clinical relevance of granulysin. *Tissue Antigens* 73:193.

Sulica A, Morel P, Metes D, Herberman RB (2001) Ig-binding receptors on human NK cells as effector and regulatory surface molecules. *International Reviews Immunology* 20:371.

Chapter 12

Cerutti A, Rescigno M (2008) The biology of intestinal immunoglobulin A responses. *Immunity* 28:740.

Faria AM, Weiner HL (2006) Oral tolerance: therapeutic implications for autoimmune diseases. *Clinical and Developmental Immunology* 13:143.

Forrester JV, Xu H, Lambe T, Cornall R (2008) Immune privilege or privileged immunity? *Mucosal Immunology* 1:372.

Green D.R., Ferguson T.A. (2001) The role of Fas ligand in immune privilege. *Nature Reviews Molecular Cell Biology* 2:917.

Iweala OI, Nagler CR (2006) Immune privilege in the gut: the establishment and maintenance of non-responsiveness to dietary antigens and commensal flora. *Immunological Reviews* 213:82.

Lechler R, Chai JG, Marelli-Berg F, Lombardi G (2001) The contributions of T-cell anergy to peripheral T-cell tolerance. *Immunology* 103:262.

Mathis D, Benoist C (2004) Back to central tolerance. *Immunity* 20:509.

Mueller DL (2010) Mechanisms maintaining peripheral tolerance. *Nature Immunology* 11:21.

Peron JP, de Oliveira AP, Rizzo LV (2009) It takes guts for tolerance: the phenomenon of oral tolerance and the regulation of autoimmune response. *Autoimmunity Reviews* 9:1.

Stein-Streilein (2008) Immune regulation and the eye. *Trends in Immunology* 29:548.

't Hart BA, van Kooyk Y (2004) Yin-Yang regulation of autoimmunity by DCs. *Trends in Immunology* 25:353.

Tsuji NM, Kosaka A (2008) Oral tolerance: intestinal homeostasis and antigen-specific regulatory T cells. *Trends in Immunology* 29:532.

Chapter 13

Anthony RM, Rutitzky LI, Urban JF Jr, Stadecker MJ, Gause WC (2007) Protective immune mechanisms in helminth infection. *Nature Reviews Immunology* 7:975.

Beeson JG, Osier FH, Engwerda CR (2008) Recent insights into humoral and cellular immune responses against malaria. *Trends in Parasitology* 24:578.

Berzofsky JA, Ahlers JD, Belyakov IM (2001) Strategies for designing and optimizing new generation vaccines. *Nature Review Immunology* 1:209.

Cooper AM (2009) Cell-mediated immune responses in tuberculosis. *Annual Review of Immunology* 27:393.

Cox FE (2002) History of human parasitology. *Clinical Microbiology Review* 15:595.

Dörner T, Radbruch A (2007) Antibodies and B cell memory in viral immunity. *Immunity* 27:384.

Garcia LS (2010) Malaria. *Clinics in Laboratory Medicine* 30: 93.

Gilboa E (2007) DC-based cancer vaccines. *The Journal of Clinical Investigation* 117:1195.

Isaacs JD, Jackson GS, Altmann DM (2006) The role of the cellular prion protein in the immune system. *Clinical & Experimental Immunology* 146:1.

Kutzler MA, Weiner DB (2008) DNA vaccines: ready for prime time? *Nature Reviews Genetics* 9:776.

Kwiatkowski DP (2005) How malaria has affected the human genome and what human genetics can teach us about malaria. *American Journal of Human Genetics* 77:171.

Lanier LL (2008) Evolutionary struggles between NK cells and viruses. *Nature Reviews Immunology* 8:259.

Lorenzo ME, Ploegh HL, Tirabassi RS (2001) Viral immune evasion strategies and the underlying cell biology. *Seminars in Immunology* 13:1.

McMichael AJ, Borrow P, Tomaras GD, Goonetilleke N, Haynes BF (2010) The immune response during acute HIV-1 infection: clues for vaccine development. *Nature Reviews Immunology* 10:11.

Pantaleo G, Graziosi C, Fauci AS (1993) New concepts in the immunopathogenesis of human immunodeficiency virus infection. *New England Journal of Medicine* 328:327.

Pierce SK, Miller LH (2009) World Malaria Day 2009: what malaria knows about the immune system that immunologists still do not. *Journal of Immunology* 182:5171.

Pieters J (2008) Mycobacterium tuberculosis and the macrophage: maintaining a balance. *Cell Host Microbe* 3:399.

Reddy A, Fried B (2007) The use of *Trichuris suis* and other helminth therapies to treat Crohn's disease. *Parasitology Research* 100:921.

Romani L (2005) Immunity to fungal infections. *Nature Reviews Immunology* 4:1.

Sacks D, Sher A (2002) Evasion of innate immunity by parasitic protozoa. *Nature Immunology* 3:1041.

Sadler AJ, Williams BR (2008) Interferon-inducible antiviral effectors. *Nature Reviews Immunology* 8:559.

Stoiber H, Soederholm A, Wilflingseder D, Gusenbauer S, Hildgartner A, Dierich MP (2008) Complement and antibodies: a dangerous liaison in HIV infection? *Vaccine* 26 Suppl 8:I79.

http://www.cdc.gov/ncird/index.html

http://www.niaid.nih.gov

http://www.who.int

http://www.vaccinealliance.org
http://www.unaids.org

Chapter 14

Bui JD, Schreiberc RD (2007) Cancer immunosurveillance, immunoediting and inflammation: independent or interdependent processes? *Current Opinion in Immunology* 19:203.
Carter P (2001) Improving the efficacy of antibody-based cancer therapies. *Nature Reviews Cancer* 1:118.
Dougan M, Dranoff G (2009) Immune therapy for cancer. *Annual Review of Immunology* 27:83.
Gerber DE (2008) Targeted therapies: a new generation of cancer treatments. *American Family Physician* 77:311.
Gilboa E (2007) DC-based cancer vaccines. *Journal of Clinical Investigation* 117:119.
Hirohashi Y, Torigoe T, Inoda S, Kobayasi J, Nakatsugawa M, Mori T, Hara I, Sato N (2009) The functioning antigens: beyond just as the immunological targets. *Cancer Science* 100:798.
Igney FH, Krammer PH (2002) Death and anti-death: Tumour resistance to apoptosis. *Nature Reviews Cancer* 2:277.
Khan N, Afaq F, Mukhtar H (2010) Lifestyle as risk factor for cancer: Evidence from human studies. *Cancer Letters* 293:133.
Plati J, Bucur O, Khosravi-Far R (2008) Dysregulation of apoptotic signaling in cancer: molecular mechanisms and therapeutic opportunities. *Journal of Cellular Biochemistry* 104:1124.
Stagg J, Johnstone RW, Smyth MJ (2007) From cancer immunosurveillance to cancer immunotherapy. *Immunological Reviews* 220:82.
Vousden KH, Lu X (2002) Live or let die: The cell's response to p53. *Nature Reviews Cancer* 2:594.
Whiteside TL (2010) Immune responses to malignancies. *Journal of Allergy & Clinical Immunology* 125(2 Suppl 2):S272.
Zitvogel L, Apetoh L, Ghiringhelli F, Kroemer G (2008) Immunological aspects of cancer chemotherapy. *Nature Reviews Immunology* 8:59.
http://www.cancer.org/docroot/home/index.asp
http://www.oncolink.upenn.edu
http://www.lymphomainfo.net/info/websites.html
http://www.cancer.gov/

Chapter 15

Atassi MZ, Casali P (2008) Molecular mechanisms of autoimmunity. *Autoimmunity* 41:123.
Bettini M, Vignali DA (2009) Regulatory T cells and inhibitory cytokines in autoimmunity. *Current Opinion in Immunology* 21:612.
Chervonsky AV (2010) Influence of microbial environment on autoimmunity. *Nature Immunology* 11:28.
Ermann J, Fathman CG (2001) Autoimmune diseases: genes, bugs and failed regulation. *Nature Immunology* 2:759.
Knip M, Siljander H (2008) Autoimmune mechanisms in type 1 diabetes. *Autoimmunity Reviews* 7:550.
Lleo A, Battezzati PM, Selmi C, Gershwin ME, Podda M (2008) Is autoimmunity a matter of sex? *Autoimmunity Reviews* 7:626.
Martinon F, Tschopp J (2004) Inflammatory caspases: linking an intracellular innate immune system to autoinflammatory diseases. *Cell* 117:561.
McInnes IB, Schett G (2007) Cytokines in the pathogenesis of rheumatoid arthritis. *Nature Reviews Immunology* 7:429.
Sfriso P, Ghirardello A, Botsios C, Tonon M, Zen M, Bassi N, Bassetto F, Doria A (2010) Infections and autoimmunity: the multifaceted relationship. *Journal of Leukocyte Biology* 87:385.
St Clair EW (2009) Novel targeted therapies for autoimmunity. *Current Opinion in Immunology* 21:648.
Veldhoen M (2009) The role of T helper subsets in autoimmunity and allergy. *Current Opinion in Immunology* 21:606.
Wentworth JM, Fourlanos S, Harrison LC (2009) Reappraising the stereotypes of diabetes in the modern diabetogenic environment. *Nature Reviews Endocrinology* 5:483.
http://www.lupus.org/
http://www.niaid.nih.gov
http://www.nlm.nih.gov/medlineplus/rheumatoidarthritis.html

Chapter 16

Abraham SN, St John AL (2010) Mast cell-orchestrated immunity to pathogens. *Nature Reviews Immunology* 10:440.
Blank U, Rivera J (2004) The ins and outs of IgE-dependent mast-cell exocytosis. *Trends in Immunology* 25:266.
Frew AJ (2010) Allergen immunotherapy. *Journal of Allergy & Clinical Immunology* 125(2 Suppl 2):S306.
Hamilton RG (2010) Clinical laboratory assessment of immediate-type hypersensitivity. *Journal of Allergy and Clinical Immunology* 125: S284.
Holgate ST, Arshad HS, Roberts GC, Howarth PH, Thurner P, Davies DE (2009) A new look at the pathogenesis of asthma. *Clinical Science (London, England:1979)* 118:439.
Kränke B, Aberer W (2009) Skin testing for IgE-mediated drug allergy. *Immunology & Allergy Clinics of North America* 29:503.

Lemanske RF Jr, Busse WW (2010) Asthma: clinical expression and molecular mechanisms. *Journal of Allergy & Clinical Immunology* 125(2 Suppl 2):S95.

Okada H, Kuhn C, Feillet H, Bach JF (2010) The 'hygiene hypothesis' for autoimmune and allergic diseases: an update. *Clinical & Experimental Immunology* 160:1.

Rook GA (2009) The hygiene hypothesis and the increasing prevalence of chronic inflammatory disorders. *Transactions of the Royal Society for Tropical Medicine and Hygiene* 101:1072.

Sandor M, Weinstock JV, Wynn TA (2003) Granulomas in schistosome and mycobacterial infections: a model of local immune responses. *Trends in Immunology* 24:44.

Schnyder B, Pichler WJ (2009) Mechanisms of drug-induced allergy. *Mayo Clinic Proceedings* 84:268.

Sicherer SH, Sampson HA (2010) Food allergy. *Journal of Allergy & Clinical Immunology* 125(2 Suppl 2):S116.

Simons R, Estelle F (2010) Anaphylaxis. *Journal of Allergy & Clinical Immunology* 125 (2 Suppl 2): S161.

Siracusa MC, Perrigoue JG, Comeau MR, Artis D (2010) New paradigms in basophil development, regulation and function. *Immunology & Cell Biology* 88:275.

Stone KD, Prussin C, Metcalfe DD (2010) IgE, mast cells, basophils, and eosinophils. *Journal of Allergy & Clinical Immunology* 125(S2):S73.

http://www.aaaai.org

http://www.niams.nih.gov/

http://www.foodallergy.org/

http://www.worldallergy.org/

Chapter 17

Alter HJ, Klein HG (2008) The hazards of blood transfusion in historical perspective. *Blood* 112:2617.

Ball LM, Egeler RM, EBMT Paediatric Working Party (2008) Acute GvHD: pathogenesis and classification. *Bone Marrow Transplantation* 41 (Suppl 2):S58-64.

Barbosa FT, Jucá MJ, Castro AA, Duarte JL, Barbosa LT (2009) Artificial oxygen carriers as a possible alternative to red cells in clinical practice. *Sao Paulo Medical Journal* 127:97.

D'Adamo PJ, Kelly GS (2001) Metabolic and immunologic consequences of ABH secretor and Lewis subtype status. *Alternative Medicine Review* 6:390.

Daniels G (2001) A century of human blood groups. *Wiener Klinische Wochenschrift* 113:781.

Dierselhuis M, Goulmy E (2009) The relevance of minor histocompatibility antigens in solid organ transplantation. *Current Opinion Organ Transplantation* 14:419.

Dodd RY (2010) Emerging pathogens in transfusion medicine. *Clinics in Laboratory Medicine* 30: 499.

Eder AF, Chambers LA (2007) Noninfectious complications of blood transfusion. *Archives of Pathology & Laboratory Medicine* 131:708.

Ferrara JL, Levine JE, Reddy P, Holler E (2009) Graft-versus-host disease. *Lancet* 373:1550.

Gökmen MR, Lombardi G, Lechler RI (2008) The importance of the indirect pathway of allorecognition in clinical transplantation. *Current Opinion in Immunology* 20:568.

Gorantla VS, Barker JH, Jones JW Jr, Prabhune K, Maldonado C, Granger DK (2000) Immunosuppressive agents in transplantation: mechanisms of action and current anti-rejection strategies. *Microsurgery* 20:420.

Hosoi E (2008) Biological and clinical aspects of ABO blood group system. *Journal of Medical Investigations* 55:174.

Makrigiannakis A, Karamouti M, Drakakis P, Loutradis D, Antsaklis A (2008) Fetomaternal immunotolerance. *American Journal of Reproductive Immunology* 60:482.

Rogers NJ, Lechler RI (2001) Allorecognition. *American Journal of Transplantation* 1:97.

Srinivasan P, Huang GC, Amiel SA, Heaton ND (2007) Islet cell transplantation. *Postgraduate Medical Journal* 83:224.

Storry JR, Olsson ML (2009) The ABO blood group system revisited: a review and update. *Immunohematology* 25:48.

Yang YG, Sykes M (2007) Xenotransplantation: current status and a perspective on the future. *Nature Reviews Immunology* 17:519.

http://www.bshi.org.uk

Index